T0297067

Birkhäuser

Current Challenges in Statistical Seismology

Edited by
Qinghua Huang
Matthew Gerstenberger
Jiancang Zhuang

Previously published in *Pure and Applied Geophysics* (PAGEOPH), Volume 173, No. 1, 2016

Editors
Qinghua Huang
Department of Geophysics
Peking University
Beijing, China

Jiancang Zhuang
Institute of Statistical Mathematics
Tokyo, Japan

Matthew Gerstenberger
GNS Science
Avalon, New Zealand

ISBN 978-3-319-28966-3 ISBN 978-3-319-28967-0 (eBook)
DOI 10.1007/978-3-319-28967-0

Library of Congress Control Number: 2015960741

Springer Cham Heidelberg New York Dordrecht London

Cover illustration: Photo courtesy Zhanhui Li

Cover design: deblik, Berlin

Printed on acid-free paper

This book is published under the trade name Birkhäuser

The registered company is Springer International Publishing AG, Switzerland

www.springer.com

Contents

Pure Appl. Geophys. 173 (2016), 1–3

DOI 10.1007/s00024-015-1222-7

Pure and Applied Geophysics

Current Challenges in Statistical Seismology

Qinghua Huang,[1] Matthew Gerstenberger,[2] and Jiancang Zhuang[3]

1. Introduction

Statistical seismology is a subject that aims to bridge the gap between physical and statistical models (Vere-Jones *et al.* 2005). It has developed rapidly during the last several decades and has been highlighted every 2 years by the International Workshop on Statistical Seismology (StatSei). This workshop has now been held nine times in many different countries around the globe. Several past special volumes related to seismicity analysis and earthquake statistics have been developed based on the StatSei and related workshops (e.g., Vere-Jones et al. 2005; Rhoades *et al.* 2010; Tsaklids *et al.* 2011; Console *et al.* 2012; Papadopoulos 2012). In this issue, we have collected articles following the 8th International Workshop on Statistical Seismology (StatSei8: http://www.geophy.pku.edu.cn/statsei8/) which was held in Beijing, China, in 2013.

Many significant achievements have been accomplished during last several decades. One is the formulation of conditional intensity models for quantifying time-varying seismicity rates. A particular example of this is the ETAS model developed by Ogata (1988). It has become a de facto standard model, or null hypotheses, for other models and ideas to be compared to. An advantage of using conditional intensity models is that the evaluation of their forecasting performance can be done in a measured and statistical way using the framework of probability gain. This means that improvements in understanding of earthquake clustering can be quantified by developing new ideas into models and then comparing them to ETAS, or other models, using statistical hypothesis testing. Ultimately, rigorous testing of forecast models is necessary in order to improve our ability to forecast seismic hazard (Jordan *et al.* 2011).

Both probabilistic earthquake forecasting and binary earthquake prediction are some of the most challenging problems in the subject of geophysics. Currently, many scientists believe that individual earthquakes cannot be deterministically predicted (e.g., Geller *et al.* 1997) due to our inability to observe many of the fundamental processes of the system and also its inherent randomness. Therefore, in order to best quantify our state of knowledge, the statistical seismology community has placed a larger focus on probabilistic forecasting. In order to provide more reliable earthquake forecast models, the challenge is to construct models that can give increased information gain with respect to a reference model, such as ETAS or complete randomness. This requires not only increased understanding of the physical process of earthquakes, such as the preparation and rupture processes of the earthquake source and the interaction between earthquakes and tectonic environments, but also development and implementation of improved statistical methods for testing and validating physical hypotheses based on observed data.

With rapid development of observation technologies, more and more observational data are obtained. For example, GPS observation of surface displacement, InSAR observation of the co-seismic deformation, ionospheric observations, etc., bring statistical seismology into the big data era. Additionally, the understanding of what an earthquake itself is has been extended by the discovery of slow earthquakes, tremors, and VLF earthquakes. These new observations provide new theories and

[1] Department of Geophysics, School of Earth and Space Sciences, Peking University, Beijing 100871, China. E-mail: huangq@pku.edu.cn

[2] GNS Science, Lower Hutt, New Zealand.

[3] Institute of Statistical Mathematics, Tokyo 1908562, Japan.

approaches to help us understand seismicity. However, there is no free lunch. To make use of these new observations, statistical seismologists are challenged by tasks of developing new methods to analyze these data more efficiently and new models to connect them to the earthquake process and tectonic environments.

This special issue emerged after the recent 8th International Statistical Seismology (StatSei8) workshop in Beijing. The articles within have been collected to report on the exciting new research in statistical seismology methods and applications; the issue contains a collection of the newest methods, techniques and outputs related to statistical analysis of earthquake occurrence and earthquake probability forecasting, and to ultimately helps to define future research directions in the field.

Papers submitted to this topical issue can be loosely categorized into the following topics:

1. *The connection between seismicity and physics* In this class of articles, Bebbington *et al.* (2016), Iwata (2016), and Leptokaropoulos *et al.* (2016), tested the seismicity rate of aftershocks or background events after big events, revealing the complexity of the seismicity changes due to the effect of Coulomb stress changes. Wang *et al.* (2016) evaluate the influence of the 2011 M9 Tohoku-Oki mega earthquake on the seismicity in eastern China. Zhang (Shenjian) *et al.* (2016a) and Zhang (Shengfeng) *et al.* (2016b) discussed the b-value changes before and after the Wenchuan earthquakes and explain such variation by changes in the stress field and fault healing.
2. *Data quality* Seismology is often an observational scientific subject and research in the field is dependent on not only the availability of high-quality data, but also on a thorough understanding of what the data quality is. Panzera *et al.* (2016) provided a revised Iceland catalog, which is extremely useful for analyzing seismicity in this region. Mignan and Chen (2016) showed the detected seismicity scaled in space in the Taiwan region.
3. *Statistical analysis of seismicity* In that category, Chen and Shearer (2016b) compared the foreshocks in the California region and in ETAS synthetic catalogs. Li *et al.* (2016); Telesca *et al.* (2016); and Sarlis *et al.* (2016) analyzed seismicity in different regions around the world by using different statistical methods. For example, Li *et al.* (2016) used the p value and K-S statistics as goodness-of-fit test in estimating the completeness magnitude threshold. Different from other papers in this category, Ford and Labak (2016) analyzed the aftershock activity due to underground nuclear explosions.
4. *Seismicity-based probability forecasting of seismicity and seismic hazard evaluation* Articles in this category focus on developing statistical models and methods for earthquake forecasts. Yu *et al.* (2016); Zhang *et al.* (2016c); and Chang *et al.* (2016) discussed two forecast algorithms, PI and LURR, and their applications. Wu *et al.* (2016) developed a probability model where seismicity rate depends on quiescence and activation of small earthquakes. Nava *et al.* (2016) discussed the application of the Bayesian estimate and forecast procedures to a semi-periodic model, while Bayrak and Türker (2016) illustrated how to use Bayesian estimates to evaluate seismic hazard.
5. *Alternative datasets and testing of precursors* This class of articles moves beyond using only observed seismicity and attempts to find correlations between precursory non-seismic observations and seismicity, with a potential outcome of improved earthquake forecasting models. Fujinawa and Noda (2016); Jiang *et al.* (2016) studied the correlation between the seismoelectric wave field and the occurrence of earthquakes. Using the Molchan error diagram, Eleftheriou *et al.* (2016) and Chen *et al.* (2016a) evaluate the precursory information in the thermal infra-red spectrum emitted by the Earth and in the observation of mobile gravity fields, respectively.

In all papers presented, a good understanding of both the physical and statistical aspects of the problems investigated is necessary. To improve our understanding of how the Earth works, it is necessary to couple our understanding of the physics of the process with robust statistical methods. Doing so, and understanding the uncertainties involved will ultimately help address the important demands in seismology and the social and engineering sciences; importantly this includes the testing of models of earthquake forecasts, earthquake early warning, and seismic hazard assessments. Statistical seismology has presented many challenges to statisticians and geophysicists, but there are undoubtedly many more exciting discoveries yet to come.

References

Bayrak Y and T Türker (2016) *The Determination of Earthquake Hazard Parameters Deduced from Bayesian Approach for Different Seismic Source Regions of Western Anatolia*. Pure Appl Geophys (This issue).

Bebbington M, D Harte and C Williams (2016) *Cumulative Coulomb stress triggering as an explanation for the Canterbury (New Zealand) aftershock sequence: Initial conditions are everything?* Pure Appl Geophys (This issue).

Chang LY, CC Chen, YH Wu, TW Lin, CH Chang and CW Kan (2016) *A strategy for a routine pattern informatics operation applied to Taiwan*. Pure Appl Geophys (This issue).

Chen S, CS Jiang and JC Zhuang (2016) *Statistical evaluation of efficiency and possibility of earthquake predictions with gravity field variation and its analytic signal in western China*. Pure Appl Geophys (This issue).

Chen XW and PM Shearer (2016) *Analysis of foreshock sequences in California and implications for earthquake triggering*. Pure Appl Geophys (This issue).

Console R, K Yamaoka and J Zhuang (2012) *Implementation of Short- and Medium-Term Earthquake Forecasts*. Int J Geophys, 217923. doi:10.1155/2012/217923.

Eleftheriou A, C Filizzola, N Genzano, T Lacava, M Lisi, R Paciello, N Pergola, F Vallianatos and V Tramutoli (2016) *Long term RST analysis of anomalous TIR sequences in relation with earthquakes occurred in Greece in the period 2004–2013*. Pure Appl Geophys (This issue).

Ford SR and P Labak (2016). *An Explosion Aftershock Model with Application to On-Site Inspection*. Pure Appl Geophys (This issue).

Fujinawa Y and Y Noda (2016) *Characteristics of seismoelectric wave fields associated with natural microcracks*. Pure Appl Geophys (This issue).

Geller RJ, DD Jackson, YY Kagan and F Mulargia (1997) *Earthquakes cannot be predicted*. Science, *275* (5306), 1616.

Iwata T. (2016) *A variety of aftershock decay in the rate- and state-friction model due to the effect of secondary aftershocks: Implications from real aftershock sequences*. Pure Appl Geophys (This issue).

Jiang F, XB Chen, Y Zhan, GZ Zhao, H Yang, LQ Zhao, L Qiao and LF Wang (2016) *Shifting correlation between earthquakes and electromagnetic signals: a case study of the 2013 Minxian-Zhangxian ML6.5 (MW6.1) earthquake in Gansu, China*. Pure Appl Geophys (This issue).

Jordan TH, Y-T Chen, P Gasparini, R Madariaga, I Main, W Marzocchi, G Papadopoulos, G Sobolev, K Yamaoka and J Zschau (2011) *Operational earthquake forecasting: state of knowledge and guidelines for implementation*. Annals Geophysics, *54* (4), 316–391.

Leptokaropoulos KM, EE Papadimitriou, B Orlecka-Sikora and VG Karakostas (2016) *Evaluation of Coulomb Stress Changes from Earthquake Productivity Variations in Western Corinth Gulf, Greece*. Pure Appl Geophys (This issue).

Li HC, CH Chang and CC Chen (2016) *Quantitative analysis of seismicity before Large Taiwanese Earthquakes Using G-R Law*. Pure Appl Geophys (This issue).

Mignan A and CC Chen (2016) *The spatial scale of detected seismicity*. Pure Appl Geophys (This issue).

Nava FA, CB Quinteros, E Glowacka and J Frez (2016) *A Bayesian Assessment of Seismic Semi-Periodicity Forecasts*. Pure Appl Geophys (This issue).

Ogata Y (1988) *Statistical models for earthquake occurrences and residual analysis for point processes (in Applications)*. J Am Stat Associ, *83* (401), 9–27.

Panzera F, JD Zechar, KS Vogfjörd and DAJ Eberhard (2016) *A revised earthquake catalogue for South Iceland*. Pure Appl Geophys (This issue).

Papadopoulos G. (editor) (2012). *7th International Workshop in Statistical Seismology, 2011*. Res Geophys, 2 (1).

Rhoades D, M Savage, E Smith, M Gerstenberger and D Vere-Jones (2010) *Introduction. Special Issue: Seismogenesis and Earthquake Forecasting: the Frank Evison Symposium*. Pure Appl Geophys, *167*, 619–621.

Sarlis NV, ES Skordas, S-R G Christopoulos and PA Varotsos (2016) *Statistical significance of the minimum of the order parameter fluctuations of seismicity before major earthquakes in Japan*. Pure Appl Geophys (This issue).

Telesca L, M Lovallo, SK Aggarwal, PK Khan and BK Rastogi (2016) *Visibility graph analysis of 2003–2012 earthquake sequence in Kachchh region, Western India*. Pure Appl Geophys (This issue).

Tsaklids GM, EE Papadimitriou and N Limnios (2011) *Statistical tools for earthquake and mining seismology: preface to the topical issue*. Acta Geophysica, *59*, 657–658. doi: 10.2478/s11600-0022-4.

Vere-Jones D, Y Ben-Zion and R Zuniga (2005) *Statistical seismology*. Pure Appl Geophys, *162* (6–7), 1023–1026.

Wang LF, J Liu, J Zhao and JG Zhao (2016) *Tempo-spatial impact of the 2011 M9 Tohoku-Oki earthquake on Eastern China*. Pure Appl Geophys (This issue).

Wu YH, CC Chen and HC Li (2016) *Conditional probabilities for Large Events Estimated by Small Earthquake Rate*. Pure Appl Geophys (This issue).

Yu HZ, FR Zhou, QY Zhu, XT Zhang and YX Zhang (2016) *Development of a combination approach for seismic hazard evaluation*. Pure Appl Geophys (This issue).

Zhang SJ and SY Zhou (2016) *The spatial and temporal variation of the b-value in Southwest China*. Pure Appl Geophys (This issue).

Zhang, SF, ZL Wu and CS Jiang (2016) *The central China North-South Seismic Belt: Seismicity, Ergodicity, And Five-Year PI Forecast in Testing*. Pure Appl Geophys (This issue).

Zhang SF, ZL Wu and CS Jiang (2016) *Signature of fault healing in an aftershock sequence? The 2008 Wenchuan earthquake*. Pure Appl Geophys (This issue).

Pure Appl. Geophys. 173 (2016), 5–20

DOI 10.1007/s00024-015-1062-5

Pure and Applied Geophysics

Cumulative Coulomb Stress Triggering as an Explanation for the Canterbury (New Zealand) Aftershock Sequence: Initial Conditions Are Everything?

Mark Bebbington,[1] David Harte,[2] and Charles Williams[2]

Abstract—Using 2 years of aftershock data and three fault-plane solutions for each of the initial M7.1 Darfield earthquake and the larger ($M > 6$) aftershocks, we conduct a detailed examination of Coulomb stress transfer in the Canterbury 2010–2011 earthquake sequence. Moment tensor solutions exist for 283 of the events with $M \geq 3.6$, while 713 other events of $M \geq 3.6$ have only hypocentre and magnitude information available. We look at various methods for deciding between the two possible mechanisms for the 283 events with moment tensor solutions, including conformation to observed surface faulting, and maximum ΔCFF transfer from the Darfield main shock. For the remaining events, imputation methods for the mechanism including nearest-neighbour, kernel smoothing, and optimal plane methods are considered. Fault length, width, and depth are arrived at via a suite of scaling relations. A large (50–70 %) proportion of the faults considered were calculated to have initial loading in excess of the final stress drop. The majority of faults that accumulated positive ΔCFF during the sequence were 'encouraged' by the main shock failure, but, on the other hand, of the faults that failed during the sequence, more than 50 % of faults appeared to have accumulated a *negative* ΔCFF from all preceding failures during the sequence. These results were qualitatively insensitive to any of the factors considered. We conclude that there is much unknown about how Coulomb stress triggering works in practice.

Key words: Coulomb failure stress, Canterbury earthquake sequence, Fault plane solutions.

1. Introduction

The Canterbury earthquake sequence was notable for being very active across multiple faults where previously there was little observed activity. Further, the sequence displayed considerable spatial migration to the east, had considerable longevity compared to other main shock-aftershock sequences, and also consisted of four rather late major aftershocks each with a distinct aftershock sequence (Christophersen *et al.* 2013). Was it a 'normal' main shock–aftershock phenomenon, or is something more organized going on?

A number of hypotheses have been advanced for the possibly anomalous nature of the sequence:

- That the region is seismogenically immature (Fry and Gerstenberger 2011; Ristau *et al.* 2013), with a tendency to produce more and larger events than in more mature settings.
- The region has multiple receiver fault orientations (Hainzl *et al.* 2010b); however, the GeoNet (www.geonet.org.nz) data tends to indicate quite a limited range of fault orientations in the Canterbury sequence.
- The faults in the Christchurch region were 'primed' prior to the Darfield event for a long and energetic sequence.

We investigate this last possibility, which is based on a triggering argument.

It is generally accepted that aftershocks are triggered by changes in static or dynamic stress resulting from earlier events, although it is difficult to isolate their relative contributions (Freed 2005). However, static stress changes, while limited in their effect to a few fault lengths from the rupture, are permanent rather than transient (assuming elastic behaviour), and explain the observed stress-shadowing effect in aftershocks (Toda *et al.* 2012). Dynamic stress changes appear to be more important in distant co-seismic triggering rather than aftershock production (Hill 2008; Richards-Dinger *et al.* 2010), and hence we will concentrate on static stress changes as an explanation for the aftershock sequence.

[1] IAE, Massey University, Private Bag 11222, Palmerston North 4442, New Zealand. E-mail: m.bebbington@massey.ac.nz
[2] GNS Science, 1 Fairway Drive, Avalon, Lower Hutt 5010, New Zealand.

The Coulomb failure function (CFF) stress is produced by the dislocation on the master (or source) fault based on Okada's solution (OKADA 1992), where the stress is resolved onto a receiver fault to estimate the stress perturbation of that fault (KING *et al.* 1994; STEIN *et al.* 1997). Recall that Coulomb stress changes (ΔCFF) are calculated as $\Delta\text{CFF} = \Delta\tau + \mu \times (\Delta\sigma_n + \Delta P)$, where $\Delta\tau$ is the shear stress in the direction of slip on the receiver fault plane, $\Delta\sigma_n$ is the normal stress change (positive for extension), μ is the friction coefficient and ΔP is the pore pressure change. Assuming a homogeneous and isotropic region, and that the pore pressure change is related to the normal stress, this becomes the constant effective friction model

$$\Delta\text{CFF} = \Delta\tau + \mu' \times \Delta\sigma_n, \quad (1)$$

where μ' is the *apparent coefficient of friction*, usually assumed to lie between 0 and 0.8 (HARRIS 1998).

In addition to the dependence on the source rupture model, the Coulomb stress change on a receiver fault depends on its location, geometry and rake relative to the source fault. This naturally leads to the question of what these should be (STEACY *et al.* 2005; HAINZL *et al.* 2010b). Two alternative paradigms have emerged (HAINZL *et al.* 2010b). The first is that aftershocks occur on optimally oriented planes (KING *et al.* 1994), the orientation of which is a function of both the source parameters and the regional stress field. The second, slightly different idea, is that regions have a predominant fault orientation, and that aftershocks occur in accordance with this (TODA *et al.* 2011), thus the aftershock mechanisms are largely independent of the main shock mechanism except in so far as the latter is also in line with the regional mechanism. However, neither of these has been entirely successful in explaining aftershock production, particularly relative to using the actual mapped faults (PARSONS *et al.* 2012a, b). We intend to exploit the moment tensor catalogue from GeoNet in a way analogous to the latter paradigm.

A natural question at this point is 'what is an aftershock?' Aftershocks are usually defined in a space-time-magnitude window of the 'main shock', in accordance with various statistical models such as the ETAS model (OGATA and ZHUANG 2006), but this is a subtly different definition to that of triggering. An event can be defined to be 'triggered' if the static stress drop is greater than the sum of the ΔCFF contributions from earlier events in the sequence, but not all triggered events are necessarily aftershocks (PARSONS *et al.* 2012a). It has been shown (STEACY *et al.* 2014) that the use of ΔCFF improves the probabilistic forecasting in a hybrid STEP/ΔCFF model, a necessary preliminary check on our hypothesis. More complex frictional models (CATALLI *et al.* 2008; HAINZL *et al.* 2010a; CATALLI and CHAN 2012), based on the rate-state model (DIETERICH 1994) suggest a decay in the predicted number of aftershocks due to the evolution of the state variable.

The definition of triggered events generates some curiosity about what percentage of the stress drop is accounted for by the ΔCFF contributions. Let us suppose that the stress on each fault has a loading cycle, which is released in the form of earthquakes. This is a common assumption (e.g., STEIN *et al.* 1997) in static triggering. In general, one possibility is that faults are at a random point in their cycle, and tectonic loading and inter-event transfer results in what one might consider 'typical' aftershock sequences, separated by 'typical' periods of quiescence. However, if the stress loading pre-Darfield was coherent rather than random across the faults, then the unexpected occurrence and scale of the Canterbury earthquake sequence could have been a consequence of the fault segments being 'in phase', and thus failing more in unison. This would have major implications for determining the expected activity (both persistence and event magnitudes) of an evolving earthquake sequence in a previously quiescent part of the country.

The unusual nature of the Canterbury earthquake sequence (well-located, with multiple large events in a previously quiescent area) provides an ideal opportunity to quantify localized temporal correlation between fault loading cycles, and examine the hazard implications.

2. Method

Let the ith event in the sequence occur at time t_i. We will calculate the cumulative Coulomb failure function

$$C_{\Delta \mathrm{CFF}}(j) = \sum_{i:t_i < t_j} \Delta \mathrm{CFF}(i \rightarrow j) \tag{2}$$

on each fault j up to the time of failure t_j, where $\Delta\mathrm{CFF}(i \rightarrow j)$ is the Coulomb failure function (King *et al.* 1994; Stein *et al.* 1997) from the event i resolved on the fault j. In order to calculate this using the method of Okada (1992), we need the location (x, y, z), mechanism (strike, dip, and rake), the dimension of the faults (both source and receiver), and the slip on the source fault. Note that every fault is first a receiver and, after its rupture, a source.

In calculating the ΔCFF we will use an apparent coefficient of friction $\mu' = 0.4$ (Steacy *et al.* 2014). The initial stress state will be immediately before the Darfield main shock, the source model for the latter being taken as one of the available multiple segment models (Atzori *et al.* 2012; Beavan *et al.* 2012; Elliott *et al.* 2012). As tectonic stress loading can be assumed to be negligible during the Canterbury sequence, the cumulative ΔCFF provides a potential explanation of the entire (years-long) earthquake sequence. Our investigation will involve looking at the ΔCFF evolution of each fault-presumptive until its failure. If ΔCFF reaches a maximum at the time of failure, we interpret this to mean that stress triggering is a plausible explanation of the failure mechanism.

Assuming static stress transfer to be a suitable model, the initial (immediately before the Darfield event) loading state of the fault can be estimated (Stein *et al.* 1997) as

$$X_j = (\Delta\sigma - C_{\Delta \mathrm{CFF}}(j))/\Delta\sigma, \tag{3}$$

where $\Delta\sigma$ is the stress drop in the event (Cotton *et al.* 2013; Shaw 2013). If $X_j > 1$, it represents a stress condition that is higher than the level of stress drop in the subsequent earthquake. A level of $X_j < 0$ then represents stress shadowing. We will calculate $\Delta\sigma$, assuming uniform slip on a rectangular plane of length L and width W, as

$$\Delta\sigma = \frac{8}{3\pi}\frac{M_0}{W^2 L} \tag{4}$$

(Kanamori and Anderson 1975), where M_0 is the seismic moment. The values $\{X_j\}$ can then be examined against various hypotheses such as randomness (i.e., a uniform distribution) that would indicate no coherence.

3. Data Overview

The evolution of the Canterbury earthquake sequence has been described by Bannister and Gledhill (2012). The sequence (Fig. 1) shows a west to east migration. A closer look, considering only the longitude and time (Fig. 2) reveals four significant sub-sequences [following the M7.1 Darfield, M6.3 (February) Christchurch, M6.4 (June) Christchurch and M6.1 Pegasus Bay events], each with its own contribution to the overall aftershock sequence (Fig. 3). All events are of shallow depth, most being less than 20 km. Many have a catalogued depth of 5 km, which is an indicator of considerable depth uncertainty.

The GeoNet moment tensor catalogue[1] provides two possible fault solutions (i.e. strike, rake, and dip). We use all the 287 available solutions of the Canterbury sequence until 31 December 2011. This provides 287 solutions with $M \geq 3.6$, including those of the four major ($M > 6$) events during this period. The GeoNet catalogue contains another 713 earthquakes of $M \geq 3.6$ without moment solutions during this period. This magnitude cut-off is comfortably above the completeness magnitude of 3.0 determined by Shcherbakov *et al.* (2012), as demonstrated in Fig. 4. We note that a number of large events closely following the major events have missing moment tensor solutions. This is due to the difficulty in separating the signals received concurrently at these times from the large number of events.

The overriding principle in our analysis will be to have a homogeneous (with respect to the error structure) data set. Hence, although portions of the catalog have been relocated (e.g. Bannister *et al.* 2011; Syracuse *et al.* 2012), or had stress drops calculated (Kaiser *et al.* 2013), we will use the original data in the GeoNet catalogue. The location bias appears to be locally homogeneous (Syracuse *et al.* 2012), and relocation requires reprocessing the seismogram data, which is beyond the scope of this paper.

The tectonic stress regime for the Canterbury region is predominantly strike-slip, with σ_1 horizontal

[1] http://info.geonet.org.nz/display/appdata/Earthquake+Resources.

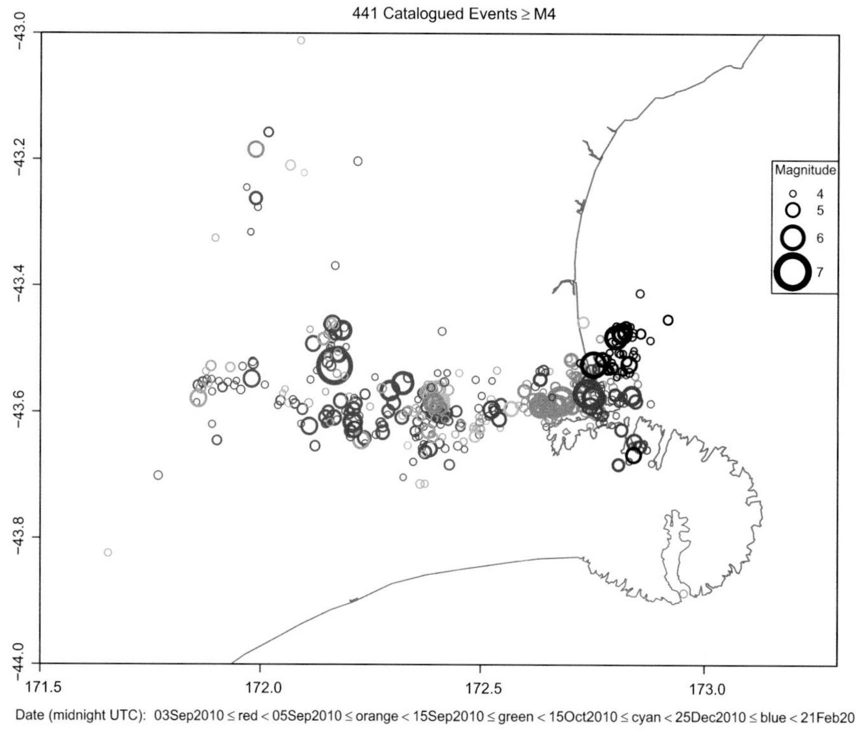

Figure 1
Epicentral plot of the Canterbury earthquake sequence, with *colour* indicating time

and oriented at 115° (Sibson *et al.* 2011). This implies that σ_2 is vertical and σ_3 is horizontal and at right angles to σ_1; however, the mixture of strike-slip and reverse faulting throughout this region implies that $\sigma_2 \approx \sigma_3$. The regional stress field is important when computing ΔCFF on optimally oriented receiver faults, but it is not relevant when the receiver fault mechanism is specified.

4. Fault Mechanisms, Dimensions, and Slip

For our purposes we need the location, mechanism, fault dimensions, and slip for every event considered. Our data fall into three classes: major events (e.g., M7.1 Darfield, the M6.3 February 22 and M6.4 June 6 Christchurch, and M6.1 December 23 Pegasus Bay events) with detailed slip models available, catalog events with moment tensor solutions (a choice of two fault solutions), and catalog events with only location and magnitude. If we can variously select, impute, and estimate the missing information, we will be able to calculate the cumulative ΔCFF (from the beginning of the Darfield sequence) on each subsequently active receiver fault. We will now discuss the available options.

4.1. Major ($M > 6$) Events

A number of detailed finite source models, giving dimensions, mechanisms, and slip, exist for the four major ($M > 6$) events in the sequence. While Beavan *et al.* (2012) provide solutions for all four, the Atzori *et al.* (2012) solutions omit Pegasus Bay, and Elliott *et al.* (2012) provide fault models only for the Darfield and first Christchurch event. In order to have three complete sets of models, we will use the Atzori *et al.* (2012) solution [as the earlier solutions are closer than the Beavan *et al.* (2012) ones] for the second Christchurch event in the Elliott *et al.* (2012) model, and the Beavan *et al.* (2012) solution

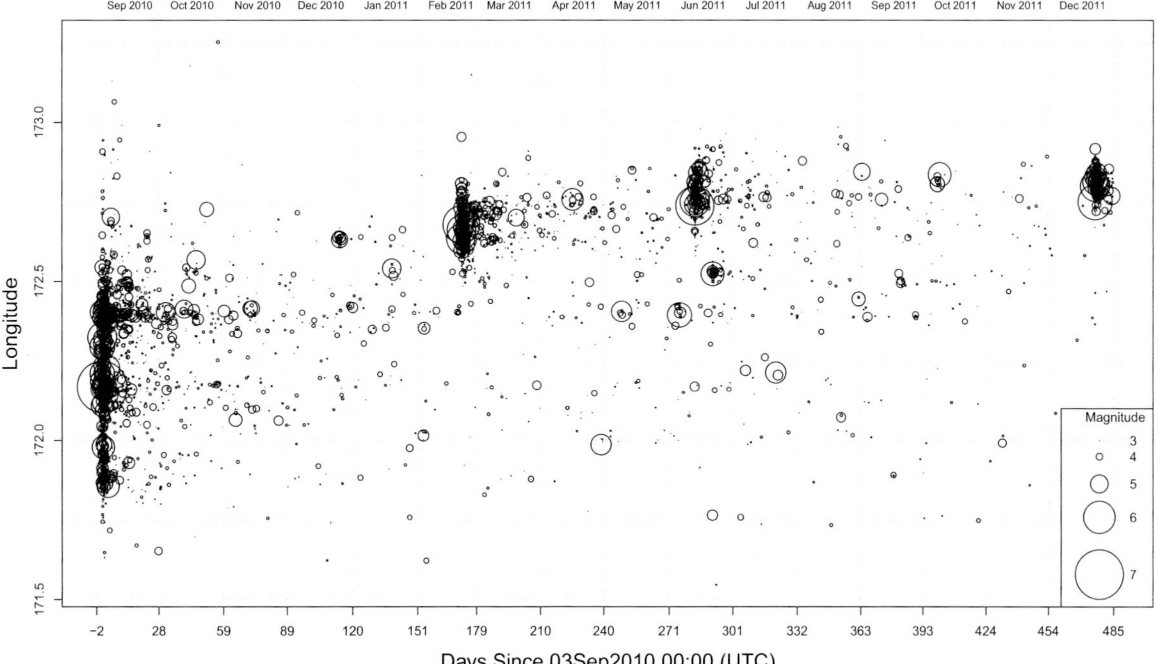

Figure 2
Longitude (showing the easterly migration of events) over time during the Canterbury sequence. There were four significant sub-sequences, each with their own contribution to the overall aftershock sequence. The last sub-sequence, which started on 23 Dec 2011, is located slightly north of the rest, whereas the first three sub-sequences were spread over very similar latitudes

for Pegasus Bay in all the models. The three resulting models are summarized in Table 1.

Steacy *et al.* (2004) found that slip solutions incorporating the correct rupture geometry, but greatly simplified slip produce stress fields consistent with the aftershock distribution when very near-fault events are excluded. Hence, and because of the number of computations required for individual receiver fault orientations, we will consider the fault planes to be rectangular, with uniform slip. So, in the case of the non-uniform slip Atzori *et al.* (2012) solutions, the average slip is obtained via the relation (Aki and Richards 1980; Stirling *et al.* 2012)

$$M_0 = \mu L W \bar{D}, \tag{5}$$

where $\mu = 3 \times 10^{11}$ dyn/cm^2 is the rigidity modulus, and L, W and $\bar{D}$ are fault length, width and average slip (in cm).

We will treat all other events as occurring on single rectangular planes.

4.2. *Choosing Mechanisms for Events with Moment Tensor Solutions*

For the remaining 283 events with moment tensor solutions, we need to select the fault plane from the two solutions. There are a number of ways in which we can choose between the two possibilities:

- The dominant mechanism in the Canterbury sequence is strike-slip, striking generally E–W and dipping nearly vertically (Sibson *et al.* 2011) and consistent with the CMT moment solution mechanism for the Darfield event (Gledhill *et al.* 2011). There are also some reverse faults striking NE–SW (Sibson *et al.* 2011). Hence, a *decision rule* is to choose between the mechanisms according to the priority rules (cf. Furlong 2013):

1. A strike-slip event is identified with the highest priority if one of the two solutions has a rake between −45° and 45°, or less than −135°, or

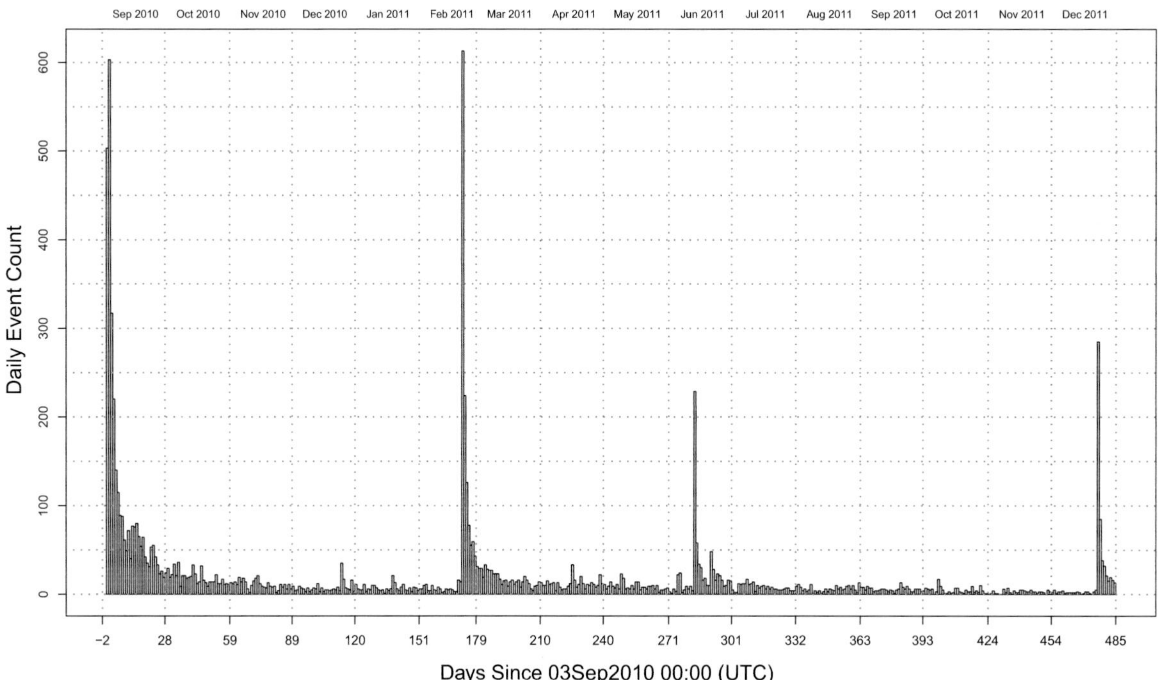

Figure 3
Event counts for each day. There were four significant sub-sequences, each with their own contribution to the overall aftershock sequence

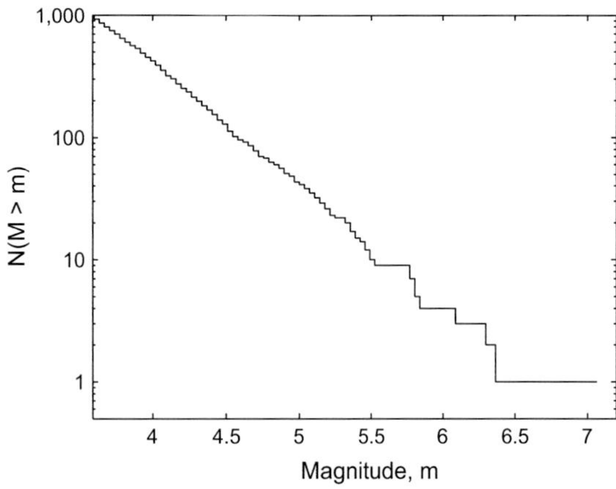

Figure 4
Frequency-magnitude plot for the earthquakes used in our analysis

more than 135°, and a strike between 45° and 135°, or between 225° and 315°.

2. A reverse faulting event if the rake is between 45° and 135° with a strike between 0° and 90° or between 180° and 270°.
3. The steepest dipping of any remaining strike-slip mechanisms
4. The steepest dipping of any remaining reverse mechanisms
5. The steepest remaining mechanism

- If we consider the sequence as the aftershocks of the Darfield event, we can select the mechanism that provides the higher ΔCFF from that event, including regional stress. The regional stress orientations are taken from Sibson *et al.* (2011). Compressive stress values are assumed to be $\sigma_1 =$ 10 MPa, $\sigma_2 = 0.5$ MPa and $\sigma_3 = 0.1$ MPa (Steacy *et al.* 2014). The easiest way to do this is to use the Harvard CMT solution for the Darfield event (strike = 268, dip = 87, rake = −166), with a fault length of 29.5 km, average slip of 2.5 m (Quigley *et al.* 2012), and width of 12 km (Gledhill *et al.* 2011).
- Steacy *et al.* (2004) noted that orientations are much more likely to be consistent with more detailed slip models, so we can choose the mechanism that has the greater ΔCFF from each of the solutions in Table 1. Steacy *et al.* (2014)

Table 1

Major event source solutions: number of fault planes

Event	Atzori *et al.* (2012)	Beavan *et al.* (2012)	Elliott *et al.* (2012)
Darfield 2011.10.4	8	7	8
Christchurch 2011.2.22	2	3	2
Christchurch 2011.6.13	1	2	Use Atzori *et al.* (2012)
Pegasus Bay 2011.12.23	Use Beavan *et al.* (2012)	1	Use Beavan *et al.* (2012)

Table 2

Count of mechanisms for the 283 events with moment tensor solutions

Method	Number		
	Strike-slip	Normal	Reverse
Decision rule	249	2	32
Max ΔCFF from Darfield main shock (Harvard CMT)	248	2	33
Max ΔCFF from Darfield main shock (Atzori *et al.* 2012)	247	5	31
Max ΔCFF from Darfield main shock (Beavan *et al.* 2012)	248	2	33
Max ΔCFF from Darfield main shock (Elliott *et al.* 2012)	247	5	31

note that the aftershocks are more consistent with the eight rupture plane models of Atzori *et al.* (2012) and Elliott *et al.* (2012) than the seven plane solution of Beavan *et al.* (2012).

The count of mechanisms for each method is shown in Table 2.

We can see on first inspection that it appears to make little difference which rule we use, but the more detailed look in Fig. 5 is very revealing. Selecting the mechanism on the basis of the greater CFF transfer results in a considerable number of N–S striking faults, where we expect to see almost none, regardless of the main shock representation.

4.3. *Imputing Mechanisms for Events Without Moment Tensor Solutions*

For these 713 events, we have only hypocentre and magnitude information.

- Although the magnitudes considered were lower than is the case here, Hardebeck (2006) found that for events within 5 km of each other, the mechanisms were identical to within observation error. This suggests that the mechanisms can be imputed by spatially smoothing the known mechanisms. Note that in order to avoid problems with the strike having modes at 90 and 270, the existing strike (s_0), dip (d_0) and rake (r_0) have to be reparameterized:

$$\begin{aligned} d_1 &= d_0 I_{[0,180)}(s_0) + (180 - d_0) I_{[180,360)}(s_0) \\ r_1 &= r_0 I_{[0,180)}(s_0) - r_0 I_{[180,360)}(s_0) \\ s_1 &= s_0 I_{[0,180)}(s_0) + (s_0 - 180) I_{[180,360)}(s_0) \end{aligned}$$

where

$$I_A(x) = \begin{cases} 1, & x \in A \\ 0, & \text{otherwise.} \end{cases}$$

Both the new strike s_1 and dip d_1 are between 0° and 180°, while the rake remains between −180° and 180°. After smoothing the transformation is inverted to return the dip to between 0° and 90°. We will consider two possibilities; either the mechanism is assigned to be that of the nearest known mechanism, or a Gaussian kernel smoother with bandwidth 1.1, 0.8, 2.4 km for strike, dip, and rake, respectively, is used. The bandwidths were determined via leave-one-out cross-validation for the decision rule-based GeoNet mechanisms. For the CMT-optimal GeoNet mechanisms, the bandwidth was similarly determined to be 1.3, 1.8, 2.1 km for strike, dip, and rake, respectively.
- An alternative is to assign these events an optimally oriented plane (King *et al.* 1994). In practice we shall search over a grid with strike and rake from 0° to 350° in 10° increments, and dip from 30° to 90° in 5° increments. We will use the Harvard CMT and Atzori *et al.* (2012) Darfield solutions for the main shock in this instance, which we take to represent the regional fault dynamics.

The resulting distribution of imputed mechanisms are shown in Fig. 6, with the decision rule-based GeoNet mechanisms presented for reference. We see, first of all, that the optimal planes are distributed completely differently from the known mechanisms. The preponderance of shallow dips is of particular concern,

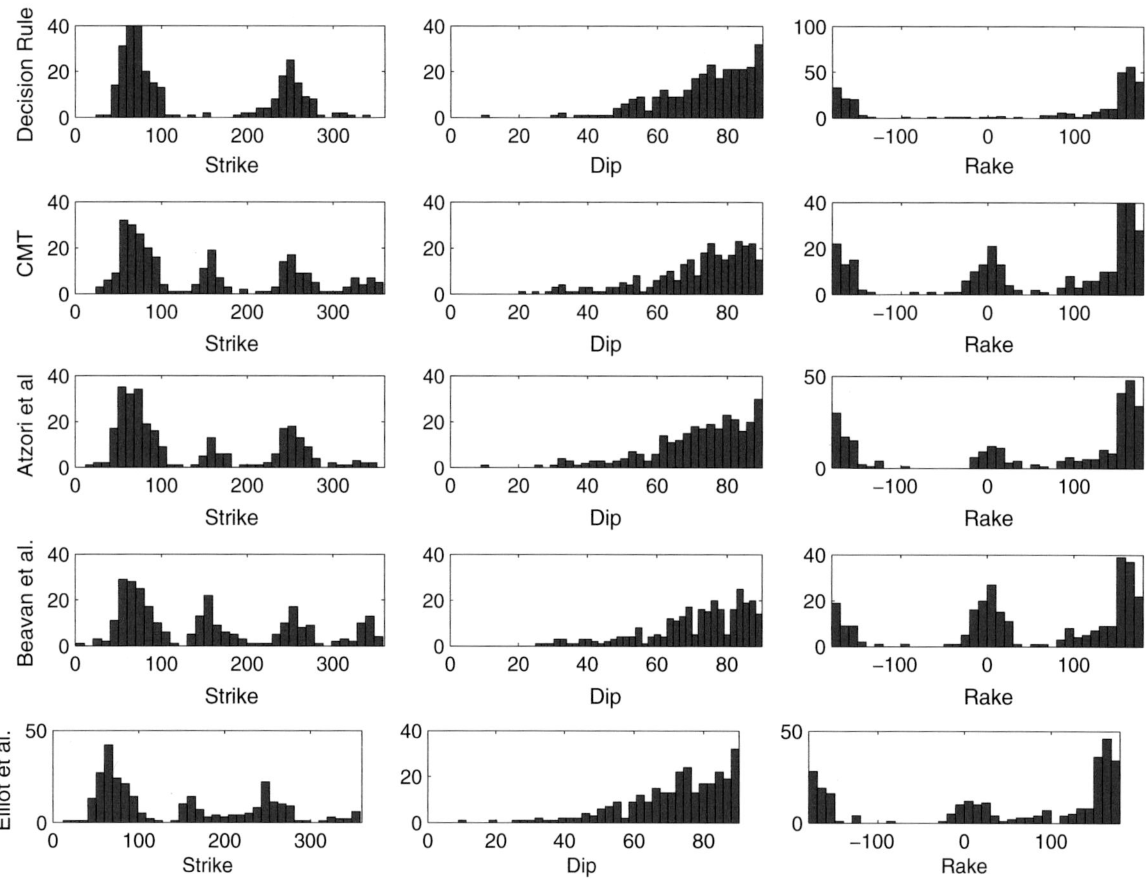

Figure 5
Strike, dip, and rake of the selected mechanism from the 283 GeoNet solutions, for the different selection rules

as is the concentration of strikes near 360° from the Atzori *et al.* (2012) solution. The kernel smoothed solutions are also clearly different from the known mechanisms. The best of the lot appear to be the nearest-neighbor solutions, particularly to the decision rule-based GeoNet mechanisms.

4.4. Fault Dimensions and Slip for $M \leq 6$ Events

While the major events in the sequence have dimensions and slips assigned, the remaining 996 $M \geq 3.6$ events do not. We thus need to estimate length L, width W, and average slip $\bar{D}$ from the known M_W. Stirling *et al.* (2013) surveyed the available scaling relations. For the strike-slip dominated, plate boundary crustal environment in Canterbury, where the total slip rate is 2.5–7 mm/year (Atzori *et al.* 2012), the relations recommended by Stirling *et al.* (2013) for slip <10 mm/year are

$$M_W = \log A + 3.98 \tag{6}$$

(Hanks and Bakun 2008) or

$$M_W = 5.56 + 0.87 \log L \tag{7}$$

(Wesnousky 2008) for strike-slip events, and

$$M_W = 4.18 + (2/3)\log W + (4/3)\log L \tag{8}$$

(Stirling *et al.* 2012) for dip-slip events, where M_W is obtained via

$$M_0 = 10^{16.05+1.5M_W}\,\mathrm{dyn\,cm} = 10^{9.11+1.5M_W}\,\mathrm{Nm} \tag{9}$$

(Yen and Ma 2011).

For normal faulting events

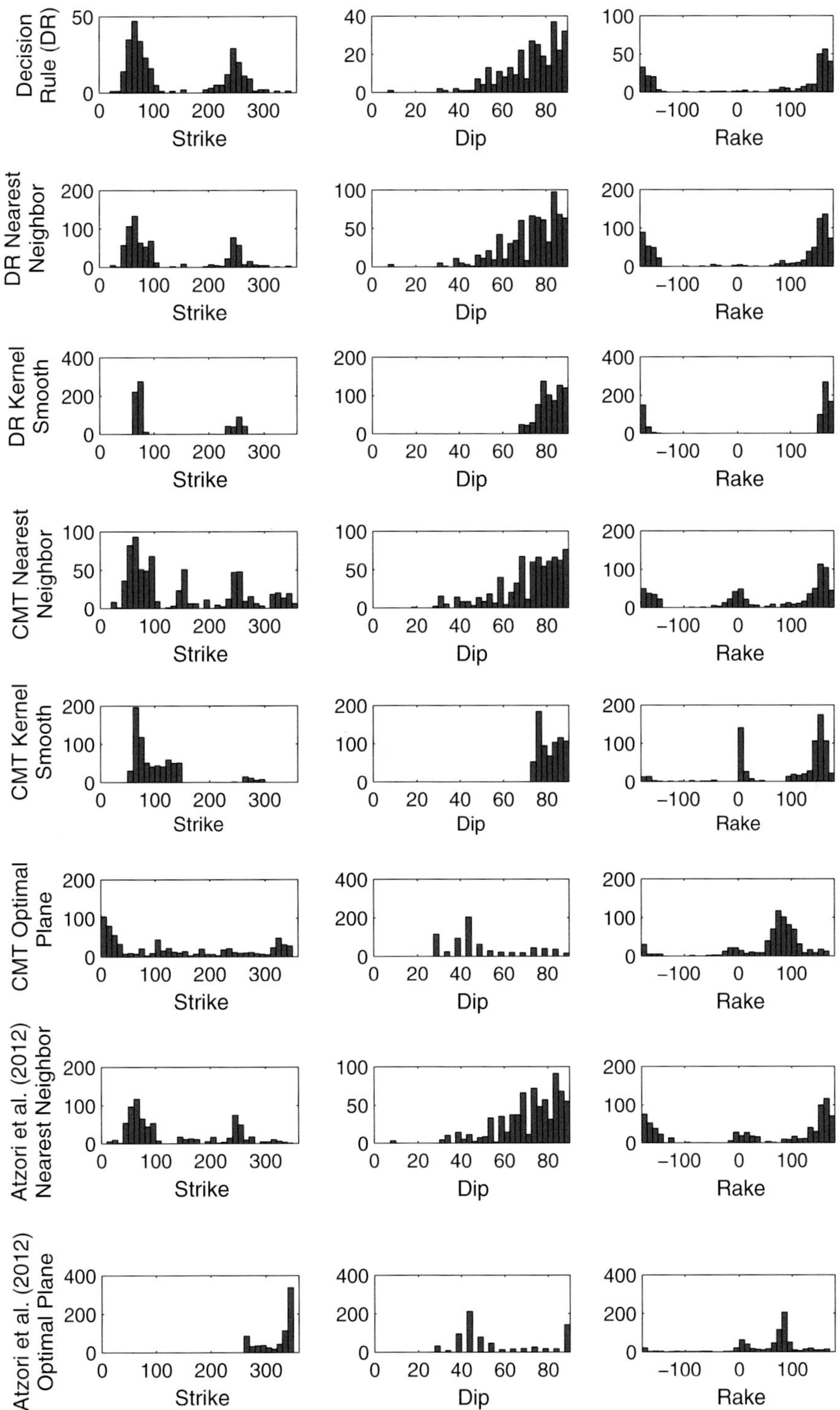

Figure 6
Strike, dip, and rake of the imputed mechanisms for the 713 events without them, for the different selection rules

$$M_W = 6.12 + 0.47 \log L \qquad (10)$$

(Wesnousky 2008), while for reverse faults Stirling *et al.* (2013) recommend

$$M_W = 4.11 + 1.88 \log L \qquad (11)$$

(Wesnousky 2008), and Eq. (8).

However, these relations are tuned to large ($M \sim 6.5$) earthquakes, and so we need to check that they scale sensibly to smaller events. For a typical (in the Canterbury sequence) event $M_W = 4.0$ strike-slip rupture, Eqs. (7), (10), and (11) produce fault lengths of 160 m, 3 cm, and 874 m, respectively, of which only the latter appears reasonable. Using Eq. (8) with the assumption of a square fault (Convertito *et al.* 2013) yields a fault length and width of 813 m for a $M_W = 4$ earthquake. Note that in this case the Wells and Coppersmith (1994) scaling relations give a fault length of 832 m, width of 1862 m, and an average slip of 9 mm.

Instead, let us consider the relations of Leonard (2010), where

$$\begin{aligned}\log M_0 &= (3.0 \log L + 6.09) I_{(0,3.4]}(L) \\ &\quad + (2.5 \log L + 7.85) I_{(3.4,45)}(L) \\ &= 1.5 \log A + 6.09 \qquad (12)\end{aligned}$$

for strike-slip faults, and

$$\begin{aligned}\log M_0 &= (3.0 \log L + 6.10) I_{(0,5.5]}(L) \\ &\quad + (2.5 \log L + 7.96) I_{>5.5}(L) \\ &= 1.5 \log A + 6.10 \qquad (3)\end{aligned}$$

for normal and reverse faults.

With the relation (9), we obtain from (12) a fault length of 1,023 m and width of 316 m for a strike-slip fault with $M_W = 4$. In this case we do not need to assume a square fault, and for this reason we will adopt this scaling. We will assume that the hypocenter is the centroid of the fault plane, reducing the width (keeping the area constant) as necessary to ensure that the top of the fault plane is not above ground. The average slip is then calculated from (5), yielding 13 cm for our hypothetical $M_W = 4$ strike-slip event. We note that the Stirling *et al.* (2012) fault patch would slip 6.5 cm.

4.5. Experimental Design

We have considered a number of factors that can be influential in the model of Coulomb stress triggering, and have outlined what form the variations can take in each case (these are known as 'levels'). The various factors and their levels are summarized in Table 3.

We have included the means of selection among the moment tensor solutions, even though only the decision rule appears in line with ground truth, and the means of determining the fault dimensions, which again we have discussed above. In order to be consistent, if imputed events are on optimal planes, then mechanisms must be chosen from the GeoNet pairs on the basis of maximum ΔCFF from the source model. We thus arrive at a total of 20 possible scenarios, as listed in Table 3.

5. Results

The primary quantity of interest is the initial (immediately before the Darfield event) loading state of the faults, expressed on a 0–1 scale, where 1 is the

Table 3

The various design factors in our experiment

Factor		No. of levels	Details of levels
I	Source models of $M > 6$ events	4	Harvard CMT, Atzori *et al.* (2012), Beavan *et al.* (2012), Elliott *et al.* (2012)
II	Method of selecting between GeoNet mechanisms	2	Decision rule, maximum ΔCFF from Darfield source model
III	Include events without mechanisms	2	Yes, no
IV	How mechanisms are imputed for events without them	2	Nearest neighbour, optimal plane (cannot then use Decision Rule)

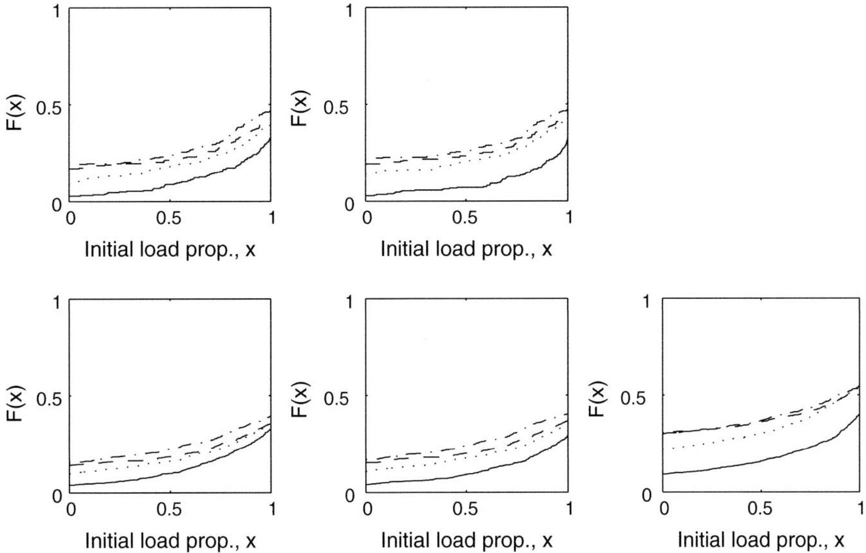

Figure 7
Distribution $F(x)$ (cumulative proportion less than x) of initial stress loading. *Top panels* do not include imputed events, *bottom panels* do. *Left panels* use the decision rule to select from mechanism pairs, other panels use the mechanism with greater ΔCFF from the Darfield source model. The *bottom-centre* and *bottom-left panels* impute mechanisms using the nearest-neighbour rule, the *bottom-right panel* calculates optimal planes. In all panels, *solid line* is the CMT source model, *dashed line* is that of Atzori *et al.* (2012), *dotted line* Beavan *et al.* (2012), and *dot-dash line* Elliott *et al.* (2012)

stress drop. This is shown in Fig. 7. We see that while the CMT source model leads to the conclusion that almost all faults considered are positively loaded, the Atzori *et al.* (2012) and Elliott *et al.* (2012) source models suggest that approximately one in five faults (two in five if ruptures occur on optimal planes) have a negative initial (immediately before the Darfield event) load, i.e., will accumulate more stress in the sequence prior to its failure than the stress drop, indicative of lying in a stress shadow. The Beavan *et al.* (2012) model lies between these extremes. In all models a large (50–70 %) proportion begin with a loading that exceeds the nominal breaking stress. In this sense, the initial load condition certainly encourages the sequence, but the idea implicit in (3) is that the stress increases (not monotonically) until it exceeds the stress drop on the fault, at which stage the earthquake occurs. This was used by Stein *et al.* (1997) to calculate the hazard. What Fig. 7 shows is that regardless of the details of the source model used and the allocation of aftershock mechanisms, > 50 % of the faults appear to have had stresses in excess of this critical level immediately before the Darfield event. There appears to be something missing from the theory.

It is thus of interest to see how well the peak stress correlates with the failure time. Figure 8 shows the time as a proportion of failure time (0 = Darfield, 1 = failure time of the fault concerned) at which the maximum ΔCFF was recorded on the fault. We see that under the source models of Atzori *et al.* (2012), Beavan *et al.* (2012), and Elliott *et al.* (2012) approximately 60 % of the faults have their peak loading at time 0, i.e., immediately prior to the Darfield earthquake. Using the CMT model, on the other hand, this is reduced to 10 % without imputed mechanisms, or 30 % with imputed mechanisms. In all cases, approximately 10 % of failures occur at the point of maximum load. This insensitivity is remarkable. The greater degree of peak loading in the Beavan *et al.* (2012) model, especially under the optimal planes imputation, probably reflects the omission of the Charing Cross reverse fault from the Beavan *et al.* (2012) solution.

The total loading (proportion of fault stress drop due to stress transfer from the preceding earthquakes in the sequence) is shown in Fig. 9. We see that the range of loading is least for the CMT source model, and greatest for the Atzori *et al.* (2012) and Elliott *et al.* (2012) source models. More than half of the faults receive a cumulative ΔCFF less than zero,

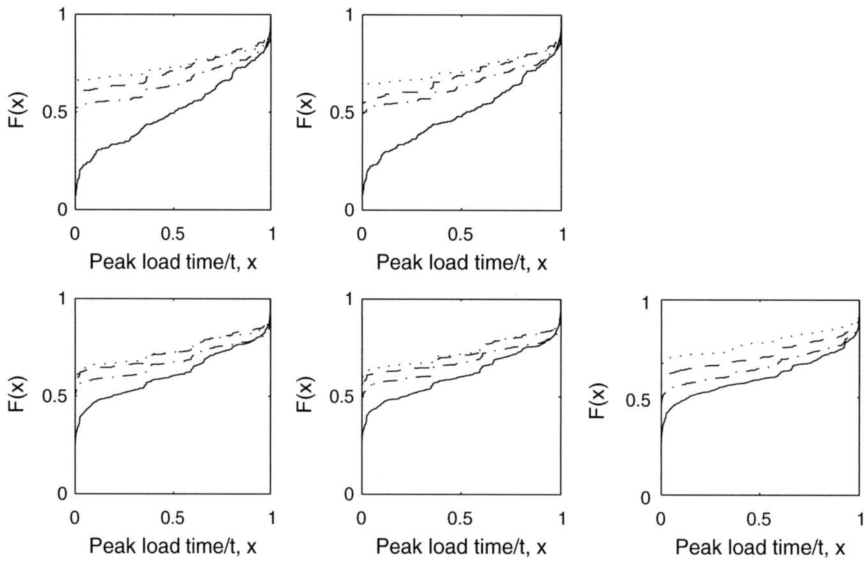

Figure 8
Distribution $F(x)$ (cumulative proportion less than x) of peak loading time as normalized between 0 (=Darfield main shock) and 1 (=fault failure time). Otherwise as for Fig. 7

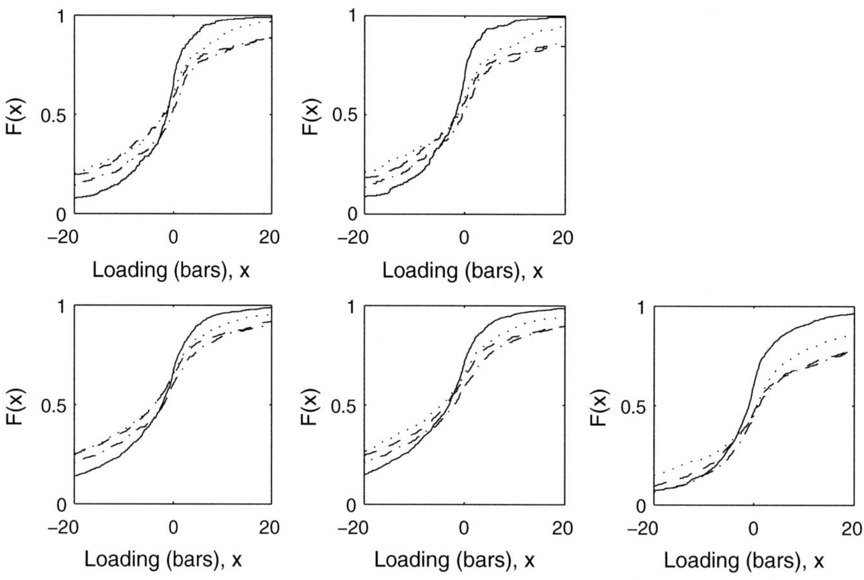

Figure 9
Distribution $F(x)$ (cumulative proportion less than x) of total (Darfield to fault failure) loading. Otherwise as for Fig. 7

regardless of the source model or data, except for the optimal plane solutions (non-CMT). This is compensated for by the initial 'overloading' shown in Fig. 7. Again, the results are insensitive to the inclusion of imputed mechanisms, or the method used to select between pairs of mechanisms.

In the case where the total ΔCFF is positive, the proportion of the received stress deriving from the Darfield source model is shown in Fig. 10. We see that all the source models result in approximately 50 % of the faults having a contribution from Darfield which is positive and less than the total

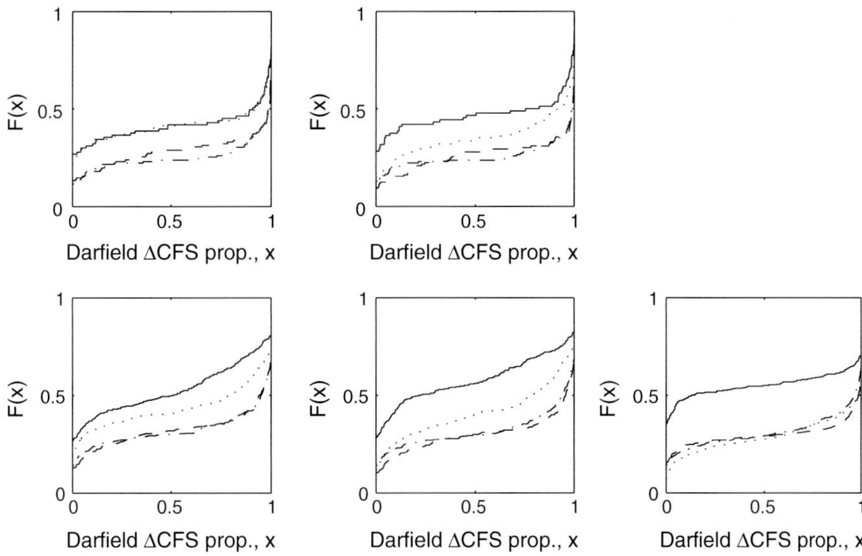

Figure 10
Distribution $F(x)$ (cumulative proportion less than x) of ΔCFF derived from Darfield normalized by the total ΔCFF from all preceding earthquakes. Otherwise as for Fig. 7

ΔCFF. The differences are seen with the CMT source model producing negative contributions in 30 % of cases, while ATZORI *et al.* (2012) and ELLIOTT *et al.* (2012) source-model based contributions are negative in only 10–15 % of cases. At the other end, the ATZORI *et al.* (2012) and ELLIOTT *et al.* (2012) models have contributions from Darfield in excess of the total ΔCFF (implying that the sum of the other contributions is negative) in 30–35 % of cases, while CMT only does so in 20% of cases. As usual, the BEAVAN *et al.* (2012) model is in the middle. With the absence of the triggering Charing Cross event in the BEAVAN *et al.* (2012) source model, using the decision rule without imputed events produces the same result as the CMT model in this respect. Interestingly, the results from the optimal planes solutions for the three detailed source mechanisms are almost identical in this regard.

6. Discussion

The main finding is that contrary to the paradigm outlined in STEIN *et al.* (1997), CFF does not increase until it reaches the stress drop $\Delta\sigma$ and failure occurs. Instead, and regardless of the modeling details, more than half of the faults in the Canterbury sequence appear to have had a loading in excess of the stress drop prior to the initiation of the sequence. In a study of the Landers aftershock sequence, MEIER *et al.* (2014) found that 27 % of aftershocks receive greater positive stress from aftershocks than from main shocks, and that while 85 % of aftershocks were encouraged by main shocks, adding in the aftershocks reduces this to 79 %. Hence, the role of aftershocks in stress triggering is not proven, and it is unlikely that events below the magnitude cutoff of $m = 3.6$ are the missing element. So, in a sense, our hypothesis that the Canterbury region was ‘primed’ for a long energetic sequence is consistent with the results. What we do not have is an explanation of how such ‘overstress’ is able to accumulate. It is possible that the region lies in an old stress shadow, for which the existence of a different triggering mechanism has been deduced by MEIER *et al.* (2014).

Figure 9 shows that, depending on the source model, 70–90 % of aftershocks that are triggered receive a positive ΔCFF contribution from the Darfield event. The effect may be magnitude-dependent: STEACY *et al.* (2014) determined that 100 % of $M \geq 5.5$ and 88 % of $M \geq 5$ aftershocks in the first 2 years of the sequence occurred in positive stress lobes (optimally oriented faults) of the

Darfield event, treated as a single fault plane. Also, for both the Atzori *et al.* (2012) and Elliott *et al.* (2012) models, 98 % of $M \geq 4$ events are consistent with static stress triggering (Steacy *et al.* 2014). We note, however, these figures are for optimally oriented faults, which appear from Fig. 6 to be differently distributed than those events for which tensor solutions exist. Also, we have used the catalog depth of the aftershocks, rather than an optimal depth: Catalli and Chan (2012) noted that the most important sensitivity factor is the depth at which stress changes are calculated. However, measurement uncertainties are not negligible (Cattania *et al.* 2014), and many of the catalog events are at a 'default' depth of 5 km.

Hainzl *et al.* (2010b) treated fault mechanisms as having a distribution, rather than a specified orientation, obtaining good agreement with the spatial distribution of Landers aftershocks. Hence, we can compare the relative locations of Canterbury aftershocks with those of Landers. While Hardebeck *et al.* (1998) showed that 85 % of 1992 $M_W 7.3$ Landers earthquake aftershocks at distances of 5–75 km were consistent with static stress triggering, the corresponding figure for the 1994 $M_W 6.4$ Northridge earthquake was only 60 %. The 1999 $M_W 7.6$ Chi-Chi earthquake had 83 % of its $M \geq 4$ aftershocks in regions where the Coulomb stress increased by at least 0.1 bar or 61 % for $M \geq 2$ aftershocks (Chan and Stein 2009). In this sense, the Canterbury sequence appears to have been more concentrated than the comparators, providing further support for the 'synchronization' hypothesis.

What are the implications for earthquake forecasting? The usual method to produce a point process intensity (forecastable) model from ΔCFF changes uses a rate-state formulation (Dieterich 1994) to convert the stress change into a rate of events (Catalli *et al.* 2008; Hainzl *et al.* 2010a; Catalli and Chan 2012). While this has been shown to be consistent with stress transfer modeling (Borovkov and Bebbington 2003), the result is strongly dependent (Catalli *et al.* 2008; Hainzl *et al.* 2009) on the estimated constitutive parameter (Bebbington 2008). However, this has been shown not to be a particularly informative method of forecasting in the Canterbury sequence (Steacy *et al.* 2014), or indeed elsewhere (Catalli *et al.* 2008; Rhoades *et al.* 2010; Parsons *et al.* 2012a) as it assumes a certain uniformity in the receiver faults. If an exponential function [cf. Bebbington and Harte (2003)] of the stress loading relative to the stress drop presents no technical issues, it does mean that stresses exceeding the critical stress drop lead quickly to simulated failure. However, some of the faults appear, prior to the Darfield event, to have had decades or more worth of tectonic stress stored on them in excess of the stress drop. Modelling this requires that something additional be added to the formulation.

There are, of course, a number of unquantifiable uncertainties in the method. These include using a simple one-plane rectangular slip model, which is adequate for ΔCFF calculation in the far-field. While in the near-field it will introduce potentially large errors, these are unavoidable; we cannot limit ourselves just to known slip models (four events total) or to only far-field transfer, as we are investigating the *evolution* of an earthquake sequence. As we have considered a number of options, including those most favorable for positive stress transfer, we consider it unlikely that the near-field errors are biased sufficiently and consistently enough to produce our results in qualitative terms. While the calculation of stress drop is dependent on fault length and width, we have little option but to impute these given the data and objectives, and again it is difficult to conceive of the resulting errors being consistent enough to cause the results. We have, of course, omitted transient deformations (e.g., afterslip and viscoelastic relaxation) from our investigation, which may yet provide at least partial answers to some of the conundrums uncovered here.

7. Conclusion

We have shown that according to the static triggering model, the majority of faults that failed in the Canterbury earthquake sequence were at a 'superloaded' (higher than failure strength) level of stress loading prior to initiation of the sequence by the Darfield event. While this may explain the severity of the aftershock sequence, it poses as yet unanswered questions about the static stress triggering hypothesis.

Acknowledgments

The authors gratefully acknowledge discussions with GS seismologists Matt Gerstenberger, Annemarie Christopherson, Bill Fry, and John Ristau. Two anonymous referees provided helpful feedback.

References

Aki, K., and P. G. Richards. 1980. Quantitative Seismology: Theory and Methods. San Francisco: WH Freeman.

Atzori, S., C. Tolomei, A. Antonioli, J. P. Merryman Boncori, S. Bannister, E. Trasatti, P. Pasquali, and S. Salvi. 2012. *The 2010–2011 Canterbury, New Zealand, seismic sequence: Multiple source analysis from INSAR data and modeling*. J. Geophys. Res. *117*:B08305.

Bannister, S., and K. Gledhill. 2012. *Evolution of the 2010–2012 Canterbury earthquake sequence*. NZ J. Geol. Geophys. *55*:295–304.

Bannister, S., B. Fry, M. Reyners, J. Ristau, and H. Zhang. 2011. *Fine-scale relocation of aftershocks of the 22 February mw 6.2 Christchurch earthquake using double-difference tomography*. Seismol. Res. Lett. *82*:839–845. doi:10.1785/gssrl.82.6.839.

Beavan, J., M. Motagh, E. J. Fielding, N. Donnelly, and D. Collett. 2012. *Fault slip models of the 2010-2011 Canterbury, New Zealand, earthquakes from geodetic data and observations of postseismic ground deformation*. NZ J. Geol. Geophys. *55*:207–211.

Bebbington, M. 2008. *Estimating rate- and state-friction parameters using a two-node stochastic model for aftershocks*. Tectonophysics *457*:71–85. doi:10.1016/j.tecto.2008.05.017.

Bebbington, M., and D. S. Harte. 2003. *The linked stress release model for spatio-temporal seismicity: formulations, procedures and applications*. Geophysical Journal International *154*:925–946. doi:10.1046/j.1365-246X.2003.02015.x.

Borovkov, K., and M. Bebbington. 2003. *A stochastic two-node stress transfer model reproducing Omori's law*. Pure and Applied Geophysics *160* (8):1429–1445. doi:10.1007/s00024-003-2354-8.

Catalli, F., and *C. H. Chan*. 2012. *New insights into the application of the Coulomb model in real-time*. Geophys. J. Int. *188*:583–599. doi:10.1111/j.1365-246X.2011.05276.x.

Catalli, F., M. Cocco, R. Console, and L. Chiaraluce. 2008. *Modeling seismicity rate changes during the 1997 Umbria-Marche sequence (central Italy) through a rate- and state-dependent model*. J. Geophys. Res. *113*:B111301.

Cattania, C., S. Hainzl, L. Wang, F. Roth, and B. Enescu. 2014. *Propagation of Coulomb stress uncertainties in physics-based aftershock models. J. Geophys. Res. 119*:7846–7864. doi:10.1002/2014JB011183.

Chan, C. H., and R. S. Stein. 2009. *Stress evolution following the 1999 Chi-Chi, Taiwan, earthquake: consequences for afterslip, relaxation, aftershocks and departures from Omori decay*. Geophys. J. Int. *177*:179–192.

Christophersen, A., D. A. Rhoades, S. Hainzl, E. G. C. Smith, andM. C. Gerstenberger. 2013. The Canterbury sequence in the context of global earthquake statistics, GNS Science Consultancy Report 2013/196, GNS Science, Lower Hutt.

Convertito, V., F. Catalli, and A. Emolo. 2013. *Combining stress transfer and source directivity: the case of the 2012 Emilia seismic sequence*. Sci. Rep. *3*:3114.

Cotton, F., R. Archuleta, and M. Causse. 2013. *What is sigma of the stress drop?* Seismol. Res. Lett. *84*:42–48.

Dieterich, J. H. 1994. *A constitutive law for rate of earthquake production and its application to earthquake clustering*. J. Geophys. Res. *99*:2601–2618. doi:10.1029/93JB02581.

Elliott, J. R., E. Nissen, P. C. England, J. A. Jackson, S. Lamb, Z. Li, M. Oehlers, and B. E. Parsons. 2012. *Slip in the 2010–2011 Canterbury earthquakes, New Zealand*. J. Geophys. Res. *117*:B03401.

Freed, A. M. 2005. *Earthquake triggering by static, dynamic, and postseismic stress transfer*. Ann. Rev. Earth Planet. Sci. *33*:335–367.

Fry, B., and M. C. Gerstenberger. 2011. *Large apparent stresses from the Canterbury earthquakes of 2010 and 2011*. Seismol. Res. Lett. *82*:833–838. doi:10.1785/gssrl.82.6.833.

Furlong, K. P. 2013. The Intraplate Earthquake Cycle: Strain and Displacement Behaviour During the the Canterbury, NZ Earthquake Sequence, Technical Report NEHRP Award G12AP20031, Pennsylvania State University.

Gledhill, K., J. Ristau, M. Reyners, B. Fry, and C. Holden. 2011. *The Darfield (Canterbury, New Zealand)* $M_W 7.1$ *earthquake of September 2010: A preliminary seismological report*. Seismol. Res. Lett. *82*:378–386. doi:10.1785/gssrl.82.3.378.

Hainzl, S., G. B. Brietzke, and G. Zoller. 2010a. *Quantitative earthquake forecasts resulting from static stress triggering*. J. Geophys. Res. *115* (B11311). doi:10.1029/2010JB007473.

Hainzl, S., G. Zoller, and R. Wang. 2010b. *Impact of the receiver fault distribution on aftershock activity*. J. Geophys. Res. *115* (B05315). doi:10.1029/2008JB006224.

Hainzl, S., B. Enescu, M. Cocco, J. Woessner, F. Catalli, R. Wang, and F. Roth. 2009. *Aftershock modeling based on uncertain stress calculations*. J. Geophys. Res. *114* (B05309). doi:10.1029/2008JB006011.

Hanks, T. C., and W. H. Bakun. 2008. $M - \log A$ *observations of recent large earthquakes*. Bull. Seismol. Soc. Amer. *98*: 490–494.

Hardebeck, J. 2006. *Homogeneity of small-scale earthquake faulting, stress and fault strength*. Bull. Seismol. Soc. Amer. *96*:1675–1688.

Hardebeck, J. L., J. J. Nazareth, and E. Hauksson. 1998. *The static stress triggering model: Constraints from two southern California aftershock sequences*. J. Geophys. Res. *103*:24427–24437.

Harris, R. A. 1998. *Introduction to special section: Stress triggers, stress shadows, and implications for seismic hazard*. J. Geophys. Res. *103*:24347–24358.

Hill, D. P. 2008. *Dynamic stresses, Coulomb failure, and remote triggering*. Bull. Seismol. Soc. Amer. *98*:66–92.

Kaiser, A. E., A. Oth, and R. A. Benites. 2013. Separating source, path and site influences on ground motion during the Canterbury earthquake sequence, using spectral inversions. Paper no. 18 (8 p.) in: Same risks, new realities: New Zealand Society for Earthquake Engineering Technical Conference, April 26–28, 2013, Wellington.

Kanamori, H., and D. L. Anderson. 1975. *Theoretical basis of some empirical relations in seismology*. Bull. Seismol. Soc. Amer. *65*:1073–1095.

King, G. C. P., R. S. Stein, and J. Lin. 1994. *Static stress changes and the triggering of earthquakes*. Bull. Seismol. Soc. Amer. *84*:935–953.

Leonard, M. 2010. *Earthquake fault scaling: Self-consistent relating of rupture length, width, average displacement, and moment release*. Bull. Seismol. Soc. Amer. *100*:1971–1988. doi:10.1785/0120090189.

Meier, M. A., M. J. Werner, J. Woessner, and S. Wiemer. 2014. *A search for evidence of secondary static stress triggering during the 1992 $M_W 7.3$ Landers, California, earthquake sequence*. J. Geophys. Res. *119*:3354–3379. doi:10.1002/2013JB010385.

Ogata, Y., and J. C. Zhuang. 2006. *Space-time ETAS models and an improved extension*. Tectonophysics *413*:13–23. doi:10.1016/j.tecto.2005.10.016.

Okada, Y. 1992. *Internal deformation due to shear and tensile faults in a half-space*. Bull. Seismol. Soc. Amer. *82*:1018–1040.

Parsons, T., Y. Ogata, J. C. Zhuang, and E. L. Geist. 2012a. *Evaluation of static stress change forecasting with prospective and blind tests*. Geophys. J. Int. *188*:1425–1440. doi:10.1111/j.1365-246X.2011.05343.x.

Parsons, T., E. H. Field, M. T. Page, and K. Milner. 2012b. *Possible earthquake rupture connections on mapped California faults ranked by calculated Coulomb linking stresses*. Bull. Seismol. Soc. Amer. *102*:2667–2676.

Quigley, M., R. J. Van Dissen, N. J. Litchfield, P. Villamor, B. Duffy, D. J. A. Barrell, K. Furlong, T. Stahl, E. Bilderback, and D. Noble. 2012. *Surface rupture during the 2010 $M_W 7.1$ Darfield (Canterbury) earthquake: Implications for fault rupture dynamics and seismic-hazard analysis*. Geology *40*:55–58. doi:10.1130/G32528.1.

Rhoades, D. A., E. E. Papadimitriou, V. G. Karakostas, R. Console, and M. Murru. 2010. *Correlation of static stress changes and earthquake occurrence in the North Aegean region*. Pure Appl. Geophys. *167*:1049–1066. doi:10.1007/s00024-010-0092-2.

Richards-Dinger, K., R. S. Stein, and S. Toda. 2010. *Decay of aftershock density with distance does not indicate triggering by dynamic stress*. Nature *467*:583–586.

Ristau, J., C. Holden, A. Kaiser, C. Williams, S. Bannister, and B. Fry. 2013. *The Pegasus Bay aftershock sequence of the $M_W 7.1$ Darfield (Canterbury), New Zealand earthquake*. Geophys. J. Int. *195*:444–459.

Shaw, B. E. 2013. *Earthquake surface slip-length data is fit by constant stress drop and is useful for seismic hazard analysis*. Bull. Seismol. Soc. Amer. *103*:876–893.

Shcherbakov, R., M. Nguyen, and M. Quigley. 2012. *Statistical analysis of the 2010 $M_W 7.1$ Darfield earthquake aftershock sequence*. NZ J. Geol. Geophys. *55*:305–311.

Sibson, R., F. Ghisetti, and J. Ristau. 2011. *Stress control of an evolving strike-slip fault system during the 2010–2011 Canterbury, New Zealand, earthquake sequence*. Seismol. Res. Lett. *82*:824–832. doi:10.1785/gssrl.82.6.824.

Steacy, S., A. Jimenez, and C. Holden. 2014. *Stress trigeering and the Canterbury earthquake sequence*. Geophys. J. Int. *196*:473–480. doi:10.1093/gji/ggt380.

Steacy, S., D. Marsan, S. S. Nalbant, and J. McCloskey. 2004. *Sensitivity of static stress calculations to the earthquake slip distribution*. J. Geophys. Res. *109* (B04303). doi:10.1029/2002JB002365.

Steacy, S., S. S. Nalbant, J. McCloskey, C. Nostro, O. Scotti, and D. Baumont. 2005. *Onto what planes should Coulomb stress perturbations be resolved?* J. Geophys. Res. *110* (B05S15). doi:10.1029/2004JB003356.

Steacy, S., M. C. Gerstenberger, C. Williams, D. A. Rhoades, and A. Christophersen. 2014. *A New hybrid Coulomb/statistical model for forecasting aftershock rates*. Geophys. J. Int. *196*:918–923. doi:10.1093/gji/ggt404.

Stein, R. S., A. Barka, and J. H. Dieterich. 1997. *Progressive failure on the North Anatolian fault since 1939 by earthquake stress triggering*. Geophys. J. Int. *128*:594–604. doi:10.1111/j.1365-246X.1997.tb05321.x.

Stirling, M., G. H. McVerry, M. C. Gerstenberger, N. J. Litchfield, R. J. Van Dissen, K. Berryman, P. Barnes, L. Wallace, P. Villamor, R. Langridge, G. Lamarche, S. Nodder, M. Reyners, B. Bradley, D. A. Rhoades, W. D. Smith, A. Nicol, J. Pettinga, K. Clark, and K. Jacobs. 2012. *National seismic hazard model for New Zealand: 2010 update*. Bull. Seismol. Soc. Amer. *102*:1514–1542. doi:10.1785/0120110170.

Stirling, M., T. Goded, K. Berryman, and N. J. Litchfield. 2013. *Selection of earthquake scaling relationships for seismic-hazard analysis*. Bull. Seismol. Soc. Amer. *103*:1–19. doi:10.1785/0120130052.

Syracuse, E. M., R. A. Holt, M. K. Savage, J. H. Johnson, C. H. Thurber, K. Unglert, K. N. Allan, S. Karaliyadda, and M. Henderson. 2012. *Temporal and spatial evolution of hypocentres and anisotropy from the Darfield aftershock sequence: implications for fault geometry and age*. NZ J. Geol. Geophys. *55*:287–293.

Toda, S., R. S. Stein, and J. Lin. 2011. *Widespread seismicity excitation throughout central Japan following the 2011 $M = 9.0$ Tohoku earthquake and its interpretation by Coulomb stress transfer*. Geophys. Res. Lett. *38* (L00G03).

Toda, S., R. S. Stein, G. C. Beroza, and D. Marsan. 2012. *Aftershocks halted by static stress shadows*. Nature Geoscience *5*:410–413. doi:10.1038/NGEO1465.

Wells, D. L., and K. J. Coppersmith. 1994. *New empirical relationships among magnitude, rupture length, rupture width, rupture area, and surface displacement*. Bull. Seismol. Soc. Amer. *84*:974–1002.

Wesnousky, S. G. 2008. *Displacement and geometrical characteristics of earthquake surface ruptures: Issues and implications for seismic-hazard analysis and the process of earthquake rupture*. Bull. Seismol. Soc. Amer. *98*:1609–1632.

Yen, Y. T., and K. F. Ma. 2011. *Source-scaling relationship for M 4.6–8.1 earthquakes, specifically for earthquakes in the collision zone of Taiwan*. Bull. Seismol. Soc. Amer. *101*:464–481.

(Received September 26, 2014, revised February 3, 2015, accepted February 26, 2015, Published online March 12, 2015)

Pure Appl. Geophys. 173 (2016), 21–33

DOI 10.1007/s00024-015-1151-5

Pure and Applied Geophysics

A Variety of Aftershock Decays in the Rate- and State-Friction Model Due to the Effect of Secondary Aftershocks: Implications Derived from an Analysis of Real Aftershock Sequences

Takaki Iwata[1,2]

Abstract—The model based on rate- and state-dependent friction law reproduces the temporal decay of an aftershock sequence with the p value of the Omori–Utsu law equal to 1, if we simply assume a constant stress rate over time. However, because p values vary in real aftershock sequences, this model requires some modification. This study examined the effect of secondary aftershocks on the variety of the p value. A large aftershock causes a stepwise stress increase in the aftershock area, and the expected seismicity rate derived from the friction law also increases abruptly. These multiple increases in the seismicity rate during its decay following a mainshock could cause variation in the apparent p value. In this study, a model incorporating this idea is applied to two aftershock sequences observed in Japan and is shown to substantially modify the modeling of aftershock activity.

Key words: Aftershock decay, secondary aftershock, point process, rate- and state-dependent friction law.

1. Introduction

Power-law decay of aftershock activity over time is widely and empirically accepted and described by the Omori–Utsu law (Utsu 1961). In contrast to this empirical law, Dieterich (1994) suggests a physics-based model that relates the rate- and state-dependent friction law (Dieterich 1994; Ruina 1983) and the seismicity rate. Since his suggestion, this model has been applied frequently to explain/examine the observed temporal and/or spatial distribution of aftershocks (e.g., Toda *et al.* 1998, 2003, 2012; Catalli *et al.* 2008; Hainzl *et al.* 2009, 2010; Cocco *et al.* 2010; Steacy *et al.* 2014).

In applying the Dieterich model, we often assume that the stress rate is constant over time. Under this simple assumption, the decay of the seismicity rate after a sudden increase in stress (stepwise stress increase) only resembles the Omori–Utsu law when the power-law exponent (i.e., the p value) is equal to 1 (Dieterich 1994), whereas in general cases, the p value ranges from 0.9 to 1.8 (Utsu *et al.* 1995). Therefore, the Dieterich model is not adaptable to the variety in the p value.

To resolve this inconsistency between the physics-based model and empirical law, a possible solution is to examine the time-varying rate of stress change (Dieterich 1994; Helmstetter and Shaw 2009), which probably results from afterslips or fault creeping.

Dieterich (1994) makes another suggestion to eliminate the inconsistency. We frequently observe secondary aftershocks (i.e., aftershocks caused by previous aftershocks) in an aftershock sequence. Consideration of secondary aftershocks is crucial to modeling the occurrence of aftershocks (e.g., Utsu 1970; Ogata 1988). Dieterich (1994) computes the seismicity rate based on his model with the simple assumption that each aftershock generates spatially constant stress changes in a study region, and showed that the computed seismicity rate apparently decays with $p \neq 1$ (see Fig. 7 of Dieterich 1994).

Ziv and Rubin (2003) pointed out that the inclusion of secondary aftershocks in the rate- and state-friction model has no effect on the decay of aftershock activity. On the other hand, Marsan (2006) shows that there is a case where the decay pattern of an aftershock activity depends on whether or not the effect is considered (see Fig. 3 of Marsan 2006).

Ziv and Rubin (2003) and Marsan (2006) have derived their results from a theoretical point of view

[1] The Institute of Statistical Mathematics, 10-3 Midori-cho, Tachikawa, Tokyo 190-8562, Japan. E-mail: tiwata@tokiwa.ac.jp
[2] Tokiwa University, 1-430-1 Miwa, Mito, Ibaraki 310-8585, Japan.

and/or numerical simulation; it is also crucial to discuss the effectiveness of secondary aftershocks on the basis of an analysis of real datasets. Thus, this study introduces a model incorporating the secondary aftershocks into the framework of the rate- and state-friction model and examines the performance of the introduced model with its application to real aftershock sequences.

In addition to this modification of the aftershock model, another interest of this study is the comparison of the modified model with the Epidemic Type Aftershock Sequence (ETAS) model (Ogata 1988), which is currently considered as a standard model of seismic activity and has many applications (e.g., Zhuang *et al.* 2012, and references therein). As stated above, the Dieterich model is physics-based, while the ETAS model is purely statistical. There is another contrastive feature between the two models. In the ETAS model, seismicity rate is represented as a summation of rates of secondary aftershocks caused by each of triggering events; that is, the model is additive. Meanwhile, the effect of the secondary aftershock in the Dieterich model is multiplicative as will be shown in Sect. 3 (Eqs. 1 and 3); the rate immediately after a triggering event is represented as a product of the rate just before the triggering event and the triggering effect caused by the event. The comparison of the performance or goodness-of-fit of such contrastive models may deepen our understanding on aftershocks.

As the first step of the application of the Dieterich model with secondary aftershock effect to real datasets, only temporal evolution of a seismicity rate is investigated in this study. To consider the spatial variation of seismicity change in the Dieterich model with the secondary aftershock effect, it is required to evaluate the spatial pattern of stress changes generated by each triggering events as precise as possible. However, such precise evaluation is challenging if moderate or small events are contained in an examined dataset as triggering events.

Because of this inconsideration of the spatial variation, the seismicity change due to the secondary aftershock effect is assumed to be constant in space, which implies that the stress changes given by a triggering event are spatially homogeneous. This assumption provides strong simplification to the model, but even examining the simplified model will reveal advantage/disadvantage of the physics-based model and is helpful to make further developments of aftershock modeling.

2. *Data*

In this study, we analyzed the aftershock sequences of the 2004 Mid-Niigata Prefecture (Chuuetsu) earthquake and the 1995 Kobe earthquake. The aftershocks were taken from the earthquake catalog compiled by the Japan Meteorological Agency (JMA). We compiled datasets of the aftershocks that had occurred within rectangular areas (Fig. 1) with their longer sides parallel to the strike direction of the mainshocks: 212° for the Mid-Niigata earthquake determined from the National Institute for Earth Science and Disaster Prevention (NIED) centroid moment tensor (CMT) solutions (National Research Institute for Earth Science and Disaster Prevention 2004) and 230° for the Kobe earthquake determined from the Harvard CMT solution (Dziewonski *et al.* 1996).

To determine the threshold magnitude for the analysis below, the temporal variation in earthquake (aftershock) detection capability was estimated through a method developed in Iwata (2008, 2012, 2013). In this method, a statistical model representing a magnitude–frequency distribution covering the entire magnitude range (Ogata and Katsura 1993) is introduced. Then, the temporal variation in the magnitude at which 50 % of earthquakes are expected to be detected contained in the statistical model is evaluated with a piecewise linear function and an appropriate smoothness constraint on the basis of a Bayesian approach. With another parameter of the statistical model of Ogata and Katsura (1993), we can derive the temporal variation in completeness magnitude, such as the magnitude at which the detection probability is 99.9 % (Iwata 2013). The *b* value of the Gutenberg–Richter law, which was used in a numerical simulation as well be described in Sect. 5.2, is estimated simultaneously. In Fig. 2, the estimated temporal profiles of magnitudes at which 50 and 99.9 % of earthquakes are expected to be detected are drawn. From this feature of the earthquake detection capability, aftershocks with $M \geq 3.0$ and 2.8 after 0.5 day (see also Sect. 4) were used for the Mid-Niigata and Kobe sequences, respectively.

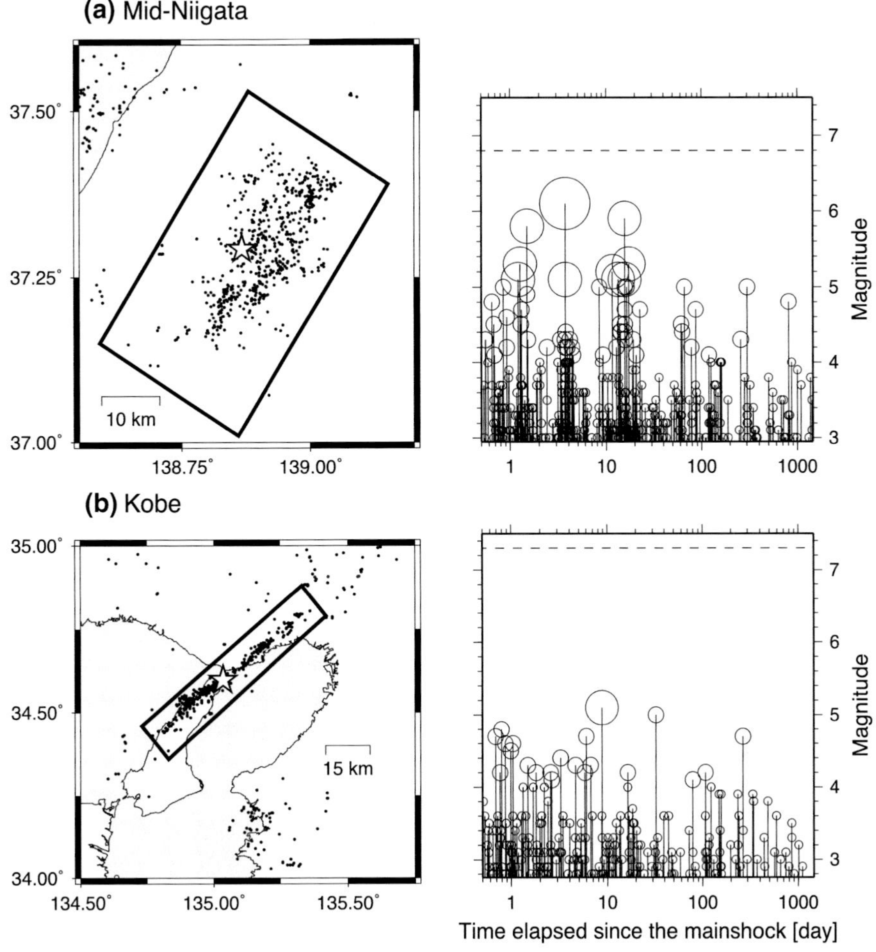

Figure 1

Locations of shallow (depth < 30 km) earthquakes (*left*) and their magnitude–time plot (*right*) for **a** $M \geq 3.0$ following the 2004 Mid-Niigata Prefecture earthquake and **b** $M \geq 2.8$ following the 1995 Kobe earthquake. This plot includes the aftershocks occurring in the 4 years after the mainshocks, indicated by the *stars*. The *rectangles* are regarded as the aftershock regions of the two destructive mainshocks in this study

The two aftershock sequences differ in one aspect. In the aftershock sequence of the Kobe earthquake, the magnitudes of the mainshock and the largest aftershock differed greatly (Fig. 1b). As a general tendency, the number of triggered earthquakes increases with the magnitude of a parent earthquake (e.g., Utsu 1971; Ogata 2001). Hence, the aftershock activity caused by the mainshock was dominant and the secondary aftershocks had little effect in the Kobe sequence. Compared with the Kobe case, in the Mid-Niigata sequence, some of the aftershocks had magnitudes close to that of the mainshock (Fig. 1a); the activity of the secondary aftershocks was effective in the sequence. Application of the model described in the next section to such two contrasting aftershock sequences can provide some implications for modification of the Dieterich model with the inclusion of the secondary aftershock effect.

3. Seismicity Rate Derived from the Rate- and State-Dependent Friction Law with the Effect of Secondary Aftershocks

By applying the rate- and state-dependent friction law to seismicity, Dieterich (1994) derived the temporal evolution of the seismicity rate $\lambda(t)$, which is the expected occurrence rate of earthquakes in unit

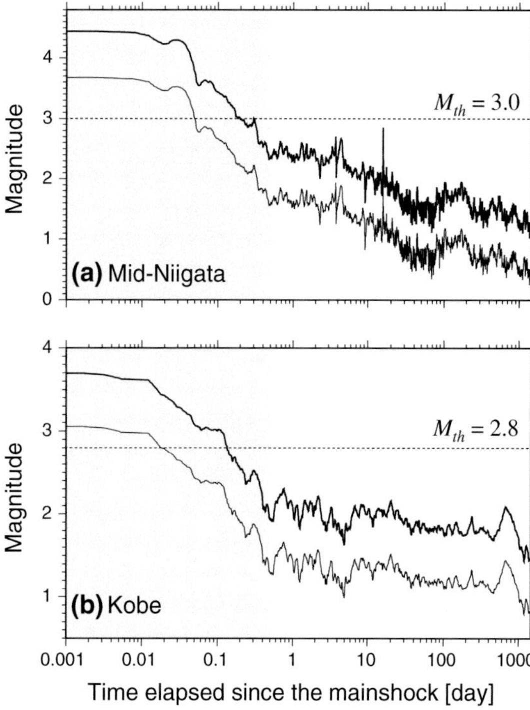

Figure 2
Estimated temporal profiles of magnitudes at which 50 % (*thin lines*) and 99.9 % (*bold lines*) of earthquakes are expected to be detected for the aftershock sequences of **a** the 2004 Mid-Niigata and **b** 1995 Kobe earthquakes. The periods until 0.5 day after the occurrence of the mainshock are shaded in *gray*. *Horizontal dotted lines* indicate the thresholds of magnitude used in this study

time as shown below. If there is a constant increase in stress over time and the stress steps are given, the temporal evolution of the state variable $\gamma(t)$ is obtained using the following two formulas:

$$\gamma_{2i} = \gamma_{2i-1} \exp\left[-\frac{\Delta \mathrm{CFF}_i}{A\sigma}\right], \tag{1}$$

$$\gamma_{2i+1} = \left[\gamma_{2i} - \frac{t_a}{A\sigma}\right] \exp\left[-\frac{\Delta t_i}{t_a}\right] + \frac{t_a}{A\sigma}. \tag{2}$$

The state variable $\gamma(t)$ changes from γ_{2i-1} to γ_{2i} because of the ith sudden stress step of $\Delta \mathrm{CFF}_i$ and then evolves into γ_{2i+1} at the time of the next (i.e., $(i+1)$th) stress step. Δt_i is the time interval between the ith and $(i+1)$th stress steps, and t_a and $A\sigma$ are parameters in the formulation of DIETERICH (1994). Using these two formulas iteratively, we can compute $\gamma(t)$ for multiple stress steps. Then, the estimated state variable $\gamma(t)$ is converted into the temporal evolution of the seismicity rate $\lambda(t)$ using the relation

$$\lambda(t) = \frac{r}{\gamma(t)} \cdot \frac{t_a}{A\sigma}, \tag{3}$$

where r is the reference seismicity rate, which corresponds to the seismicity rate in the steady state, and can be estimated from the average seismicity over a long period of time (e.g., STEIN 1999).

To eliminate the parameter redundancy in the above formulas, we multiply Eqs. (1) and (2) by $A\sigma/t_a$ and replace $(A\sigma/t_a)\gamma_n$ with γ'_n. Then, the two equations can be rewritten as follows:

$$\gamma'_{2i} = \gamma'_{2i-1} \exp\left[-\frac{\Delta \mathrm{CFF_i}}{A\sigma}\right], \tag{1'}$$

$$\gamma'_{2i+1} = \left[\gamma'_{2i} - 1\right] \exp\left[-\frac{\Delta t}{t_a}\right] + 1, \tag{2'}$$

and Eq. (3) is equivalent to $\lambda(t) = r/\gamma'(t)$, where $\gamma'(t) = (A\sigma/t_a)\gamma(t)$.

In this study, as noted in Sect. 1, we assume that all aftershocks cause spatially constant stress changes in a study region. For the simplicity of the analysis, the amount of stress step $\Delta \mathrm{CFF}_i$ caused by the ith event is also assumed to only depend on its magnitude M_i and to be proportional to an exponential function of the magnitude: $\Delta \mathrm{CFF}_i = a\exp(dM_i)$. In this formulation, the seismicity rate is determined uniquely if the value of the parameter vector $\boldsymbol{\theta} = [r, t_a, a/(A\sigma), d]$ is specified.

4. *Application of the Seismicity Model to the Observed Aftershock Sequences*

To evaluate the effect of secondary aftershocks on the variation in aftershock decay, the model introduced above was applied to the aftershock sequences that followed the Kobe and Mid-Niigata Prefecture earthquakes. We searched for the optimal value of the parameter vector $\boldsymbol{\theta}$ that provides the best fit to the observed aftershock sequences using the maximum likelihood method. According to DALEY and VERE-JONES (2003), the log-likelihood ($\ln L$) of a point process for a study period [S, T] is given as follows:

$$\ln L = \sum_{S \le t_i \le T} \ln \lambda(t_i|\boldsymbol{\theta}) - \int_S^T \lambda(t|\boldsymbol{\theta})\mathrm{d}t, \tag{4}$$

where t_i is the occurrence time of the ith event. We chose the value of the parameter vector that maximizes $\ln L$ as the best estimate.

This study examined the 4-year sequences of the aftershocks; therefore, T was set at 4 years since the occurrence of the mainshock. To take into account a significant portion of the undetected aftershocks immediately after their mainshock (e.g., KAGAN 2004), S was set at 0.5 day. Note that the seismic activity during the study period is greatly influenced by the earthquakes that occurred before the study period (between 0 and 0.5 day after the occurrence of the mainshock). Therefore, although the function $\lambda(t|\boldsymbol{\theta})$ was fitted to the data during the period $[S, T]$, the earthquakes occurring during the period between the mainshock and S were considered in the computation of $\lambda(t|\boldsymbol{\theta})$ (see OGATA 2006 for more details).

We applied the model described in Sect. 3 without any constraint on the four parameters (hereafter Model A) to the two aftershock sequences. In addition, to examine the effect of secondary aftershocks on the nature of aftershock decay, a model without the effect of secondary aftershocks was applied. The second model (hereafter Model B) only considered the effect of the stress step caused by the mainshock; that is, we assumed $\Delta\mathrm{CFF}_i = 0$ for all of the earthquakes except the mainshock. In this case, d was not optimized and was fixed at an arbitrary value (and chosen to be 1.0 in this study) because only ΔCFF caused by the mainshock was optimized based on its magnitude.

The goodness-of-fit of the two models was compared using the Akaike Information Criterion (AIC) (AKAIKE 1974); AIC $= -2$(maximum $\ln L$) $+$ 2(number of parameters). The model with the smaller AIC value is considered to offer the better fit to the dataset. Instead of AIC, the Bayesian information criterion (BIC) (SCHWARZ 1978) is sometimes applied to the problem of model comparison. Even if BIC is used in this study, the results of the model comparison shown below do not change.

Tables 1 and 2 show the parameter values (i.e., the maximum likelihood estimate, MLE) and the AIC values estimated using the maximum likelihood method, respectively. Table 1 also presents the standard error of the MLE calculated from the Fisher information matrix or Hessian matrix (e.g., OGATA 1983).

For both sequences, the AIC value associated with Model A was smaller than that associated with Model B, and the differences in the AIC values are 181.6 and 30.2 for the Mid-Niigata and Kobe sequences, respectively. Since a difference in the AIC values of around 1.5–2 approximately corresponds to the 5 % significance level (e.g., ZHENG and VERE-JONES 1994), the computed differences are very significant, suggesting that the consideration of secondary aftershocks would effectively improve the Dieterich model.

Table 1

Estimated parameters of the Dieterich model and their standard errors for the examined two models

	r (day^{-1})	t_a (day)	$a/(A\sigma)$	d
Mid-Niigata				
Model A	$(6.35 \pm 3.26) \times 10^{-3}$	$(4.88 \pm 2.65) \times 10^{2}$	$(8.57 \pm 4.69) \times 10^{-3}$	$(9.44 \pm 1.07) \times 10^{-1}$
Model B	$(1.36 \pm 4.90) \times 10^{-6}$	$(3.29 \pm 11.82) \times 10^{7}$	∞	1.00 (fixed)
Kobe				
Model A	$(2.87 \pm 4.36) \times 10^{-5}$	$(2.97 \pm 4.43) \times 10^{5}$	$(1.36 \pm 2.41) \times 10^{-3}$	1.16 ± 0.39
Model B	$(8.51 \pm 27.18) \times 10^{-7}$	$(3.28 \pm 10.47) \times 10^{7}$	∞	1.00 (fixed)

Table 2

AIC values for the examined two models based on the Dieterich model in this study and for the ETAS model

	Model A	Model B	ETAS
Mid-Niigata	−375.5	−193.9	−446.7
Kobe	−50.9	−20.7	−80.0

Note that in Model B, the values of $a/A\sigma$ is unable to be estimated numerically, either for the Mid-Niigata or for the Kobe sequence. In this model, this parameter controls the amplitude of the seismicity rate enhanced by the mainshock, and it plays a prominent role in the determination of the rate immediately after the mainshock. However, the model was fitted with the data 0.5 day after the mainshock, resulting in almost no constraint on that parameter. Consequently, $a/A\sigma$ approaches to a large value to make $\gamma(0)$ equal to zero and the seismicity rate immediately after the mainshock is divergent to infinity.

5. Discussion

5.1. Model Fitting and Interpretation on the Estimated Parameters

As mentioned above, the consideration of secondary aftershocks makes a significant contribution towards modifying the seismicity model. Figure 3 illustrates the seismicity rate deduced from the two Dieterich models examined here. As a reference model, the Omori–Utsu law is fitted to the sequences, and the seismicity rate based on the empirical law is also plotted. Using the maximum likelihood method, the p value of the Omori–Utsu law is estimated at 1.35 for the Mid-Niigata sequence, which is significantly different from 1. As seen in Fig. 3a, Model A (the Dieterich model incorporating the effect of secondary aftershocks) provides a seismicity rate matching the Omori–Utsu law, while Model B (the Dieterich model without the effect of secondary aftershocks) does not, indicating that consideration of secondary aftershocks can cause the variety of the p value. As seen in Fig. 4a, the curve corresponding to Model A fits the curves of the observation and the Omori–Utsu law much better than does the curve of Model B.

For the Mid-Niigata sequence, t_a, which is equivalent to $A\sigma/\dot{\tau}$ (Dieterich 1994), was estimated at 470 days as shown in Table 1. The shear strain rate around this region is 0.5–1.5 $\times 10^{-7}$ year^{-1}, as derived from GPS data (Sagiya *et al.* 2000; Noda and Matsu'ura 2010). Assuming that the shear modulus

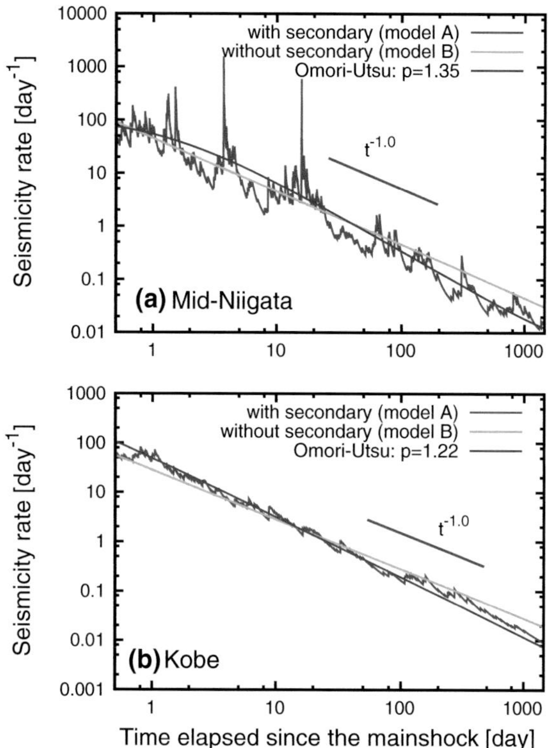

Figure 3
Seismicity rates of the aftershocks derived from the two types of the Dieterich model (*red* and *green*), as a function of the time elapsed since the mainshock, following the **a** 2004 Mid-Niigata and **b** 1995 Kobe earthquakes. The seismicity rate estimated using the Omori–Utsu law (*blue*) is also drawn

is 40 GPa, the corresponding stress rate $\dot{\tau}$ is 4–12 $\times 10^4$ Pa/year; $A\sigma$ was estimated as 0.005–0.015 MPa, which is comparable to values estimated in other studies (Hainzl *et al.* 2009, and references therein).

Since the late 1980s, the JMA data have completely detected earthquakes of $M \approx 2$ or above in the Mid-Niigata region (Nanjo *et al.* 2010), and 31 earthquakes of $M \geq 3$ were recorded within the study region during the 15 years before the Mid-Niigata earthquake. This observed number of earthquakes corresponds to a seismicity rate of 5.65×10^{-3} day^{-1}, which is similar to the value of r listed in Table 1. The estimated parameters have plausible values, and the estimated curves well match with the observed curves and the empirical law, as presented above. Therefore, the modeling for the Mid-Niigata sequence is quite reasonable.

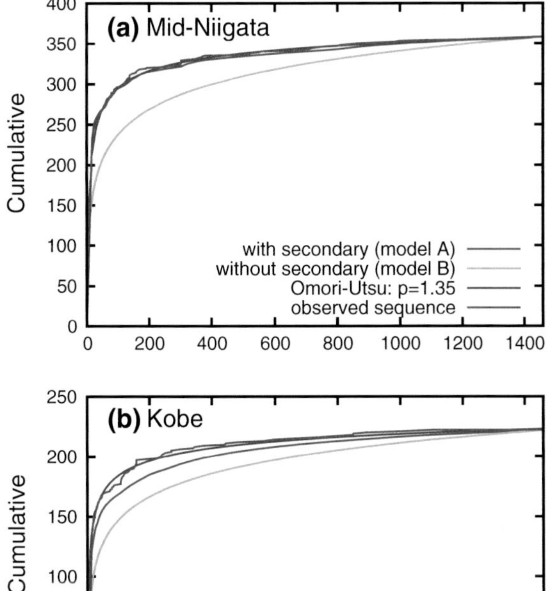

Figure 4
Cumulative number of aftershocks derived from the two types of the Dieterich model (*red* and *green*), as a function of the time elapsed since their mainshock, following the **a** 2004 Mid-Niigata and **b** 1995 Kobe earthquakes. The observed cumulative number of aftershocks (*purple*) and the expected number estimated from the Omori–Utsu law (*blue*) are also drawn

For the Kobe sequence, the modification that considers secondary aftershocks is significant but still inadequate. Comparing Models A and B, the seismicity rate deduced from Model A better fits the curve of the Omori–Utsu law, but it deviates slightly from the curve of the Omori–Utsu law in the late stage of the aftershock sequence (Fig. 3b). This feature was clarified by the plot of the cumulative number of aftershocks (Fig. 4b). The curve corresponding to Model A is much closer than that of Model B to the curves of the observation and the Omori–Utsu law; however, the Model A curve shows a slower decay (i.e., a smaller p value) than the observation and the Omori–Utsu law.

In the same manner as for the Mid-Niigata sequence, the value of $A\sigma$ is calculated as 1.25 MPa based on $t_a = 1.14 \times 10^5$ days and a shear strain rate of 0.5×10^{-7} year^{-1} around the aftershock area of the Kobe earthquake (Sagiya *et al.* 2000). This calculated value of $A\sigma$ is much greater than 0.35 MPa, as estimated by Toda *et al.* (1998). In the Kobe area, all earthquakes of $M \approx 2$ or above have been recorded since the early 1980s (Nanjo *et al.* 2010). Ten events of $M \geq 2.8$ occurred within the study region in the 15 years before the Kobe earthquake, corresponding to an observed seismicity rate of 1.83×10^{-3} day^{-1}. This rate is significantly different from the estimated reference rate shown in Table 1.

As mentioned in Sect. 2, the Mid-Niigata sequence had a high secondary aftershock activity, which provides enough flexibility to accommodate the variation in the p value. In contrast to the Mid-Niigata case, the activity of secondary aftershocks was unremarkable in the Kobe sequence. Because of such a low activity of secondary aftershocks, the Dieterich model with the effect of secondary aftershocks is not flexible in the decay exponent and the deviation between the model and data remains (Figs. 3b, 4b). This insufficient flexibility and remaining deviation cause the discrepancy found in the estimation of the parameters.

We also should note that the MLEs of the parameters are accompanied by large estimation errors (standard errors) in Model A for the Kobe sequence and Model B for the both sequences. In these three cases, as mentioned above, the model does not fit well with the real sequence, and therefore, the accuracy of the estimation is violated.

It is complicated to make the interpretation on the estimated value of d. If the productivities of offspring generated by each of triggering events are additive, as such in the ETAS model, each of those productivities are clearly isolated. For such a model, it is possible to evaluate the dependency of the productivities on the magnitude of triggering events; in many studies, the productivity is assumed to be proportional to exp($\alpha \cdot$ magnitude) and the value of α is estimated (e.g., Guo and Ogata 1997; Helmstetter *et al.* 2005; Hainzl and Marsan 2008; Ogata 2001). Contrastively, as represented in Eq. (1), the seismicity rate in the Dieterich model enhances in a multiplicative manner; the value of the state variable γ, which is proportional to a reciprocal of the seismicity rate (see Eq. 3), is multiplied by triggering effect caused by each of events. Because of this non-linear relationship of the contributions provided by all events, each of the productivities cannot be separated. At least, the physical meanings of the parameters d and α are different and the direct comparison of the two parameters is impractical.

5.2. *Validation of the Dieterich Model with Simulated Point Processes*

As mentioned in the end of the preceding subsection, the Dieterich model with secondary aftershock effect has a multiplicative property. Suppose that several earthquakes occur during a very short period and that the decay of seismicity rate represented as in Eq. (2) is negligible. Then, the state variable decreases dramatically because of the repetition of the multiplications by the influence of stress changes, and the seismicity rate increases violently; thus, one may suspect whether the seismicity rate in Model A is stable enough to support its reality.

To examine this point, 300 sequences of which seismicity rate follows the Dieterich model with secondary aftershock effect were numerically generated through the thinning algorithm for a point process (Ogata 1981). The values of the model parameters were taken from the MLE for the Mid-Niigata sequence shown in Table 1: $r = 6.35 \times 10^{-3}$, $t_a = 4.88 \times 10^2$, $a/(A\sigma) = 8.57 \times 10^{-3}$, and $d = 9.44 \times 10^{-1}$. The magnitudes of generated earthquakes were assumed to follow the Gutenberg-Richter law with $b = 0.90$ that has been estimated in the evaluation of the earthquake detection capability in Sect. 2. Only the magnitude of the first earthquake is fixed at 6.8, which is the same as the real mainshock. The case where the values of the parameters are the same as the MLE for the Kobe sequence was not examined because the MLE in this case is somewhat unrealistic, as described in the preceding subsection.

For the visualization, 20 sequences are randomly taken from the 300 sequences and their cumulative curves are plotted in Fig. 5. In this figure, unstable behavior of the seismicity rates is not found and those rates approaches to the reference (steady) seismicity rate 6.35×10^{-3} during some periods. The temporal patterns of the other 280 sequences are similar.

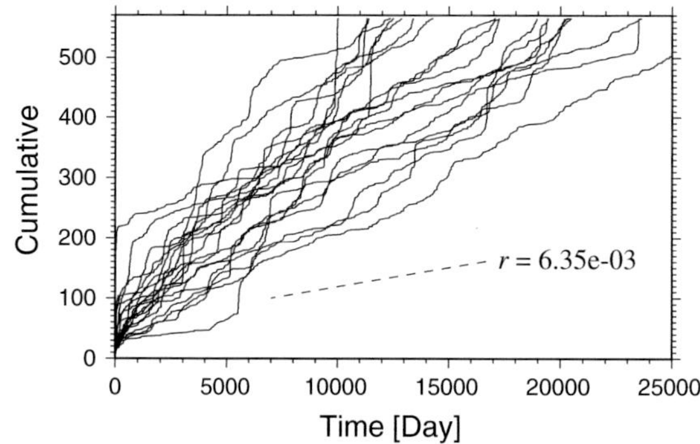

Figure 5
Cumulative curves of 20 sequences randomly taken from the 300 simulated sequences under the assumption that the seismicity rate follows the Dieterich model with the secondary aftershock effect

In relation to the multiplicative property, another concern is the influence of missing events on the estimation of the parameters. Figure 2 bolsters the completeness of the catalog during the period of $0.5(= S) < t$, but the completeness is suspicious in the time interval [0, S]. Recall that earthquakes that occur in this interval were considered as triggering events (see Sect. 4). Thus, if some events in this interval are missing, such missing may affect the estimation of the parameters.

To inspect the possibility of this effect, another numerical experiment was conducted. In a similar manner used in Iwata (2014), the influence of the detection capability visualized in Fig. 2 was imposed on the 300 sequences generated in the previous simulation; events contained in each of the sequences were randomly deleted with a detection probability derived from the magnitude of each event and the detection capability at the occurrence time of the event. Then, the parameters were estimated through the same method as described in Sect. 4. As a reference, the case without any deleted events from the sequences was also examined. The results of the estimations is summarized in Table 3, and they have only a minor difference; the influence of missing events seems not be serious.

It is common that we only use events of which magnitudes are larger than the completeness magnitude of a catalog. Within this common way, the parameter estimation in this study has been supported by the numerical experiment above. However, several recent studies (Schoenberg *et al.* 2010; Wang *et al.*

Table 3

Mean and standard deviation of the MLEs of the parameters for the 300 generated sequences

	r (day^{-1})	t_a (day)	$a/(A\sigma)$	d
Not deleted	$(7.01 \pm 1.77) \times 10^{-3}$	$(4.71 \pm 1.37) \times 10^{2}$	$(8.83 \pm 3.45) \times 10^{-3}$	$(9.49 \pm 0.69) \times 10^{-1}$
Deleted	$(7.02 \pm 1.77) \times 10^{-3}$	$(4.71 \pm 1.36) \times 10^{2}$	$(8.83 \pm 3.45) \times 10^{-3}$	$(9.49 \pm 0.69) \times 10^{-1}$

"Not deleted" corresponds to the case where no event was deleted from the sequences, and "Deleted" corresponds to the case where events were randomly deleted with with a probability derived from the magnitude of each event and the detection capability at the occurrence time of the event

2010; Harte 2013) reveal missing small earthquakes below a completeness magnitude make the biased estimation of the parameters of the ETAS model. Marsan (2005) also points out that small earthquakes that have been thought to be negligible have a significant role in stress triggering through a model including some physical aspects. Therefore, similarly to the case of the ETAS model, it would be necessary to explore the possibility of a bias in fitting of the Dieterich model caused by missing small events in future studies.

5.3. Comparison of the Dieterich Model with the ETAS Model

In this subsection, we compare the goodness-of-fit of the Dieterich and ETAS models to the data. For this comparison, the AIC values of the two types of the Dieterich model examined in this study and the ETAS model are listed in Table 2. For both sequences, the ETAS model has better performance than the Dieterich model.

This feature is quite reasonable for the Kobe sequence. This is because, as discussed above and shown in Fig. 4b, the Dieterich model including the secondary aftershock effect still deviates from the real sequence.

To explore the reason of the superiority of the ETAS model for the Mid-Niigata sequence, it is useful to compare the observed seismicity rate with expected rates of the ETAS and Dieterich models. For the visualization of the observed seismicity rate, it is necessary to divide a study period into subintervals. Then, counting the number of events in each subintervals provides the observed rate, but the seismicity rate obtained in this manner much depends on the choice of the time length of the subintervals. The choice of the time length is not a simple task, in particular, to handle a seismicity rate with marked change in a study period.

Instead, a residual analysis of a point process (e.g., Ogata 1988; Daley and Vere-Jones 2003) is frequently done for the comparison of a model and data. For the calculation of the residual, firstly, our ordinary time scale t is converted to the transformed time κ on the basis of an examined point process model as follows:

$$\kappa = \int_S^t \lambda(t|\hat{\boldsymbol{\theta}})\mathrm{d}t, \quad (5)$$

where $\lambda(t|\boldsymbol{\theta})$ is the seismicity rate of a model, $\boldsymbol{\theta}$ denotes the model parameter(s), and $\hat{\boldsymbol{\theta}}$ is MLE of $\boldsymbol{\theta}$. S indicates the beginning of an analyzed period in the ordinary time scale. If an examined model fits well with data, the occurrence pattern of observed events scaled in κ is expected to be a stationary Poisson process, and therefore, the curve of a cumulative number of events, such as depicted in Figs. 6a and 7a, closely follows a straight line (dotted lines in those figures).

Then we count the number of events $N(\kappa)$ contained in the interval $(\kappa - h, \kappa)$, which is expected to be a random variable obeying a Poisson distribution with mean h for any value of κ. Finally, using the formula suggested by Shimizu and Yuasa (1984), $N(\kappa)$ is converted into the residual $\xi(\kappa)$ that will approximately follow a standard normal distribution if a given model is appropriate (see Ogata 1988 for more details).

Figures 6a and 7a depict the cumulative number of observed events versus the transformed time scale, and Figs. 6b and 7b present the plot of the residual $\xi(\kappa)$ with $h = 8$ for the Dieterich and ETAS models, respectively. In the calculation of the residual $\xi(\kappa)$,

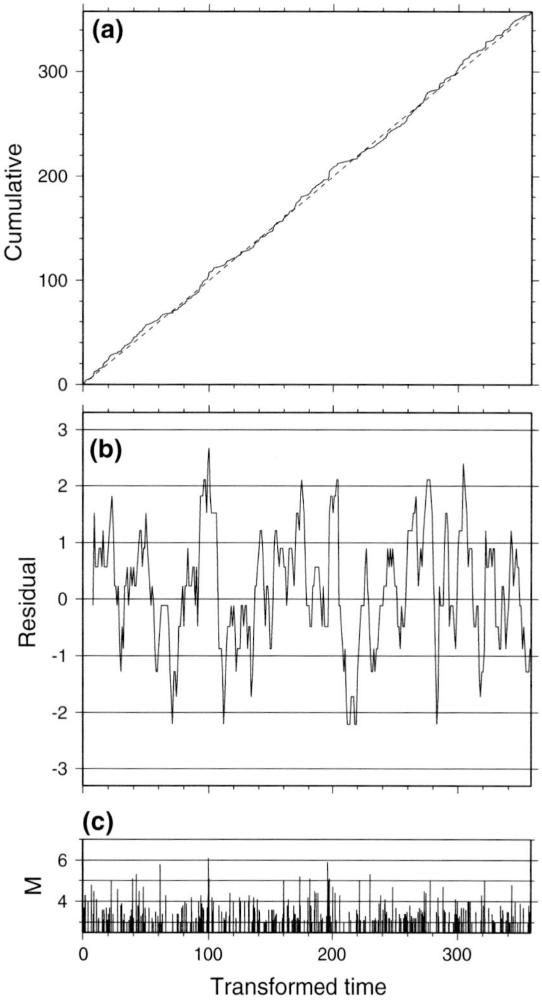

Figure 6
a Cumulative number, **b** residual $\xi(\kappa)$, and **c** magnitude of the events of the Mid-Niigata sequence versus the transformed time on the basis of the Dieterich model

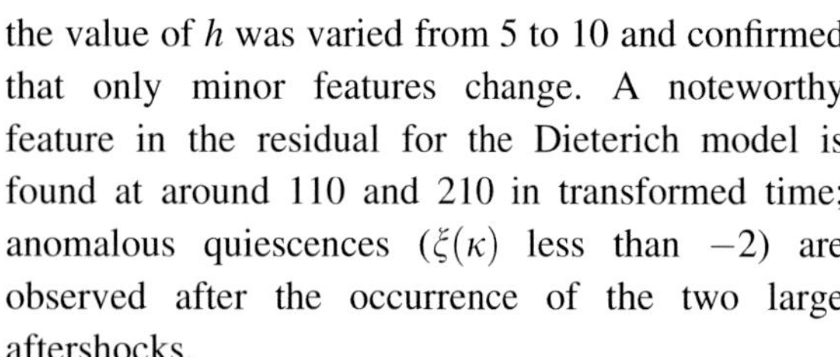

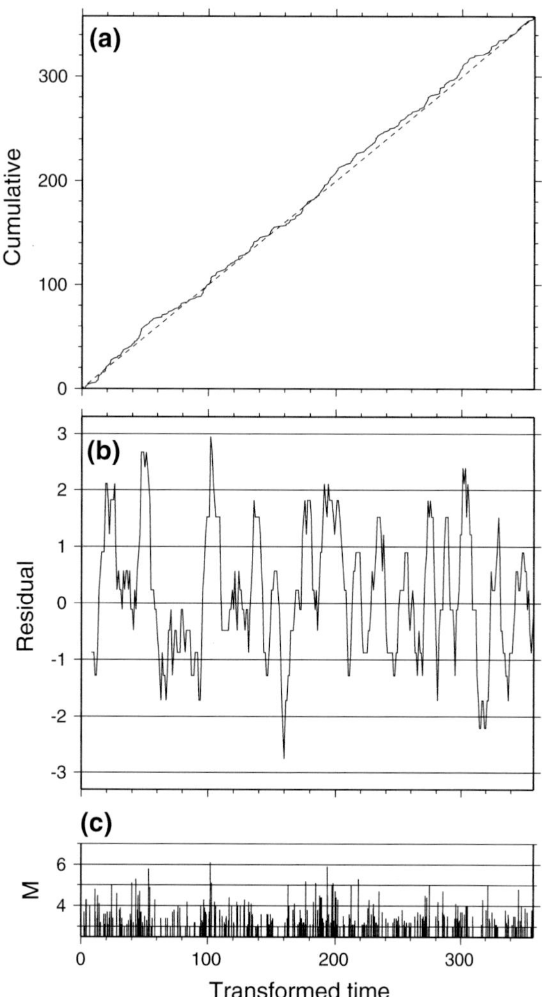

Figure 7
a Cumulative number, **b** residual $\xi(\kappa)$, and **c** magnitude of the events of the Mid-Niigata sequence versus the transformed time on the basis of the ETAS model

the value of h was varied from 5 to 10 and confirmed that only minor features change. A noteworthy feature in the residual for the Dieterich model is found at around 110 and 210 in transformed time; anomalous quiescences ($\xi(\kappa)$ less than -2) are observed after the occurrence of the two large aftershocks.

These quiescences are attributable to the multiplicative property of the Dieterich model as mentioned previously. Suppose that an examined seismicity rate $\lambda(t|\boldsymbol{\theta})$ has the form of $\theta_1 \cdot \lambda'(t|\boldsymbol{\theta}')$, where $\boldsymbol{\theta} = (\theta_1, \boldsymbol{\theta}')$. Then, the expected number of events from the optimized seismicity rate $\int_S^T \lambda(t|\hat{\boldsymbol{\theta}})\mathrm{d}t$ is equal to the observed number during a period $[S, T]$ due to the characteristics of the MLE; the derivation of this characteristics is shown in "Appendix". The seismicity rate of the Dieterich model examined in this study matches this case because the reference seismicity r corresponds to θ_1 (see Eq. 3).

Let us think the case when a large aftershock accompanied by secondary aftershocks occurs. If the seismicity rate expected from the Dieterich model is sufficiently enhanced by the large event in this case, then the rate will reach at an extremely large value because the sufficiently high seismicity rate immediately after the large event is multiplied by the stress change given by the following events. However, as we have noted, it is necessary that the expected number of events from the model is identical with the observed one; if such an extremely high seismicity rate is adopted, the expected number of events from the model becomes unreasonably large and the necessary condition is violated. To prevent such a violation, the maximum likelihood method constraints the seismicity rate immediately after a large aftershock to be a suppressed level. Then, the occurrence of successive events multiplies the rate one by one and the rate gradually approaches a peak value. In other words, there is a lag between the occurrence of the large shock and the time when the expected rate of the Dieterich model takes the maximum value while the real rate would be most active immediately after the large shock; when the maximum rate in the model appears, the decay of the activity already starts in a real sequence. Hence, the quiescence (i.e., overestimation of the seismicity rate in the model) is found as visualized in Fig. 6b. This would be a major reason of the worse performance of the Dieterich model.

As mentioned above, the Dieterich model is inferior to the ETAS model, but this is not the case throughout the study period. For instance, in the residual of the ETAS model (Fig. 7b), the significant activation and quiescence are observed at around 50 and 160, respectively whereas those anomalies are not seen in the Dieterich model. As a total performance, the Dieterich model is worse than the ETAS model, but it is better during some periods, suggesting the possibility to explore the mixture of the Dieterich and ETAS models for the improvement of an aftershock modeling.

6. Conclusion

In this study, the effect of secondary aftershock on aftershock decay in the rate- and state-friction model is investigated on the basis of a real dataset analysis. Two contrasting aftershock sequences are examined, and to some extent, the model including the secondary aftershock effect succeeds in reproducing the variety of the observed aftershock decay.

However, admittedly, the introduced model in this study is simple, and in particular, the assumption that the stress changes caused by aftershocks are constant in space is unrealistic. In real earthquake mechanics, the stress change has strong heterogeneity in space, and the consideration of the heterogeneity is indispensable for a more realistic modeling of earthquake triggering. Additionally, the dependence of the decay parameter on the magnitude of a mainshock has been discussed, and the spatial heterogeneity may perform an important role for the dependency in the context of rate- and state-friction law (Hainzl and Marsan 2008). Thus, as the next step of aftershock modeling along this direction, incorporating unconsidered factors in this study, such as spatial variability of stress changes and time-varying stress changes, is plausible; such a future development may resolve the remaining inconsistency between the real aftershock decay and rate- and state-friction model we have found in the case of Kobe sequence.

Acknowledgments

The author thanks Shinji Toda for useful discussion and JMA for the permission to use their earthquake catalog. The author also wishes to thank Matthew Gerstenberger (the associate editor) and two anonymous reviewers for their helpful comments to improve the manuscript significantly. This study was partially received support from the Grant-in-Aid 25330053 for Scientific Research (C), by Japan Society for the Promotion of Science. The figures were drawn using the Generic Mapping Tools (Wessel and Smith 1998).

Appendix

The result shown below has been mentioned in Imoto (2000) and Daley and Vere-Jones (2003).

If a seismicity rate with a form of $\lambda(t|\boldsymbol{\theta}) = \theta_1 \cdot \lambda'(t|\boldsymbol{\theta}')$. where $\boldsymbol{\theta} = (\theta_1, \boldsymbol{\theta}')$ is given, the log-likelihood shown in Eq. (4) is rewritten as

$$\ln L = \sum_{S \le t_i \le T} \ln \lambda(t_i|\boldsymbol{\theta}) - \int_S^T \lambda(t|\boldsymbol{\theta}) \mathrm{d}t$$
$$= \sum_{S \le t_i \le T} \ln \lambda'(t_i|\boldsymbol{\theta}') + n \ln \theta_1 - \theta_1 \int_S^T \lambda'(t|\boldsymbol{\theta}') \mathrm{d}t, \tag{6}$$

and n denotes the number of observed events in $[S, T]$. Taking the partial derivative of $\ln L$ with respect to θ_1 and equating it to zero yields

$$\frac{n}{\theta_1} - \int_S^T \lambda'(t|\boldsymbol{\theta}') \mathrm{d}t = 0 \tag{7}$$

because $\lambda'(t|\boldsymbol{\theta}')$ does not contain θ_1. The MLE of $\theta_1 (= \hat{\theta_1})$ is the solution of this equation. That of $\boldsymbol{\theta}' (= \hat{\boldsymbol{\theta}}')$ is determined independently from θ_1. Thus, Eq. (7) holds even for $\boldsymbol{\theta}' = \hat{\boldsymbol{\theta}}'$ and consequently

$$\int_S^T \lambda(t|\hat{\boldsymbol{\theta}}) \mathrm{d}t = \hat{\theta_1} \int_S^T \lambda(t|\hat{\boldsymbol{\theta}}') = n. \tag{8}$$

References

Akaike, H. (1974), *A New look at the statistical model identification*, IEEE Trans. Autom. Control *19*, 716–723.

Catalli, F., Cocco, M., Console, R., and Chiaraluce, L. (2008), *Modeling seismicity rate changes during the 1997 Umbria-Marche sequence (central Italy) through a rate- and state-dependent model*, J. Geophys. Res. *114*, B11301, doi:10.1029/2007JB005356.

Cocco, M., Hainzl, S., Catalli, F., Enescu, B., Lombardi, A. M., and Woessner, J. (2010), *Sensitivity study of forecasted aftershock seismicity based on Coulomb stress calculation and rate-and state-dependent frictional response*, J. Geophys. Res. *115*, B05307, doi:10.1029/2009JB006838.

Daley, D. J., and Vere-Jones D. (2003), *An Introduction to the Theory of Point Processes*, vol. I, 2nd ed., Springer, New York.

Dieterich, J. (1994), *A constitutive law for rate of earthquake production and its application to earthquake clustering*, J. Geophys. Res. *99*, 2601–2618.

Dziewonski, A. M., Ekström, G., and Salganik, M. P. (1996), *Centroid-moment tensor solutions for January–March 1995*, Phys. Earth Planet. Int. *93*, 147–157.

Guo, Z., and Ogata, Y. (1997), *Statistical relations between the parameters of aftershocks in time, space, and magnitude*, J. Geophys. Res. *102*, 2857–2873.

Hainzl, S., and Marsan, D. (2008), *Dependence of the Omori–Utsu law parameters on main shock magnitude: Observation and modeling*, J. Geophys. Res. *113*, B10309, doi:10.1029/2007JB005492.

Hainzl, S., Enescu, B., Cocco, M., Woessner, J., Catalli, F., Wang, R., and Roth F. (2009), *Aftershock modeling based on uncertain stress calculations*, J. Geophys. Res. *114*, B05309, doi:10.1029/2008JB006011.

Hainzl, S., Zöller, G., and Wang, R. (2010), *Impact of the receiver fault distribution on aftershock activity*, J. Geophys. Res. *115*, B05315, doi:10.1029/2008JB006224.

Harte, D. (2013), *Bias in fitting the ETAS model: a case study based on New Zealand seismicity*, Geophys. J. Int. *192*, 390–412.

Helmstetter, A., Kagan, Y. Y., and Jackson, D. D. (2005), *Importance of small earthquakes for stress transfers and earthquake forecasting*, J. Geophys. Res. *110*, B05S08, doi:10.1029/2004JB003268.

Helmstetter, A., and Shaw, B. E. (2009), *Afterslip and aftershocks in the rate-and-state friction law*, J. Geophys. Res. *114*, B01308, doi:10.1029/2007JB005077.

Imoto, M. (2000), *A quality factor of earthquake probability models in terms of mean information gain*, Zisin (J. Seismol. Soc. Japan) *53*, 79–81 (in Japanese with English abstract and figure captions).

Iwata, T. (2008), *Low detection capability of global earthquakes after the occurrence of large earthquakes: Investigation of the Harvard CMT catalogue*, Geophys. J. Int. *174*, 849–856.

Iwata, T. (2012), *Revisiting the global detection capability of earthquakes during the period immediately after a large earthquake: considering the influence of intermediate-depth and deep earthquakes*, Res. Geophys. *2*, 24–28.

Iwata, T. (2013), *Estimation of completeness magnitude considering daily variation in earthquake detection capability*, Geophys. J. Int. *194*, 1909–1919.

Iwata, T. (2014), *Decomposition of seasonality and long-term trend in seismological data: a Bayesian modelling of earthquake detection capability*, Aust. N. Z. J. Stat. *56*, 201–215.

Kagan, Y. Y. (2004), *Short-term properties of earthquake catalogs and models of earthquake source*, Bull. Seismol. Soc. Am. *94*, 1207–1228.

Marsan, D. (2005), *The role of small earthquakes in redistributing crustal elastic stress*, Geophys. J. Int. *163*, 141–151.

Marsan, D. (2006), *Can coseismic stress variability suppress seismicity shadows? Insights from a rate-and-state friction model*, J. Geophys. Res. *111*, B06305, doi:10.1029/2005JB004060.

Nanjo, K. Z., Ishibe, T., Tsuruoka, H., Schorlemmer, D., Ishigaki, Y., and Hirata, N. (2010), *Analysis of the completeness magnitude and seismic network coverage of Japan*, Bull. Seismol. Soc. Am. *100*, 3261–3268.

National Research Institute for Earth Science and Disaster Prevention (2004), http://www.fnet.bosai.go.jp/.

Noda, A., and Matsu'ura, M. (2010), *Physics-based GPS data inversion to estimate three-dimensional elastic and inelastic strain fields*, Geophys. J. Int. *182*, 513–530.

Ogata, Y. (1981), *On Lewis' simulation method for point processes*, IEEE Trans. Inf. Theory, *IT-27*, 23–31.

Ogata, Y. (1983), *Estimation of the parameters in the modified Omori formula for aftershock frequencies by the maximum likelihood procedure*, J. Phys. Earth, *31*, 115–124.

Ogata, Y. (1988), *Statistical models for earthquake occurrence and residual analysis for point processes*, J. Am. Stat. Assoc., *83*, 9–27.

Ogata, Y. (2001), *Exploratory analysis of earthquake clusters by likelihood-based trigger models*, J. Appl. Probab., *38A*, 202–212.

Ogata, Y. (2006), *Statistical analysis of seismicity—Updated version (SASeis2006)* Computer Science Monograph, *33*, The Institute of Statistical Mathematics, Tokyo.

OGATA, Y. (2011), *Significant improvements of the space-time ETAS model for forecasting of accurate baseline seismicity*, Earth Planets Space, *63*, 217–229.

OGATA, Y., and KATSURA, K. (1993), *Analysis of temporal and spatial heterogeneity of magnitude frequency distribution inferred from earthquake catalogues*, Geophys. J. Int. *113*, 727–738.

RUINA, A. (1983), *Slip instability and state variable friction laws*, J. Geophys. Res. *84*, 10,359–10,370.

SAGIYA, T., MIYAZAKI, S., and TADA, T. (2000), *Continuous GPS array and present-day crustal deformation of Japan*, Pure Appl. Geophys. *157*, 2303–2322.

SCHOENBERG, F. P., CHU, A., and VEEN, A. (2010), *On the relationship between lower magnitude thresholds an bias in epidemic-type aftershock sequence parameter estimates*, J. Geophys. Res. *115*, B04309, doi:10.1029/2009JB006387.

SCHWARZ, C. (1978), *Estimating the dimension of a model*, Annals of Statistics, *6*, 461–464.

SHIMIZU, R., and YUASA, M. (1984), *Normal approximation for asymmetric distributions*, Proc. Inst. Stat. Math. *32*, 141–158.

STEACY, S., GERSTENBERGER, M., WILLIAMS, C., RHOADES, D., and CHRISTOPHERSEN, A. (2014). *A new hybrid Coulomb/statistical model for forecasting aftershock rates*, Geophys. J. Int. *196*, 918–923.

STEIN, R. S. (1999), *The role of stress transfer in earthquake occurrence*, Nature *402*, 605–609.

TODA, S., STEIN, R., REASENBERG, P., DIETERICH, J., and YOSHIDA, A. (1998), *Stress transferred by the 1995* $M_w = 6.9$ *Kobe, Japan, shock: Effect on aftershocks and future earthquake probabilities*, J. Geophys. Res. *103*, 24543–24565.

TODA, S., and STEIN, R. (2003), *Toggling of seismicity by the 1997 Kagoshima earthquake couplet: A demonstration of time-dependent stress transfer*, J. Geophys. Res. *108*, 2567, doi:10.1029/2003JB002527.

TODA, S., STEIN, R., BEROZA, G. C., and MARSAN, D. (2012), *Aftershocks halted by static stress shadows*, Nat. Geosci. *5*, 410–413.

UTSU, T. (1961), *A statistical study on the occurrence of aftershocks*, Geophys. Mag. *30*, 521–605.

UTSU, T. (1970), *Aftershocks and earthquake statistics (II)—Further investigation of aftershocks and other earthquake sequences based on a new classification of earthquake sequences—*, J. Fac. Hokkaido Univ., Ser. VII *3*, 197–266.

UTSU, T. (1971), *Aftershocks and earthquake statistics (III)—Analyses of the distribution of earthquakes in magnitude, time and space with special consideration to clustering characteristics of earthquake occurrence (1)—*, J. Fac. Hokkaido Univ., Ser. VII *3*, 379–441.

UTSU, T., OGATA, Y., and MATSU'URA, R. S. (1995), *The centenary of the Omori formula for a decay law of aftershock activity*, J. Phys. Earth *43*, 1–33.

WANG, Q., JACKSON, D. D., and ZHUANG, J. (2010), *Missing links in earthquake clustering models*, Geophys. Res. Lett. *37*, L21307, doi:10.1029/2010GL044858.

WESSEL, P., and W. H. F. SMITH (1998), *New, improved version of the Generic mapping Tools* released, EOS Trans. AGU *79*, 579.

ZHENG, Z., and VERE-JONES, D. (1994), *Further applications of the stochastic stress release model to historical earthquake data*, Tectonophysics *229*, 101–121.

ZHUANG, J., HARTE, D., WERNER, M. J., HAINZL, S., and ZHOU, S. (2012), *Basic models of seismicity: temporal models*, CORSSA: Community Online Resource for Statistical Seismicity Analysis, doi:10.5078/corssa-79905851.

ZIV, A., and RUBIN, A. M. (2003), *Implications of rate-and-state friction for properties of aftershock sequence: Quasi-static inherently discrete simulations*, J. Geophys. Res. *108*, 2051, doi:10.1029/2001JB001219.

(Received October 2, 2014, revised June 28, 2015, accepted July 11, 2015, Published online August 7, 2015)

Pure Appl. Geophys. 173 (2016), 35–47

DOI 10.1007/s00024-015-1121-y

Pure and Applied Geophysics

Tempo-Spatial Impact of the 2011 M9 Tohoku-Oki Earthquake on Eastern China

Lifeng Wang,[1] Jie Liu,[1] Jing Zhao,[1] and Jingui Zhao[2]

Abstract—We investigate in this study the impact of the Tohoku-Oki earthquake on Eastern China, and particularly focus on postseismic relaxation processes. We first invert for postseismic slip on the fault plane based on the GPS measurements of GEONET in Japan. Then, we use a layered rheological model to theoretically investigate the deep viscoelastic relaxation process. The Tohoku-Oki mainshock produced significant strain changes in Eastern China, dominantly east–west-oriented extension with a level close to or higher than the tectonic strain rates at the east border of China. The strain due to the postseismic stress relaxations has similar patterns as those produced by the mainshock, but with smaller magnitudes. The Tohoku-Oki earthquake impacts Eastern China for decades, but dominantly in the first 2–3 years after the mainshock and caused an apparent displacements and decrease of seismicity rate in Northeast China. For a long-term of 100 years, the Tohoku-Oki earthquake produces about 10 % of the tectonic strain rates in Eastern China, due to viscoelastic relaxation at the deep depth.

Key words: GPS, postseismic relaxation.

1. Introduction

The 2011 M9 Tohoku-Oki earthquake was a devastating event, which occurred in the subduction zone between the Eurasian and Pacific plates, and caused secondary disasters, tsunami, and nuclear leaking. The coseismic slip reached tens of meters (Ozawa *et al.* 2011, 2012; Pollitz *et al.* 2011), and the measured displacements on the seafloor reached 50 meters (Fujiwara *et al.* 2011).

The occurrence of the 2011 M9 Tohoku-Oki earthquake imposed significant influence in the surrounding regions. It produced coseismic displacements of 14.5–57.7 mm in Korea (Hwang *et al.* 2012) and of ~32 mm in Northeast China (Wang *et al.* 2011; Yang *et al.* 2011). This amount is higher than the average secular displacement rate due only to tectonic loading. Triggering effect from the point of view of Coulomb stress change has long been an important aspect in seismic hazard analysis (e.g., Jia *et al.* 2012; Peng *et al.* 2012). Though some studies analyzed Coulomb stress change produced by the Tohoku-Oki earthquake on the major faults in China (Chen *et al.* 2012; Cheng *et al.* 2014), it is still lacking a relatively comprehensive picture of its impacts in both temporal and spatial space. Our study here focuses on the impact of the Tohoku-Oki earthquake on East China from a large tempo-spatial scale and puts emphasis on the postseismic phase.

In response to coseismic stress disturbance, the early postseismic relaxation mainly concentrates near the fault plane, where coseismic stress perturbation is large, though the relaxation type could be different due to local rheological properties (Tse and Rice 1986; Dieterich 1979; Rundle and Jackson 1977; Pollitz 1992; Bürgmann and Dresen 2008; Wang *et al.* 2010; Masterlark and Wang 2002; Peltzer *et al.* 1998). The early postseismic deformation is in general consistent with the motion of the mainshock and can be simulated by slip/creep on the fault plane. In this study, we utilize the continuous GPS measurements of GEONET to invert for the early postseismic slip. Furthermore, we apply a layered viscoelastic relaxation model to theoretically investigate the long-term effect of this big event. Based on the constructed models, we calculate the displacement/strains field and analyze the spatial and temporal impacts of the Tohoku-Oki earthquake.

Electronic supplementary material The online version of this article (doi:10.1007/s00024-015-1121-y) contains supplementary material, which is available to authorized users.

[1] China Earthquake Networks Center, Beijing, China. E-mail: chsxwlf@hotmail.com

[2] Taiyuan University of Technology, Taiyuan, China.

 Reprinted from the journal

2. Postseismic Modeling

2.1. GPS-Measured Postseismic Displacements of GEONET

The GEONET of Japan includes ~1000 stations and covers the whole Japanese island with an average distance of ~25 km (see '+' in Fig. 1a). The daily data of GEONET are processed in the reference frame of ITRF2005 by Geospatial Information Authority (GSI) of Japan (Yamagiwa *et al.* 2006). These continuous GPS measurements provide valuable information for both secular and transient deformations in the region. Plots 1b–1d and 1e–1g in Fig. 1 show the GPS-measured displacements, respectively, at station 93022 and 940023, after correcting for the secular deformations that are determined from the long-term linear component in the displacement time series over 10 years before the occurrence of the Tohoku-Oki earthquake. The plots display the significant postseismic deformation with a time-decaying rate following the Tohoku-Oki earthquake.

In order to construct postseismic slip model that is temporally stable, we firstly filter the displacement time series by averaging each displacement value with its six neighbors in the time series (i.e., $D_i = (d_{i-3} + d_{i-2} + d_{i-1} + d_i + d_{i+1}+d_{i+2}+d_{i+3})/7$ with d and D representing, respectively, the measurements and the filtered displacements). We consider all of the three components (E-W, N-S, and vertical) at 456 GPS sites, where the measurements are relatively stable with time (no apparent fluctuations relating to instrumental problems or local large aftershocks that displayed in Fig. 1a). Based on the filtering procedure, the influence of random noise and outliers that usually exist in the GPS measurements (e.g., vanDam *et al.* 1994; Vey *et al.* 2002; Heki 2001) can be decreased to some extent. We show the filtered data by the black-dashed lines in Fig. 1.

2.2. Postseismic Slip Model

For the postseismic slip modeling, we use the fault geometry provided by Pollitz *et al.* (2011), which is curved according to the subduction zone. Considering postseismic relaxation that occurs mainly surrounding the coseismic rupture where coseismic stress perturbation is high, we utilize a modified fault plane that is enlarged by ~200 km along strike and 80 km along downdip based on Pollitz *et al.* (2011). The fault plane is discretized as 1800 patches with the size of 20 × 20 km, and embedded in a homogenous halfspace. Okada's code (Okada 1992) is adopted to construct the matrix of Green's function. Given the fault geometry, the fault slip inversion is a linear problem, and can be solved by either least-square or Bayesian approach (Wang *et al.* 2012). We utilize the least-square method in this study, and fix the rake angle as 89° (the dominant direction of coseismic rupture) in the inversion so that to stabilize the inversion procedure and capture the major deformation processes. The checkerboard test to the inversion approach and the used observation network is provided in the Appendix (Fig. S1). The result indicates that the observation network can in general resolve the slip cluster of 100 × 100 km, but with a relatively lower resolution near the trench.

Utilizing the filtered displacements ("GPS-measured postseismic displacements of GEONET"), we invert for the postseismic slip which occurred over 10, 30, 60, 90, 180, and 360 days after the Tohoku-Oki earthquake. Meanwhile, we also invert for the postseismic slip occurred in the 2nd postseismic year (03.2012–03.2013). The resulting slip is shown in Fig. 2, and the corresponding displacement residuals of the models are provided in the Appendix (Fig. S2). As seen, the postseismic slip consistently concentrates along downdip of the coseismic rupture in the first 2 years after the Tohoku-Oki earthquake.

2.3. Postseismic Viscoelastic Relaxation Model

The major focus of this study is the impact of the Tohoku-Oki earthquake on China mainland, which is minimal ~1000 km from the hypocenter. For such a large spatial scale, we utilize the stratified model of PREM (Dziewoński and Anderson 1981), and adapt the model by including a viscoelastic layer at the depth of 40 km, the Moho depth in North Japan (Zhao and Hasegawa 1993; Katsumata 2010) (see Fig. 3). This depth is also consistent with the average Moho depth of China (Xiong *et al.* 2011). In the

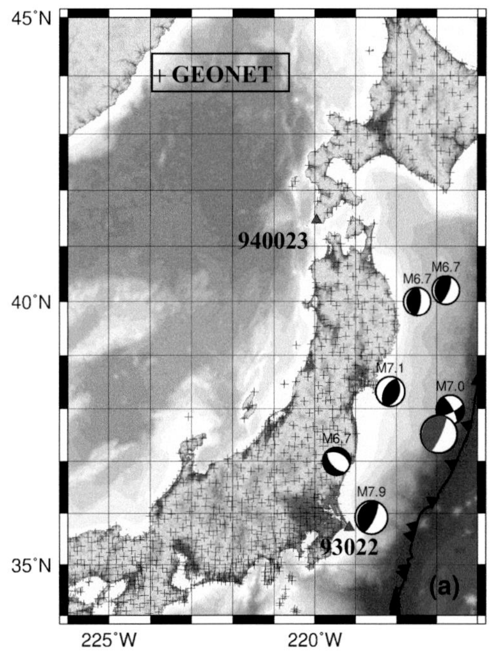

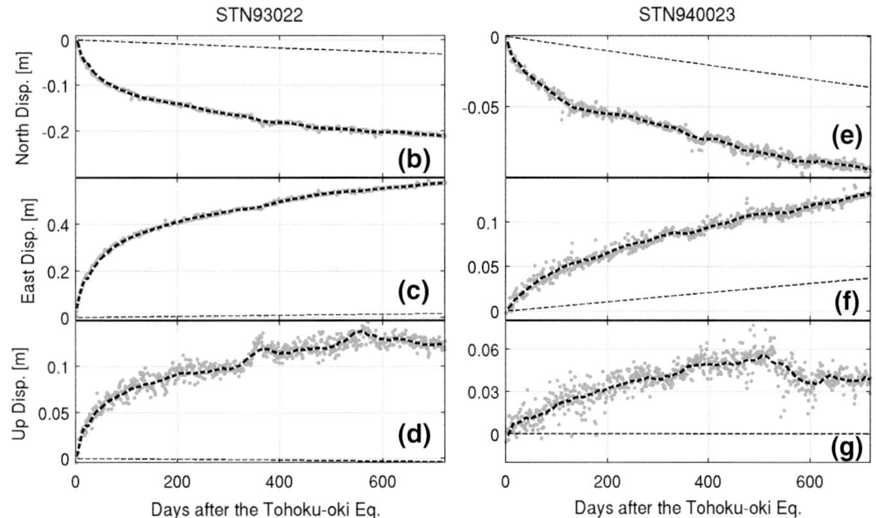

Figure 1

GPS stations (marked by '*plus*') of GEONET in Japan (**a**) and postseismic displacements (after correcting for the secular components) at two sites [marked by *red triangles* in panel (**a**)] shown as illustrations (**b**–**g**). In panel (**a**), the Tohoku-Oki earthquake is displayed by the *red-white* beach-ball and its large aftershocks are marked by the *black-white* beach-balls. In panel (**b**–**g**), the *gray dots* indicate the measurements; the *black curves* show the results based on the filtering procedure. The thin *dashed lines* indicate the displacements calculated from the VER-model ("Postseismic viscoelastic relaxation model")

viscoelastic layer, we consider the Maxwell rheology with the viscosity of 9.3×10^{18}Pa s that has been applied for North Japan (Suito *et al.* 2002). This value is also close to the optimized viscosity of 2×10^{19}Pa s in the postseismic viscoelastic relaxation model for the 2011 M9 Tohoku-Oki earthquake (Diao *et al.* 2013). Utilizing the VISCO1D (Pollitz 1992) that designed for a spherically stratified elastic–viscoelastic medium, we calculate the displacement and strain fields produced by the

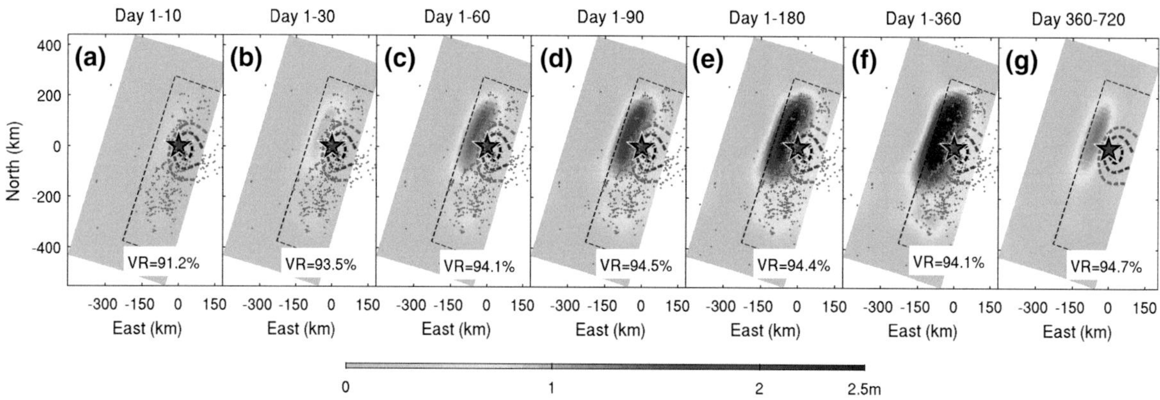

Figure 2
Inverted postseismic slip for 10 days (**a**), 30 days (**b**), 60 days (**c**), 90 days (**d**), 180 days (**e**), 360 days (**f**) and the second years (**g**) after the Tohoku-Oki earthquake. The *dashed rectangle* marks the fault plane used for the coseismic slip modeling (Pollitz *et al.* 2011). The *gray dots* show the aftershocks (M $\geq$ 5.0) near the fault plane ($\pm$25 km). The *contour lines* indicate the major coseismic rupture inverted from GEONET measurements in this study

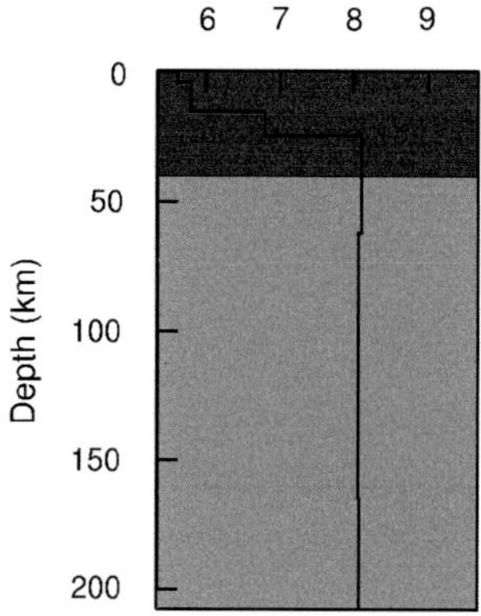

Figure 3
Stratified medium used in the viscoelastic relaxation model

viscoelastic relaxation (VER) in response to the mainshock (Pollitz *et al.* 2011). We illustrate the surface displacements due to viscoelastic relaxation by gray dashed lines in Fig. 1b–e, which indicate rather small values comparing with the observed postseismic displacements in the first 2 years.

3. *Impact of the Tohoku-Oki Earthquake on Eastern China*

3.1. *Overview of Tectonic Blocks and Seismicity in Eastern China*

From the late Cenozoic Era on, tectonic deformation of China mainland is characterized by the movements of the tectonic blocks; the major tectonic blocks (displayed by black solid lines in Fig. 4) in Eastern China from north to south include Xing'an East Mongolia block, Liaodong block, Yanshan block, Ordos block, North China Plain block, Yellow sea block, and South China block. The boundaries between the blocks have the highest gradient of deformation and are characterized by the seismic zones (Shanxi seismic zone, Huabei Plain Seismic Zone, Tancheng-Luhuo seismic zone; Yanshan seismic zone, white marked in Fig. 4), where most of the strong tectonic activities occur there. It has been reported that all earthquakes of M $\geq$ 8 and most (80–90 %) of earthquakes of M7.0–7.9 in China occur in these seismic zone along boundaries of active tectonic blocks (Deng *et al.* 2003; Zhang *et al.* 2003).

Studies on triggering effect of the Tohoku-Oki earthquake on Eastern China have been carried out by several groups. Cheng *et al.* (2014) found that the Tohoku-Oki earthquake produced small Coulomb stress change on the major faults in Eastern China, and thus do not significantly change the regional stress field. From the point of view of dynamic triggering, Parsons et al. did not find triggered events when searching the seismicity recorded over 45 min after the occurrence of the Tohoku-Oki earthquake (Parsons *et al.* 2014). We focus in this study tempo-spatial scale of the impacts imposed by this big event, particularly in the postseismic phase.

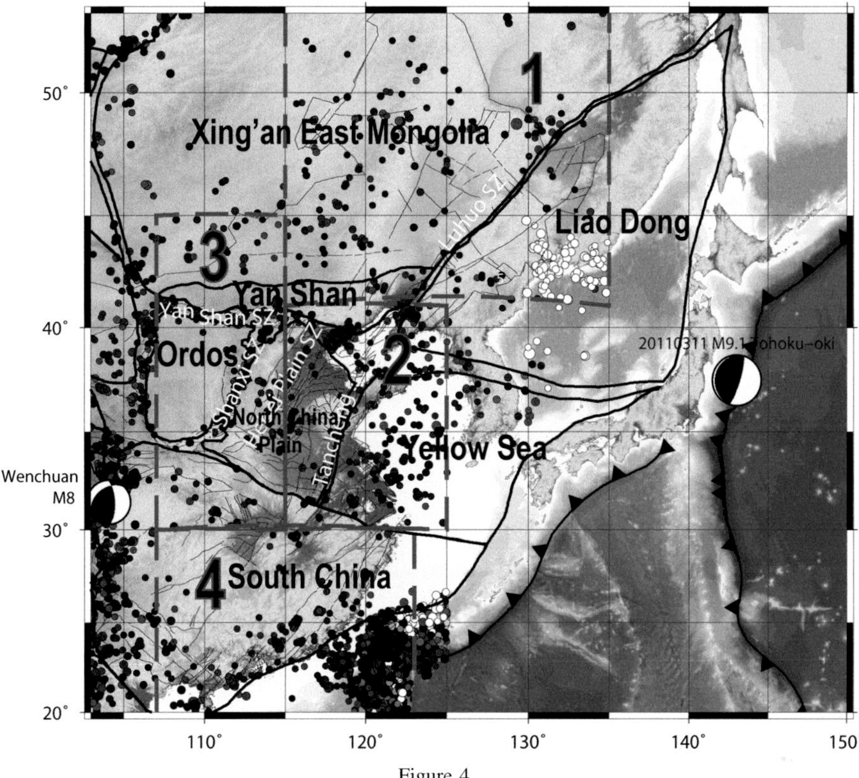

Figure 4
Tectonic settings in Eastern China and earthquake (M ≥ 4.5) since 1970. The main tectonic blocks are displayed by *solid black lines* and *black-bold labeled*. The major seismic zones between the blocks are *white-labeled*. The four zones for investigating seismicity rate are marked by *blue dashed line* and indexed by *numbers*

3.2. *Temporal and Spatial Impacts From the Point of View of Displacement Fields*

The national GPS network of China that including ~260 continuous GPS stations and ~1000 campaign stations has provided valuable data for studying contemporary tectonics in China (e.g., Burchfiel *et al.* 2008; Liu *et al.* 2009; Zhang *et al.* 2008). We display its measured secular displacement rates (per year; in the reference frame of ITRF 2005) (Wu *et al.* 2009) in East China in Fig. 5a. The displacement rates indicate the east-southeastwards tectonic motion with a mean rate of ~20 mm/years in Eastern China.

Based on the slip models (Fig. 2) presented in "Postseismic slip model," we calculate their produced displacements in Eastern China. Meanwhile, we also calculate the coseismic displacements, and quantitatively investigate the influence from the Tohoku-Oki mainshock. We decompose the displacements produced by the Tohoku-Oki earthquake to parallel/perpendicular to the GPS-measured secular displacements in Eastern China, and find that the components perpendicular to the secular displacements are rather small and close to zero. Therefore, we only analyze the components parallel to the secular displacements. The projected displacements are shown in Fig. 5b–d, respectively, along three profiles marked in Fig. 5a. The stations along each profile are selected in the latitude range of ±1°.

As seen, the displacement produced by the Tohoku-Oki mainshock is comparable to or even higher than the annual motion in east border of China, and decrease to the West (with increasing distance to the Tohoku-Oki epicenter). In Southeast China, the effect of the Tohoku-Oki earthquake is small. Meanwhile, postseismic relaxation of the Tohoku-Oki earthquake produced significantly deformation in Eastern China. The postseismic displacements in the first year reach ~30 % of the annual

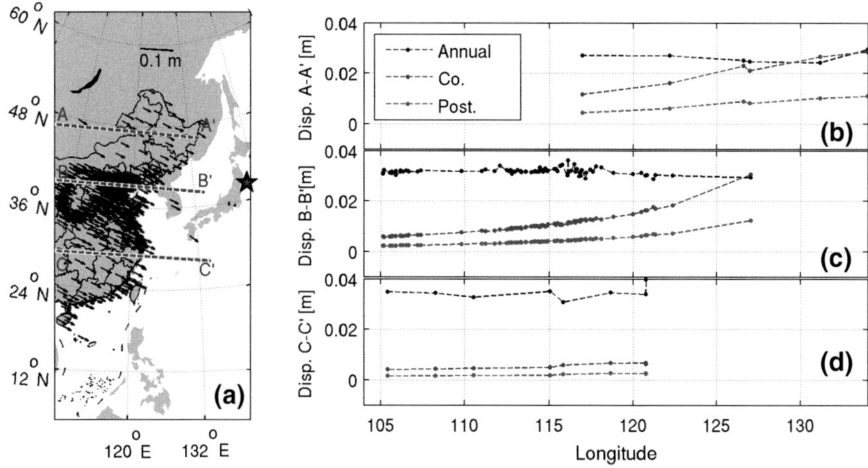

Figure 5
GPS-measured secular displacements in Eastern China (**a**), and the co-/post-seismic displacements projected to the direction of the secular displacements in Eastern China, respectively, along profile A–A' (b), B–B' (**c**) and C–C' (**d**). In (**a**), the *star* indicates the epicenter of the Tohoku-Oki earthquake. In panels (**b–d**), the *black dots* with *dashed line* show the secular displacements. The postseismic displacements are calculated from the inverted postseismic slip in "Postseismic slip model"

displacements in Northeast China. Similar as the coseismic displacement field, the postseismic displacements decrease further to the South and to the West in China mainland.

For the long temporal scale, viscoelastic relaxation at the deep depth is important, and also affects a large spatial space. Based on the layered viscoelastic relaxation model described in "Postseismic viscoelastic relaxation model," we calculate the displacement (rate) in Eastern China. The results shown in Fig. 6 indicate that the viscoelastic relaxation in the first year (Fig. 6a) following the Tohoku-Oki earthquake mainly influence Japanese Island and produce GPS-measureable displacements only in Northeast China. The cumulative displacements due to viscoelastic relaxation increase with time and spread to a wide area. The deep relaxation occurred over 100 years after the Tohoku-Oki earthquake produced displacements higher than 100 mm in Eastern China (see Fig. 6b). In Fig. 6c, we illustrate the displacement rates of 2 mm/y for the 10th, 50th, and 100th year after the Tohoku-Oki earthquake, respectively. The results demonstrate that viscoelastic relaxation has a slow decay process, and imposes long-lasting impact of hundred years in Northeast China.

3.3. Characteristics of the Strain Fields Produced by the Tohoku-Oki Earthquake

We investigate here the impact of the Tohoku-Oki earthquake on Eastern China from the point of view of horizontal strain field. Our analysis focuses on the area that is well covered by the national GPS network of China. In this calculation, we firstly interpolate the displacement measurements at the regular grid points utilizing the least-square collocation approach (Moritz 1978) and then calculate the strain field based on the partial differential relation (Jiang *et al.* 2000). We demonstrate in Fig. 7 the principal strain fields calculated, respectively, based on the GPS-measured secular displacement rates in Eastern China [panel (a)], based on the coseismic slip model (Pollitz *et al.* 2011) of the Tohoku-Oki earthquake [panel (b)], based on the postseismic slip model (Fig. 2) of the 1st postseismic year of the Tohoku-Oki earthquake [panel (c)], and based on the VER-model for the 50th postseismic year [panel (d)].

Figure 7a indicates that the tectonic strain rates (displayed by arrows) which are calculated from the secular displacement rates in Eastern China are dominated by NW–SE oriented extension, but with local heterogeneities, particularly in the block

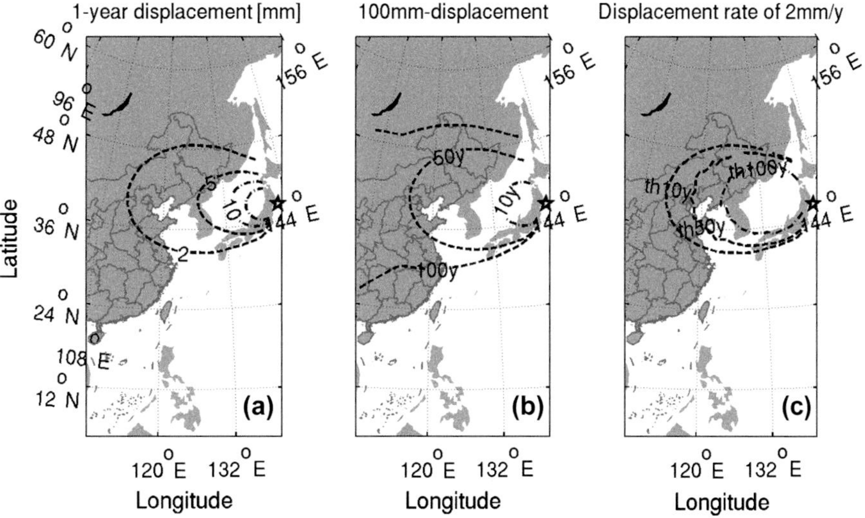

Figure 6
Contour plots of postseismic displacements due to deep viscoelastic relaxation based on a spherical model. **a** Displacements (in mm) in the first year following the Tohoku-Oki earthquake; **b** displacements of 100 mm occurred over 10, 50, and 100 years following the mainshock; **c** displacement rates of 2 mm/years in the 10th year, the 50th year, and the 100th year after the mainshock. The *star* shows the epicenter of the Tohoku-Oki earthquake

margins where tectonic activities are significant and the tectonic strain rates are high. The values are mostly higher than the average noise level of 0.3×10^{-8} given the usual displacement error of 3 mm in the horizontal GPS measurements (Zhu and Shi 2007). In both coseismic and postseismic phases, the Tohoku-Oki earthquake imposes dominantly EW extensional strain in Eastern China. The EW strain components are color-coded in Fig. 7. The EW component in the tectonic strain rates of Eastern China displays the intertwined compression and extension zones. The Tancheng-Luhuo and Shanxi seismic belts, which involve the major faults in Eastern China, are mainly located in the compression zone. Thus, the EW extension produced by the Tohoku-Oki earthquake might play a role to relax the strain accumulated during tectonic loading along the two major seismic belts.

4. Discussion

4.1. Ambiguities in Postseismic Slip Model of the Tohoku-Oki Earthquake

Whether the postseismic slip of the Tohoku-Oki earthquake overlaps the major coseismic rupture area has been argued. Some researchers (Ozawa *et al.* 2011; Perfettini and Avouac 2014) favor for large overlapping, some (Diao *et al.* 2013; Fukuda *et al.* 2013) do not. Our model (Fig. 2) presents disjoint locations between the major postseismic slip and coseismic rupture. Nevertheless, when we apply a smaller fault plane that is the same as the for the coseismic slip model of Pollitz *et al.* (2011) in the inversion, the postseismic slip is largely overlapping with the coseismic rupture zone (Fig. S3 in the Appendix). These slip models can also well explain the observations with Variance Reduction (VR) higher than 90 %.

Such a difference suggests that the obtained shallow slip in the model using the smaller fault plane is likely artifact due to limited slip area considered. This can also be confirmed by the fact that slip in Fig. 2 extends slightly deeper than the coseismic fault plane (dashed rectangle). In comparison, the modeling results based on the larger fault plane of Fig. 2 favor more for the physical mechanism of postseismic relaxation. The different modeling results also reflect the non-uniqueness of the fault slip inversion depending on the model setups and other factors (e.g., Lohman and Simons 2005; Page *et al.* 2009; Wang *et al.* 2012).

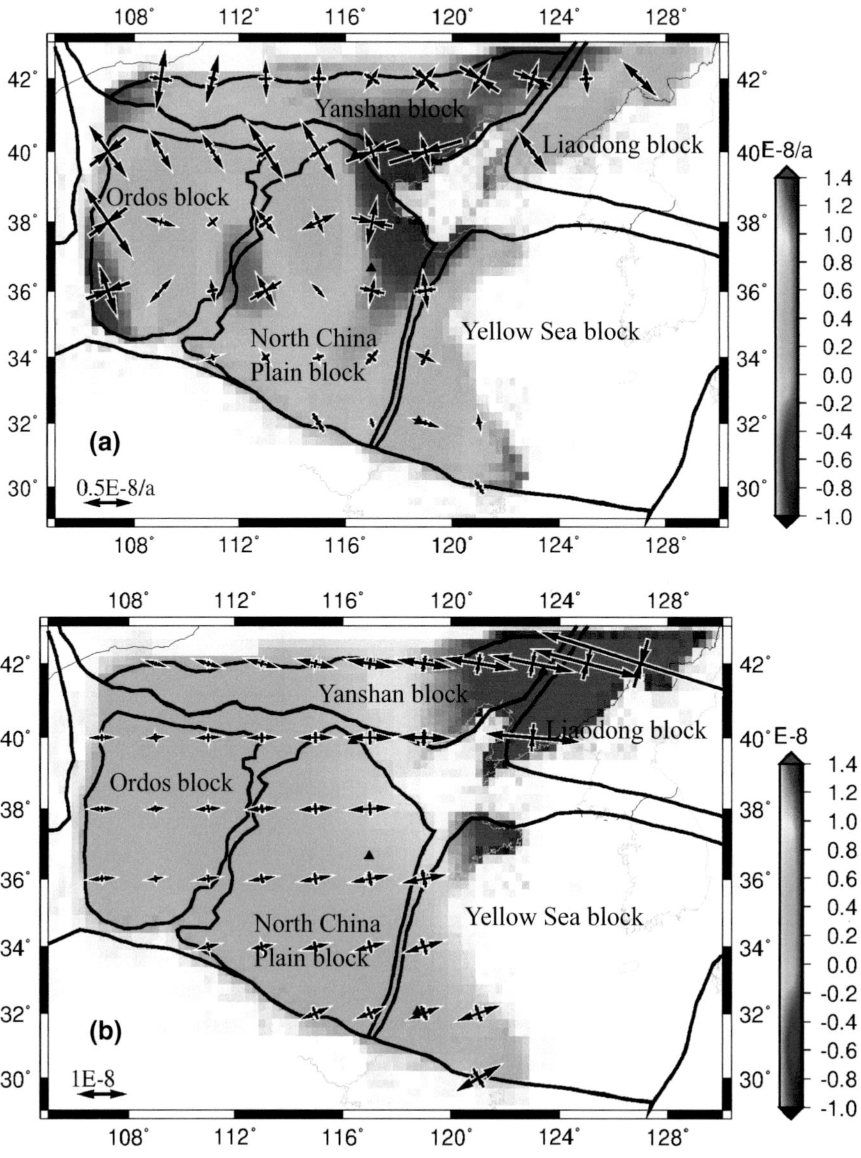

Figure 7
Color-coded EW extensional strain in Eastern China, produced by the tectonic loading according to the GPS-measured secular displacement rates (**a**), produced by the Tohoku-Oki mainshock (**b**), produced by the postseismic relaxation in the 1st year following the mainshock (**c**), and produced by the viscoelastic relaxation in the 50th postseismic year (**d**). The *solid black lines* display the boundary of geological blocks and the *arrows* show the principal strains

In the model of either Fig. 2 or Fig. S3, we have not accounted for viscoelastic relaxation. To quantitatively investigate the significance of the early viscoelastic relaxation, we model the postseismic slip using the displacements after correcting for the effect of viscoelastic relaxation. The results (Fig. S4 in the Appendix) provide that the postseismic slip excluding the contribution of viscoelastic relaxation is similar as that based on the entire measured surface displacements for the 1st postseismic year (Fig. 2f), but has apparently smaller magnitude for the 2nd postseismic year. It indicates that the deep viscoelastic relaxation increasingly contributes to the entire postseismic deformation. The reduction in apparent deep slip when excluding the effect of viscoelastic relaxation suggests that deep postseismic slip and

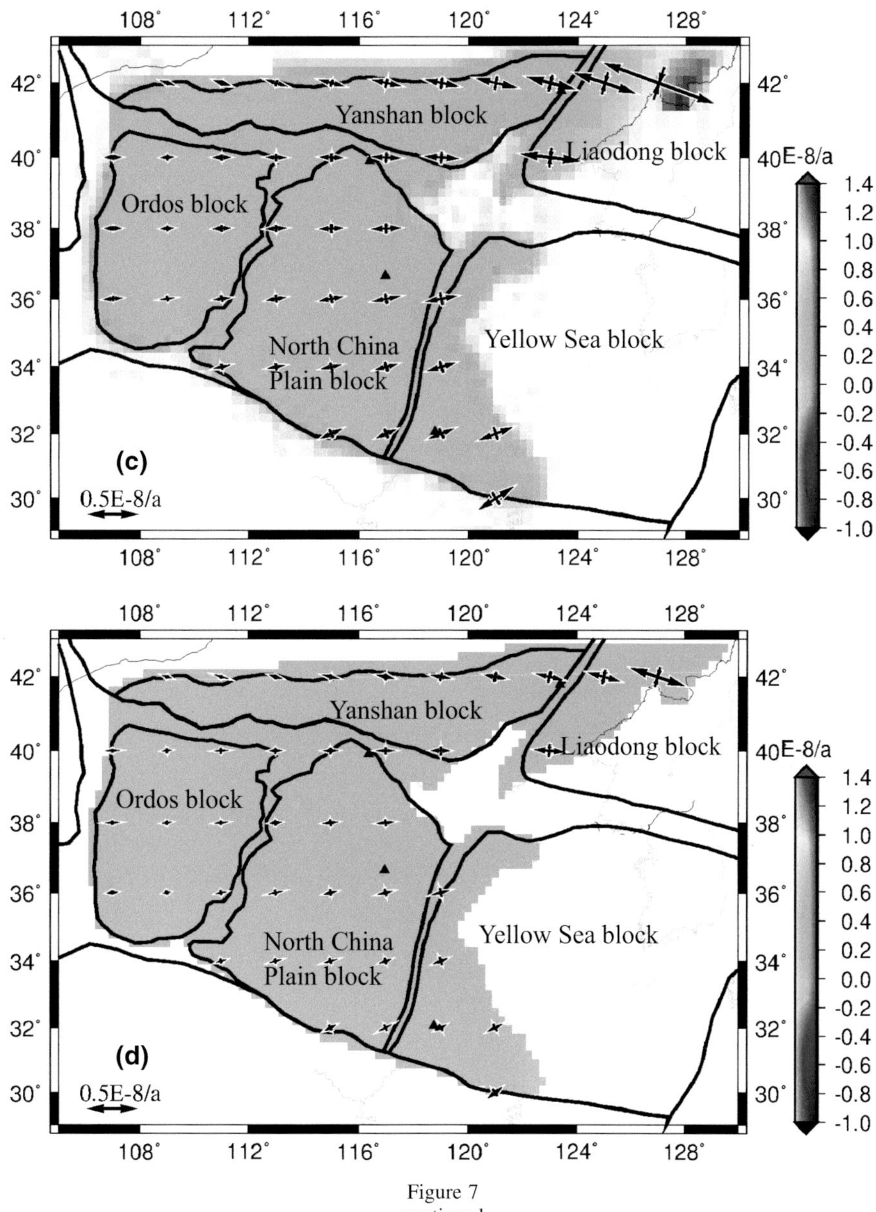

Figure 7
continued

viscoelastic relaxation make also similar signals in the surface displacements, as suggested in other study (Wang *et al.* 2009; Diao *et al.* 2013).

4.2. *Seismic Rate Before and After the Tohoku-Oki Earthquake in Eastern China*

To quantitatively investigate the impact of the Tohoku-Oki earthquake, we analyze the seismicity before and after this mega-thrust event in Eastern China. We delineate the earthquake sequence using epidemic type aftershock sequence (ETAS) model (Ogata 1988, 1999; Zhuang *et al.* 2005), which has been widely used to evaluate the characteristics of seismicity. The ETAS model expresses the seismicity rate $\lambda(t)$ as

$$\lambda(t) = \mu + \sum_{t_i < t} \frac{K e^{\alpha(M_i - M_c)}}{(t - t_i + c)^p}, \qquad (1)$$

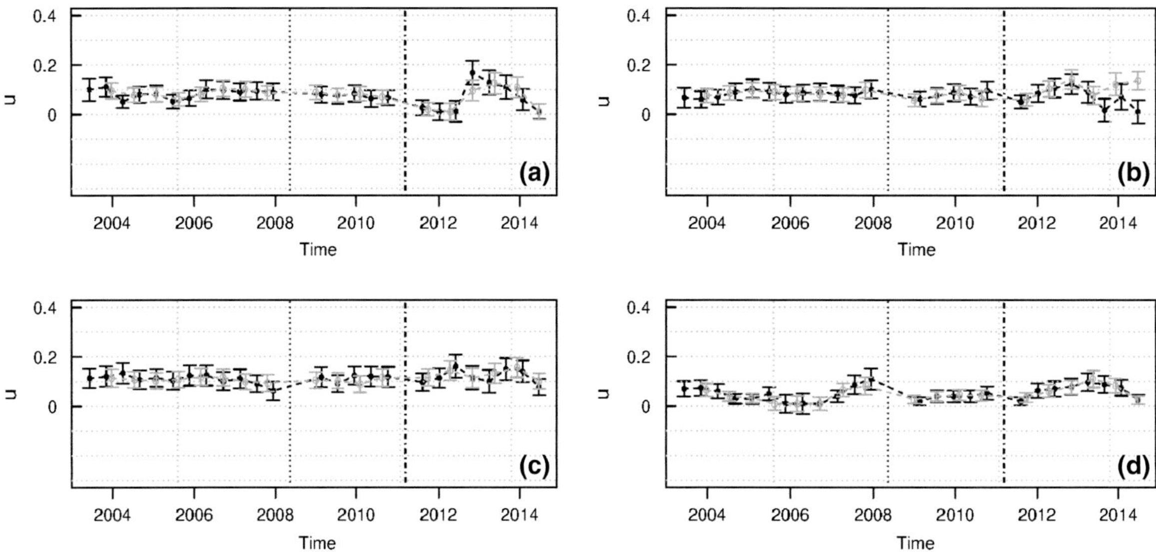

Figure 8
Seismicity rates (*dots* with *error bars*) in Eastern China estimated from ETAS model for the four zones (panel **a** for zone 1, **b** for zone 2, **c** for zone 3, and **d** for zone 4) marked in Fig. 4. The vertical *dashed-dotted line* marks the time of the Tohoku-Oki earthquake and the *dotted line* marks the time of the 2008 M8 Wenchuan earthquake. The *black color* indicates the results based on the time window of 300 days with overlapping of 150 days; while the *gray color* displays the results for the time window of 400 days with overlapping of 200 days

where μ represents the background seismicity rate and the second term at the right side describes the contribution from every preceding earthquake of magnitude M_i ($M_i \geq M_c$) at time t_i ($t_i < t$). α, c, K, and p are constants to characterize the contribution of the earthquakes, and M_c is the lower limit of the magnitude in the earthquake catalog. Therefore, ETAS has an explicit form to separate the background seismicity (μ parameter) that is related to the tectonic loading (e.g., Ide 2013) and the secondary seismicity (aftershocks) that is related to the local stress perturbations due to the preceding earthquakes. In this study, we target to analyze the changes in the stress state due to the occurrence of the Tohoku-Oki earthquake, which might co- and post-seismically interact with the regional tectonic loading, and thus use μ parameter in the ETAS model to quantify the seismicity rate.

We estimate the background seismicity rate of selected four zones (see Fig. 4) over 10 years (2004–2014) in time windows of 300 days, overlapping by 150 days. We use the earthquake catalog provided by China Earthquake Networks Center and define the four zones that, respectively, cover Northeast China (zone1, in which the North part of Tancheng-Luhuo seismic zone is included), Southeast China (zone 4), the middle part of East China (zone 2, in which the South part of Tancheng-Luhuo seismic zone is located; zone 3, in which Shanxi and Yanshan seismic zone is included). We select the earthquakes of M $\geq$ 3.0 (above the completeness level of earthquake records of China, Mignan *et al.* 2013) above the 40 km depth, the Moho depth of East China (Teng *et al.* 2014). Besides, we only consider the data inside the border of China mainland, which is well covered by the seismic network of China.

The estimated background seismicity rates are shown by dots with error bars in Fig. 8. The uncertainties are quantified with the two-sigma rule using a Hessian matrix (Okutani and Ide 2011). The results indicate (1) that the seismicity of Northeast China (zone 1) appears as a clear low rate lasting about 2 years after the occurrence of the Tohoku-Oki earthquake; (2) such a decrease in seismicity rates of zone 2 is not apparent, so do the seismicity rates of zone 3 and 4. Therefore, the Tohoku-Oki earthquake likely imposed an apparent effect on the seismicities in Northeast China, where this big event produced highest displacement and strain amount than other area of China.

We also tested to calculate the seismicity rates in time windows of 400 days (overlapping by 200 days). The results indicated in gray color in Fig. 8 do not show apparent difference from those based on a time window of 300 days, indicating the calculation procedure is stable. Besides, although our analyzed period covers the 2008 M8 Wenchuan earthquake (see Fig. 4), we did not find relevant changes in the seismicity rates for the analyzed zones. It suggests that the M9 mega-event in Japan, releasing 30 times more energy than the Wenchuan earthquake, might impact a much larger region.

5. Conclusion

In this study, we model the postseismic relaxation processes of the 2011 M9 Tohoku-Oki earthquake, and analyze the influence of this big event on Eastern China. We apply the fault slip inversion to investigate the early postseismic slip based on the GPS measurements of GEONET in Japan, and utilize a viscoelastic relaxation model to theoretically investigate the long-term relaxation process. Based on the constructed postseismic models, we calculate the strain fields in Eastern China, and compare them with the tectonic strain rates derived from the GPS-measured secular displacement rates. The results demonstrate that the 2011 Tohoku-Oki mainshock produced strain changes in Eastern China comparable or higher than the tectonic loading rate. The strain changes are dominantly east–west oriented. Similar strain patterns are obtained for the postseismic phase, but with smaller magnitudes. The impacts of the Tohoku-Oki earthquake on Eastern China are dominantly limited in the first 2–3 years after the mainshock, and cause the apparent decrease of seismicity rate in Northeast China. Nevertheless, this big event imposes a long-term impact in Eastern China for nearly 100 years, with a level of $\sim 10\ \%$ of the tectonic strain rate.

Acknowledgments

We acknowledge Geospatial Information Authority (GSI) of Japan for providing access to the GPS measurements of GEONET, Yanqiang Wu for providing the GPS date in Eastern China, and the anonymous reviewers for constructive suggestions. This work is supported by NSFC-41204065 project and Key Project of the National Eleventh-Five Year Research Program of China (2012BAK19B02).

References

Bürgmann, R. & Dresen, G., 2008. *Rheology of the lower crust and upper mantle: evidence from rock mechanics, geodesy, and field observations*, Ann. Rev. Earth Planet. Sci., *36*, 531–567.

Burchfiel, B.C., Royden, L.H., Hilst, R.D.v.d., Hager, B.H., Chen, Z., King, R.W., Li, C., Lu, J., Yao, H. & Kirby, E., 2008. *A geological and geophysical context for Wenchuan earthquake of 12 May 2008, Sichuan, People's Republic of China*, GSA Today, *18*.

Chen, W., Gan, W., Xiao, G., Liang, S. & Sheng, C., 2012. *The impact of 2011 Tohoku-Oki earthquake in Japan on crustal deformation of Northeastern region in China*, Seismology and Geology, 425–439.

Cheng, J., Liu, M., Gan, W., Xu, X., Huang, F. & Liu, J., 2014. *Seismic impact of the Mw 9.0 Tohoku earthquake in Eastern China*, Bull. Seis. Soc. Am, *104*, 1258–1267.

Deng, Q., Zhang, P., Ran, Y., Yang, X., Min, W. & Chen, L., 2003. *Active tectonics and earthquake activities in China*, Earth Science Frontiers, *10*, 66–73.

Diao, F., Xiong, X., Wang, R., Zheng, Y., Walter, T.R., Weng, H. & Li, J., 2013. *Overlapping post-seismic deformaiton processes: afteslip and viscoelastic relaxation following the 2011 Mw 9.0 Tohoku (Japan) earthquake*, Geophys. J. Int.

Dieterich, J.H., 1979. *Modeling of rock friction: 2. Simulation of preseismic slip*, J. Geophys. Res., *84*, 2161–2168.

Dziewoński, A. & Anderson, D.L., 1981. *Preliminary reference earth model*, Phys. Earth Planet. Inter., *25*, 297–356.

Fujiwara, T., Kodaira, S., No, T., Kaiho, Y., Takahashi, N. & Yoshiyuki, K., 2011. *The 2011 Tohoku-oki earthquake: Displacement reaching the Trench axis*, Science, *334*.

Fukuda, J., Kato, A., Kato, N. & Aoki, Y., 2013. *Are the frictional properties of creeping faults persistent? Evidence from rapid afterslip following the 2011 Tohoku-oki earthquake*, Geophys. Res. Lett.

Heki, K., 2001. *Seasonal Modulation of Interseismic Strain Buildup in Northeastern Japan Driven by Snow Loads*, Science, *293*, 89–92.

Hwang, J.-s., Yun, H.-S., Huang, H., Jung, T.-J., Lee, D.-H. & We, K.-J., 2012. *The 2011 Tohoku-Oki earthquake's influence on the Asian plates and Korean geodetic network*, Chinese J. Geophys., 1884–1893.

Ide, S., 2013. *The proportionality between relative plate velocity and seismicity in subduction zones*, Nat Geosci., *6*, 780–784. doi:10.1038/ngeo1901.

Jia, K., Zhou, S. & Wang, R., 2012. *Stress interactions within the strong earthquake from 2001 to 2010 in the Bayankala Block of Eastern Tibet*, Bull. Seis. Soc. Am, *102*, 2157–2164.

Jiang, Z., Zhang, X. & Chen, B., 2000. *Characteristics of recent horizontal movement and strain-stress field in the crust of North China*, Chinese J. Geophys., *43*, 657–665.

Katsumata, A., 2010. *Depth of the Moho discontinuity beneath the Japanese islands estimated by traveltime analysis*, J. Geophys. Res., *115*.

Liu, J.Y., Chen, Y.I., Chen, C.H., Liu, C.Y., Chen, C.Y., Nishihashi, M., Li, J.Z., Xia, Y.Q., Oyama, K.I., Hattori, K. & Lin, C.H., 2009. *Seismoionospheric GPS total electron content anomalies observed before the 12 May 2008 Mw7.9 Wenchuan earthquake*, J. Geophys. Res., *114*.

Lohman, R. & Simons, M., 2005. *Some thoughts on the use of InSAR data to constrain models of surface deformation: Noise structure and data downsampling*, Geocham. Geophys. Geosyst., *6*.

Masterlark, T. & Wang, H.F., 2002. *Transient stress-coupling between the 1992 Landers and 1999 Hector Mine, California, earthquakes*, Bull. Seism. Soc. Am., *92*, 1470–1486.

Mignan, a., Jiang, C., Zechar, J.d., Wiemer, S., Wu, Z. & Huang, Z., 2013. *Completeness of the Mainland China Earthquake Catalog and Implications for the Setup of the China Earthquake Testing Center*, Bull. Seis. Soc. Am., *103*, 845–859.

Moritz, H., 1978. Least-Squares Collocation, *Review of Geophysics and Space Physics,* 16, 421–429.

Ogata, Y., 1988. *Statistical models for earthquake occurrences and residual analysis for point processes*, J. Am. Stat. Assoc., *83*, 9–27.

Ogata, Y., 1999. *Seismicity analysis through point-process modeling: A review,* Pure Appl. Geophys., *155*, 471–507.

Okada, Y., 1992. *Internal deformation due to shear and tensile faults in a half-space*, Bull. Seism. Soc. Am., *82*, 1018–1040.

Okutani, T. & Ide, S., 2011. *Statistic analysis of swarm activities around the Boso Peninsula, Japan: Slow slip events beneath Tokyo Bay?,* Earth and Planetary Space, *63*, 419–426.

Ozawa, S., Nishimura, T., Munekane, H., Suito, H., Kobayashi, T., Tobita, M. & Imakiire, T., 2012. *Preceding, coseismic, and postseismic slips of the 2011 Tohoku earthquake, Japan*, J. Geophys. Res., *117*.

Ozawa, S., Nishimura, T., Suito, H., Kobayashi, T., Tobita, M. & Imakiire, T., 2011. *Coseismic and postseismic slip of the 2011 magnitude-9 Tohoku-Oki earthquake*, Nature.

Page, M., Custodio, S., Archuleta, R. & Carlson, J.M., 2009. *Constraining earthquake source inversions with GPS data: 1. Resolution-based removal of artifacts*, J. Geophys. Res, *114*.

Parsons, T., Segou, M. & Marzocchi, W., 2014. *The global aftershock zone*, Tectonophysics, *618*, 1–34.

Peltzer, G., Rosen, P., Rogez, F. & Hudnut, K., 1998. *Poroelastic rebound along the Landers 1992 earthquake surface rupture*, J. Geophys. Res., *103*, 30,131–130,146.

Peng, Y., Zhou, S., Zhuang, J. & Shi, J., 2012. *An approach to detect the abnormal seismicity increase in Southwestern China triggered co-seismically by 2004 Sumatra Mw9.2 earthquake*, Geophys. J. Int, 1734–1740.

Perfettini, H. & Avouac, J.P., 2014. *The seismic cycle in the area of the 2011 Mw 9.0 Tohoku-Oki earthquake,* J. Geophys. Res.

Pollitz, F., 1992. *Postseismic relaxation theory on the spherical earth,* Bull. Seism. Soc. Am, *82*, 422–453.

Pollitz, F., Burgmann, R. & Banerjee, P., 2011. *Geodetic slip model of the 2011 M9.0 Tohoku earthquake*, Geophys. Res. Lett, *38*.

Rundle, J.B. & Jackson, D.D., 1977. *A three-dimensional viscoelastic model of a strike-slip fault*, Geophys. J. R. astr. Soc., *49*, 575–591.

Suito, H., Iizuka, M. & Hirahara, K., 2002. *3-D viscoelastic FEM modeling of crustal deformation in Northeast Japan*, Pageoph, *159*, 2239–2259.

Teng, J., Deng, Y., Badal, J. & Zhang, Y., 2014. *Moho depth, seismicity and seismogenic structure in China mainland*, Tectonophysics, *627*, 108–121.

Tse, S.T. & Rice, J.R., 1986. *Crustal earthquake instability in relation to the depth variation of frictional slip properties*, J. Geophys. Res., *91*, 9452–9472.

vanDam, T.M., Blewitt, G. & Heflin, M.B., 1994. *Atmospheric pressure loading effects on Global Positioning System coordinate diterminations*, J. Geophys. Res., *99*, 23939–23950.

Vey, S., Calais, E., Llubes, M., Florsch, N., Woppelmann, G., Hinderer, J., Amalvict, M., Lalancette, M.F., Simon, B., Duquenne, F. & Hasse, J.S., 2002. *GPS measurements of ocean loading and its impact on zenith tropospheric delay estimates: a case study in Brittany, France*, J. Geod., 419–427.

Wang, L., Hainzl, S., Özeren, M.S. & Ben-Zion, Y., 2010. *Postseismic deformation induced by brittle rock damage of aftershocks*, J. Geophys. Res., *115*.

Wang, L., Hainzl, S., Zöller, G. & Holschneider, M., 2012. *Stress- and aftershock- constrained joint inversions for co- and post- seismic slip applied to the 2004 M6.0 Parkfield earthquake*, J. Geophys. Res., *117*.

Wang, L., Wang, R., Roth, F., Enescu, B., Hainzl, S. & Ergintav, S., 2009. *Afterslip and viscoelastic relaxation following the 1999 M7.4 İzmit earthquake, from GPS measurements*, Geophys. J. Int., *178*, 1220–1237.

Wang, M., Li, Q., Wang, F., Zhang, R., Wang, Y., Shi, H., Zhang, P. & Shen, Z., 2011. *Far-field coseismic displacements associated with the 2011 Tohoku-Oki earthquake in Japan observed by Global Positioning System*, Chinese Sci. Bull., *56*.

Wu, Y., Jiang, Z., Yang, G., Fang, Y. & Wang, W., 2009. *The application and method of GPS strain calculation in whole mode using least square collection in sphere surface*, Chinese J. Geophys, *52*, 1707–1714.

Xiong, X., Gao, R., Zhang, X., Li, Q. & Hou, H., 2011. *The Moho Depth of North China and Northeast China Revealed by Seismic Detection*, Acta Seismologica Sinica, *32*, 36–56.

Yamagiwa, A., Hatanaka, Y., Yutsudo, T. & Miyahara, B., 2006. *Real-time capability of GEONET system and its application to crust monitoring*, Bull. Geogr. Surv. Inst., 27–33.

Yang, S., Nie, Z., Jia, Z. & Peng, M., 2011. *Far-field Coseismic Surface Displacement Caused by the Mw9.0 Tohoku Earthquake*, Geomatics and Information Science of Wuhan University, *36*, 31–54.

Zhang, P., Deng, Q., Zhang, G., Ma, J., Gan, W., Min, W., Mao, F. & Wang, Q., 2003. *Active tectonic blocks and strong earthquakes in the continent of China*, Science in China series D: Earth Science, *46*.

Zhang, P., Xu, X., Wen, X. & Ran, Y., 2008. *Slip rates and recurrence interval of Longmen Shan active fault zone, and tectonic implications for eht mechanism of the May 12 Wenchuan earthquake, 2008, Sichuan, China*, Chinese J. Geophys., *V51*, 1066–1073.

Zhao, D. & Hasegawa, A., 1993. *P wave tomographic imaging of the crust and upper mantle beneath the Japan islands,* J. Geophys. Res., *98*, 4333–4354.

ZHU, S. & SHI, Y., 2007. *Error analysis of strian rates resulted from errors of GPS measurement*, Journal of Geodesy and Geodynamics, *27*, 52–57.

ZHUANG, J., CHANG, C.-P., OGATA, Y. & CHEN, Y.-I., 2005. *A study on the background and clustering seismicity in the Taiwan region by using point process models*, J. Geophys. Res., *110*.

(Received October 6, 2014, revised May 31, 2015, accepted June 6, 2015, Published online June 30, 2015)

Pure Appl. Geophys. 173 (2016), 49–72

DOI 10.1007/s00024-015-1057-2

Pure and Applied Geophysics

An Evaluation of Coulomb Stress Changes from Earthquake Productivity Variations in the Western Gulf of Corinth, Greece

K. M. Leptokaropoulos,[1,2] E. E. Papadimitriou,[2] B. Orlecka–Sikora,[1] and V. G. Karakostas[2]

Abstract—Spatial and temporal evolution of the stress field in the seismically active and well-monitored area of the western Gulf of Corinth, Greece, is investigated. The highly accurate and vast regional catalogues were used for inverting seismicity rate changes into stress variation using a rate/state-dependent friction model. After explicitly determining the physical quantities incorporated in the model (characteristic relaxation time, fault constitutive parameters, and reference seismicity rates), we looked for stress changes across space and over time and their possible association with earthquake clustering and fault interactions. We focused our attention on the Efpalio doublet of January 2010 ($M = 5.5$ and $M = 5.4$), with a high aftershock productivity, and attempted to reproduce and interpret stress changes prior to and after the initiation of this seismicity burst. The spatial distribution of stress changes was evaluated after smoothing the seismological data by means of a probability density function (PDF). The inverted stress calculations were compared with the calculations derived from an independent approach (elastic dislocation model) and this comparison was quantified. The results of the two methods are in good agreement (up to 80 %) in the far field, with the inversion technique providing more robust results in the near field, where they are more sensitive to the uncertainties of coseismic slip distribution. It is worth mentioning that the stress inversion model proved to be a very sensitive stress meter, able to detect even small stress changes correlated with spatio–temporal earthquake clustering. Data analysis was attempted from 1975 onwards to simulate the stress changes associated with stronger earthquakes over a longer time span. This approach revealed that only $M > 5.5$ events induce considerable stress variations, although in some cases there was no evidence for such stress changes even after an $M > 5.5$ earthquake.

Key words: Seismicity rate changes, Rate/state friction, Static Coulomb stress changes, Gulf of Corinth, Efpalio January 2010 seismic sequence.

[1] Seismology and Physics of the Earth's Interior, Institute of Geophysics, Polish Academy of Sciences, 01452 Warsaw, Poland. E-mail: kleptoka@igf.edu.pl; kleptoka@geo.auth.gr; orlecka@igf.edu.pl

[2] Department of Geophysics, School of Geology, Aristotle University of Thessaloniki, 54124 Thessaloniki, Greece. E-mail: ritsa@geo.auth.gr; vkarak@geo.auth.gr

1. Introduction

The study area constitutes part of the western Corinth rift, central Greece, which is one of the most rapidly deforming continental extension areas in the Mediterranean, accommodating intense seismic activity (Fig. 1). This rift has been generalised as an east–west oriented asymmetric half-graben with a north–south striking extension controlled by a series of en echelon north-dipping normal faults bounding the southern coast, along with minor south-dipping antithetic faults along its northern boundary (Roberts and Jackson 1991; Armijo *et al.* 1996; Bell *et al.* 2008). Fault plane solutions of the strongest ($M \geq 6.0$) earthquakes determined in recent decades by waveform modelling (Taymaz *et al.* 1991; Braunmiller and Nabelek 1996; Baker 1997; Kiratzi and Louvari 2003) along with solutions of moderate events that occurred during the Efpalio sequence in January 2010 (Karakostas *et al.* 2012) verify the pattern of east–west striking normal faults dipping to the north. This faulting type is also dominant in fault plane solutions of microearthquakes registered by dense temporary networks (Hatzfeld *et al.* 1990; Rigo *et al.* 1996, among others). Several devastating earthquakes have been reported from ancient times in association of the major regional faults (Papazachos and Papazachou 2003) as the ones recorded during the instrumental era, causing extensive damage and severe casualties. Events of moderate magnitude ($5.0 \leq M \leq 6.0$) are fairly frequent in the area.

The purpose of this study is to investigate seismicity rate changes in the domains of space and time and, from these changes, to reveal the associated stress-field variations. Usually, the impact of stressing history on reference seismicity rates in specific areas is studied aiming to forecast future seismicity

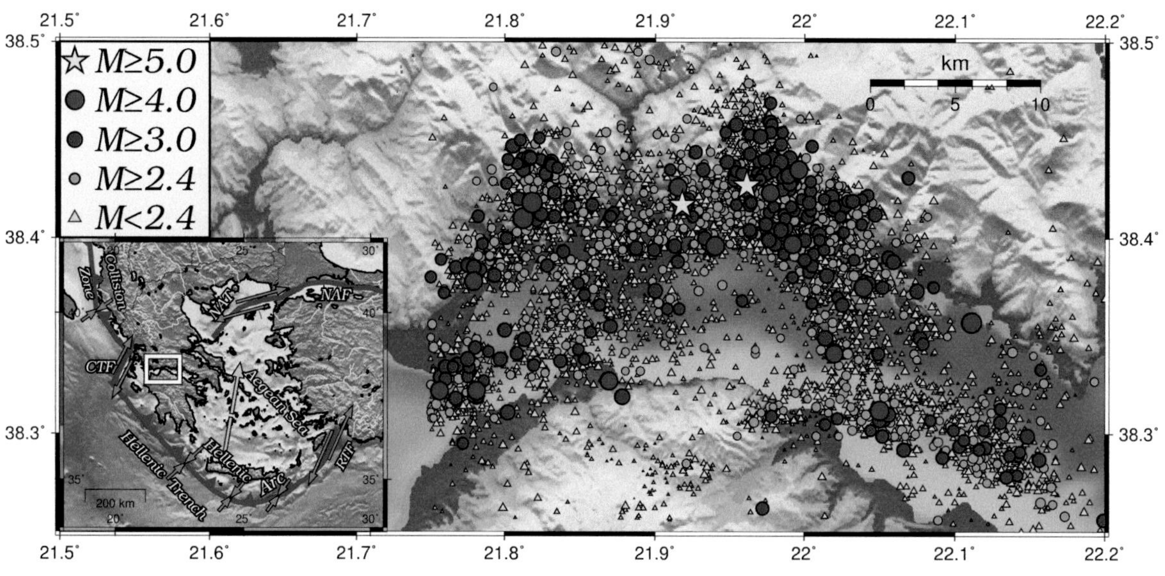

Figure 1
Morphological map of the study area with seismicity occurring between August 2008 and December 2012. *Asterisks* denote the epicentres of the two 2010 Efpalio main shocks. The *inlaid figure* shows the broader Aegean region and its major tectonic features, with the *white box* indicating the study area

rates (see the Appendix for the rate/state forward modelling). The inverse procedure, i.e., the calculation of stress changes from earthquake-occurrence rates obtained from catalogues (achieving adequate spatial and temporal resolution), first led to successful results by Dieterich *et al.* (2000), despite the nonlinearity of earthquake-rate changes with respect to stress and time. These authors proposed and applied two methods (boundary element and elastic dislocation models) on data from the Kilauea volcanic region, with results yielding sufficient agreement with independent estimates of stress changes. The short-term stress fluctuations could not be investigated because of random seismicity rate changes and possible catalogue inconsistencies. The long-term stressing rate evolution and sudden stress steps appeared to be well resolved.

Following the same methodology Toda and Matsumura (2006) studied a large-scale silent slip in the Tokai region. Their purpose was to investigate whether this phenomenon was uniquely associated with the expected Tokai earthquake or with sustained ordinary activity that was repeatedly occurring in the region. They calculated stress changes inverted from microseismicity ($M \geq 1.5$), which allowed the detection of only moderate to large slippage, rather than short-term motions. These inverted stress–change values strongly depended upon slip direction and fault orientation and could be interpreted only after revisiting the regional seismotectonic setting. They resulted in a new constraint of plate–coupling for the Tokai region, proposing a slip distribution on the plate interface. Ghimire *et al.* (2008) estimated the spatio–temporal evolution of Coulomb stress from the analysis of seismicity rate changes within the subducted Pacific slab in Hokkaido. The pattern of stress changes inverted from the seismicity rate changes was comparable with the pattern resulted from dislocation models calculations. Their inversion analysis also revealed that stressing events with $M_W < 7.0$ appear to have a minimal impact on Coulomb stress change in the Pacific slab and that deep-focused, large earthquakes did not significantly change Coulomb stress in shallower layers.

The methodology of Dieterich *et al.* (2000) is followed for calculating the stress field evolution in space and time and the associated fault interaction and earthquake clustering. The spatial and temporal evolution of stress field changes are evaluated for datasets corresponding to different increments of time and areas of major interest, such as those encompassing the failed fault segments. To achieve this task, high-accuracy seismicity catalogues at low magnitude thresholds for a long time interval are necessary. Data availability is the reason for the selection of the study area, despite the fact that stronger earthquakes recently occurred in other areas of the Greek territory, less well monitored. We focused our attention on the most recent catalogue since August 2008, which is characterised by high epicentral accuracy and low completeness magnitude. A second reason for this selection was a swarm–like seismic sequence, which started on 18 January 2010 with the first main event of $M = 5.5$ (Fig. 1). Four days later, the intense aftershock activity culminated in a second main event ($M = 5.4$), and then continued for several weeks with intense but lower magnitude seismicity. The spatio–temporal evolution of the sequence, along with seismotectonic implementations such as coseismic shifts and crustal structure, have already been studied by several researchers (Jansky *et al.* 2011; Ganas *et al.* 2012; Karakostas 2012; Kostelecky and Dousa2012; Novotny *et al.* 2012; Sokos *et al.* 2012). Our aim was to take advantage of the large number of well-recorded earthquakes before and after the 2010 sequence for calculating the spatial and temporal distribution of stress variations. We performed a more detailed analysis close in space and time to the activated fault and attempted to associate earthquake clustering with stress-field variation. The stress changes calculated from the inversion of seismicity rates were compared with the stress changes due to the coseismic slips of the two main events, calculated from the elastic dislocation model, and quantification of this comparison was demonstrated. By prolonging the time of calculations, seismicity rates from 1975 onwards were considered to pursue stress variations over a longer timescale, including a larger number of strong earthquakes.

2. *Selection of Catalogue and Datasets*

The Gulf of Corinth is one of the best monitored areas of the Aegean region, as the adequately dense Hellenic Unified Seismological Network (HUSN) and the local morphology guarantee quite satisfactory azimuthal coverage. This, in turn, results in low completeness magnitude and adequate hypocentral accuracy. The analysis was performed in two periods: from 2008 to 2012 ($\sim$4.5 years) and one from 1975 onwards ($\sim$38 years), with two dinstictive datasets exhibiting different properties. The first data set comprises the catalogue compiled from the recordings of the HUSN between August 2008 and December 2012 (dataset 1, available at http://geophysics.geo.auth.gr/ss/), which demonstrated homogeneity with respect to magnitude estimation and detection level. We excluded from this catalogue earthquakes located at a depth of more than 20 km, which corresponded to either poorly determined hypocentres or to intermediate-depth events in the subducting Eastern Mediterranean plate underneath Peloponnese at this site. Given that these events did not exceed a magnitude of 4.0, they should have negligible influence in the stress calculations. For the analysis performed for longer time, the second dataset was used, taken from the catalogue of the Geophysics Department of Aristotle University of Thessaloniki (dataset 2, available at http:/geophysics.geo.auth.gr/ss/). This dataset comprises earthquakes that have occurred since 1975 in an area extended by 0.3° to the east for including a larger number of strong ($M \geq 6.0$) events. Information on the two datasets is provided in Table 1.

The reliable determination of the completeness magnitude, M_C, is of major importance for any analysis based on seismicity rates. For this determination the modified goodness of fit test (MGFT) proposed by Leptokaropoulos *et al.* (2013) was applied. Although

Table 1

Information about the datasets used in this study

Dataset No.	Boundaries		Duration	M_C	No. of events $M \geq M_C$
	Long (°E)	Lat (°W)			
1	21.75–22.20	38.20–38.50	08/2008–12/2012 (4.4 years)	2.4	988
2	21.75–22.50	38.20–38.50	1975–2012 ($\sim$38 years)	3.5	1613

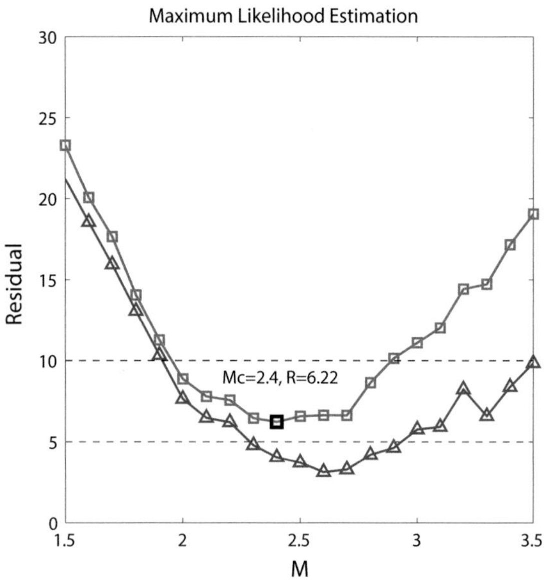

Figure 2
Modified goodness of fit test (MGFT) for M_C determination for dataset 1 during 2010. The residuals yielded as a function of the minimum magnitude from applying the original and modified goodness of fit tests are shown in *green* and *red*, respectively

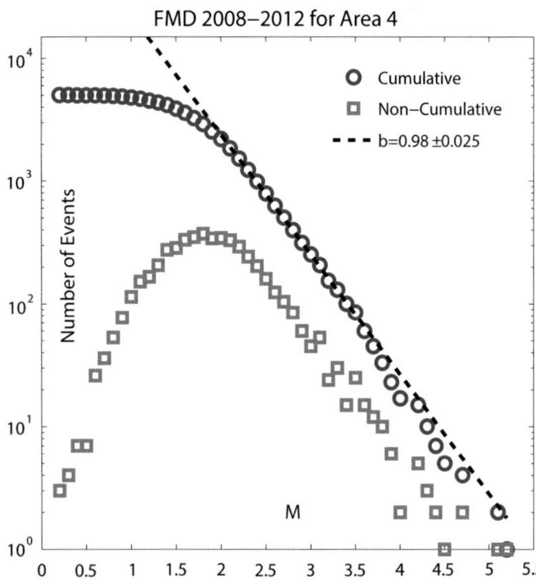

Figure 3
Frequency-magnitude distribution of the events recorder by HUSN from August 2008 to December 2012. The *b* value and its standard error, σ_b, was calculated by estimating the maximum likelihood (Aki 1965) for 1- and 2-year duration subsets and did not show significant fluctuations. On the contrary, M_C fluctuated between 1.9 and 2.4 depending on the selected dataset

a relatively constant detection level of seismicity was achieved since August 2008, a temporal analysis of M_C should be carried out in order to secure the stability of the completeness threshold. For this purpose, annual and two yearly datasets were compiled and the final selection was the highest of the resulting M_C values. The M_C found equal to 2.4 for dataset 1 (Fig. 2) and was observed, as expected, during the aftershock sequence of January 2010, because a fraction of smaller-magnitude earthquakes cannot be distinguished within the coda of larger events (Woessner and Wiemer, 2005). This magnitude is 0.1–0.5 units higher than the M_C calculated for the other tested datasets corresponding to periods before and after the sequence. The *b* value was estimated in the range $0.96 < b < 1.03$ for all subsets, with an average value of $b = 0.98 \pm 0.025$ (Fig. 3). The M_C for dataset 2 (1975–2013) was found equal to 3.5 for 10-year lasting datasets overlapping by 5 years.

3. *Stress–Inversion Methodology*

According to Dieterich (1994) the rate of earthquake production, *R*, can be expressed as:

$$R = \frac{r}{\gamma \dot{S}_r} \tag{1}$$

with:

$$d\gamma = \frac{1}{A\sigma}[dt - \gamma \Delta S] \tag{2}$$

where γ is the state variable, t, is the time, A is a dimensionless fault constitutive parameter, σ is the effective normal stress, r is the steady state earthquake rate at the reference stressing rate $\dot{S}_r$, and ΔS stands for the changes of the modified Coulomb stress function *S*, defined as:

$$S = \tau - [\mu - \alpha]\sigma \tag{3}$$

In Eq. 3, τ is the shear stress, σ is the normal stress, μ is the coefficient of friction and α is a constitutive parameter (Linker and Dieterich 1992). According to this formulation, a causative relationship exists between the stress-field evolution and the deviation of earthquake-production rates from their unperturbed background state. Despite the inherent uncertainties embodied in this approach connected with the determination of parameter values, random seismic fluctuation and fault orientation (Stein 1999; Harris 2000), its application usually yields satisfactory results. A very important precondition is that the local seismicity should be continuously high and well-monitored. Thus, the information about seismicity rates contained in the earthquake catalogues may be interpreted as a stress meter (Toda and Matsumura 2006). Dieterich *et al.* (2000) developed and applied two methods to estimate stress perturbations from seismicity rate changes. In this study we apply the second one, which uses the solution of Eq. 1 as an initial assumption of the constant stressing rate. This solution provides the spatial distribution of stress changes, ΔS, for a stress event:

$$\Delta S = A\sigma \ln\left[\frac{\dot{S}\left(\exp\left(N_2\dot{S}_r t_1/N_1 A\sigma\right) - 1\right)}{\dot{S}_r\left(\exp\left(\dot{S}t_2/A\sigma\right) - 1\right)}\right] \quad (4)$$

where $\dot{S}_r$ and $\dot{S}$ are the background stressing rate and the Coulomb stressing rate on the fault, respectively, N_1 is the count of earthquakes within the time interval, t_1, immediately before the stress event and N_2 is the count of earthquakes within the time interval, t_2, immediately after the stress event. Therefore, ΔS is estimated from the observed time–dependent seismicity, by counting the number of earthquakes occurred during specified time intervals (t_1, t_2). By using the term 'stress event' one can refer to an earthquake (Helmstetter and Shaw 2006; Mallman and Zoback 2007; Ghimire *et al.* 2008), a magmatic intrusion or eruption (Dieterich *et al.* 2000, 2003), or a silent slip event (slow creep that may lead to dynamic instability—Toda and Matsumura 2006 and references therein). For the application of this formulation in our study area, it is assumed that the stressing rate remains constant during the time our study covers and that it is independent of sudden stress events (earthquakes), and identical faulting mechanism since it concerns a small crustal volume (Ghimire *et al.* 2008). The latter assumption is justified by the fact that the average focal mechanism of 31 events in the area exhibits slight variations (258° ± 22° strike, 41° ± 11° dip and −80° ± 22° rake), implying $\dot{S}_r = \dot{S}$. Note that the parameter $A\sigma$ and stressing rate are interconnected in the following equation:

$$t_a = \frac{A\sigma}{\dot{S}_r} \quad (5)$$

where t_a stands for the characteristic relaxation time (or aftershock decay period) necessary to elapse between the stress perturbation and the seismic activity returning to its steady state level.

4. Parameterization

In this section, we describe the process of determining the parameter values of the rate/state model, i.e., the reference and background seismicity rates, the stressing rate, the characteristic relaxation time, and the product Aσ.

4.1. Reference and Background Seismicity Rates

An evaluation of the reference seismicity rate is necessary for a comparison with the stress changes calculated by the elastic dislocation modelling and for estimating an average background seismicity rate for estimating next the t_a from the Omori–Utsu law. The definitions of reference and background seismicity rates are adopted from Cocco *et al.* (2010) as follows: the reference seismicity rate refers to a time-independent, spatially smoothed seismicity rate calculated by using a non-declustered catalogue. The background seismicity rate is considered a time-independent average seismicity rate calculated in a predefined time window from a declustered catalogue.

The reference seismicity rates were estimated in terms of earthquake probabilities by spatially smoothing the seismicity using a probability density function (PDF). A normal grid was superimposed on the study area and the PDF for $M \geq M_C$ earthquakes was determined at the centre of each cell. The PDF used in the present study has the form (Silverman 1986):

$$f(x,y) = \frac{1}{nh^2}\sum_{i=1}^{n} k\left(\frac{x - X_i}{h}, \frac{y - Y_i}{h}\right) \quad (6)$$

where k is a Gaussian Kernel:

$$k(x,y) = \frac{1}{2\pi}e^{-\frac{x^2+y^2}{2}} \quad (7)$$

and X_i and Y_i are the epicentral coordinates (longitude and latitude, respectively), x and y the geographical

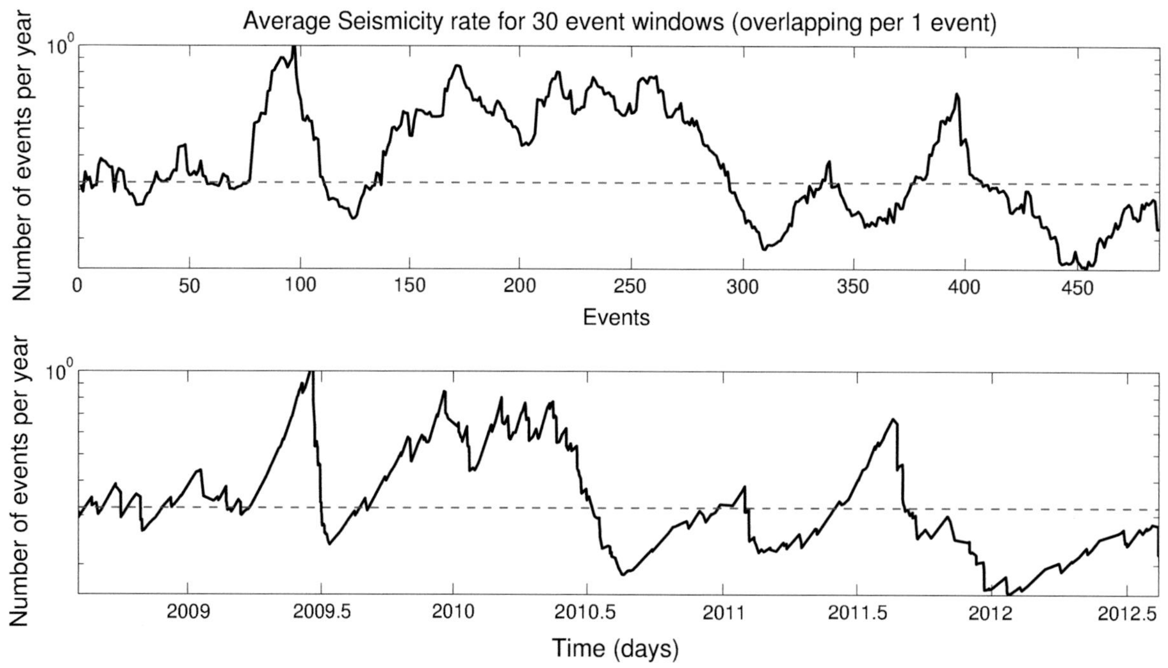

Figure 4
Background seismicity rate from the declustered catalogue in the domains of event (*upper frame*) and time (*lower frame*) for 30-event subsets (overlapping per one event). The averaged background seismicity rate was found to equal 0.32 events per day in the period just before the January 2010 sequence (*dashed line*)

coordinates of each cell centre, n is the number of events inside each cell, and h is the bandwidth (or smoothing parameter) that determines the degree of smoothing.

To estimate background seismicity rates, a declustered catalogue was constructed for the first dataset following Reasenberg's (1985) approach. The seismicity rates were temporally smoothed for 30-event windows and an average background rate equal to 0.48 ± 0.17 events per day was found. Given that declustering methods often apply subjective and arbitrary criteria, it is likely that a fraction of the aftershocks set still remains in the declustered dataset. For this reason, an even more conservative evaluation of the background rate was decided, taken from the declustered seismicity from August 2008 to 15 January 2010, a period without noticeable (enhanced) seismic activity in the study area. In this way we ensured that even less dependent events were included in the dataset and calculated a background seismicity rate equal to 0.32 ± 0.04 events per day (Fig. 4). This rate varies in space and time, but it is assumed to be representative because the unperturbed regional rate is relatively stable during the period before the seismicity enhancement (Fig. 4), as its standard error (0.04) is less than one-quarter of the respective error for the entire dataset (0.17). It is worth to note here that the study area is relatively small, and thus variations in the spatial rate do not significantly influence the results. This is because most events are concentrated in specified areas, which mostly contribute to the value of the calculated background seismicity rate, and where all next calculations are performed.

4.2. *Characteristic Relaxation Time* (t_a)

The characteristic relaxation time is estimated in two different ways. The first approach is based upon a temporal variation of the inter-event time between successive events since August 2008 (Fig. 5), and the second estimates t_a from the parameters of the Omori–Utsu law for the January 2010 sequence (dataset 1, Fig. 6):

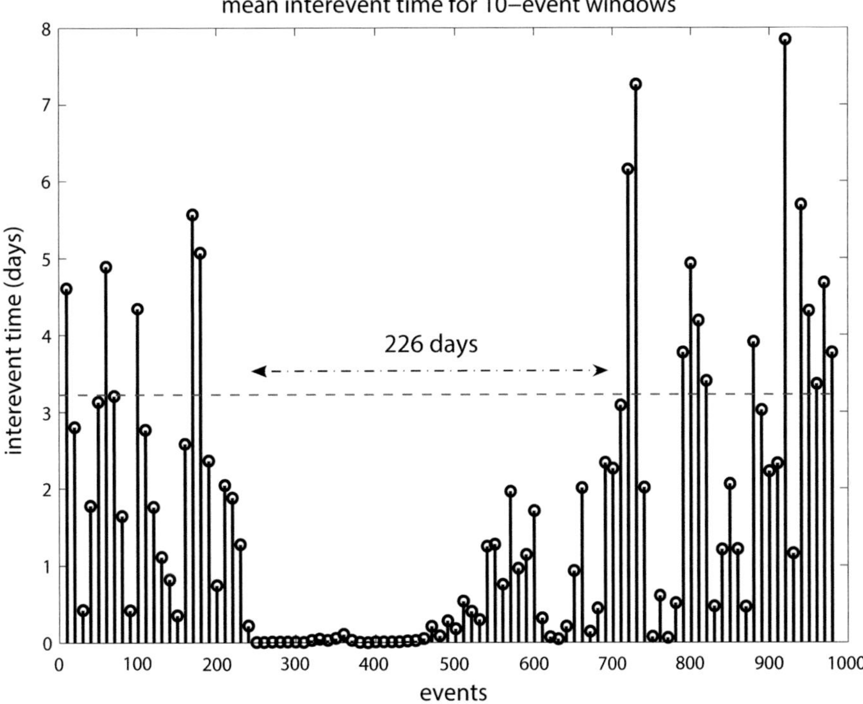

Figure 5
The temporal variation of inter-event times reveals that the seismicity rate returns to the background rate approximately 226 days after the first main shock. These times were yielded by averaging the times of overlapping ten-event windows. The *dashed line* shows the average inter-event time, which is the inverse of the average background seismicity rate as estimated in the previous section

$$\lambda(t) = \frac{K}{(c + t)^p} \qquad (8)$$

The parameters used in the Omori-Utsu decay law (Fig. 6) were estimated as $K = 87.5 \pm 4.5$, $p = 1.05 \pm 0.05$, and $c = 0$, implying a characteristic relaxation time of 223 days (166–307 days at a 95 % confidence level). The calculations were then repeated for the second catalogue (dataset 2, i.e., events since 1975, with $M \geq 3.5$) by fitting the decay law to the aftershock sequence that followed the Aigion earthquake of 15 June 1995 ($M = 6.4$, 38.36°N, 22.20°E). The average background seismicity rate was estimated at 0.1 events per day and the parameters of the Omori–Utsu law were found equal to $K = 45.2 \pm 11.7$, $p = 1.12 \pm 0.09$, and $c = 0.725 \pm 0.16$. These values led to a characteristic relaxation time of 229 days with the 95 % confidence bounds lying between 115 and 473 days. This result is in very good agreement with the ones yielded from the previously mentioned methods for the 2010 sequence (226 days from inter-event time plots and 223 days from the Omori–Utsu law fitting to the Efpalio sequence data). Considering the above we adopted $t_a = 225$ days (or 7.5 months) in the following calculations.

4.3. $A\sigma$ and Stressing Rate ($\dot{S}_r$)

The fault constitutive parameter, A, exhibits a broad potential value range. Harris and Simpson (1998) found acceptable values of A between 10^{-4} and 10^{3}, although it often ranges from 0.005 to 0.015 (Dieterich 1994; Scholz 1998). The effective normal stress, σ, depends on depth, regional stress, fault orientation and pore pressure (Hainzl *et al.* 2010). Usually, the combined parameter $A\sigma$ is considered as a product that describes the frictional resistance of the fault segments. Uncertainties relating to the determination of $A\sigma$ arise from the fact that its values in the Earth's crust are very difficult to estimate. Laboratory experiments also provide a wide range of observed values depending on rock type and environmental

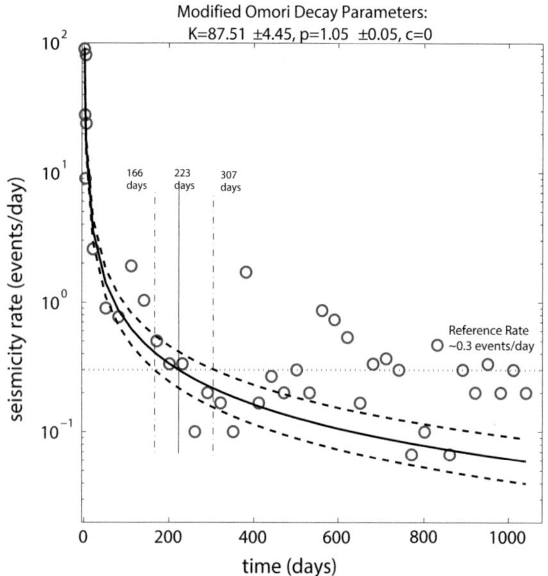

Figure 6
Seismicity rate decay over time (*circles*) and Omori-Utsu law fitted to the data (*solid line*) with its 95 % confidence interval (*dashed lines*) after the 18 January 2010 sequence (dataset 1). The first five points (*solid circles*) show the daily seismicity rates for the first 5 days after the main shock. The rest of the *circles* represent the daily seismicity rates smoothed over 30-days non-overlapping periods. The *horizontal dotted line* represents the background seismic activity as derived from the declustered dataset 1, intersecting with the Omori-Utsu law *curve* at $t_a \sim 223$ days

conditions. The role of $A\sigma$ was analysed by Catalli *et al.* (2008) who showed an increased total number of triggered events in a given time interval after a main shock when $A\sigma$ decreases, in agreement with Belardinelli *et al.* (2003). In the present study, t_a, along with $\dot{S}_r$ was used to estimate $A\sigma$. $\dot{S}_r$ was assumed to be spatially uniform and constant over time. The long term slip rates on the major fault segments were constrained based on GPS data analysis (McClusky *et al.* 2000; Flerit *et al.* 2004; Reilinger *et al.* 2006) and considering 60 % of the geodetic slip value to account for the seismic part of the secular tectonic motion (Ambraseys and Jackson 1990). A stressing rate of $\sim$0.06 bars/year was estimated, equivalent to $\sim 1.68\ 10^{-4}$ bars/day. This value, along with $t_a = 225$ days, was inserted into Eq. 5 to obtain the product $A\sigma$. The stressing rate and characteristic time values yield to $A\sigma = 0.04$ bars ($\sim$0.03–0.05 bars considering the t_a uncertainties), which is relatively low, but still inside the proposed accepted values ($A\sigma = 0.01$–9 bars, Harris and Simpson, 1998) and also in agreement with recent studies (e.g., Hainzl *et al.*, 2013: 0.0016–0.16 bars). Maccaferi *et al.* (2013) accepted a value of 0.05 bars for their analysis of the extensional regime of Iceland, which is fairly similar to the one estimated in this study.

5. *Results and Implications*

After achieving data elaboration and determining parameter values we applied the inversion technique presented above. In this section we will demonstrate the results and their association with various implications for clustering, tectonics, and stress changes. These implications are discussed in the following subsections.

- The temporal evolution of stress field changes inverted from seismicity rate variations between August 2008 and December 2012 (dataset 1).
- A comparison between the inverted stress changes associated with the coseismic slip of the two strongest events of the 2010 sequence ($M = 5.5$ and $M = 5.4$) and the respective ΔCFF (changes in the Coulomb failure function) calculated from the elastic dislocation model (dataset 1).
- The spatial evolution of stress changes inverted from the seismicity rate variations before and after the two main shocks of January 2010 (dataset 1).
- The stress changes close to the fault segments associated with the 2010 doublet (dataset 1, excluding events located approximately one fault length further from the ruptured segments).
- The stress changes associated with spatio–temporal earthquake clustering (dataset 1).
- Temporal analysis of the stress field variations inverted from the seismicity rate changes since 1975 (dataset 2).

5.1. *The Temporal Evolution of the Stress Field Derived from Seismicity Rate Changes*

The stress changes calculated by inverting seismicity rate variations (Eq. 4) and their evolution through time were examined. We illustrated our result in two ways: firstly, we followed an equal–

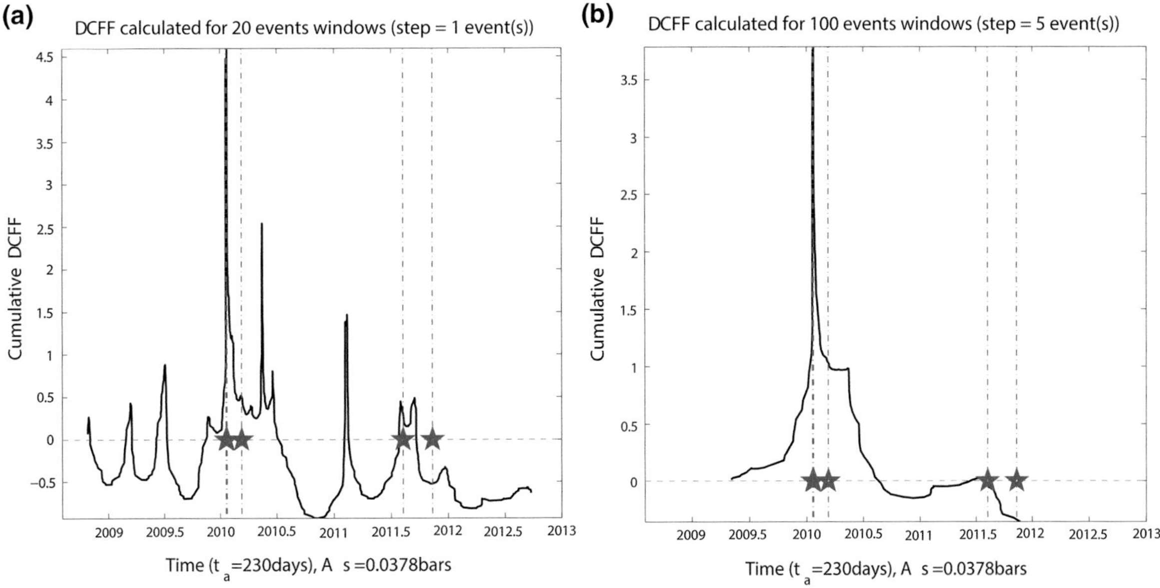

Figure 7
Temporal variation of stress changes inverted from the seismicity rate considering overlapping datasets of 20 (*left-hand frame*) and 100 (*right-hand frame*) events. The size of the event window determines the resolution of ΔCFF: small windows reveal more detailed stress variations, whereas broader windows demonstrate stress changes associated only with the aftershock sequence that followed the 18 January 2010 $M = 5.5$ event. *Blue asterisks* denote the $M \geq 4.5$ events

event–number approach (Fig. 7) and then an equal–time–step approach (Fig. 8). In the first approach we calculated the seismicity rates for unequal moving time windows encompassing a constant predefined number of earthquakes. Since all the datasets contained the same number of events, positive ΔCFF occurred when a time window had a shorter duration than the previous one. These stress changes were plotted versus time (the continuous line in Fig. 7). We first selected a 20–event window overlapping per single event (Fig. 7, left-hand frame). The occurrence times of the strongest shocks ($M \geq 4.5$) are denoted with asterisks. Stress changes associated with these events are clearly demonstrated, particularly for the January 2010 doublet, when the highest ΔCFF values were computed. There are also some distinctive stress steps that do not seem to be connected with an $M \geq 4.5$ event, but are rather a product of a swarm-like activity (we will return to this point in the section concerning earthquake clustering). Taking this approach, it is clear that each positive stress step is followed by a negative one due to the depression of the seismicity rates. This was expected according to this concept because we compared the seismicity rate within each time (or equivalently, event) step with the one that occurred immediately beforehand. Therefore, instead of using a uniform and not explicitly determined constant background rate as reference activity, we compared the differences in the seismicity rate observed in subsequent increments of time. When a dataset corresponded to a lower rate than the preceding one (equivalently, when there was a slope decrease of the cumulative events curve; see Fig. 8), this was interpreted as a stress drop in Fig. 7. The tuning parameter for illustrating stress changes in this case was the size of the time (or event) window. We then selected a broader 100–event window overlapping per five events (Fig. 7, right-hand frame). In this case, the temporal range of the results and their resolution was reduced, but it is clear that the only significant stress step is related to the strongest earthquake occurrence. This stress step was followed by a gradual decrease of stress in long term, which almost stabilised after August 2010. No minor stress changes are distinguished in this figure.

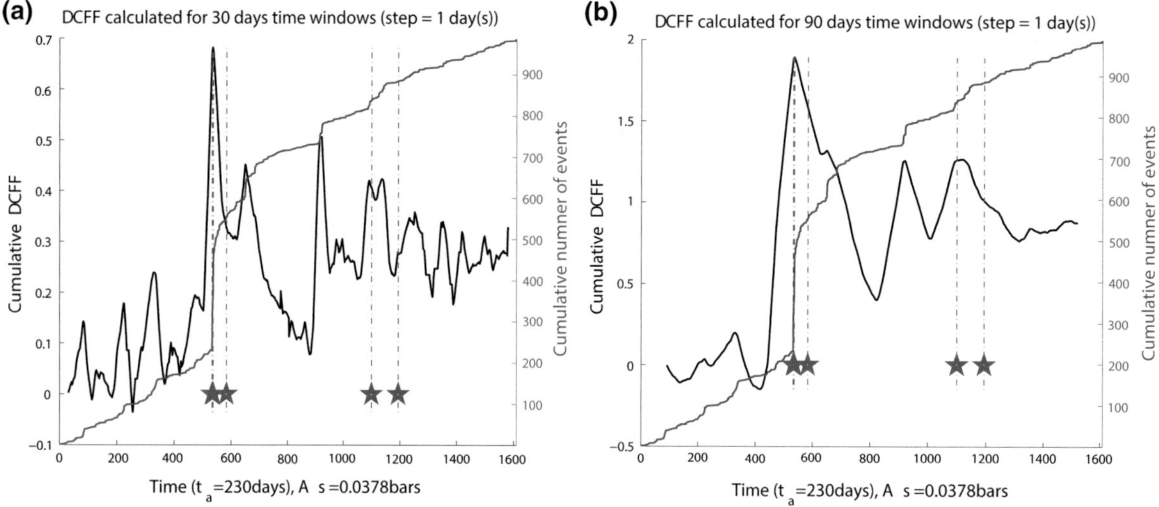

Figure 8
Stress changes inverted from seismicity rate variation considering overlapping datasets of 30-days (*left-hand frame*) and 90-days (*right-hand frame*) time windows. *Blue asterisks* denote the $M \geq 4.5$ events

In the second approach we calculated stress changes by considering fixed time windows before and after successive calculation points (Fig. 8). Therefore, ΔCFF was positive when the dataset that followed a certain calculation point included more events than the one preceding it. The left-hand frame of Fig. 8 shows the stress derived from 30-day time windows, whereas the right-hand frame displays the respective results yielded for 90-day time windows. The cumulative number of events as a function of time is also plotted on the same figures. The stress changes associated with larger magnitude events are obvious, but there are still considerable stress steps caused by seismic enhancement; these are not directly connected with an event of a significantly larger magnitude. These changes are evident before and after the January 2010 doublet. The application of the 90-day time window revealed that, in addition to the stress jumps associated with the main shocks, there were also two notable positive stress steps, one before and one after the January 2010 seismicity burst, which are connected with swarm-like activity rather than a typical main shock—aftershock sequence. Nevertheless, there is also an $M = 4.5$ earthquake (close to event number 1200) that does not seem to induce remarkable stress changes; this is shown in both cases, where the time window is equal to 30 and 90 days, respectively.

5.2. Comparison with ΔCFF Derived from the Elastic Dislocation Model

The stress results inverted from the seismicity rate changes were compared with the stress changes calculated with the elastic dislocation model. This latter approach was firstly applied to calculate the Coulomb stress changes caused by the coseismic slip of the two strongest earthquakes. Then, the reference and seismicity rates of the small–magnitude events for different increments of time after the main shock were spatially smoothed by applying the selected PDF (Eqs. 6 and 7). The differences between the earthquake-occurrence rates before and after the main shock were compared and used as input data in the stress–inversion algorithm in order to provide an independent estimation of the stress changes. Finally, we investigated the quantitative correlation between the results derived from the two methods. The stressing rate is assumed to be constant throughout the study period; therefore, long-term changes in tectonic loading are not investigated here. Even if some variation in the stressing rate does exist, it is expected to be too small to influence substantial changes.

The above analysis is based upon the observation that even small changes in static stress result in considerable seismicity rate changes (Harris 2000;

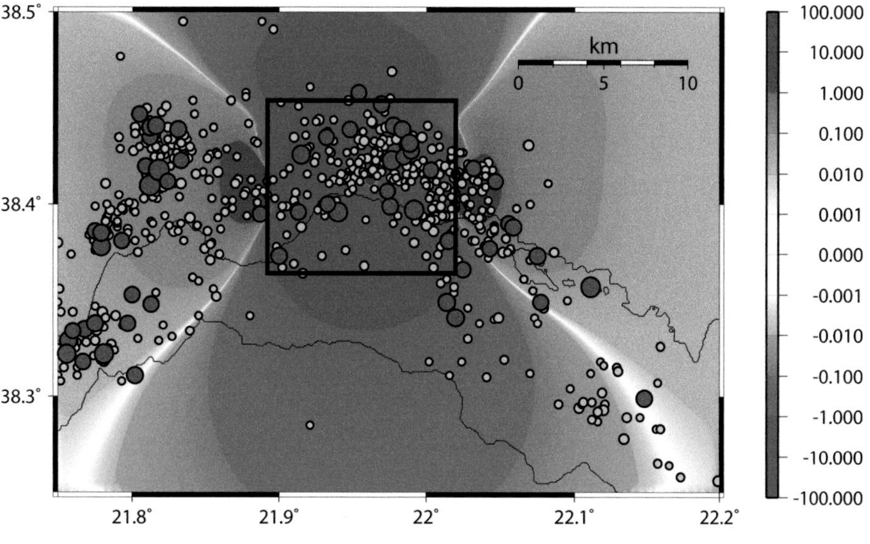

Figure 9
Stress pattern due to the combined coseismic slip of the two main shocks, resolved according to the faulting type of the first main shock at a depth of 9 km. *Grey circles* show the epicentres of subsequent events above $M_C = 2.4$ that occurred up to December 2012, and the strongest of them ($M > 3.5$) are depicted by purple circles. The *black box* indicates the approximate area located at a distance closer than one fault length ($\sim$5.5–6 km) across the ruptured zone. Colour scale is in *bars*

Steacy et al. 2005 and references therein). It was, then, important to examine in what proportion of the area an agreement could be found in the sign of ΔCFF derived from the two methods. The stressing rate and characteristic relaxation time values do not affect the spatial distribution of stress changes, but only their absolute value. Therefore, the results we sought were not sensitive to fluctuations in the values of those parameters. Instead, we focused on the agreement of the ΔCFF sign and the fixed parameter values mentioned in the previous sections: $\dot{S}_r = 0.06$bar/year, $A\sigma = 0.04$ bars and $t_a =$ 225 days. The only parameter that does affect the spatial pattern of the derived stress changes is the bandwidth and for this reason the influence of the bandwidth fluctuation was examined. Note that the epicentral error in the catalogue is 3–5 km and, therefore, the bandwidth selection was made according to this criterion.

Using the elastic dislocation approach to calculate the changes in Coulomb stress due to the coseismic slip of the two main shocks, we adopted the rupture model proposed by Karakostas (2012). The calculations were performed at a depth of 9 km, which is approximately the average depth determined for regional (7.8 ± 3.5) and relocated (8.9 ± 1.6) catalogues. Figure 9 shows the distribution of ΔCFF in the study area after the combined influence of both main shocks, along with their aftershocks. The epicentral distribution evidences important spatial clusters beyond the eastern and western tips of the ruptured fault segment, where the largest positive ΔCFFs were observed. Nevertheless, a significant fraction of seismicity is located inside the negative ΔCFF lobes, mostly comprising onto—fault aftershocks.

In Fig. 10 the agreement percentage between the two stress–estimation methods is plotted as a function of bandwidth, h. The average sign agreement is generally not sensitive to bandwidth fluctuations of between 0.01° and 0.05°, demonstrating an almost stable value between 60 and 65 %. This means that nearly two-thirds of the coseismic stress changes are compatible with the seismicity rate variations observed after the main shocks. Positive stress changes are better forecast, though, with the sign agreement in such areas reaching up to 78 % and being directly proportional to the bandwidth value. On the contrary, negative stress changes are inversely proportional to the smoothing parameter. The sign agreement in this case falls below 60 % for $h > 0.03°$.

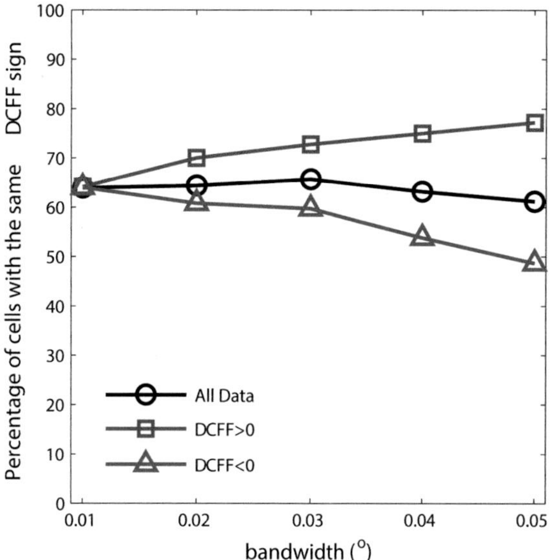

Figure 10
Percentage of cells having a common sign of ΔCFF derived from both methods. The elastic dislocation model takes into account the combined influence of both main shocks. The reference seismicity rate was derived from the events occurring from August 2008 to 18 January 2010. The *lines* demonstrate the results derived from the local catalogue (dataset 1) up to the end of 2012. The *red* and *blue* lines indicate the percentage of cells with positive and negative ΔCFF, calculated using the elastic dislocation model and the stress inversion technique, respectively

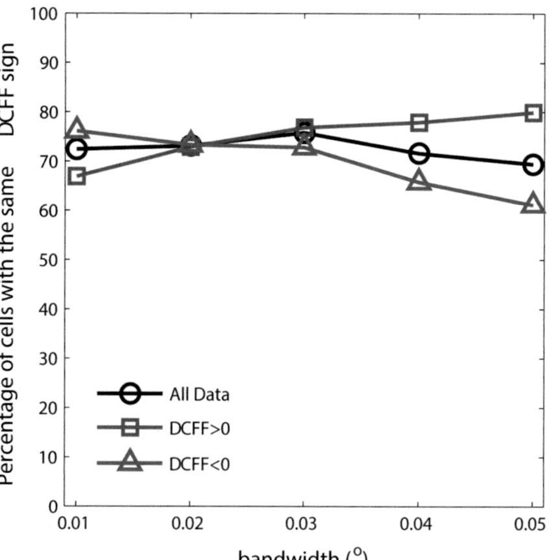

Figure 11
The percentage of cells having a common ΔCFF sign derived from both methods outside the *square area* indicated in Fig. 9. The elastic dislocation model takes into account the influence of both main shocks. The reference seismicity rate was derived from the events occurring from August 2008 to 18 January 2010. The *solid lines* demonstrate the results derived from the local catalogue (dataset 1) up to the end of 2012. The *red* and *blue* lines indicate the percentage of cells with positive and negative ΔCFF, calculated using the elastic dislocation model and the stress inversion technique, respectively

Our next step was to deal with off–fault seismicity rate changes, where the elastic dislocation model calculated ΔCFF with significantly higher accuracy than in the near field, where lack of slip details influenced the stress changes pattern. These stress values were compared with the values calculated by inverting changes in the earthquake rate after excluding the area inside the box shown in Fig. 9, for investigating the distribution of off–fault aftershocks alone. Therefore, the consistency between the two methods is based upon results that do not suffer from significant uncertainties. Figure 11 illustrates that a higher correlation than previously is achieved in this way: the two methods provide the same ΔCFF sign for approximately 75 % of the area beyond the near field (indicated by the box in Fig. 9). Generally, as shown in Figs. 10 and 11, lower bandwidth values lead to stronger agreement for negative ΔCFF cells, whereas higher values of h bring out a higher percentage of agreement for positive ΔCFF areas. Nevertheless, higher bandwidth values should be avoided, because they over-smooth the calculated ΔCFF, which means that local fluctuations can no longer be distinguished, a fact that may lead to erroneous interpretation. Silverman (1986) suggests that bandwidth should be estimated as:

$$h \sim \sigma N^{-1/6} \qquad (9)$$

where $\sigma = \sqrt{\frac{S_x + S_y}{2}}$ and N is the sample size. S_x and S_y stand for the sample variances of the epicentral geographical distribution of the coordinates (X_i and Y_i). This equation yields $h \sim 0.03$, which, as shown in Figs. 10 and 11, leads to a higher percentage of sign agreement and balances the differences between positive and negative ΔCFF cells. This bandwidth value also leads to the highest correlation between the observed and synthetic seismicity rates yielded from rate/state forward modelling (see the Appendix for details).

The comparison of the stress results implies that, in general, there is a good agreement between

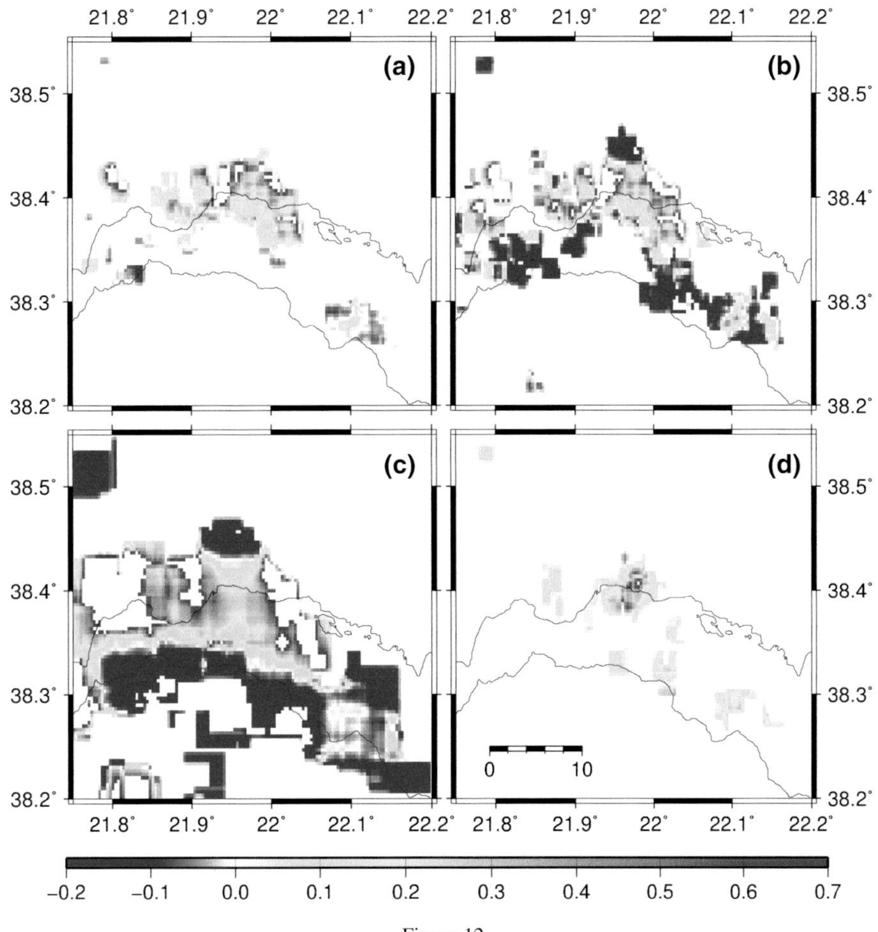

Figure 12
Spatial variation of Coulomb stress derived from seismicity rate changes in different time windows before and after the January 2010 doublet: **a** 1 year before and 1 year after, $h = 0.02°$, **b** 532 days before and 1,077 days after (entire dataset), $h = 0.02°$, **c** 532 days before and 1,077 days after (entire dataset), $h = 0.04°$, and **d** 100 days before and 100 days after, $h = 0.02°$. The *white cells* represent areas with insufficient data. The spatial step in all cases is 0.001°. Colour scale is in *bars*

the elastic dislocation and inversion calculations (>50 % in almost all cases), especially in the far field (reaching up to 80 %). It is of major importance that significant seismicity rates must be available before and after the stress events for the analysis to provide conspicuous results. Approximately three quarters of the stress changes in the far field as calculated by the elastic dislocation model can be reproduced by applying the stress–inversion methodology.

5.3. The Spatial Distribution of Stress Changes

The stress changes due to coseismic slip of the two strongest earthquakes ($M \geq 5.4$) in the study area since August 2008 were retrieved from seismicity both before and after their occurrence. In doing so we selected various time windows preceding and following the $M = 5.5$ earthquake on 18 January 2010 and mapped the stress as calculated from Eq. 4. This application was performed on a dense normal grid was superimposed onto the study area with cells of 0.001° sides (Fig. 12a–d). The stress values in each cell were smoothed using a Gaussian filter with a radius of 0.02° in all cases except in Fig. 9c, where the applied bandwidth value was $h = 0.04°$. In the approaches described below, the condition that each cell contained at least two events before and after the main shock was fulfilled. The parameter values used were the same as in the previous subsection and equal to $A\sigma = 0.04$ bars,

$t_a = 225$ days and $\dot{S}_r = 0.06$ bars/year. In Fig. 12a the stress changes were inverted from 1-year time intervals before and after the first main shock. Positive stress changes of up to 0.6 bars were detected close to the activated fault segments and also to the west. Minor stress changes, positive and negative, are evident in the southern part of the region, although the data in this area do not provide sufficient resolution. A more detailed stress pattern was revealed when longer time spans and the entire dataset were considered (Fig. 12b, c). Dataset 1 includes 232 events occurring in 532 days before and 755 events within 1,077 days after the 18 January earthquake. Positive stress changes of up to 0.7 bars were accommodated in approximately the same areas as previously (Fig. 12a). Stress drops of 0.2 bars were detected in contrast with Fig. 12a, which are amplified in Fig. 12c, where the bandwidth value was doubled. Persistent positive stress changes at the location of the first main shock can be attributed to the stress transfer from the second main shock and possibly to several strong ($M > 4.0$) aftershocks located to the east of the first failed fault segment. The resulting stress pattern seems consistent with the one derived from the elastic dislocation model when onto–fault areas are excluded (Fig. 9). A narrow time window of 100 days before and after the main shock was able to provide only local, low–amplitude positive stress changes, despite the fact that some of these variations were located at a considerable distance from the activated faults (Fig. 12d).

5.4. *ΔCFF Changes Close to the Fault Segments Associated with the 2010 Doublet*

We will now focus on the close vicinity of the fault segments associated with the January 2010 doublet, in an area approximately one fault length further from the rupture zone. The aim is to look for anomalies in earthquake occurrence rates prior to the sequence and their possible connection with stress changes. These anomalies were detected by means of the inter-event time distribution (the inverse quantity of the seismicity rate) and how far this time deviates from the average inter-event time resulting from the declustered catalogue (the dotted horizontal line in Fig. 13). 2 months before the 5.5 event, an $M = 3.9$ earthquake took place, followed by abundant seismicity at a rate more than ten times higher than the

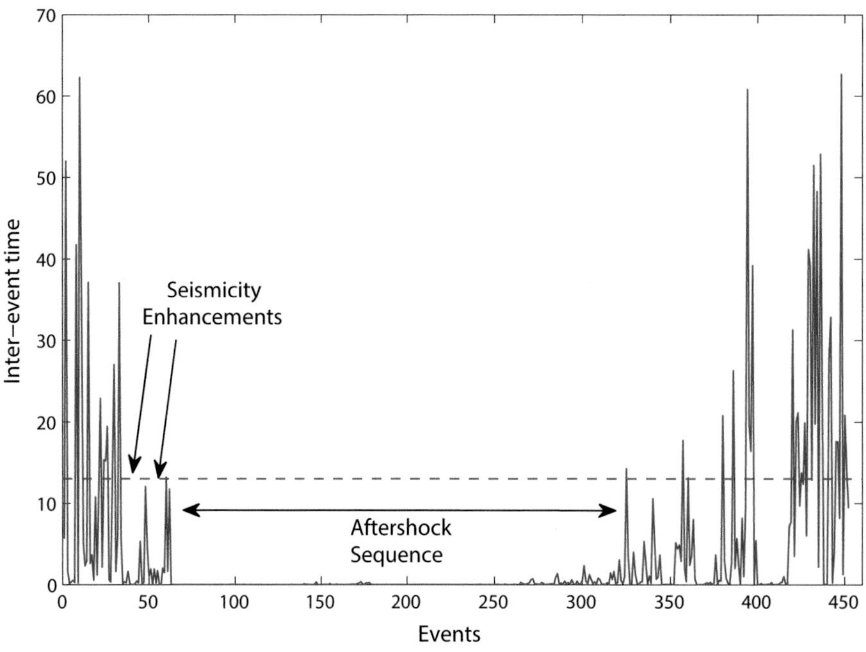

Figure 13
Seismicity rates in the close vicinity of the fault segments associated with the January 2010 doublet. Evidence of anomalous activity prior to the first strong event is shown here. We interpret this activity in terms of stress increase that finally led to the seismic burst less than 1 month later

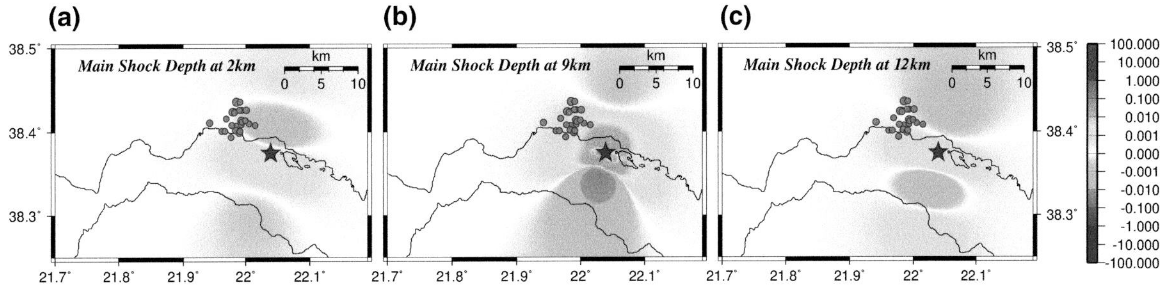

Figure 14
Coulomb stress changes caused by the $M_L = 3.9$ event resolved according to the focal mechanism of 18 January 2010, $M = 5.5$ event, considering the nucleation depth of the causative fault at 2, 9, and 12 km (**a**, **b**, and **c**, respectively). Colour scale is in *bars*

background activity. The seismic activity was depressed for 2 weeks and then a second burst occurred 25 days before the beginning of the Efpalio sequence, with the strongest earthquake of magnitude $M_{\max} = 3.1$. After this seismic enhancement, only two earthquakes occurred during the 25 days before the beginning of the sequence on 18 January.

Both clusters occurred very close to the epicentre of the second main shock. We will now attempt to derive stress changes from these rate enhancements and to associate those changes with the Efpalio sequence. The regional data relocated by Karakostas (2012) indicate an average focal depth of 9 km. The calculations performed at this focal depth produced a stress pattern more consistent with the location of the cluster that preceded the $M = 5.5$ main shock, which is also located inside a lobe of positive ΔCFF. The stress field variation is resolved according to the $M = 5.5$ earthquake of 18 January 2010. If $h = 2$ km is adopted for the $M = 3.9$ event (Fig. 14a), most of the events in the cluster are located in a negative ΔCFF lobe. At a depth of 12 km, which is at the lower boundary of the regional seismogenic layer which reaches roughly 15 km (Karakostas 2012), the locations are shared between positive and negative stress changes (Fig. 14c). Obviously, the depth selection of 9 km (Fig. 14b) appears to reproduce the most consistent stress pattern, because all the earthquakes of the cluster are located in an area of increased stress.

5.5. Stress Changes Associated with Spatio–Temporal Clustering

Earthquake clustering in the western Gulf of Corinth happens quite frequently, attracting the interest of several researchers. Mesimeri *et al.* (2013) identified that there have been 18 earthquake clusters in northwest Peloponnese since 1980 and classified them into three categories (main shock—aftershocks, swarms, and swarm-like sequences) according to their history of moment release and the time of occurrence of the main event. Karagianni *et al.* (2013) studied spatio-temporal earthquake clustering in the western Gulf of Corinth during 2010 and 2011 and demonstrated the evolution of the swarms using space–time plots. They suggested that this activity could be associated with fluid-flow phenomena in the region. Taking advandage of the availability and adequate number of observed clusters, we attempted to identify them by the stress changes they induce (Fig. 15). The eight clusters, three before and five after the 2010 sequence, that were detected are shown in Fig. 16 and summarised in Table 2. All the latter are located inside areas of positive stress changes due to the coseismic slip of the two main events.

Table 2 evidences that most clusters exhibit swarm–like behaviour, as in several cases the largest magnitude event occurred when the activity was already in progress. Even when the strongest shock took place at the beginning of a sequence, the difference in magnitude between that and the second strongest event was less than 0.4 units, except in cluster 7, where this difference was equal to 0.6 units. This sustains an additional evidence that the activated fault segments came closer to failure by remote stress triggering rather than producing their own aftershock sequences (induced by a near-field main shock). Figure 17 shows the locations of the last five clusters in relation to ΔCFF as calculated by the application

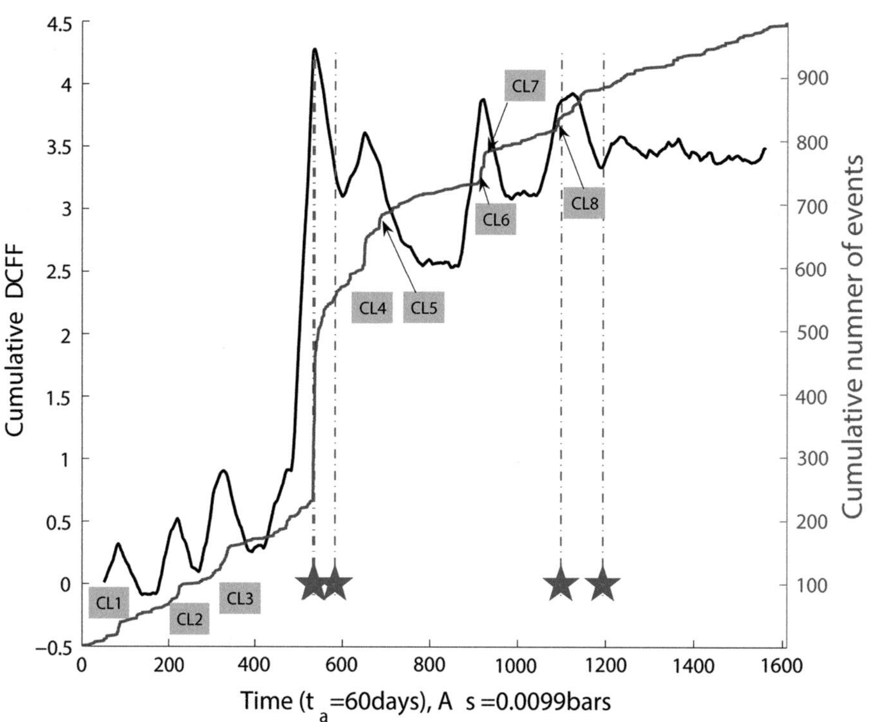

Figure 15
Cluster identification by stress changes (*black line*) associated with their occurrence. In addition to the larger stress changes connected with the January 2010 sequence, eight more clusters are shown to produce remarkable stress changes. *Blue asterisks* denote the $M \geq 4.5$ events

of the elastic dislocation model. Although several events occurred inside negative lobes, only spatio-temporal clusters characterised by increased seismicity rates were located in stress-enhanced areas. This suggests that non-clustered activity is probably related to reference seismicity, whereas enhanced–rated clusters are plausibly considered to be associated with stress–triggering.

5.6. *Temporal Variation of Stress Analysis Since 1975*

Stress calculations were performed by inverting seismicity rates since 1975 (dataset 2) for an area extended by 0.3° to the east in order to include more of the strong ($M \geq 6.0$) events. We mapped the stress evolution from 1975 onwards for various time windows corresponding to durations that are smaller than, equal to, and larger than the calculated characteristic relaxation time. Patterns that could not be distinguished using the previous approaches because of insufficient resolution were then revealed (Fig. 18). When narrow time windows were tested (left frame, Fig. 18), even small earthquake clusters appeared to produce measurable increases of stress, which were usually followed by analogous decreases. These short-term stress changes may be artefacts arising from random seismicity fluctuations and ambiguous model performance when dealing with small datasets covering small increments of time. As the duration of time windows approaches t_a (central frame, Fig. 18), these minor fluctuations are smoothed and the stress changes of all but the strongest earthquakes (followed by plenty of aftershocks) become insignificant. When the time window becomes approximately 1.5 times larger (1 year) than t_a (right-hand frame, Fig. 18), only three peaks in ΔCFF are distinguished, associated with the 1984 $M = 5.6$ event, the 1995 $M = 6.5$ event, and the 2010 doublet ($M = 5.5$, $M = 5.4$). Taking the seismicity rate change approach, the $M = 5.9$ event (18 November 1992) did not induce notable stress

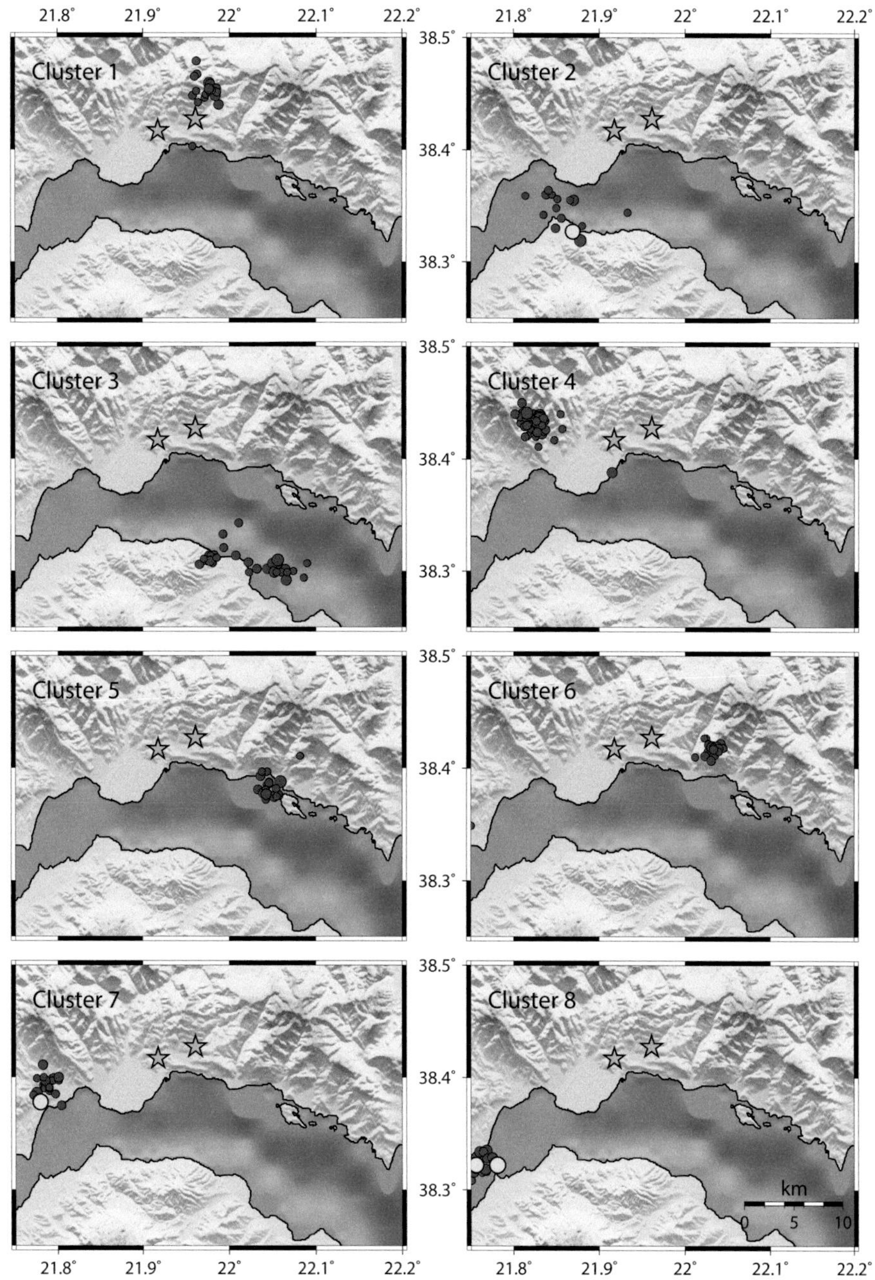

Figure 16
Seismicity clusters associated with minor ΔCFF changes. *Yellow circles* illustrate $M \geq 4.0$ events. Information about these clusters is provided in Table 2

enhancement, because of its limited number of aftershocks. HATZFELD *et al.* (1996) speculated that the reason for the depopulated aftershock sequence was the spatial change in the mechanical properties of the fault. A region of high strength was surrounded by regions of low strength, and the stress drop during the main shock did not increase stresses in the surrounding region to a level that could induce aftershocks. It is also notable that the 1995 and 2010 sequences both occurred after a relatively long-term increase of stress followed by a drop just before their initiation.

Table 2

Information on the eight clusters associated with estimated stress changes in the study area (excluding the 18 January sequence)

Cluster	No. of events	Duration (days)	1st event's magnitude	Largest magnitude	ΔM between two strongest events	M_{max} event
1	18	7.5	3.7	3.7	0.3	1st
2	15	5	4.0	4.0	0.1	1st
3	29	20	2.8	3.8	0.3	23rd
4	42	4	3.5	3.9	0.1	34th
5	19	5.5	2.9	3.8	0.3	5th
6	20	2.5	2.5	3.5	0.3	11th
7	23	3.5	2.4	4.2	0.6	2nd
8	18	7	2.5	4.3	0.1	8th

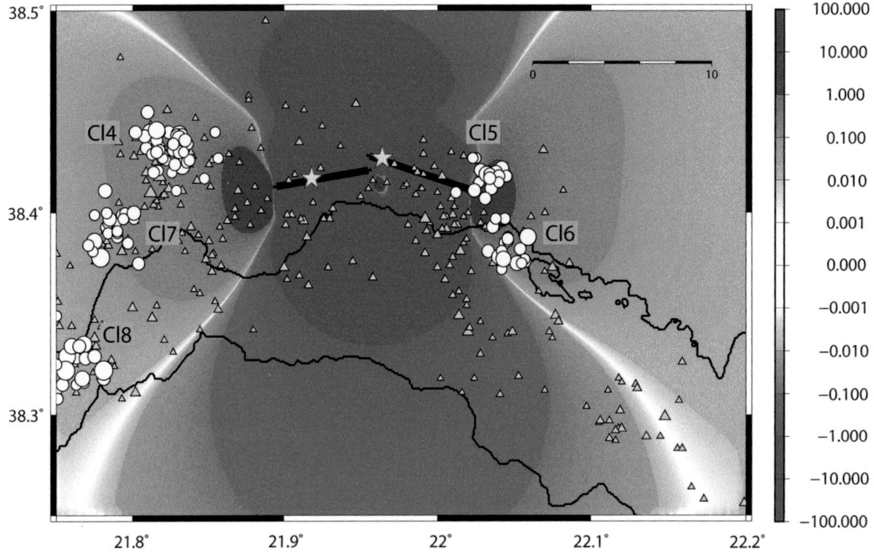

Figure 17

Map with ΔCFF caused by the January 2010 doublet. Aftershocks since April 2010 are also plotted. Non-clustered activity is indicated by *grey triangles*. The seismicity locations of the spatio-temporal earthquake clusters are depicted as *white circles*. It is shown that almost all clusters are found in stress-enhanced areas. Cluster codes correspond to the ones provided in Table 2 and shown in Fig. 16. Colour scale is in *bars*

6. Discussion and Conclusions

In this study we attempted to derive static stress changes from the variation of earthquake production rates in the western part of the Gulf of Corinth by applying DIETERICH (2000) rate/state formulation. The aim was to investigate seismicity rate changes in space and time domains and then obtain information concerning the associated stress field variations. The earthquake catalogue used started from August 2008, since when the geometry and detectability of the seismological network were considerably improved to ensure consistently high recording rates. An M_C equal to 2.4 for $a \sim 4.5$–year period was determined, whereas, for example, TODA and MATSUMURA (2006) carried out their analysis with an M_C of 1.5 for a 24-year period. Nevertheless, the purpose of this study was not to provide a forecast for a certain area, but to take advantage of one of the best monitored sites of seismic activity in the Aegean region in order to obtain an independent estimate of Coulomb stress changes.

Although several factors used in the calculations were imprecisely determined or not directly measured (the rate/state model parameters, poroelastic effects, pre-existing fault orientations, and random

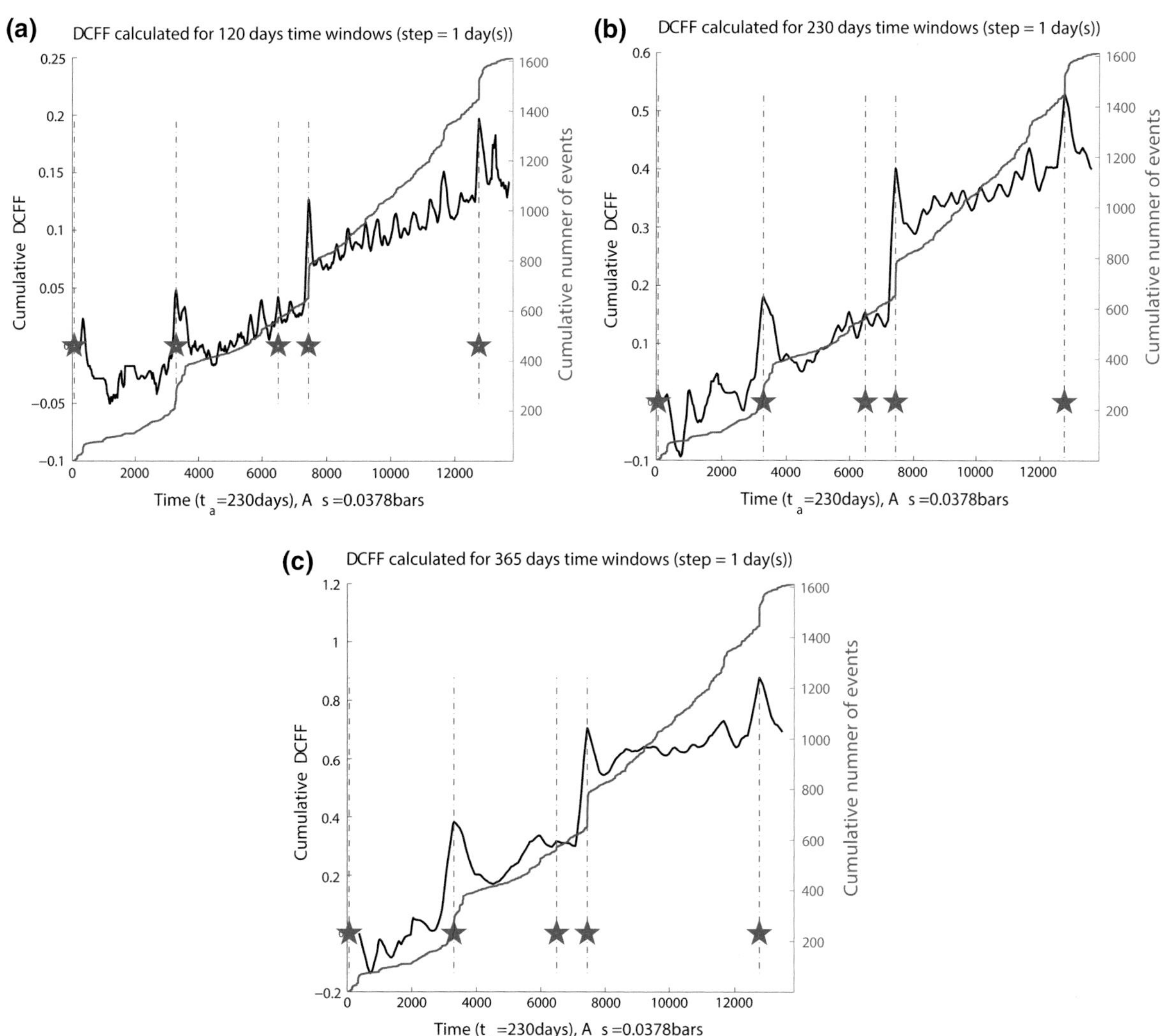

Figure 18
Stress changes inverted from the seismicity rate variation (1975–2013) considering overlapping datasets of 120-day (*left-hand frame*), 230-day (*central frame*), and 365-day (*right-hand frame*) time windows. *Blue asterisks* denote the $M \geq 5.5$ events

fluctuation of earthquake occurrence), the choice of the selected values was made after thoroughly considering all the available information and data. The background rate (constant in space and time) was considered only for estimating the characteristic relaxation time and, in turn, the parameter $A\sigma$. The temporal analysis of stress variations was performed for subsequent datasets (exhibiting either equal event numbers or equal duration), whereas the spatial analysis was performed after spatially smoothing the seismicity rates in certain areas. Therefore, the background rate was estimated only as an alternative way to determine a representative value of characteristic relaxation time for aftershock decay. This latter parameter, along with the stressing rate as derived by an analysis of GPS data, was inserted into Eq. 5 to determine the product $A\sigma$. This was done because we believe that t_a and $\dot{S}_r$ can be more robustly determined by independent approaches, rather than $A\sigma$, the determination of which involves several uncertainties and a wide range of plausible values.

The stresses were resolved onto receiver faults that had geometrical features identical to the causative fault segments. Even though this approach appears to be a simplification, the study area is small and the available focal mechanisms indicate small deviations of the assumed faulting mechanism. This is verified by Karakostas (2012), suggesting an average fault geometry derived from 31 events of the 2010 sequence as a $258° \pm 22°$ strike and a $41° \pm 11°$ dip. As far as the dip and rake uncertainties are concerned and their influence on stress changes calculations, we rely on Karakostas *et al.* (2003) who performed ΔCFF calculations for the 2001 Skyros earthquake (M_w6.4, North Aegean) by varying strike, dip, and rake angles in a range of 40°, with no remarkable changes in the stress pattern.

We have pointed out that the main aim of this study was to test whether the stress-inversion formulation is appropriate for deriving reliable results and, therefore, would enable future works to handle seismicity rate changes as an alternative but reliable way to determine stress changes in the actively deforming region of the Aegean Sea and its adjacent areas. In this context, the ΔCFF results yielded by this approach were compared with the respective results derived from the application of the elastic dislocation model. The accurate depth determination of the available earthquakes assisted in the resolution of the stress field at the respective depth layers. This led to a sufficient correlation between the independent stress-estimating methods, even in the near field, which is inherently characterised by significant complexity and inhomogeneity on many scales. The qualitative (stress pattern comparison) and quantitative (ΔCFF sign agreement of the two methods) correlation was proven to be sufficient, with 60–80 % of the results being consistent. In the far field, more than three quarters of the stress changes as calculated by the elastic dislocation model were successfully reproduced by the stress-inversion methodology. Although stress increases were adequately detected by the analysis of earthquake catalogues, stress shadows could not be robustly indicated because of the relatively low levels of the reference seismicity rates.

Another important issue for discussion concerns the time scales that were considered in this analysis. On the one hand, the stress-calculation forward modelling provides a snapshot of the stress field variation induced by the two main shocks. On the other hand, the inverse method intrinsically requires a considerable time window, accumulating sufficient data in order to estimate stress changes. However, it is generally accepted that only the strongest events have a major impact on Coulomb stress changes (e.g., Ghimire 2008). In the sequence studied, the strongest events (excluding the doublet) had magnitudes smaller than 4.6, meaning that they were approximately one unit of magnitude lower than the main shocks ($M = 5.5$ and $M = 5.4$, respectively). Because of this, we might expect that the evolution of the sequence induced only minor stress changes, such that the total stress pattern is negligibly modified. Following this assumption, we compared the stress values derived from the two methods, considering the seismic data up to the end of 2012. This was done in order to exploit as much of the available data as possible, during longer time scales than the characteristic relaxation time, to obtain robust quantitative results. Nevertheless, shorter time windows were considered as well in order to reproduce the spatial stress pattern (Fig. 12) and compare it with the one derived by using the forward modelling (Fig. 9).

Also important is the fact that the model applied here was proven to constitute a very sensitive stress meter, able to detect even small changes in stress associated with seismicity rate changes. In addition to the profound stress changes that follow a strong ($M \geq 5$) earthquake, the method adopted here was able to successfully detect swarm–like activity as well as spatio–temporal seismic clusters. This cluster identification is very important and differs significantly from a simple inspection of the number of earthquakes occurring in positive and negative lobes: space–time clusters clearly demonstrate considerable rate increases, interpreted as stress changes, and are fully consistent with the ΔCFF pattern derived from the elastic dislocation approach (Fig. 17).

As a challenge for future research, relocated catalogues should be taken into consideration, together with detailed slip models, in order to model ΔCFF in the near field. The efficiency of the seismological networks is continuously improving and, therefore, the available datasets are being enriched with data from more earthquakes at lower magnitudes. Consequently, the catalogue–based stress calculation method will become a promising tool for future studies. Accurate depth determination and focal-

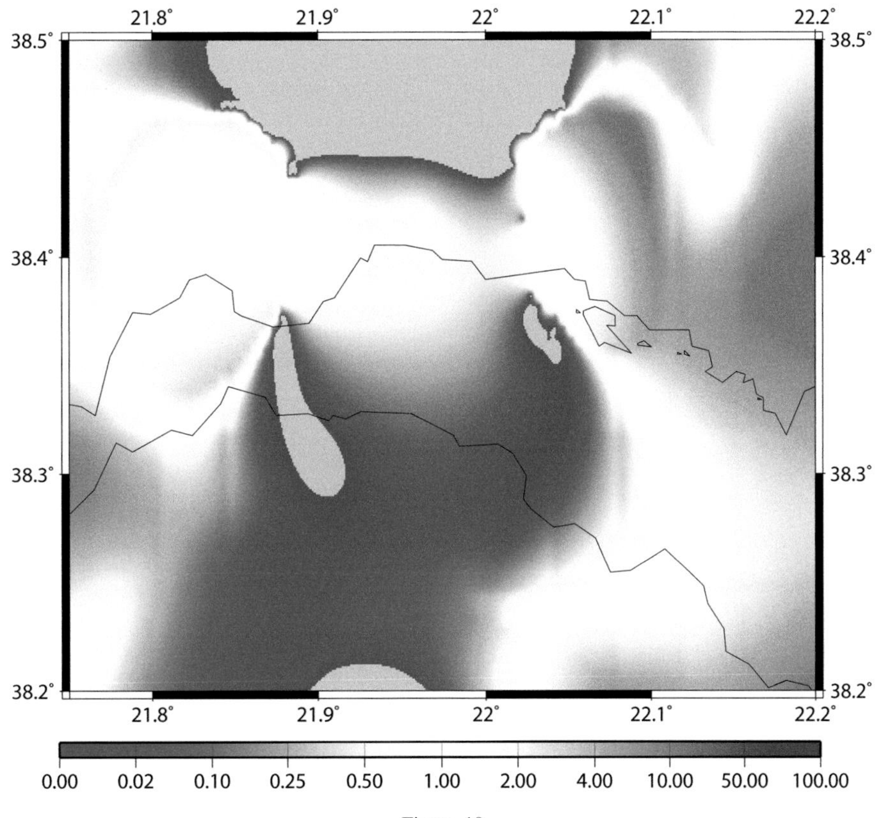

Figure 19
Ratio of expected/observed seismicity rates given the reference seismicity rate calculated from 1 August 2008 to 18 January 2010 ($M \geq 2.4$). *Red* shows overestimated excepted values in comparison with the actual ones, whereas *blue* shows seismicity rates observed to be higher than the simulated ones. *White areas* correspond to a ratio value between 0.5 and 2, suggesting sufficient model performance. Calculations were not performed in *grey areas* because of data insufficiency. The parameter values applied were: $h = 0.03°$, $t_a = 225$ days, $A\sigma = 0.4$ *bars*

plane solutions for further events may provide a very detailed three-dimensional Coulomb stress model for onto–fault aftershocks, which may help in the effort to comprehend the trigger mechanisms and the spatio–temporal evolution of aftershock sequences and swarms.

Acknowledgments

The stress tensors were calculated using the DIS3D code developed by S. Dunbar, which was later improved by Erikson (1986) and the expressions developed by G. Converse. This work was supported by the THALES Programme of the Ministry of Education of Greece and the European Union within the framework of the project entitled 'Integrated understanding of Seismicity, using innovative Methodologies of Fracture mechanics along with Earthquake and non-extensive statistical physics—Application to the geodynamic system of the Hellenic Arc. SEISMO FEAR HELLARC'. The GMT system (Wessel and Smith 1998) was used to plot some of the figures. Department of Geophysics, AUTH, contribution number 840.

Appendix

Forward Modelling

The Dieterich (1994) rate/state forward model provides seismicity rate changes, as a function of time, t, as:

$$R(t) = \frac{r}{\left[\exp\left(\frac{-\Delta S}{A\sigma}\right) - 1\right]\exp\left(\frac{-t}{t_a}\right) + 1} \tag{A1}$$

By substituting in this equation S (Eq. 3), $A\sigma$ (0.04 bars) and t_a (225 days), along with the

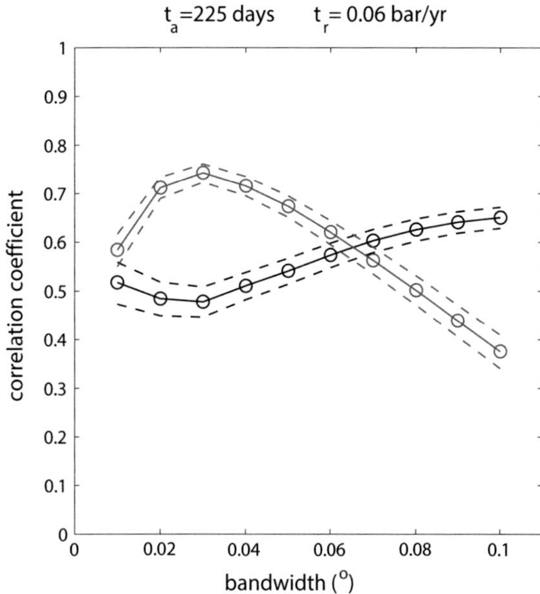

Figure 20
Quantitative evaluation of the difference between observed and synthetic seismicity rates during the inter-event time period between the two strong events (18 January 2010 and 22 January 2010, *black lines*) and until the end of 2012 (*red lines*). *Solid lines* indicate the value of the Pearson linear correlation coefficient; 95 % confidence intervals for each coefficient are also depicted (*dashed lines*)

smoothed reference seismicity rates, r (Eq. 6), we obtained the forecasted rates, R, after the two main shocks. These rates were compared with the recorded seismicity rates and found to be in good agreement (Fig. 19). Then, a quantitative analysis (Fig. 20) showed that, especially after the second strong event, the correlation coefficient between forecast and observed seismicity rates is approximately 75 % for a bandwidth equal to 0.03°. This is a significant result concerning the fact that after applying Silverman's formulation (Eq. 9), $h = 0.033°$ is obtained, which is almost identical to the corresponding value of the best correlation between real and synthetic seismicity rates ($h = 0.03°$). It is noteworthy that this best correlation resulted for $h = 0.03°$ either using forward (Fig. 20) or inverse (Figs. 11, 12) approaches. The inter-event period (time between the two main shocks) exhibits a lower correlation (50–60 %), but is not representative of the standard model performance because of its short duration ($\sim$4 days) and the abundance of onto–fault aftershocks.

References

Aki, K. (1965). *Maximum likelihood estimate of b in the formula logN = a − bM and its confidence limits*, Bull. Earthquake Res. Inst. Tokyo Univ., *43*, 237–239.

Ambraseys, N. N., and J. A. Jackson (1990), *Seismicity and associated strain of central Greece between 1890 and 1988*, Geophys. J. Int., *101*, 663–708.

Armijo, R., B. Meyer, G. C. P. King, A. Rigo, and D. Papanastassiou (1996), *Quaternary evolution of the Corinth Rift and its implications for the late Cenozoic evolution of the Aegean*, Geophys. J. Int. *126*, 11–53.

Baker, C., D. Hatzfeld, H. Lyon-Caen, E. Papadimitriou, and A. Rigo (1997), *Earthquake mechanisms of the Adriatic Sea and Western Greece: implications for the oceanic subduction–continental collision transition*, Geophys. J. Int., *131*, 559–594.

Belardinelli, M. E., A. Bizzarri, and M. Cocco (2003), *Earthquake triggering by static and dynamic stress changes*, J. Geophys. Res., *108*, B3, 2135, doi:10.1029/2002JB001779.

Bell, R. E., L. C. McNeill, J. M. Bull, and T. J. Henstock (2008), *Evolution of the offshore western Gulf of Corinth*, Bull. Geol. Soc. Amer., *120*, 156–178.

Braunmiller, J., and J. Nabelek (1996), *Geometry of continental normal faults: seismological constraints*, J. Geophys. Res., *10*, 3045–3052.

Catalli, F., M. Cocco, R. Console, and L. Chiaraluce (2008), *Modeling seismicity rate changes during the 1997 Umbria–Marche sequence (central Italy) through a rate-and-state dependent model*, J. Geophys. Res., *113*, B11301, doi:10.1029/2007JB005356.

Cocco, M., S. Hainzl, F. Catalli, B. Enescu, A. M. Lombardi, and J. Wossner (2010), *Sensitivity study of forecasted aftershock seismicity based on coulomb stress calculation and rate- and state- dependent frictional response*, J. Geophys. Res., *115*, B05307, doi:10.1029/2009JB006838.

Dieterich, J. H. (1994), *A constitutive law for rate of earthquake production and its application to earthquake clustering*, J. Geophys. Res., *99*, 2601–2618.

Dieterich, J., V. Cayol, and P. Okubo (2000), *The use of earthquake rate changes as stress meter at Kilauea volcano*, Nature, *408*, 457–460.

Dieterich, J., V. Cayol, and P. Okubo (2003), *Stress changes before and during the Pu'u 'Ō 'ō-Kūpaianaha Eruption*, U. S. Geol. Survey Professional Paper, *1676*, 187-202.

Erikson, L. (1986), *User's manual for DIS3D: A three–dimensional dislocation program with applications to faulting in the Earth*. Master's Thesis, Stanford Univ., Stanford, Calif., 167 pp.

Flerit F., R. Armijo, G. King, and M. Bertrand (2004), *The mechanical interaction between the propagating North Anatolian Fault and the back-arc extension in the Aegean*, Earth Planet. Sci. Lett., *224*, 347–362.

Ganas, A., K. Chousianitis, E. Batsi, M. Kolligri, A. Agalos, G. Chouliaras, and K. Makropoulos (2012), *The January 2010 Efpalion earthquakes (Gulf of Corinth, Central Greece): earthquake interactions and blind normal faulting*, J. Seismol., DOI 10.1007/s10950-012-9331-6.

Ghimire, S., K. Katsumata, and M. Kasahara (2008), *Spatio–temporal evolution of Coulomb stress in the Pacific slab inverted from the seismicity rate change and its tectonic interpretation in Hokkaido, Northern Japan*, Tectonophysics, *455*, 25–42.

Hainzl, S., S. Steacy, and D. Marsan (2010), *Seismicity models based on Coulomb stress calculations, Community Online Resource for Statistical Seismicity Analysis*, doi:10.5078/corssa-32035809.

Hainzl, S., Y. Ben-Zion, C. Cattania, and J. Wassermann (2013), *Testing atmospheric and tidal earthquake triggering at Mt. Hochstaufen, Germany*, J. Geophys. Res., *118*, 1–11.

Harris, R. A. (2000), *Earthquake stress triggers, stress shadows, and seismic hazard*, Curr. Science, *79*, 1215–1225.

Harris, R. A., and R. W. Simpson (1998), *Suppression of large earthquakes by stress shadows: A comparison of Coulomb and rate/state failure*, J. Geophys. Res., *103*, 24439–24451.

Hatzfeld D., G. Pedotti, P. Hatzidimitriou, and K. Makropoulos (1990), *The strain pattern in the western Hellenic arc deduced from a microearthquake survey*. Geophys. J. Int., *101*, 181–202.

Hatzfeld D., D. Kementzetzidou, V. Karakostas, M. Ziazia, S. Nothard, D. Diagourtas, A. Deshamps, G. Karakaisis, P. Papadimitriou, M. Scordilis, R. Smith, N. Voulgaris, A. Kiratzi, K. Makropoulos, M. Bouin, and P. Bernard (1996), *The Galaxidi earthquake sequence of November 18, 1992: a possible geometrical barrier within the normal fault system of the Gulf of Corinth (Greece)*, Bull. Seismol. Soc. Am., *86*, 1987–1991.

Helmstetter, A., and B. E. Shaw (2006), *Relation between stress heterogeneity and aftershock rate in the rate-and-state model*, J. Geophys. Res., *111*, B07304, doi:10.1029/2005JB004077.

Jansky J., O. Novotny, V. Plicka, J. Zahradnik, and E. Sokos (2011), *Earthquake location from P-arrival times only: problems and some solutions*, Stud. Geophys. Geod., *56*. doi:10.1007/s11200-011-9036-2.

Karagianni, E., P. Paradisopoulou, and V. Karakostas (2013), *Spatio-temporal earthquake clustering in the Western Corinth Gulf*, Bull. Seismol. Soc. Greece, XLVII, 2013.

Karakostas, V. G., Papadimitriou, E. E., Karakaisis, G. F., Papazachos, C. B., Skordilis, E. M., Vargemezis, G., and Aidona, E. (2003), *The 2001 Skyros, northern Aegean, Greece, earthquake sequence: off-fault aftershocks, tectonic implications, and seismicity triggering*, Geophys. Res. Lett., *30*, doi:10.1029/2002gl015814.

Karakostas, V., E. Karagianni, and P. Paradisopoulou (2012), *Space-time analysis, faulting and triggering of the 2010 earthquake doublet in western Corinth Gulf*, Nat. Hazards, *63*, 1181–1202.

Kostelecky, J., and J. Dousa (2012), *Results of geodetic measurements during the January 2010 Efpalio earthquakes at the western tip of the Gulf of Corinth, central Greece*, Acta Geodyn. Geomater., *9*, 291–301.

Kiratzi, A., and E. Louvari (2003), *Focal mechanisms of shallow earthquakes in the Aegean Sea and the surrounding lands determined by waveform modeling: a new database*, J. Geodyn., *36*, 251–274.

Leptokaropoulos, K. M., Karakostas, V. G., Papadimitriou, E. E., Adamaki, A. K., Tan, O., and İnan, S., (2013), *A homogeneous earthquake catalogue compilation for western Turkey and magnitude of completeness determination*, Bull. Seismol. Soc. Am., *103*, 5, 2739–2751.

Linker, J., and J. Dieterich (1992), *Effects of variable normal stress on rock friction: Observations and constitutive equations*, J. Geophys. Res., *97*, 4923–4940, doi:10.1029/92JB00017.

Maccaferri, F., E. Rivalta, L. Passarelli, and S. Jónsson (2013), *The stress shadow induced by the 1975-1984 Krafla rifting episode*, J. Geophys. Res., *118*, 1109–1121.

Mallman, E. P., and M. D. Zoback (2007), *Assessing elastic Coulomb stress transfer models using seismicity rates in southern California and southern Japan*, J. Geophys. Res., *112*, B03304, doi:10.1029/2005JB004076.

McClusky, S., S. Balassanian, A. Barka, C. Demir, S. Ergintav, I. Georgiev, O. Gurkan, M. Hamburger, K. Hurst, H. Kahle, K. Kastens, G. Kekelidze, R. King, V. Kotzev, O. Lenk, S. Mahmoud, A. Mishin, M.Nadariya, A. Ouzounis, D. Paradissis, Y. Peter, M. Prilepin, R. Reilinger, I. Sanli, H. Seeger, A. Tealeb, M. N. Toksöz, and G. Veis (2000), *Global Positioning System constraints on plate kinematics and dynamics in the eastern Mediterranean and Caucasus*, J. Geophys. Res., *105*, 5695–5719.

Mesimeri, M, E. Papadimitriou, V. Karakostas, and G. Tsaklidis (2013), *Earthquake clusters in NW Peloponnese*, Bull. Seismol. Soc. Greece, XLVII, 2013.

Novotny, O., E. Sokos, and V. Plicka (2012), *Upper crustal structure of the western Corinth Gulf, Greece, inferred from arrival times of the January 2010 earthquake sequence*, Stud. Geophys. Geod., *56*, 1007–1018, doi: 10.1007/s11200-011-0482-7.

Papazachos B. C., and C. C. Papazachou (2003), *The earthquakes of Greece*. Ziti Publication Co., Thessaloniki, pp. 304.

Reasenberg, P. A. (1985), *Second order moment of central California Seismicity, 1969–1982*, J. Geophys. Res., *90*, B7, 5479–5495.

Reilinger, R., S. McClusky, P. Vernan, S. Lawrence, S. Ergitav, R. Cakmak, H. Ozener, F. Kadirov, I. Guliev, R. Stepanyan, M. Nadariya, G. Hahubia, S. Mahmoud, K. Sakr, A. ArRajehi, D. Paradissis, A. Al-Aydrus, M. Prilepin, T. Guseva, E. Evren, A. Dmitrotsa, S. V. Filikov, F. Gomez, R. Al–Ghazzi, and G. Karam (2006), *GPS constraints on continental deformation in the Africa–Arabia–Eurasia continental collision zone and implications for the dynamics of plate interactions*, J. Geophys. Res., *111*, B05411, doi:10.1029/2005JB004051.

Rigo, A., H. Lyon-Caen, H. R. Armijo, A. Deschamps, D. Hatzfeld, K. Makropoulos, P. Papadimitriou, and I. Kassaras (1996), *A microseismic study in the western part of Gulf of Corinth (Greece): implications for large scale normal faulting mechanisms*, Geophys. J. Int., *126*, 663–668.

Roberts, S. and J. Jackson (1991), *Active normal faulting in central Greece: An overview*, Geol. Soc. Lon. Spec. Pub., *56*, 125–142.

Scholz, C. H. (1998), *Earthquakes and friction laws*, Nature, *391*, 37–42.

Silverman, B. W. (1986), *Density Estimation for Statistic and Data Analysis*. Chapman and Hall, London, pp. 9, 21.

Sokos E, J. Zahradnik, A. Kiratzi, J. Jansky, F. Gallovic, O. Novotny, J. Kostelecky, A. Serpetsidaki, and A. Tselentis (2012), *The January 2010 Efpalio earthquake sequence in the western Corinth Gulf (Greece)*. Tectonophysics, *530–531*, 299–309.

Steacy, S., J. Gomberg, and M. Cocco (2005), *Introduction to special section: Stress transfer, earthquake triggering, and time-dependent seismic hazard*,J. Geophys. Res., *110*, B05S01, doi:10.1029/2005 JB003692.

Stein, R. S. (1999), *The role of stress transfer in earthquake occurrence*, Nature, *402*, 594–604.

Taymaz, T., J. Jackson, and D. McKenzie (1991), *Active tectonics of the north and central Aegean Sea*, Geophys. J. Int., *106*, 433–490.

Toda, S., and S. Matsumura (2006), *Spatio-temporal stress states estimated from seismicity rate changes in the Tokai region, central Japan*, Tectonophysics, *417*, 53–68.

Wessel, P. and W. H. F. Smith (1998), *New improved version of the Generic Mapping Tools Released*, EOS Trans. AGU, *79*, 579.

Woessner, J., and S. Wiemer (2005), *Assessing the quality of earthquake catalogues: Estimating the magnitude of completeness and its uncertainty*, Bull. Seismol. Soc. Am., *95*, 684–698.

(Received March 25, 2014, revised February 9, 2015, accepted February 11, 2015, Published online February 28, 2015)

Pure Appl. Geophys. 173 (2016), 73–84

DOI 10.1007/s00024-015-1086-x

Pure and Applied Geophysics

Signature of Fault Healing in an Aftershock Sequence? The 2008 Wenchuan Earthquake

SHENGFENG ZHANG,[1] ZHONGLIANG WU,[1] and CHANGSHENG JIANG[1]

Abstract—We analyzed the aftershock sequence of the 2008 Wenchuan earthquake from May 12, 2008 to May 12, 2013 using the earthquake catalog of the China Earthquake Networks Center (CENC). In the analysis performed, we took under consideration the temporary variation in the completeness of the earthquake catalog just after the Wenchuan mainshock. The cutoff completeness magnitude from May 12 to June 27, 2008 was above 3.0 due to the impact of the earthquake sequence on the seismological observatory practice. It was observed that the *b* value has an increasing trend from June 27, 2008 to late April 2009, while since May 2009, the *b* value has remained stable. If these characteristics were associated with the possible signature of fault healing, the 'apparent healing time' could be pinpointed by this measure as around 1 year. Due to two strong asperities present on the rupture of the Wenchuan mainshock, the aftershock zone can be divided into two segments, namely the north and the south segment. The *b* values of the two segments seem to show different trends of temporal variation. The main contribution of the increasing trend comes from the south segment, or the 'initiation segment' of the main rupture.

Key words: Aftershock sequence, fault healing, *b* value, the 2008 Wenchuan earthquake.

1. Introduction

Fault healing after a major earthquake determines the recurrence behavior of the earthquakes with similar sizes, which plays an important role in the assessment of time-dependent seismic hazard. The healing process, which is supposed to be a combination of fracture closer, sealing, precipitation, biogenic growth, and pressure solution (XUE *et al.* 2013), reflects the physics of the fault zone as well as the geodynamic environment of the seismic source. In earthquake case studies, however, the measurements to constrain the healing process of an earthquake fault are still limited. This situation has been changed in recent years by the direct observations of fault zone trapped waves (LI 2010) and fault zone permeability (XUE *et al.* 2013) and the indirect observations such as after-slip observation by GPS measurement (ZHANG *et al.* 2013) and the deep slip variation by 'repeating earthquakes' (MA *et al.* 2014). In the related studies, independent measures, even if with limited precision and marginal reliability, are still needed to contribute to the interdisciplinary constraint of the healing process.

It has been well known that an aftershock sequence contains the information of the mainshock and, possibly, the information of fault healing. For a long time, the post-earthquake relaxation process has been quantified by the Omori's law (OMORI 1894) and its generations (UTSU 1961; UTSU *et al.* 1995; OGATA and ZHUANG 2006). In this paper, we take the 2008 Wenchuan earthquake as an example to analyze the post-earthquake temporal variation of the sequence parameters, namely the *b* value in the Gutenberg–Richter frequency–magnitude relation (GUTENBERG and RICHTER 1944). After the Wenchuan earthquake, there were several studies, directly related to the healing process, that considered the fault zone permeability (XUE *et al.* 2013) and the variation of seismic velocities within the fault zone (LI 2010). The Wenchuan earthquake sequence was selected as a case study example for two reasons. On one hand, the Wenchuan earthquake sequence shows a rapid healing process (LI 2010; XUE *et al.* 2013) during which the aftershock sequence was still in progress, providing a unique case for examining the healing process by aftershock analysis. On the other hand, the aftershock sequence of the Wenchuan earthquake showed a relatively simple spatial and temporal

[1] Institute of Geophysics, China Earthquake Administration, Beijing 100081, China. E-mail: wuzhl@ucas.ac.cn

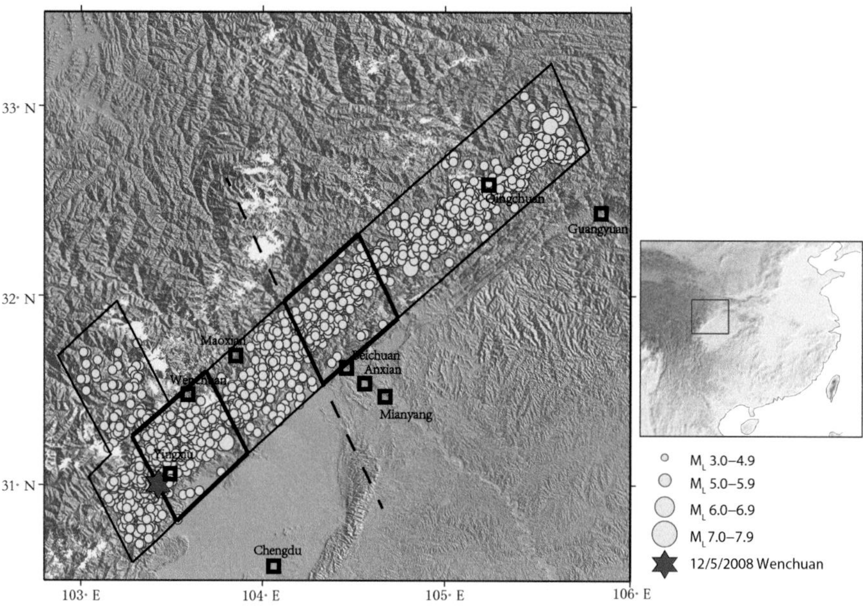

Figure 1
Distribution of the two main asperities of the Wenchuan earthquake and earthquakes for the period from 14:28 (local time, the origin time of the Wenchuan M_S 8.0 earthquake), May 12, 2008, to 14:28, May 12, 2013. The region under study is displayed in the indexing figure to the right. *Yellow dots* show the epicenters of events with $M_L \geq 3.0$, scaled by the magnitude. The *red hexagon* indicates the epicenter, or the surface projection of the initiation point of the rupture, of the Wenchuan mainshock. The *polygon* with *black thin lines* indicates the region of the Wenchuan aftershock sequence, which is consistent with the rupture zone of the mainshock. The *two boxes* with *black thick lines* indicate the main asperities, located in Wenchuan and Beichuan, respectively, causing tremendous disasters (from the results of Shen *et al.* 2009). The *black dashed line* indicates the 'boundary' between the north and the south segment of the aftershock distribution (see Fig. 8a, b, respectively)

distribution, which provides this study with convenience to a large extent.

2. Data Used for the Analysis

Figure 1 shows the location of the 2008 Wenchuan earthquake, its aftershock sequence and the local seismicity in the area. The Wenchuan earthquake had an extended rupture which spanned nearly 240 km on the surface of the Earth, with the aftershock distribution basically consistent with the rupture distribution. The rupture process exhibited two main asperities, one in Wenchuan County to the south, and the other in Beichuan County to the north, both causing tremendous disasters. Figure 1 approximates the two main asperity pictures by a polygon. Prior to the Wenchuan mainshock in 2008, the associated Longmenshan fault zone exhibited low seismicity, and the Wenchuan aftershock sequence has dominated the regional seismicity since then.

The spatial distribution of earthquakes selected from the catalog of the China Earthquake Networks Center (CENC),[1] from 14:28 (local time, the origin time of the Wenchuan earthquake) May 12, 2008 to 14:28 (local time) May 12, 2013, with a 5 year time span, appears in Fig. 1. The magnitude of earthquakes in the catalog is in the Chinese M_L scale.[2] The area under study consists of the rectangle with spatial range 30.5–33.5°N, 102.8–106.0°E. Figure 2 shows the magnitude–time distribution of earthquakes in the study area with magnitude $M_L \geq 3.0$, and Fig. 3a–d shows the temporal distribution of the cumulative number of events for different magnitude classes. From this figure, the change of the monitoring capability is clearly visible. Figure 4a–e shows the frequency–magnitude distribution. Apparently, the

[1] http://www.csndmc.ac.cn/wdc4seis@bj/earthquakes/csn_catalog_p001.jsp.

[2] Chinese Standard GB 17740-1999 General ruler for earthquake magnitude (in Chinese).

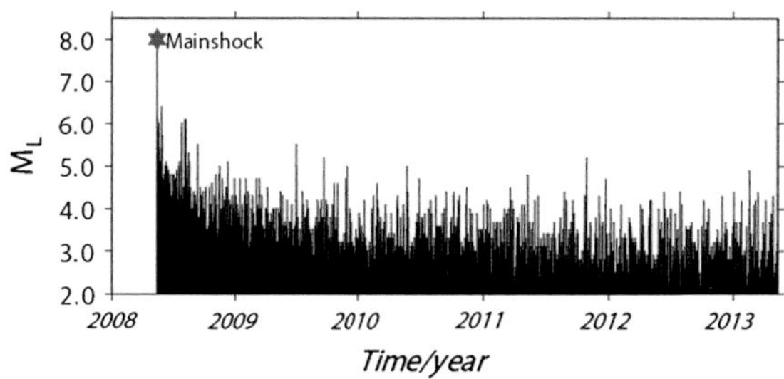

Figure 2
Temporal distribution of the earthquakes with $M_L \geq 3.0$ in the studied region as shown in Fig. 1, with the *red hexagon* showing the time of the Wenchuan mainshock. In the figure, except the mainshock which uses the Chinese surface wave magnitude, all the earthquakes use M_L

piecewise power–law-type frequency–magnitude relation has a kink around M_L 4.0. This kink further shows the reduction of the monitoring capability of local and regional seismic networks after the Wenchuan earthquake. Indeed, after a major to great earthquake, due to the sharp increase in the number of aftershocks and the limited capability of routine seismological interpretation (most of which are interactive and analyst review based), it is quite common that the effective monitoring capability decreases dramatically after the mainshock and needs a time for recovery (Iwata 2008). In the case of Wenchuan, Fig. 4d, e shows that before and after June 27, 2008, the frequency–magnitude distribution has a distinct difference that before this time the kink is clear, and after this time the kink disappears so that the frequency–magnitude distribution shows a typical GR relation. Here, the selection of the date for the transition, June 27, 2008, is both by visual trial-and-error and with the reference of Fig. 3c.

The temporal variation of the monitoring capability can be further verified by Fig. 5a which is the magnitude–index plot showing the number of earthquakes in the studied region. From the magnitude–index figure, the 'dynamic completeness magnitude' (Ogata *et al.* 1991) can be determined by examining the peak of the number (as shown by the black solid line). In Fig. 5a, the red lines together with the inserting text mark the point where the 'dynamic completeness magnitude' drops to below M_L 3.0. The associated time point (May 22, 2008) seems earlier than what is indicated from the cumulative numbers (Fig. 3) and the GR relation (Fig. 4). Choosing the

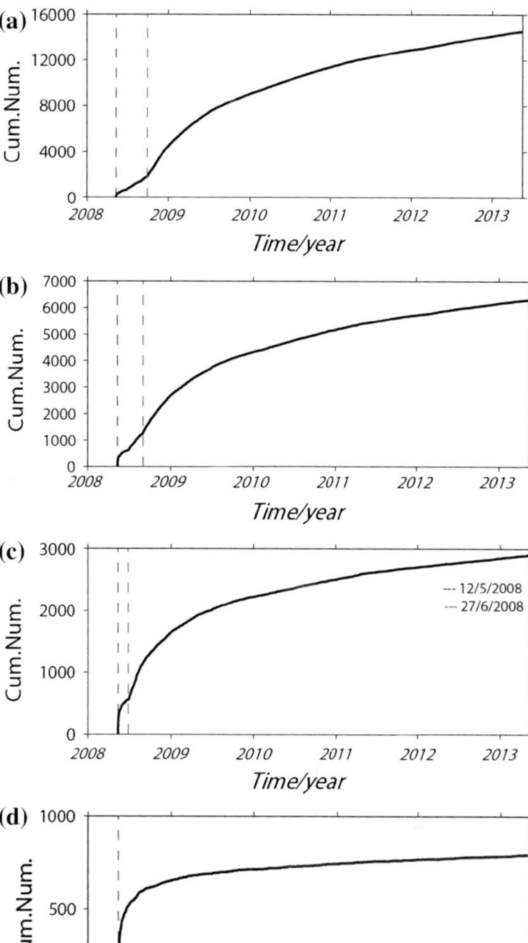

Figure 3
a Cumulative number of earthquakes with $M_L \geq 2.0$ in the spatial range as shown in Fig. 1, with the *black vertical dashed line* indicating the Wenchuan mainshock and the *red vertical dashed line* indicating the date at which the completeness had recovered to be 'normal' from the impact of the Wenchuan mainshock. **b** Cumulative number of earthquakes with $M_L \geq 2.5$. **c** Cumulative number of earthquakes with $M_L \geq 3.0$, as shown in Fig. 2, with the *red vertical dashed line* indicating the date, June 27, 2008, at which the completeness had recovered to be 'normal' from the impact of the Wenchuan mainshock. **d** Cumulative number of earthquakes with $M_L \geq 4.0$

date June 22, 2008, therefore, as the date of transition of monitoring capability is a kind of 'conservative choice'. Figure 5b further shows the completeness magnitude obtained using the maximum curvature technique (Wyss *et al.* 1999; Wiemer and Wyss 2000). The analysis takes the catalog of 1 month,

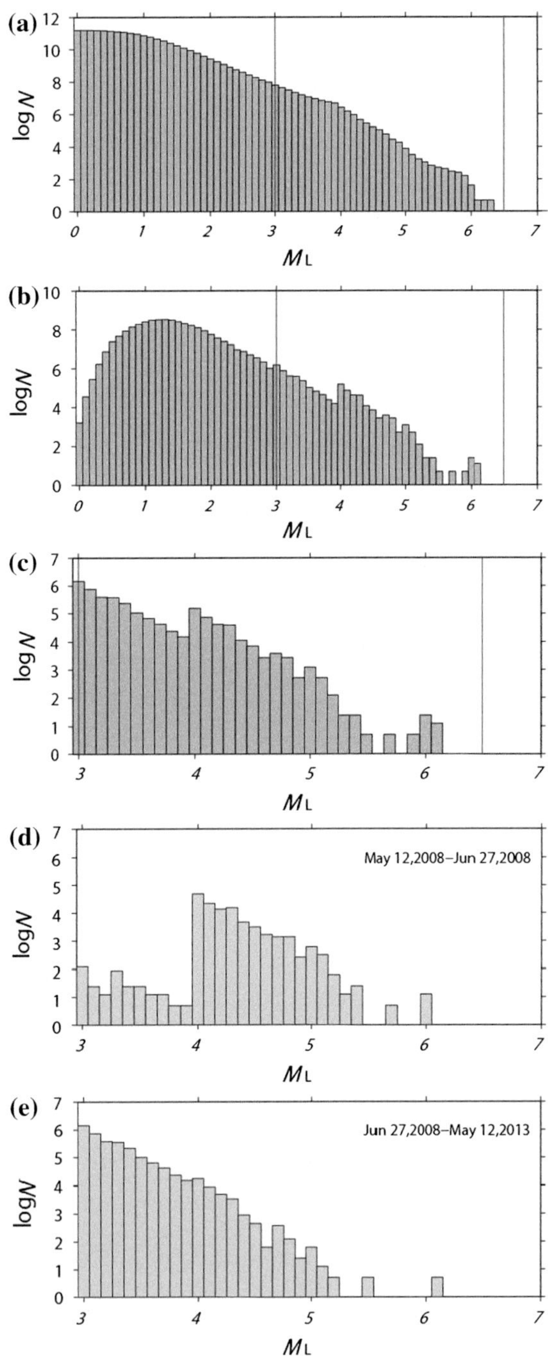

Figure 4
Frequency–magnitude distribution of the earthquakes in the region and time period as shown in Figs. 1 and 2. **a** Cumulative distribution; **b** Non-cumulative distribution. The *two vertical red lines* indicate the range of magnitude as shown in **c**–**e**. **c** Period May 12, 2008 (14:28 local time) to May 12, 2013 (14:28 local time). **d** Period May 12, 2008 (14:28 local time) to June 27, 2008 (14:28 local time), which shows clearly the kink at M_L 4.0. **e** Period June 27, 2008 (14:28 local time) to May 12, 2013 (14:28 local time) in which the kink does not exist. From **e**, the completeness magnitude of the catalog can be determined, by visual inspection, as M_L 3.0. From **e** it may be seen that the power-law distribution, or log-linear trend, exists between M_L 3.0 and 5.5, which is used to calculate *b* value

with the starting time, *Tc*, sliding with a 1 day step. It can be seen that the completeness magnitude exhibits apparent time dependence, indicating the impact of the Wenchuan earthquake. Figure 5c shows the result obtained using the entire magnitude range (EMR) technique (Ogata and Katsura 1993), which shows a similar result to that of Fig. 5a, b. The software for the catalog completeness analysis, using the maximum curvature technique and the EMR technique, is from Mignan and Woessner (2012), while the magnitude–index diagram is plotted by the software of ourselves.

After the Wenchuan earthquake, there were several studies on the relocation of the aftershock sequence (e.g., Huang *et al.* 2008; Lü *et al.* 2008; Zhu *et al.* 2008; Chen *et al.* 2009; Zhao *et al.* 2011). These relocated catalogs, despite having significantly improved precision (and improved accuracy for some cases with good ‘ground truth’ information), share the limitation that all the relocation algorithms have to eliminate some events due to the selection criteria for the relocation, which in turn affects the completeness of the catalogs in use, and thus affects the calculation of the *b* value. In this study, therefore, we only use the routine earthquake catalog of the CENC which has lower precision and accuracy but better completeness. From the relocation results, it may be seen that all the earthquakes in this region belong to ‘shallow’ events within the crust. Accordingly, in our analysis, we did not consider the variation of the hypocentral depths.

3. *Calculation of the b Value*

Related to the long history of seismological observation and interpretation in the Yunnan–Sichuan region since the 1970s, there have been several studies on the instrument-recorded seismicity and the distribution of *b* value (Yi *et al.* 2006; Wen *et al.* 2008; Jiang and Zhuang 2010), at least partly due to

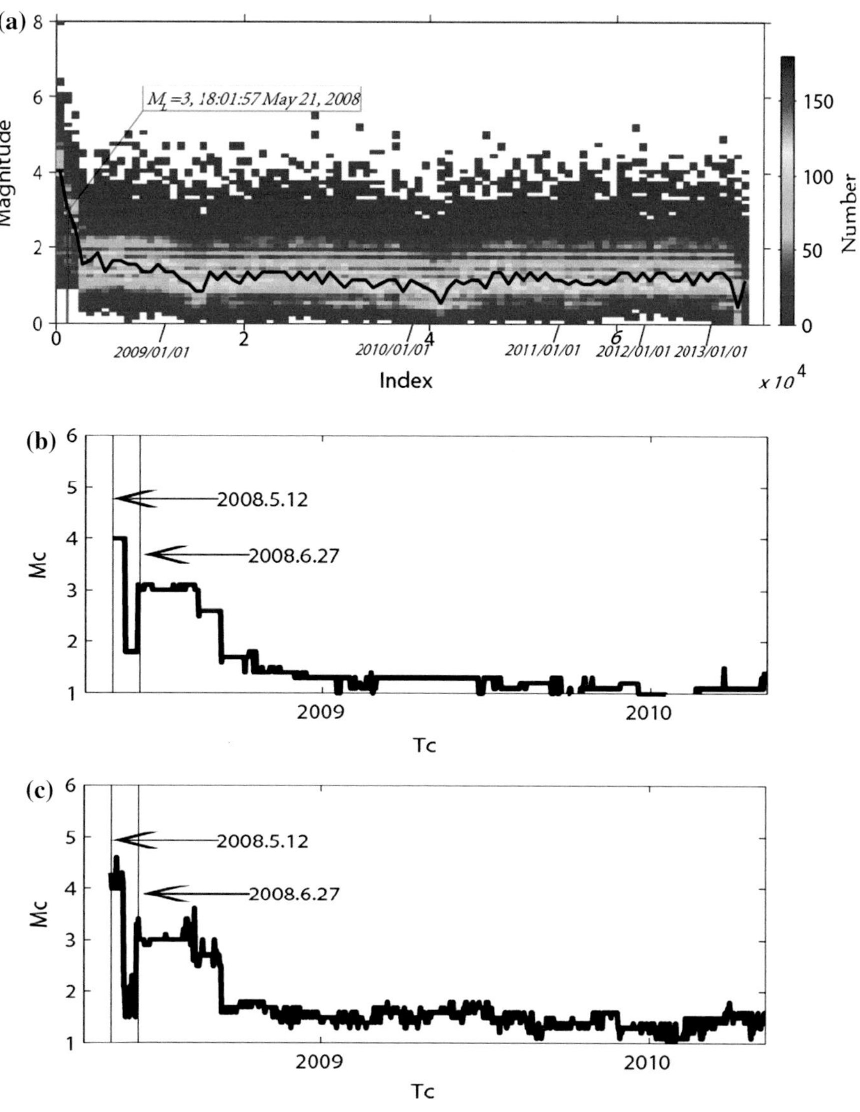

Figure 5

a Magnitude–index plot showing the density distribution of the earthquakes in the studied region. From the magnitude–index figure, the 'dynamic completeness magnitude' (Ogata *et al.* 1991) can be determined by examining the peak of the density (as shown by the *black solid line*). The starting index, No. 1, is for the Wenchuan mainshock. The index step is 800 events, and magnitude step is 0.1. In the figure, the index, together with the magnitude and origin time of the event, after which the dynamic completeness magnitude changes from above M_L 3.0 to below, was marked by the *red lines* together with the *inserting box*, which shows that May 22, 2008 is the date after which the completeness magnitude is below M_L 3.0. **b** Completeness magnitude using the maximum curvature method (Wyss *et al.* 1999; Wiemer and Wyss 2000). The analysis takes the catalog of 1 month, with the starting time, *Tc*, sliding with a 1 day step. **c** Completeness magnitude using the entire magnitude range (EMR) method (Ogata and Katsura 1993). For captions, see (**b**)

the theoretical relation between the *b* value and the stress level (Wyss 1973; Wiemer and Wyss 1997). For instance, Yi *et al.* (2006) mapped the *b* value along the Longmenshan fault zone and identified the potential asperities or locking units. Since the Wenchuan earthquake, these studies have been continued and extended (Zhao and Wu 2008; Wang *et al.* 2009; Jiang *et al.* 2011; Yi *et al.* 2013). Zhao and Wu (2008) investigated the *b* value distribution before and after the Wenchuan earthquake and discussed the relation between the mainshock rupture, the *b* value, and the strong aftershocks.

In our analysis, the b value was calculated by the widely used maximum likelihood approach (Aki 1965) that

$$b = \frac{\log_{10}e}{M_{\text{mean}} - M_{\text{min}}} \quad (1)$$

with the standard deviation depending on the number of samples, N, in the calculation

$$\sigma = \frac{b}{\sqrt{N}} \quad (2)$$

Following Utsu (1965), if the binning of magnitude is 0.1, then M_{min} is taken as:

$$M_{\text{min}} = \min(M) - 0.05 \quad (3)$$

Figure 6a shows the temporal variation of b value in the studied region, as a demonstration of how the b values are calculated and illustrated. The sliding step is selected, without overlaps with its succeeding steps, as 30 events. The time period before June 27, 2008 is not considered due to the limited monitoring capability as discussed above (see Figs. 3, 4 and 5). The horizontal axis to plot the b value is chosen, somewhat arbitrarily, as the mid point of the 30-events time window. Standard deviation calculated by Eq. (2) is marked to the data point by the error bar. From this figure, it can be seen that the b value has an apparently increasing trend which was kept till late April 2009. Considering the possible effect of the cutoff magnitude on the calculation, Fig. 6b, c shows the results with cutoff magnitude M_{L} 3.5 and M_{L} 4.0, respectively. From these series of figures, it can be seen that the variation trend of the b values, from an increasing trend to stable, is seemingly robust against the change of the cutoff magnitude.

The statistical significance of such an increasing trend seems hard to confirm, but could be measured using the AIC (Utsu 1999) or BIC (Leonard and Hsu 1999; Seher and Main 2004) criterion. As a simplification, we compare the increasing trend [as represented by the function $b(t) = b_0 + b_1t$) and the 'null hypothesis' of random fluctuation with an average level (as represented by the function $b(t) = b_2$]. Associated with Figs. 6, 7 and 8, the BIC and BIC gain were calculated using the formulation of Seher and Main (2004) that

$$\text{BIC} = L - \frac{p}{2}\ln\frac{N}{2\pi} \quad (4)$$

in which L is the maximized logarithmic likelihood, p is the number of parameters or the degrees-of-freedom of the model, and N is the number of measurements. According to Seher and Main (2004), for different models with different p values, a change in BIC by one unit is regarded as significant. In Fig. 6, the statistical significance of the increase from June 27, 2008, to April 27, 2009, can be justified by the BIC that for the linear trend, BIC gives 0.78 for (a), 1.26 for (b), and 2.12 for (c), as a reference baseline for comparison, and for the increasing trend, BIC gives 1.77 for (a), 2.36 for (b), and 3.63 for (c). This gives the BIC gain 0.97 for (a), 1.10 for (b) and 1.51 for (c). In Fig. 7, BIC gain gives 0.97 for (a), 1.01 for (b) and 1.84 for (c). These results show that the increasing trend has some statistical significance as shown by the BIC gain larger than one unit for higher cutoff magnitude, but the significance is limited to some extent for lower cutoff magnitude because the BIC gain is near to but not larger than one unit.

4. *Variation of b Values: Temporal and Spatial Features*

4.1. *Temporal Variation of the b Value of the Aftershock Sequence, and its Possible Relation with the Healing Process*

Aftershocks are generally hard to define by a strict criterion. However, for the Wenchuan sequence, since local background seismicity (that is, the decade-scale seismicity before the Wenchuan earthquake) was low, as the first-order approximation, the abruptly increased seismicity following the Wenchuan mainshock, within the region consistent with the rupture zone, could be considered as the aftershocks. As shown earlier in Fig. 1, such a region could be sketched by a polygon. Figure 7 shows the temporal variation of b values for the aftershocks as defined above, which shows a similar pattern of variation to that in Fig. 6. By examining the aftershock sequence, again it is shown that the b value has a trend of increase from June 27, 2008 to late April 2009, while since late April 2009, the b value has kept stable. As a

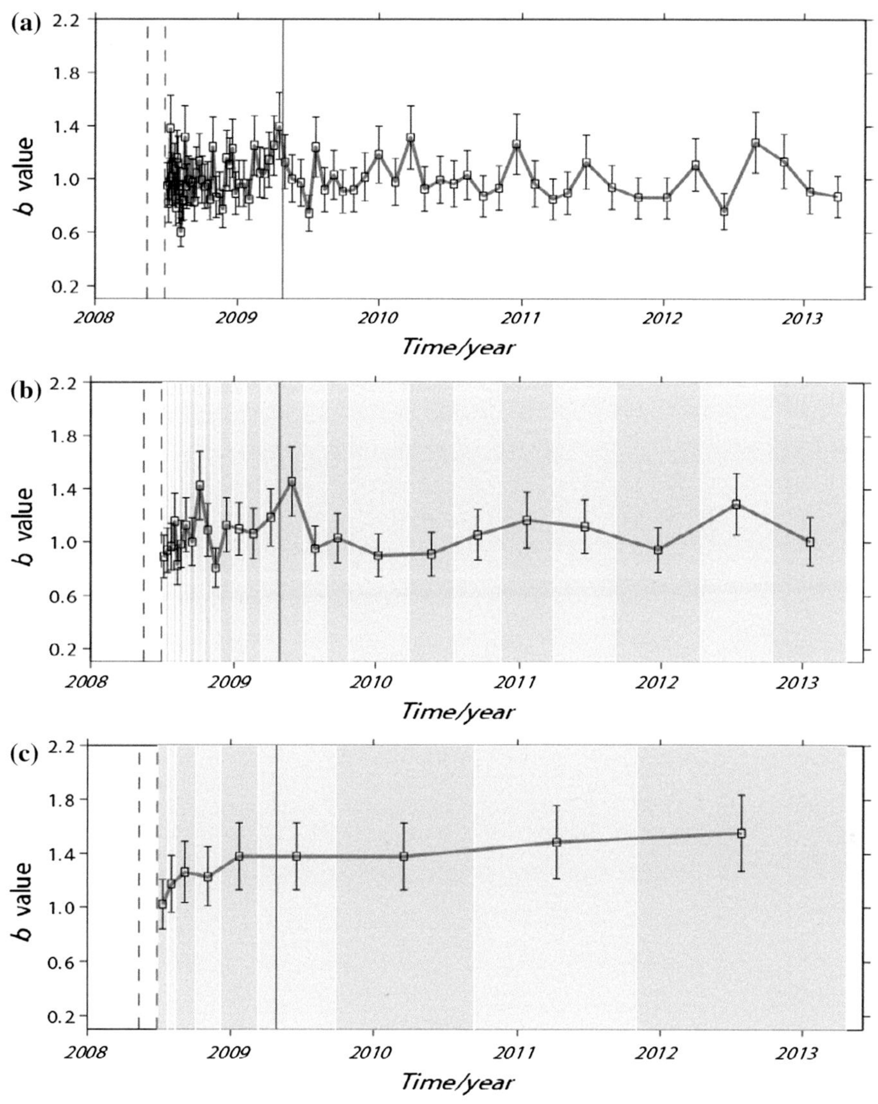

Figure 6
Temporal variation of b values for all the events in the studied area. Sliding step is 30 events, and *error bar* shows the standard deviation of the b value. The time at which the b value is shown (the *horizontal axis* in the figure) is chosen as the mid point of the time window of 30 events. The *black and red vertical dashed lines* are the time of the mainshock and June 27, 2008, between which the completeness of the earthquake catalog is above M_L 3.0. The *vertical red solid line* indicates the time crossing which the b value varies in different ways that before this time the b value has an apparently increasing trend, and after this time the b value varies around a stable level. **a** Result of b values calculated with events of $3.0 \leq M_L \leq 5.5$; **b** The same as (**a**), with events of $3.5 \leq M_L \leq 5.5$; **c** The same as (**a**), with events of $4.0 \leq M_L \leq 5.5$. In (**b**) and (**c**), the time span of each 30-events window is shown by different *vertical color bars*

reference, in the period from May 2009 to May 2013, there have been no strong aftershocks above M_L 6.0. If these characteristics were associated with the possible signature of fault healing, then the 'apparent healing time' would be pinpointed by this measure as around 1 year. In Fig. 7(c), exploiting the use of the completeness magnitude from May 12, 2008 to June 27, 2008, we had a glimpse at the variation of b value which shows that the b value had a decrease just after the mainshock which was followed by the increasing trend.

Regarding the healing process of the Wenchuan earthquake fault, fault zone permeability study (Xue *et al.* 2013) obtained the exponential decay time

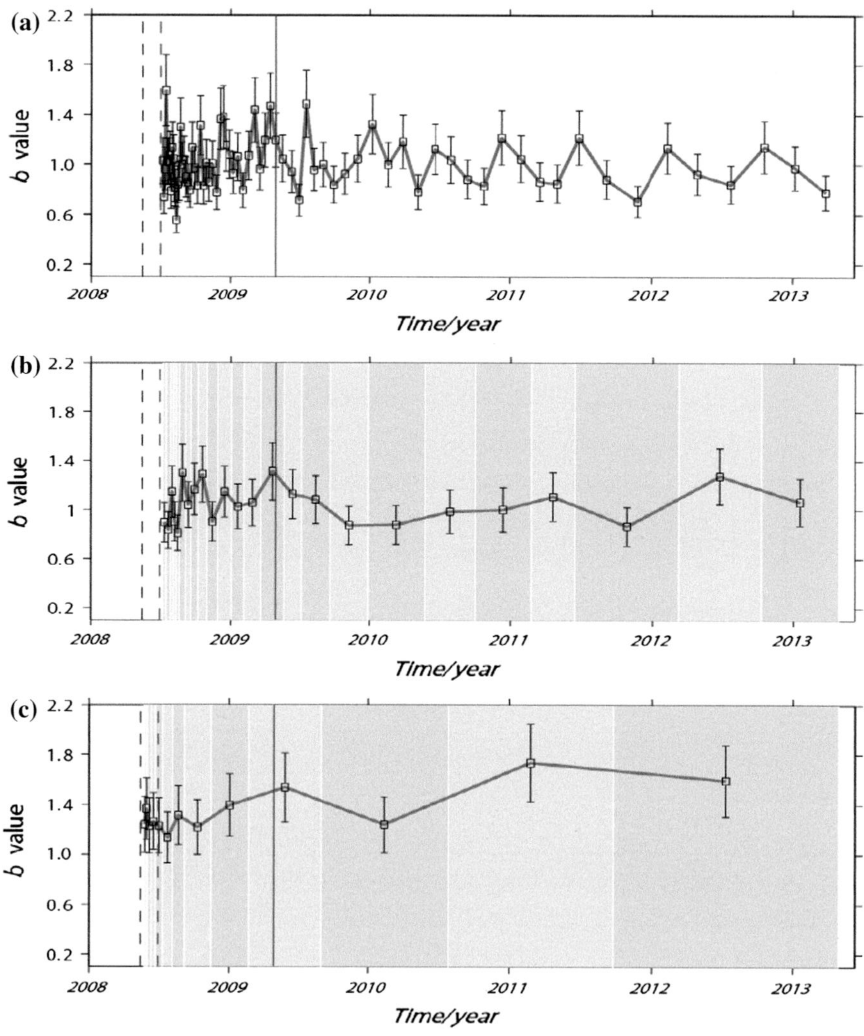

Figure 7
Temporal variation of *b* values for the aftershocks, that is, the events within the *polygon* shown in Fig. 1. For captions, see Fig. 6

0.6–2.5 years, and the variation of seismic velocities within the fault zone (Li 2010) gave that co-seismic reduction of the seismic wave velocity was about 10–15 %, while at least 5 % was restored during the first year after the mainshock, providing the orders of magnitudes of the half life period in case of an exponential decay. Our result, which indicates that the healing time is about 1 year, basically agrees with these results. Li (2010) also pointed out that the healing rate was not constant, with larger healing rate in the earlier stage after the earthquake. It seems that the healing process reflected by the variation of *b* value may reflect the early stage in which the healing rate is larger.

4.2. Spatial Pattern of the b Value

In Fig. 1, the two boxes with black thick lines are used as a proxy of the main asperities (from the results of Shen *et al.* 2009). The black dashed line indicates the 'boundary' between the north and the south segment of the aftershock distribution—actually shown in the figure is the north limit of such a boundary: the boundary can be moved in between the two strong asperities. Figure 8a, b shows the temporal variation of *b* values for the events in the north segment and the south segment, respectively. Apparently, the south segment shows a clear increasing trend from June 27, 2008 to mid 2009, while the

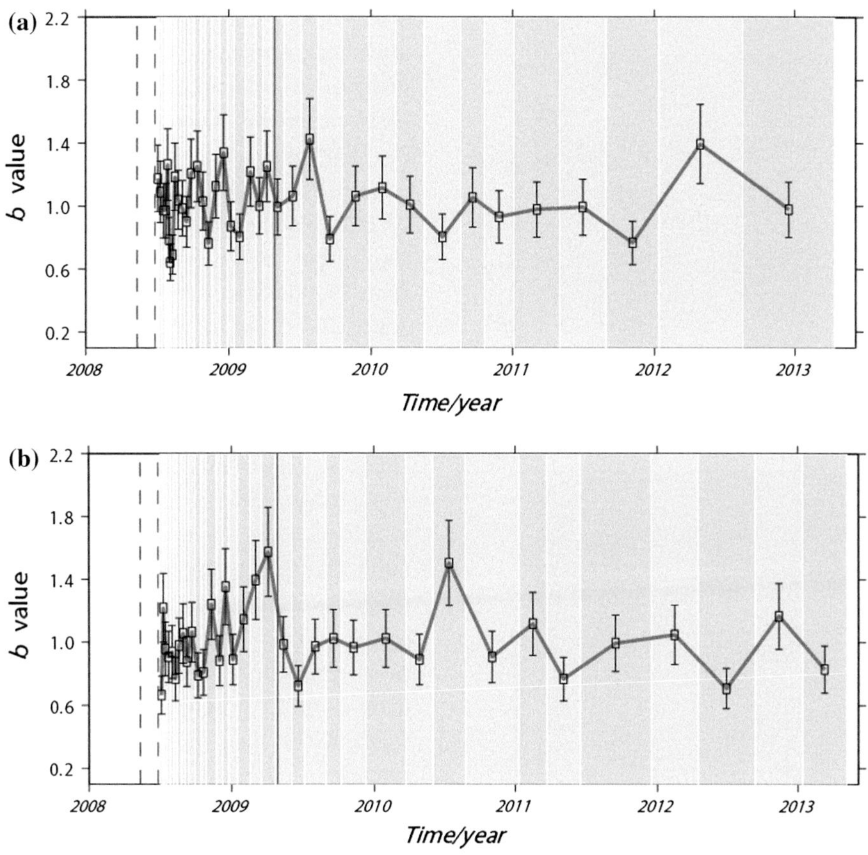

Figure 8
Temporal variation of b values for the events of $3.0 \leq M_L \leq 5.5$ in the north part (**a**) and the south part (**b**) of the aftershock zone, respectively, referring to Fig. 1. For captions, see Fig. 6

north segment does not show such a trend. In Fig. 8, BIC gain gives 0.93 for (a), 1.08 for (b), further demonstrating that the south part shows more clear trend of increasing. Changing the position of the boundary between the north and the south segment within the gap between the two strong asperities, the trend that the south segment plays a predominant role remains.

Figure 9 maps the b value with the reference of the Wenchuan main rupture zone. The whole rupture zone is divided into 24 rectangles. The rectangles in which there are less than 25 events are not shown in the figure. Figure 9a shows the b value of the events between June 27, 2008 and April 27, 2009, with $3.0 \leq M_L \leq 5.5$, in which white dots show the epicenters of strong aftershocks with $M_L \geq 6.0$ occurred in the same time period. Apparently, two strong aftershocks located to the north margin of the main rupture (seemingly related to the fracture extension), and one fell in the bin with lower b value (probably associated with the barrier). The low b value bin to the south margin seems to explain why the main rupture propagated to the north and stopped to the south. Figure 9b shows the b value during the time from April 27, 2009 to May 12, 2013, with $3.0 \leq M_L \leq 5.5$, exhibiting different patterns from Fig. 9a. During this time, there had been no strong aftershocks above M_L 6.0. Most of the segments along the rupture zone had high or 'normal' b values, while the bin accommodating the nucleation point exhibited the low b value. The comparison between Fig. 9a, b might imply that before and after the 'healing time' the spatial distribution of the b values is different.

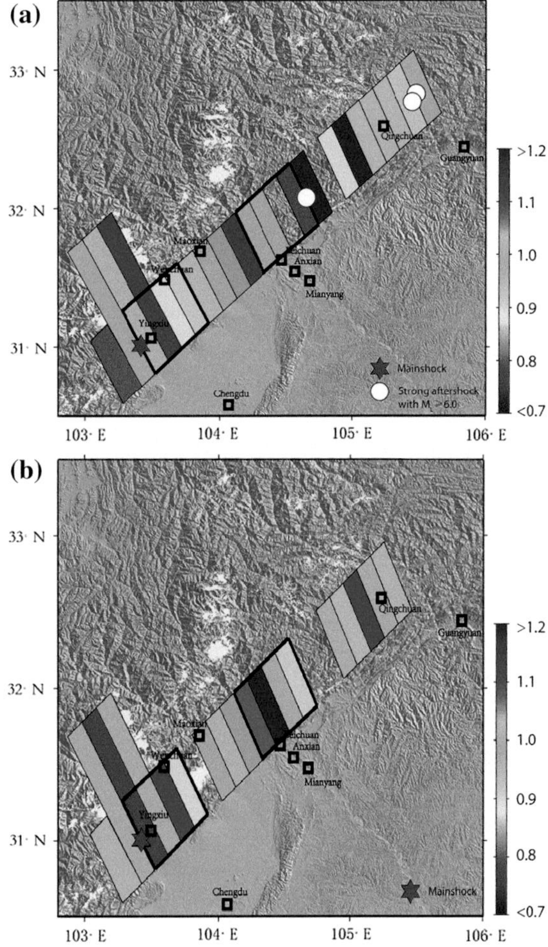

Figure 9

Mapping the b value, with the reference of the Wenchuan main rupture zone. The whole rupture zone is divided into *24 rectangles*. The *red hexagon* indicates the epicenter of the Wenchuan mainshock. **a** Mapping the b value of the events between June 27, 2008 and April 27, 2009, with $3.0 \leq M_L \leq 5.5$. *White dots* show the epicenters of strong aftershocks with $M_L \geq 6.0$ occurred in the same time period. **b** Mapping the b value during the time from April 27, 2009 to May 12, 2013, with $3.0 \leq M_L \leq 5.5$. During this time period, there had been no strong aftershocks above M_L 6.0

5. *Conclusions and Discussion*

Using the 2008 Wenchuan earthquake sequence as an example, we try to relate the temporal variation of b value to the healing process of the earthquake fault. The tricky aspect of discussing this problem lies in the entanglement of two time-lapse processes: one is the variation of monitoring capability of micro earthquakes after a major to great earthquake; the other is the relaxation process associated with the aftershock sequence which may reflect the healing process. In our analysis, it is shown that the completeness of the catalog from May 12, 2008 to June 27, 2008 was above M_L 3.0, which has to be accounted for in the calculation of the b value.

By examining the aftershock sequence, it is shown that the b value has a trend of increase from June 27, 2008 to late April 2009, while since late April 2009, the b value has kept stable. If the temporal variation of b value observed here could be related to the post-earthquake healing process of the fault zone, the healing time is approximately 1 year, which is consistent with other measures (Li 2010; Xue *et al.* 2013). In a spatial perspective, the aftershocks can be divided into two segments, the north and the south segment, respectively, with the reference of the two strong asperities of the main rupture. The b values of the two segments seem to show different trends of temporal variation that the main contributor of the increasing trend is the south segment, or the 'initiation segment' of the main rupture.

In discussing the above-mentioned observation, it is worth pointing out that although there has been a physical assumption to relate the b value to the stress level (Wyss 1973; Wiemer and Wyss 1997; Nakaya 2006; Narteau *et al.* 2009), whether this physical assumption could be extended to an aftershock sequence is still an open question. Also bear in mind that in considering the process of fault healing, the fluid permeability (Xue *et al.* 2013), the seismic wave velocity increase within the fault zone (Li 2010), and the increase of the b values (this paper) are still different processes, which have to be cautioned in the comparative discussion. Nevertheless, extending the study along this direction, more earthquake cases and more detailed observations of aftershocks, together with the comparative analysis using multidisciplinary data, seem essential.

Acknowledgments

The catalog data are taken from the China Earthquake Networks Center (CENC). Hui Jiang and Xiaoxiao Song provided helps in the processing of the catalog and the making of the figures as well as the revision. The suggestions of the referees as well as the guest editor helped much to the improvement of the paper, for which we are grateful.

References

AKI K (1965) *Maximum likelihood estimate of b in the formula logN = a-bm and its confidence limits.* Bull Earthq Res Inst Univ *43*: 137–139

CHEN J, LIU Q, LI S, GUO B, LI Y, WANG J, QI S (2009) *Seismotectonic study by relocation of the Wenchuan M_S8.0 earthquake sequence.* Chinese J Geophys *52*: 390–397 (in Chinese with English abstract)

GUTENBERG B, RICHTER CF (1944) *Frequency of earthquakes in California.* Bull Seismol Soc Am *34*: 185–188

HUANG Y, WU JP, ZHANG TZ, ZHANG DN (2008) *Wenchuan 8.0 earthquake and its aftershocks relocation.* Sci China Ser D-Earth Sci *38*: 1242–1249

IWATA T (2008) *Low detection capability of global earthquakes after the occurrence of large earthquakes: investigation of the Harvard CMT catalogue.* Geophys J Int *174*: 849–856. doi:10.1111/j.1365-246X.2008.03864.x

JIANG CS, ZHUANG JC (2010) *Evaluation of background seismicity and potential source zones of strong earthquakes in the Sichuan-Yunan region by using the spacetime ETAS model.* Chinese J Geophys *53*: 305–317 (in Chinese with English abstract)

JIANG ZQ, LIU FR, XU RP (2011) *Study on temporal and spatial distribution of seismic sequence of the M_S = 8.0 Wenchuan earthquake.* Journal of Earth Science and Environment *33*(4): 428–433 (in Chinese with English abstract)

LEONARD T, HSU J S J (1999) Bayesian Methods. Cambridge University Press, New York, 6–9

LI YG (2010) *Fault damage in the 2008 M8 Wenchuan earthquake epicentral region.* Acadmic Prospective *6*: 2–16

LÜ J, SU J, LONG F, YANG Y, ZHANG Z, Tang L, LI C (2008) *Discussion on relocation and seismo-tectonics of the M_S8.0 Wenchuan earthquake sequences.* Seismology and Geology *30*: 917–925 (in Chinese with English abstrcat)

MA XJ, WU ZL, JIANG CS (2014) *'Repeating earthquakes' associated with the WFSD-1 drilling site.* Tectonophysics *619–620*: 44–50

MIGNAN A, WOESSNER J (2012) *Estimating the magnitude of completeness for earthquake catalogs, Community Online Resource for Statistical Seismicity Analysis.* doi:10.5078/corssa-00180805. Available at http://www.corssa.org.

NAKAYA S (2006) *Spatiotemporal variation in b value within the subducting slab prior to the 2003 Tokachi-oki earthquake (M8.0), Japan.* J Geophys Res *111*: B03311. doi:10.1029/2005JB003658

NARTEAU C, BYRDINA S, SHEBALIN P, SCHORLEMMER D (2009) *Common dependence on the stress for the two fundamental laws of statistical seismology.* Nature *462*: 642–646. doi:10.1038/nature08553

OMORI F (1894) *On aftershocks of earthquakes.* J Coll Sci Imp Univ Tokyo *7*:11–200

OGATA Y, IMOTO M, KATSURA K (1991) *3-D spatial variation of b-values of magnitude-frequency distribution beneath the Kanto District, Japan.* Geophys J Int *104*:135–146

OGATA Y, KATSURA K (1993) *Analysis of temporal and spatial heterogeneity of magnitude frequency distribution inferred from earthquake catalogues.* Geophys J Int *113*:727–738

OGATA Y, ZHUANG JC (2006) *Space-time ETAS models and an improved extension.* Tectonophysics *413*:13–23

SEHER T, MAIN I G (2004) *A statistical evaluation of a 'stress-forecast' earthquake.* Geophys J Int *157*:187–193

SHEN ZK, SUN JB, ZHANG PZ, WAN YG, WANG M, BURGMANN R, ZENG YH, GAN WJ, LIAO H, WANG QL (2009) *Slip maxima at fault junctions and rupturing of barriers during the 2008 Wenchuan earthquake.* Nature Geoscience 2:718–724. doi:10.1038/NGEO636

UTSU T (1961) *A statistical study on the occurrence of aftershocks.* Geophys Mag *30*:526–605

UTSU T, OGATA Y, MATSU'URA R (1995) *The centenary of the Omori formula for a decay law of aftershock activity.* J Phys Earth *43*:1–33

UTSU T (1965) *A method for determining the value of b in a formula logN = a-bm showing magnitude-frequency relation for earthquakes.* Geophys Bull Hokkaido Univ *13*:99–103 (in Japanese with English abstract)

UTSU T (1999) *Representation and analysis of the earthquake size distribution: a historical review and some new approaches.* Pure Appl Geophys *155*: 509–535

WANG J, RUAN X, ZHEN JR, SUN YJ, JIANG HL, ZHAN XY (2009) *Study on the variation of b value in Wenchuan earthquake series.* Seismological and Geomagnetic Observation and Research *30*(2): 15–21 (in Chinese with English abstract)

WEN XZ, MA SL, XU XW, HE YN (2008) *Historical pattern and behavior of earthquake ruptures along the eastern boundary of the Sichuan-Yunnan faulted-block, southwestern China.* Phys Earth Planet Interiors *168*: 16–36. doi:10.1016/j.pepi.2008.04.013

WIEMER S, WYSS M (1997) *Mapping the frequency-magnitude distribution in asperities: An improved technique to calculate recurrence times?* J Geophys Res *102*: 15115–15128

WIEMER S, WYSS M (2000) *Minimum magnitude of complete reporting in earthquake catalogs: examples from Alaska, the western United States, and Japan.* Bull Seismol Soc Am *90*: 859–869

WYSS M (1973) *Towards a physical understanding of the earthquake frequency distribution,* Geophys J R Astron Soc *31*: 341–359

WYSS M, HASEGAWA A, WIEMER S, UMINO N (1999) *Quantitative mapping of precursory seismic quiescence before the 1989, M7.1 Off-Sanriku earthquake, Japan.* Annals of Geophysics *42*: 851–869

XUE L, LI HB, BRODSKY EE, XU ZQ, KANO Y, WANG H, MORI JJ, SI JL, PEI JL, ZHANG W, YANG G, SUN ZM, HUANG Y (2013) *Continuous permeability measurements record healing inside the Wenchuan earthquake fault zone.* Science *340*: 1555–1559. doi:10.1126/science.1237237

YI GX, WEN XZ, WANG SW, LONG F, FAN J (2006) *Study on fault sliding behaviors and strong-earthquake risk of the Longmen-shan-Minshan fault zones from current seismicity parameters.* Earthquake Research in China *22*: 117–125 (in Chinese with English abstract)

YI GX, WEN XZ, XIN H, QIAO HZ, WANG SW, GUAN Y (2013) *Stress state and major-earthquake risk on the southern segment of the Longmen Shan fault zone.* Chinese J Geophys *56*: 1112–1120. doi:10.6038/cjg20130407 (in Chinese with English abstract)

ZHANG QZ, TANG WP, LIU YP, LI J (2013) *Research of the activity along the central segment of the Longmenshan active faults and adjacent regions based on the high precision GPS monitoring*. Progress in Geophysics *28*(1): 190–198 (in Chinese with English abstract)

ZHAO B, SHI YT, GAO Y (2011) *Relocation of aftershocks of the Wenchuan M_S8.0 earthquake and its implication to seismotectonics*. Earthq Sci *24*: 107–113. doi:10.1007/s11589-011-0774-6

ZHAO YZ, WU ZL (2008) *Mapping the b-values along the Longmenshan fault zone before and after the 12 May 2008, Wenchuan, China, M_S8.0 earthquake*. Nat Hazards Earth Syst Sci *8*: 1375–1385

ZHU AL, XU XW, DIAO GL, SU JR, FENG XD, SUN Q, WNG YL (2008) *Relocation of the M_S8.0 Wenchuan earthquake sequence in part: preliminary seismotectonic analysis*. Seismology and Geology *30*(3): 759–767 (in Chinese with English abstract)

(Received August 3, 2014, revised April 8, 2015, accepted April 13, 2015, Published online May 8, 2015)

Pure Appl. Geophys. 173 (2016), 85–96

DOI 10.1007/s00024-015-1044-7

Pure and Applied Geophysics

Spatial and Temporal Variation of *b*-Values in Southwest China

Shenjian Zhang[1] and Shiyong Zhou[1]

Abstract—In our study, we used an improved Bayesian method based on a basis spline (B-spline) function to estimate the spatial and temporal variations of *b*-values in Southwest China. We propose that *b*-values combined with seismicity and tectonic background can provide a fairly clear stress condition and can be used in earthquake hazard analysis. Very low *b*-values in Southwest China appear to be related to the 2008 M_s 8.0 Wenchuan earthquake. We also suggest that the decreasing *b*-value trend in the Longmen Shan area from 2000 to 2008 may be an indicator of the 2008 M_s 8.0 Wenchuan earthquake.

Key words: *b*-value, frequency magnitude distribution, B-spline, Bayesian method, Wenchuan M_S 8.0 earthquake, Southwest China.

1. Introduction

Gutenberg and Richter (1944) proposed the empirical law (G–R law) that describes the frequency–magnitude distribution (FMD) as follows:

$$\log_{10} N = a - bM \tag{1}$$

where N is the cumulative number of earthquakes with magnitude larger than or equal to M. Two constants in this law, a and b, both have geophysical significance rather than mathematical meaning. The constant a characterizes seismic activity, or earthquake productivity, and the coefficient b describes the relative ratio of large and small events. A high *b*-value indicates a larger proportion of small events, and vice versa. The studies that focus on the FMD, especially the *b*-value, can be divided into several categories, but the most common topics are the relationship between *b*-value and seismicity or stress and the spatial or temporal variability of the *b*-value (Wiemer and Wyss 2002).

Spatial and temporal variations in *b*-values are known to reflect the stress field, for the *b*-value is inversely dependent on differential stress (Scholz 1968; Ogata *et al.* 1991; Urbancic *et al.* 1992; Ogata and Katsura 1993; Narteau *et al.* 2009). Schrolemmer *et al.* (2005) suggested that the *b*-value is also dependent on styles of faulting, as the *b*-values of thrust faults are the lowest among the three types of faulting mechanisms, which can be considered as evidence of the relationship between *b*-value and stress. Rock fracture experiments in the laboratory supported this relationship as well (Scholz 1968; Amitrano 2003). This relationship can be applied in building an earthquake hazard model using the *b*-value as an indirect "stressmeter" (Wiemer and Schorlemmer 2007).

Studies on computing the *b*-value and its uncertainty have been performed in the last decades to show how the *b*-value changes with space or time. Most of these studies used the maximum likelihood estimation (MLE) method to obtain the figures of *b*-value and its standard error (Aki 1965; Utsu 1965; Shi and Bolt 1982; Bender 1983). However, the MLE approach can hardly give a smooth estimation of *b*-values with space or time, using a moving window approach. It also has difficulty dealing with short-period turbulence and event clusters.

Ogata *et al.* (1991) suggested a Bayesian approach using a B-spline to estimate the *b*-value on space, which performs better in overcoming the shortages of the MLE method. An improved Bayesian approach based on a free-knot B-spline was proposed by Zheng and Zhou (2014) to deal with clustered earthquakes. In this study, we use the improved Bayesian method to estimate the spatial and temporal variations of *b*-values in Southwest China

[1] School of Earth and Space Sciences, Peking University, Beijing 100871, China. E-mail: sjzhang@pku.edu.cn; zsy@pku.edu.cn

where the 2008 M_s 8.0 Wenchuan earthquake took place (JIA *et al.* 2012). We suggest that the b-value in this area corresponds to other geophysical research results indicating that the 2008 M_s 8.0 Wenchuan earthquake was different from most other large earthquakes in regards to its seismogenic process and pre-event background seismicity.

2. Method

In our study, we used equally spaced knots for B-spline bases to estimate b-values in space, and free-knot spline functions to estimate b-values in time. This method performs better than the MLE method in the synthetic test (ZHENG and ZHOU 2014). As many parameters are needed for b-values in space and time (IMOTO and ISHIGURO 1986; OGATA and KATSURA 1988), we maximized the log-likelihood function to improve the goodness of the data fitting. On the other hand, we need to avoid unnecessary fluctuation of the spline function, so we used the roughness penalty as the constraint. In order to deal with the trade-off between giving a good fit to the data and avoiding too much rapid fluctuation, we applied the objective Bayesian method (AKAIKE 1980; ISHIGURO and SAKAMOTO 1983; OGATA *et al.* 1991; ZHENG and ZHOU 2014).

2.1. The Penalized Log-Likelihood

As the derivation of the G–R law (1), the probability density function of the events with magnitude M can be written as the exponential distribution

$$f(M) = \beta e^{-\beta(M-M_c)}, \quad M > M_c \tag{2}$$

where $\beta = b \ln 10$ and M_c is the magnitude of completeness, which is the lowest magnitude at which all the events are detected (WIEMER and WYSS 2000; WOESSNER and WIEMER 2005).

Here, we assume that b, as well as β, is dependent on the location. Considering the magnitude data M_i for each epicenter coordinate (x_i, y_i) with $i = 1, 2, \ldots, n$, the likelihood function can be written as

$$L(\theta) = \prod_{i=1}^{n} \beta_\theta(x_i, y_i) e^{-\beta_\theta(x_i,y_i)(M_i-M_c)}, \quad M_i > M_c \tag{3}$$

where θ is a parameter vector characterizing the function, and the parameterization of $\beta_\theta(x, y)$ is carried out by

$$\beta_\theta(x, y) = e^{\phi_\theta(x,y)}. \tag{4}$$

So, the estimated b-value is given as

$$b(x, y) = e^{\phi_\theta(x,y)} / \ln 10 \tag{5}$$

where ϕ is the 2-D B-spline function

$$\phi_\theta(x, y) = \sum_{l=0}^{L+3} \sum_{m=0}^{M+3} c_{l,m} F_l(x) G_m(y). \tag{6}$$

Here, the parameter vector $\phi = \{c_{l,m}\}$ is a set of coefficients, and the functions F_l and G_m are cubic B-spline bases. We use the recurrence method (DE BOOR 1978) to calculate the $(k-1)$th degree B-spline bases

$$B_{i,l} = \begin{cases} 1 & \text{if } t \in [t_i, t_{i+1}) \\ 0 & \text{otherwise} \end{cases} \tag{7}$$

and

$$B_{i,k} = \frac{t - t_i}{t_{i+k-1} - t_i} B_{i,k-1}(t) + \frac{t_{i+k} - t}{t_{i+k} - t_{i+1}} B_{i+1,k-1} \tag{8}$$

where t_i represents the knot sequence. We adopt equally spaced knots for space variations of the b-value and self-adaptive knots for time variations, which can reflect the short period changes we are interested in.

As the Eq. (6) shows, a large number of parameters are needed to represent the 2-D B-spline function, which usually causes unnecessary rapid fluctuation in the spline surface. In order to resolve the conflicting goals of a good data fit and less rapid wiggles, we use roughness penalties of the function as a constraint of estimation. The penalties can be given as

$$\Phi_1 = \int_s w_1 \left[\left(\frac{\partial \phi}{\partial x} \right)^2 + \left(\frac{\partial \phi}{\partial x} \right)^2 \right] dx dy \tag{9}$$

$$\Phi_2 = \int_s w_2 \left[\left(\frac{\partial^2 \phi}{\partial x^2} \right)^2 + 2 \left(\frac{\partial^2 \phi}{\partial x \partial y} \right)^2 + \left(\frac{\partial^2 \phi}{\partial y^2} \right)^2 \right] \mathrm{d}x\mathrm{d}y \tag{10}$$

where the non-negative constants w_1, w_2 are weights controlling the strength of the penalties. In this study, we assume that isotropy in (x, y) is reasonable, so two weights are enough for the penalty functions. Therefore, given suitable weights w_1, w_2, we estimate the parameter vector $\theta = \{c_{l,m}\}$ by maximizing the penalized log-likelihood function (GOOD and GASKINS 1971; OGATA *et al.* 1991)

$$\begin{aligned} Q(\theta|w_1, w_2) &= \log L(\theta) - \Phi_1(\theta|w_1) - \Phi_2(\theta|w_2) \\ &= \sum_{i=1}^{n} [\phi_\theta(t_i) - \mathrm{e}^{\phi_\theta(t_i)}(M_i - M_\mathrm{c})] \\ &\quad - \Phi_1(\theta|w_1) - \Phi_2(\theta|w_2) \end{aligned} \tag{11}$$

for the dataset $\{(x_i, y_i, M_i); i = 1, 2, \ldots, n\}$.

2.2. The Objective Bayesian Method

As we can see, suitable weights are critical for the maximization of Eq. (11). The first derivative norm represents total fluctuation, and minimizing Eq. (9) can avoid any global trends. Equation (10) is a curvature measurement, and suppressing it leads to reduced undulation. In order to obtain the optimal weights, we introduce the objective Bayesian method (AKAIKE 1980; GOOD 1965; OGATA *et al.* 1991) into our calculation.

According to this method, the exponential of the negative sum $-(\Phi_1 + \Phi_2)$ is proportional to the prior distribution $\pi(\theta|w_1, w_2)$characterized by w_1, w_2. Considering the expressions of Φ_1 and Φ_2, the penalties are quadratic with respect to the parameter vector $\theta = \{c_{l,m}\} = (c_p)_{p=1,2,\ldots,P}$ for a non-negative definite matrix $\Sigma(w_1, w_2)$ such that

$$\frac{\theta \Sigma \theta^\mathrm{T}}{2} = \Phi_1(\theta|w_1) + \Phi_2(\theta|w_2) \cdot \tag{12}$$

As the matrix Σ is degenerated here (OGATA *et al.* 1991), we divide the parameter vector $\theta = \{c_{l,m}\} = (c_p)_{p=1,2,\ldots,P}$ into $(\mathbf{c}^\mathbf{r}, c_p)$, so the distribution $\pi(\theta|w_1, w_2)$ can be written as

$$\pi(\mathbf{c}^\mathbf{r}|w_1, w_2, c_P) = \frac{\sqrt{\det \Sigma_r}}{\sqrt{2\pi}^{P-1}} \mathrm{e}^{-\theta \Sigma \theta^\mathrm{T}} \tag{13}$$

where c_P is the last element of θ, $\mathbf{c}^\mathbf{r}$ is the other elements, and Σ_r is the cofactor of the last diagonal element of Σ.

Thus, we have the Bayesian likelihood written as the integral of the posterior function

$$L(w_1, w_2, c_P) = \int L(\theta)\pi(\mathbf{c}^\mathbf{r}, \mathbf{c_P}|\mathbf{w_1}, \mathbf{w_2})\mathrm{d}\mathbf{c}^\mathbf{r}. \tag{14}$$

We then maximize the Bayesian likelihood to obtain the weights w_1, w_2. As the integral in Eq. (14) is non-analytical, we used the Gaussian approximation method (ISHIGURO and SAKAMOTO 1983; OGATA *et al.* 1991) to obtain an approximated log Bayesian likelihood (see OGATA *et al.* 1991), shown as

$$\begin{aligned} \log L(w_1, w_2, c_P) = Q(\hat{\theta}|w_1, w_2) + \frac{1}{2}\log\{\det \Sigma_r\} \\ - \frac{1}{2}\log\{\det \mathbf{H_r}\} \end{aligned} \tag{15}$$

where $\hat{\theta}$ is the parameter vector estimated by maximizing the penalized log-likelihood function Q with given weights w_1, w_2, and $\mathbf{H_r}$ is the cofactor of the last diagonal element of the Hessian matrix of Q at $\hat{\theta}$. Since both maximizations of the log Bayesian likelihood $\log L$ and the penalized log-likelihood function Q are non-linear, we adopted the Davidon–Fletcher–Powell method (FLETCHER and POWELL 1963).

We used an iteration algorithm for the whole calculation of b-values. We first set initial values of θ and weights, then we obtain $\hat{\theta}$ by maximizing Q in Eq. (11) with fixed weights. After that, we maximize $\log L$ in Eq. (15) with $\hat{\theta}$ calculated in last step to obtain optimal weights. These two steps are repeated until the b-values converge. In this case, the convergence is quite rapid with a large number of parameters, indicating the stability of our method. In our study, we also estimated the standard errors of b-values (OGATA *et al.* 1991). The variance–covariance matrix between (x, y) and (x', y') is shown as

$$C(x, y; x', y') = \sum_p \sum_{p'} h^{p,p'} F_l(x) G_m(y) F_{l'}(x') G_{m'}(y') \tag{16}$$

where $h^{p,p'}$ is the element of the inverse of the Hessian matrix of Q. So, the standard error of $\phi_{\hat{\theta}}$ at (x, y) is

$$\varepsilon(x,y) = C(x,y;x',y')^{\frac{1}{2}} \tag{17}$$

and the error of the $b_{\hat{\theta}}$ can be expressed as

$$\begin{aligned} e_{\hat{\theta}} = \frac{1}{2}\{\exp[\phi_{\hat{\theta}}(x,y) + \varepsilon(x,y)]\} \\ - \{\exp[\phi_{\hat{\theta}}(x,y) - \varepsilon(x,y)]\}. \end{aligned} \tag{18}$$

As for the temporal variation of b-values, we used the same algorithm as the spatial variation except that we adopted free-knots for the B-spline bases.

3. Data

In our study, we used a dataset from the China Earthquake Networks Center (CENC) earthquake catalog between the years 1985 and 2012. The epicenters are distributed in a rectangle defined by 97°E–107°E and 26°N–34°N. To estimate the completeness magnitude M_c, we apply the maximum curvature method (MAXC) combined with the bootstrap method and added a correction value of 0.2 (Wiemer and Wyss 2000; Woessner and Wiemer 2005). Considering the spatial variation of the minimum magnitude of completeness, we adopt an overall $M_c = 2.8$. The number of events larger than M_c in the dataset is 9,625 (Fig. 1). Figure 2 shows a plot of magnitude versus event time, and Fig. 3 is the (cumulative) frequency–magnitude distribution of earthquakes in the space–time range.

4. Results

In this study, we first drew the spatial variation of b-values in Southwest China. We divided the area into $L \times M = 10 \times 8$ rectangle sub-areas with sides of 1° in both latitude and longitude, for the knots of B-spline bases described in Sect. 2.1. We used equally spaced knots here because we wanted to obtain a relatively high space resolution in the whole area. The non-uniform grid cannot provide an enough resolution in low seismicity areas. Our result is shown in Fig. 4. Figure 4a is the b-value distribution in Southeast China and Fig. 4b is the spatial distribution of the standard error. The b-values vary between 0.5 and 1.9. There are several large zones in which the b-value is lower than the average level 1.0. However, due to the boundary effect of the b-value, standard errors around the boundary of the region being much higher than the central area, we did not analyze the area along the boundaries of the whole region. The main two areas that have low b-values are regions A and B (marked by polygons in Fig. 4a). Region A includes the Longmen Shan faulting zone, which is a SW–NE profile (marked as region C). Considering the G–R relationship and the meaning of b-values, a low b-value indicates an earthquake size distribution in which the small events are fewer than normal or the number of large ones is above average. As Figs. 1 and 4a show, most of $M \geq 6.0$ events took place in these two areas, especially in region A. In regions with a small number of large events, the b-value is significantly above 1.2. According to the relationship between the b-value and the stress condition, we can infer that the differential stress is higher in region A and B than other areas.

Further, we investigated the temporal variations of the b-value in region A. Here we used a non-uniform grid for B-spline bases (Zheng and Zhou 2014), which was given by a self-adaptive algorithm. After testing several parameter combinations, we chose the one which claims at least 30 events in each grid with a step size of 0.1. This type of knot sequence can show the short period wiggles pretty well.

We also calculated temporal variations of the minimum magnitude of completeness M_c using the MAXC method for each region and choose $M_c = 2.8, 2.6$ for region A and C. Figure 5 depicts the temporal b-values changes and the standard errors in region A. The b-values in this region range from approximatley 0.7 to 1.4, which have a complex structure.

The lowest b-value is immediately prior to the 2008 M_s 8.0 Wenchuan earthquake, the largest event in this area within the time range of concern. Lower b-values correspond to the occurrence of large events. Chang *et al.* (2007) reported that low b-values after a large earthquake may indicate that a lot of small events cannot be detected by seismic networks. This premise can be adopted in our study as well. As shown in Fig. 2, the magnitude of completeness rises to 4.0 in more than one month after the 2008 M_s 8.0

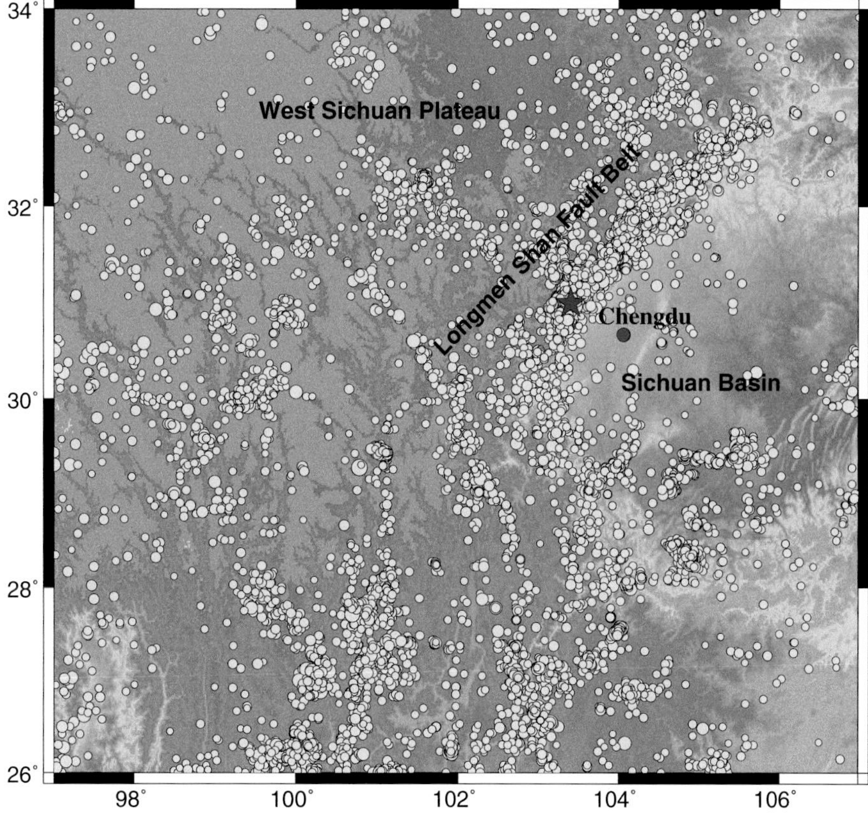

Figure 1
Epicenter distribution of earthquakes in Southwest China. Epicenters of $M \geq 2.8$ events from January 1985 through December 2012 are shown as *dots*, scaled proportional to magnitude. The *blue star* is the location of the 2008 M_s 8 Wenchuan earthquake. The *red dot* is the location of Chengdu, the capital of Sichuan Province. Major tectonic units are labeled in *black*

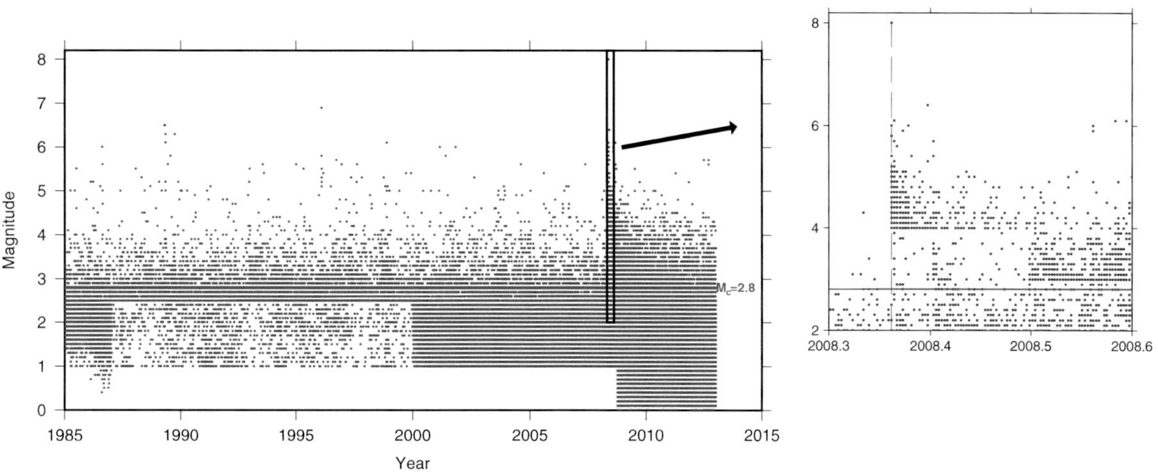

Figure 2
Magnitude versus occurrence time of events in Southwest China. The *red line* indicates M_c. The *small box* shows the relationship for three to four months around the 2008 M_s 8 Wenchuan earthquake

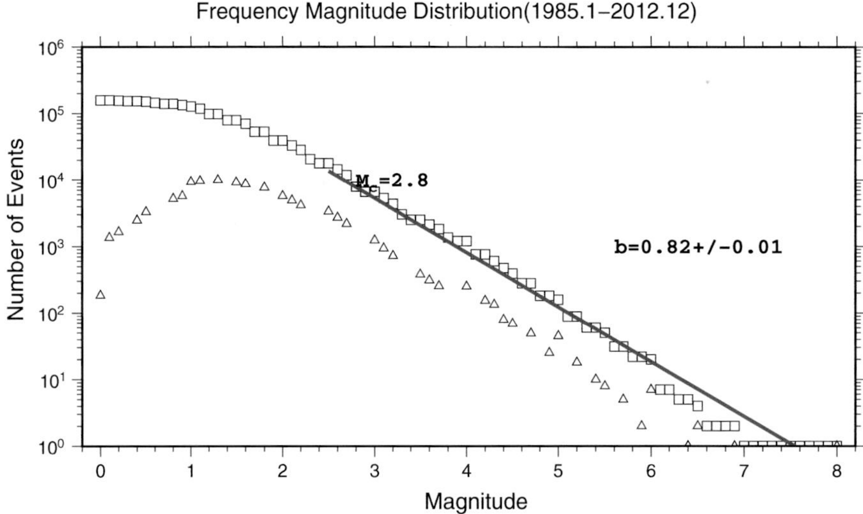

Figure 3
Frequency and magnitude distribution in Southwest China. *b*-value with its standard error are calculated via the MLE method

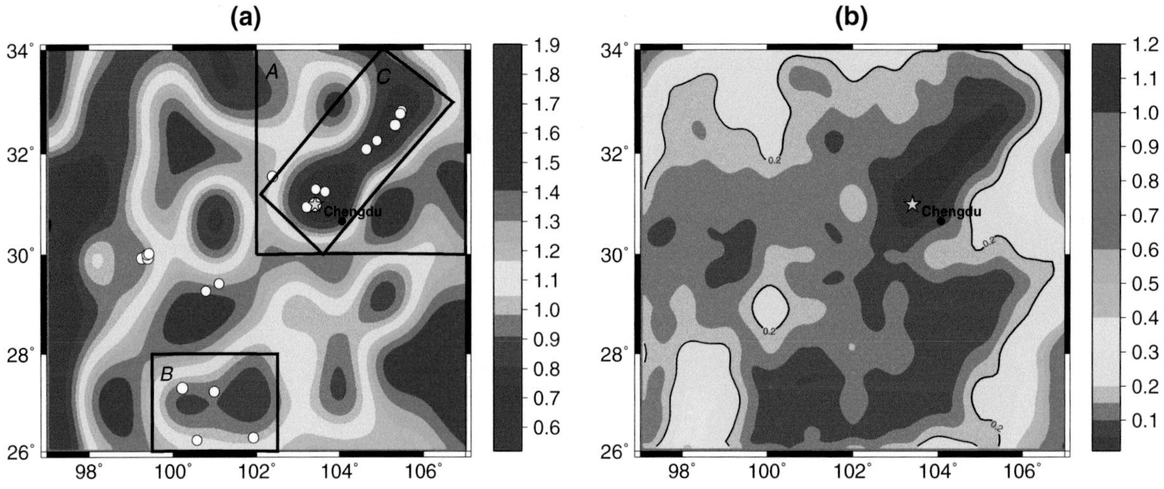

Figure 4
a Spatial variations of *b*-values in Southwest China. *Black polygons* highlight the regions with a low *b*-value. *White dots* scaled proportional to magnitude are the locations of events with $M \geq 6.0$. **b** Standard error of spatial variations of *b*-values

Wenchuan earthquake (Jia *et al.* 2014). Here, we cut off the data two months after the 2008 M_s 8.0 Wenchuan earthquake. From 1985 to 2000, *b*-values in region A were above 1.0. Only one $M \geq 6.0$ event took place in this time range. From 2000 to 2008, *b*-values show a decreasing trend with several small fluctuations. After 2005, *b*-values are below 1.0, with a drop of around 0.3 between 2000 and 2005. After the M_s 8.0 Wenchuan earthquake, the *b*-values return to average levels in less than a year, which indicates the stress condition has been normalized.

We also obtained estimated *b*-value errors for region A (drawn in Fig. 5b). In general, the standard error is <0.2 during most of the time, and <0.1 when events are clustered.

5. Discussion

In our study, we chose a raw catalog to calculate *b*-values for as many events as possible. So, we first considered the influence of the data or the catalog by

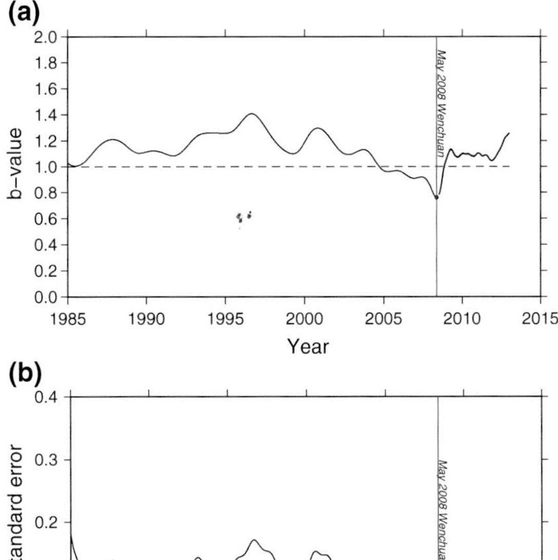

Figure 5

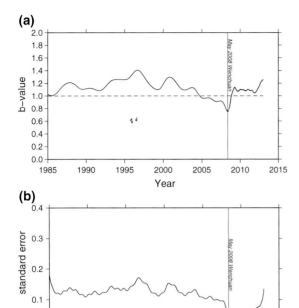

a Temporal variations of *b*-values in region A. *Red line* indicates the occurrence time of the 2008 M_s 8 Wenchuan earthquake. **b** Standard error of temporal *b*-value variations

showing the results derived from both a raw catalog and a declustered catalog. We applied the space-time ETAS model (Zhuang *et al.* 2002, 2004; Zhuang 2006) to estimate background seismic activity and decluster our catalog. In order to maximize the effect of declustering and to demonstrate this effect on the *b*-value distribution as much as possible, we adjusted the magnitude of completeness to 4.0, although the reduction of small events lead to a different spatial *b*-value pattern. As in Fig. 6a and c, the declustering effect is adequate and, as in Fig. 6b and d, the spatial *b*-value distributions of both catalog types do not show significant discrepancies. As we abandoned all the events with a magnitude <4.0, the low *b*-value of the Longmen Shan fault does not appear clearly in both *b*-value maps. But the result of the declustered catalog expresses a more clear concentrating trend of low *b*-values around the epicenter of the 2008 M_s 8.0 Wenchuan earthquake. We propose that the choice of catalog type depends on the purpose of the research being performed. A declustered catalog may perform better in estimating *b*-values but it also loses a lot of information.

We then compared the results of the improved method with the MLE method, as shown in Fig. 7. The two results share the same *b*-value trend and show an obvious decrease before the 2008 M_s 8.0 Wenchuan earthquake. However, the MLE method relies much more on the quality of data than the other method as the estimation is not smooth enough when the number of data points is small. Once the events are sufficient in number, the MLE method also reflects the rapid *b*-value fluctuations.

Besides, in Fig. 7, when we used the MLE method, we cut off the data two months after the 2008 M_s 8.0 Wenchuan earthquake. From 1985 to just prior to the 2008 M_s 8.0 Wenchuan earthquake, we used an event window with the step length of 50 events and a window length of 100 events because events in this time range do not cluster. After the 2008 M_s 8.0 Wenchuan earthquake, we calculated *b*-values from July 12th, 2008 using a time window of one tenth of a year. The temporal *b*-value variation before the large event carried out by the MLE method shows the same trend as the B-spline method and shows rapid fluctuations after the event, which is the result of a dramatic increase of the detectability of the network in this area.

In addition to a temporal comparison before and after the large event, we also conduct a spatial comparison. Result are presented in Fig. 8. Before May 12th, 2008, *b*-values in region A were relative high, especially in the West Sichuan plateau. This phenomena can be explained by the tectonic mechanism of the 2008 M_s 8.0 Wenchuan earthquake, which we will discuss later in this article. Spatial *b*-value variation after the main shock displays the aftershock distribution, from which we can clearly recognize the Longmen Shan fault (Fig. 8).

The Longmen Shan area is one of the most complicated and important tectonic zones in the eastern margin of Tibet. The M_s 8.0 Wenchuan earthquake in 2008 killed more than 80,000 people (Deng *et al.* 2010) and was the largest earthquake in China in the last few decades. This event resulted from rupture of the Yingxiu-Beichuan fault in the Longmen Shan fault belt, which lies in the eastern margin of the Tibetan plateau in Sichuan (Burchfiel

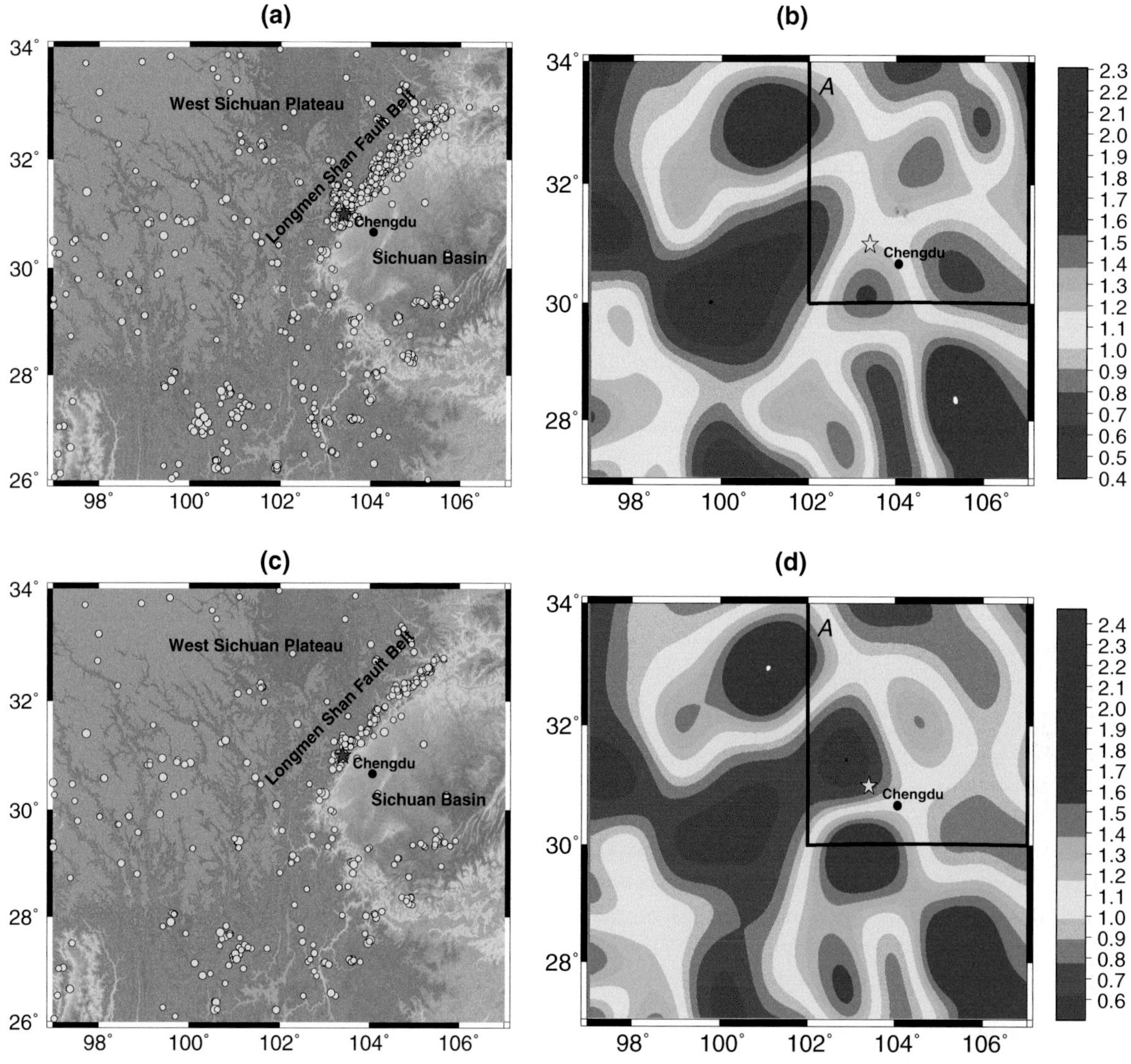

Figure 6
a Epicenter distribution of events with $M \geq 4.0$ in Southwest China from January 1985 through December 2012. **b** Spatial *b*-value variations calculated from events in **a**. **c** Epicenter distribution of a declustered catalog of events in **a** using an ETAS model. **d** Spatial *b*-value variations calculated from events in **c**

et al. 2008; ZHANG *et al.* 2009). The West Sichuan plateau, the Longmen Shan fault, and the Sichuan Basin are the three basic geological units of this area. The West Sichuan plateau is a deformation unit, in which deformation is converted to stress accumulated in the Lonemen Shan fault belt. The Longmen Shan fault zone is a locked unit with very slow deformation and fairly high stress (ZHOU 2008). The Sichuan Basin, as a support unit, blocks eastward movement of the West Sichuan plateau and the Longmen Shan fault (ZHANG *et al.* 2009). These three units are discernible in Fig. 4. For instance, high *b*-values near the south boundary of region A corresponds to the Sichuan Basin. Thus, the tectonic context of the investigated region is complex and different from typical seismic areas.

Under common conditions, low background seismicity can be found in a fault before a large event (WU and CHEN 2007; HUANG 2008; WEN *et al.* 2009). As shown in Fig. 5, however, the *b*-value of region A

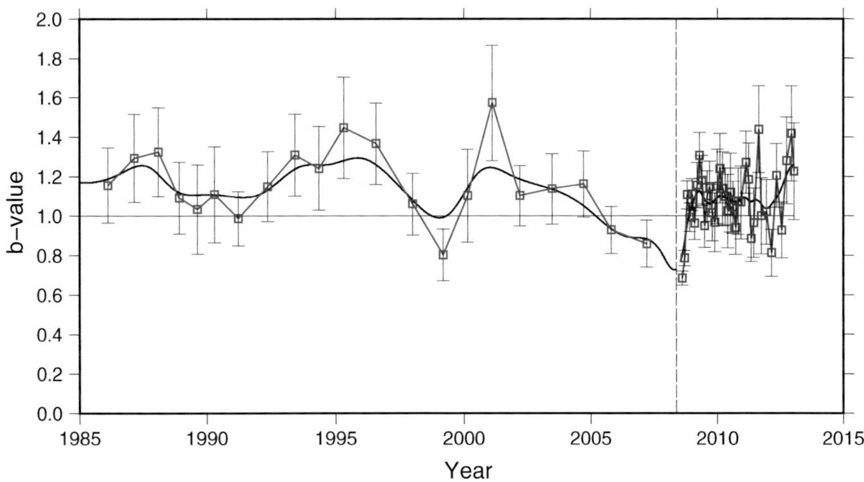

Figure 7
Temporal *b*-value variations in Region C calculated via the MLE method (*red* and *blue line*) and improved Bayesian method (*black line*). The *error bar* of the MLE method is calculated according to Shi and Bolt (1982)

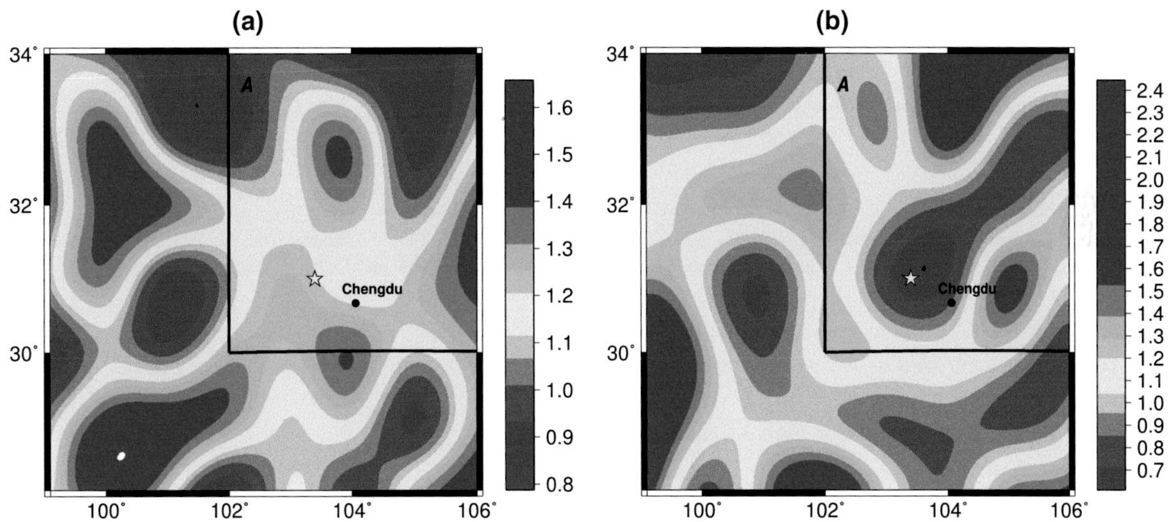

Figure 8
a Spatial *b*-value variation in Southwest China before the 2008 M_s 8 Wenchuan earthquake. **b** Spatial *b*-value variation in Southwest China after the 2008 M_s 8 Wenchuan earthquake

is relatively high, which indicates a large proportion of small events in a frequency–magnitude distribution. This phenomenon can be understood by showing the seismicity before the event. Figure 9 depicts the events in Longmen Shan area before the Wenchuan earthquake. As shown, prior to 2000, relatively smaller events occurred in all three units, especially along the Longmen Shan fault. From 2000 to 2008, there were few events along the fault, indicating the accumulation of stress. Exploring the tectonic mechanism of the Wenchuan earthquake is also helpful for understanding the *b*-value variations in this area (Zhang *et al.* 2009). During the deformation phase before the earthquake, the West Sichuan plateau, as the deformation unit, performs strike slip movement, horizontal compression and vertical uplifting. This strain and deformation results in high seismicity of small and medium events in the West Sichuan plateau (Wen *et al.* 2009; Zhang *et al.* 2009). Thus, when we draw the temporal *b*-value variations

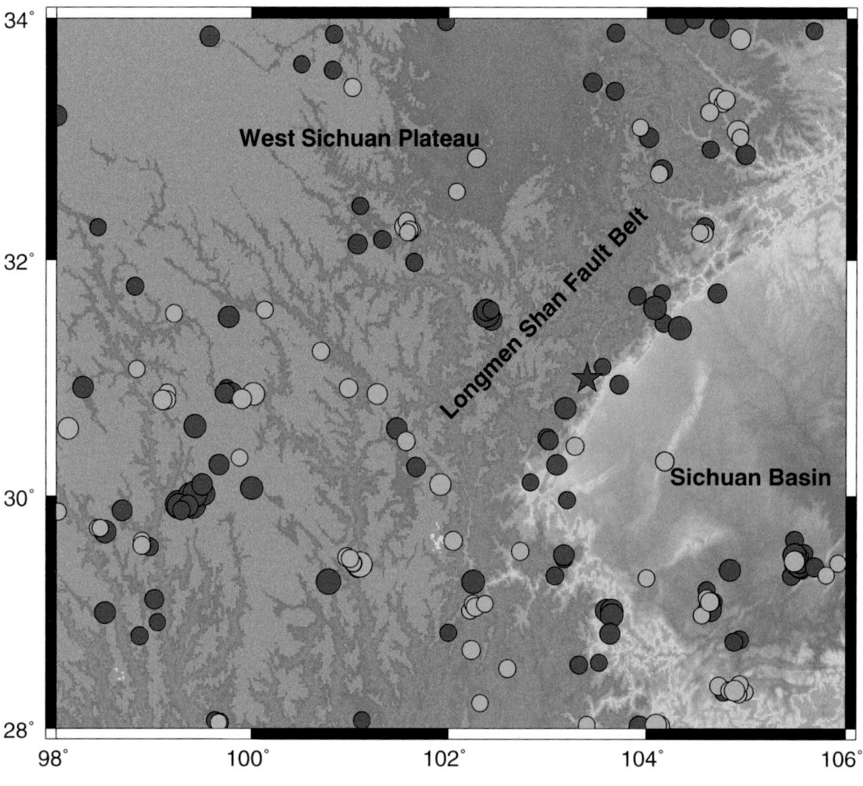

Figure 9
Epicenter distribution of events with $M \geq 4.0$ in Southwest China from January 1985 to May 2008. *Circles filled in red* represent events occurring from January 1985 through December 1999. *Circles filled in green* represent events occurring from January 2000 through May 2008. *All circles* are scaled proportional to magnitude

of region A or region C, as in Fig. 5, the *b*-values are relatively higher compared to average levels from the years 1985 to 2000. Even though the *b*-value was decreasing during 2000–2008, it is higher than the *b*-values in most of the large earthquakes (Ikeya and Huang 1997). Nevertheless, high *b*-values in Fig. 5 cannot be explained as low differential stress, because the accumulative stress in the Longmen Shan fault zone is quite high due to the locked status itself. Deformation and displacement in this unit is pretty slow or even non-existent. So, when the accumulated stress exceeded the fracture strength of the fault, the rupture was triggered and formed a 340 km long fracture belt (Zhang *et al.* 2009). In the last several years before the Wenchuan earthquake, decreasing *b*-values reached their lowest point just before the rupture took place, suggesting a rising trend of stress, consistent with the tectonic mechanism. In general, a low *b*-value accompanied by a significant decrease indicates the cumulative stress in that fault is close to the failure strength (Zheng and Zhou 2014).

6. Conclusion

In this study, we used an improved Bayesian approach with B-spline functions to estimate the spatial and temporal *b*-value variations. The results for the Longmen Shan area demonstrate that *b*-values, in either space or time, are a good indicator for seismicity and stress conditions.

The above analysis indicates that interpretation of *b*-values requires considering background seismicity and the tectonic context. The result of comprehensive analysis can provide valuable information for the estimation of a stress field and the study of earthquake hazard models. We propose that the spatial mapping of *b*-values can aid in identification of high stress areas, and temporal *b*-value variation can be used to assess earthquake risk.

Acknowledgments

The data used in this study are available from the China Earthquake Data Center (CEDC) url: http://data.earthquake.cn/index.html. The study is supported by the Seismological Research Project of China (Grant No. 201208009) and the National Natural Science Foundation of China (Grant No. 41274052). The authors would like to thank Dr. Changsheng Jiang for supplementing the catalog data.

References

Akaike, H (1980). *Likelihood and the Bayes procedure.* Trabajos de estadistica y de investigacion operativa *31*(1):143–166.

Aki, K. (1965) *Maximum likelihood estimate of b in the formula logN = a-bM and its confidence limits.* Bull. Earthq. Res. Inst. *43*:237–239.

Amitrano, D (2003). *Brittle–ductile transition and associated seismicity: experimental and numerical studies and relationship with the b value.* J Geophys Res Solid Earth *108*(B1):2044. doi:10.1029/2001JB000680

Bender, B (1983). *Maximum likelihood estimation of b-values for magnitude grouped data.* Bull. Seism. Soc. Am. *73*:831–851.

Burchfiel, B.C., Royden, L.H., van der Hilst, R.D., *et al.* (2008). *A geological and geophysical context for the Wenchuan earthquake of 12 May 2008, Sichuan, People's Republic of China.* GSA Today. *18*(7): 4–11, doi:10.1130/GSATG18A.1

Chang, C. H., Y. M. Wu, L. Zhao, and F. T. Wu (2007). *Aftershocks of the 1999 Chi-Chi, Taiwan, earthquake: The first hour,* Bull. Seism. Soc. Am., *97*, 1245–1258, doi:10.1785/0120060184.

De Boor, Carl (1978). A practical guide to splines. Springer-Verlag, New York.

Deng, K., et al. (2010). *Evidence that the 2008 M-w 7.9 Wenchuan Earthquake Could Not Have Been Induced by the Zipingpu Reservoir.* Bulletin of the Seismological Society of America *100*(5B): 2805–2814.

Fletcher, R. and M. J. D. Powell (1963). *A rapidly convergent descent method for minimization.* Comput J 6(2):163–168

Good, I. J. (1965). The estimation of probabilities: an essay on modern Bayesian methods. MIT Press, Cambridge, MA.

Good, I. J. and R. A. Gaskins (1971). *Nonparametric roughness penalties for probability densities.* Biometrika. *58*:255–277.

Gutenberg, R., and C. F. Richter (1944). *Frequency of earthquake in California.* Bull Seismol Soc Am *34*(4):185–188

Huang, Q. H. (2008). *Seismicity changes prior to the M(s)8.0 Wenchuan earthquake in Sichuan, China.* Geophysical Research Letters *35*(23).

Ikeya, M. and Q. H. Huang (1997). *Earthquake frequency and moment magnitude relations for mainshocks, foreshocks and aftershocks: Theoretical b values.* Episodes *20*(3):181–184.

Imoto, M. and M. Ishiguro (1986). *A Bayesian approach to the detection of changes in the magnitude–frequency relation of earthquake.* J. Phys. Earth. *34*:441–455.

Ishiguro, M. and Y. Sakamoto (1983). *A Bayesian approach to binary response curve estimation,* Ann. Inst. Statist. Math. *35*:115–137.

Jia, K., *et al.* (2012). *Stress Interactions within the Strong Earthquake Sequence from 2001 to 2010 in the Bayankala Block of Eastern Tibet.* Bulletin of the Seismological Society of America *102*(5): 2157–2164.

Jia, K., *et al.* (2014). *Possibility of the Independence between the 2013 Lushan Earthquake and the 2008 Wenchuan Earthquake on Longmen Shan Fault, Sichuan, China,* Seismological Research Letters, *85*, 60–67, doi:10.1785/0220130115

Narteau, C., *et al.* (2009). *Common dependence on stress for the two fundamental laws of statistical seismology.* Nature, 462, 642–645. doi:10.1038/nature08553.

Ogata, Y., *et al.* (1991). *3-D spatial variation of b -values of magnitude-frequency distribution beneath the Kanto District,* Japan. Geophys J Int *104*(1):135–146

Ogata, Y. and K. Katsura (1988). *Likelihood analysis of spatial in homogeneity for marked point patterns,* Ann. Inst. Statist. Math. *40*:29–39.

Ogata, Y. and K. Katsura (1993). *Analysis if temporal and spatial heterogeneity of magnitude frequency distribution inferred from earthquake catalogs.* Geophys J Int. *113*:727–738

Scholz, C. H. (1968). *The frequency–magnitude relation of microfracturing in rock and its relation to earthquakes.* Bull Seismol Soc Am *58*(1):399–415

Schorlemmer, D., *et al.* (2005). *Variation in earthquake-size distribution across different stress regimes,* Nature, *437*, 539–542.

Shi, Y. and B. A. Bolt (1982). *The standard error of the Magnitude-frequency b value.* Bull. Seism. Soc. Am. *72*:1677–1687.

Urbancic, T. I., *et al.* (1992). *Space–time correlations of b-values with stress release.* Pageoph *139*:449–462

Utsu, T. (1965). *A method for determining the value of b in a formula logN = a − bM showing the magnitude frequency for earthquakes.* Geophys. Bull. Hokkaido Univ. *13*, 99–103.

Wen, X. Z., *et al.* (2009). *The background of historical and modern seismic activities of the occurrence of the 2008 Ms8.0 Wenchuan Sichuan earthquake* Chinese J. Geophys. (in Chinese). *52*(2):444–454.

Wiemer, S. and D. Schorlemmer (2007). *ALM: an asperity-based likelihood model for California.* Seismol Res Lett *78*(1):134–140. doi:10.1785/gssrl.78.1.134

Wiemer, S. and M. Wyss (2000). *Minimum magnitude of complete reporting in earthquake catalogs: examples from Alaska, the Western United States, and Japan.* Bull Seismol Soc Am *90*(4):859–869.

Wiemer, S. and M. Wyss (2002). *Mapping spatial variability of the frequency–magnitude distribution of earthquakes.* Adv Geophys *45*:259–302

Woessner, J. and S. Wiemer (2005). *Assessing the quality of earthquake catalogs: Estimating the magnitude of completeness and its uncertainties.* Bull Seismol Soc Am *95*(2):684–698

Wu, Y. M. and C. C. Chen (2007). *Seismic reversal pattern for the 1999 Chi-Chi, Taiwan, M_w7.6 earthquake.* Tectonophysics. *429*(1–2):12–132.

Zhang, P. Z., Wen X Z, Xu X W, *et al.* (2009). *Tectonic model of the great Wenchuan earthquake of May 12, 2008, Sichuan, China (in Chinese).* Chinese Sci Bull (Chinese Ver). *54*(7): 944–953.

Zheng, Y. and S. Zhou (2014). *The spatiotemporal variation of the b-value and its tectonic implications in North China*. Earthquake Science, *3*.

Zhou, S. Y. (2008). *Seismicity simulation in Western Sichuan of China based on the fault interactions and its implication on the estimation of the regional earthquake risk*. Chinese Journal of Geophysics-Chinese Edition *51*(1):165–174.

Zhuang J., Ogata Y. and Vere-Jones D. (2002). *Stochastic declustering of space–time earthquake occurrences*. Journal of the American Statistical Association, *97*:369–380.

Zhuang J., Ogata Y. and Vere-Jones D. (2004). *Analyzing earthquake clustering features by using stochastic reconstruction*. Journal of Geophysical Research, *109*, No. B5, B05301, doi:10.1029/2003JB002879.

Zhuang J. (2006) *Second-order residual analysis of spatiotemporal point processes and applications in model evaluation*. Journal of the Royal Statistical Society: Series B (Statistical Methodology), *68* (4):635–653. doi:10.1111/j.1467-9868.2006.00559.x.

(Received September 30, 2014, revised January 12, 2015, accepted January 16, 2015, Published online March 1, 2015)

Pure Appl. Geophys. 173 (2016), 97–116

DOI 10.1007/s00024-015-1115-9

Pure and Applied Geophysics

A Revised Earthquake Catalogue for South Iceland

FRANCESCO PANZERA,[1,2] J. DOUGLAS ZECHAR,[3] KRISTÍN S. VOGFJÖRD,[1] and DAVID A. J. EBERHARD[3]

Abstract—In 1991, a new seismic monitoring network named SIL was started in Iceland with a digital seismic system and automatic operation. The system is equipped with software that reports the automatic location and magnitude of earthquakes, usually within 1–2 min of their occurrence. Normally, automatic locations are manually checked and re-estimated with corrected phase picks, but locations are subject to random errors and systematic biases. In this article, we consider the quality of the catalogue and produce a revised catalogue for South Iceland, the area with the highest seismic risk in Iceland. We explore the effects of filtering events using some common recommendations based on network geometry and station spacing and, as an alternative, filtering based on a multivariate analysis that identifies outliers in the hypocentre error distribution. We identify and remove quarry blasts, and we re-estimate the magnitude of many events. This revised catalogue which we consider to be filtered, cleaned, and corrected should be valuable for building future seismicity models and for assessing seismic hazard and risk. We present a comparative seismicity analysis using the original and revised catalogues: we report characteristics of South Iceland seismicity in terms of *b value* and magnitude of completeness. Our work demonstrates the importance of carefully checking an earthquake catalogue before proceeding with seismicity analysis.

1. *Introduction*

In a recent report from the London Workshop on the Future of Statistical Sciences (MADIGAN *et al.* 2014), the responsibilities of a statistician were defined as:

- to design the acquisition of data in a way that minimizes bias and confounding factors and maximizes information content;
- to verify the quality of the data after those are collected; and
- to analyse data in a way that produces insight or information to support decision-making

As statistical seismologists, our work tends to emphasize the third point, with little thought given to the others: we often treat earthquake catalogues as collections of perfect measurements, forgetting (or at least neglecting) that a hypocentre location is the uncertain result of an unsolved inversion problem; and we are almost never involved in planning data collection or seismic network design. With this article, we address the second responsibility, which directly supports the third responsibility, future data analyses. In particular, we consider the quality of the data in the earthquake catalogue generated by the SIL seismic network in Iceland.

The SIL network was planned and developed in the framework of the Nordic SIL project in 1988–1994 (STEFÁNSSON *et al.* 1993; BÖÐVARSSON *et al.* 1996, 1999). The network was originally installed and operated in South Iceland in 1990 to monitor seismicity in the South Iceland Seismic Zone (SISZ), a sinistral transform zone that crosses southern Iceland. In 1993, it was expanded to the seismic zone in northern Iceland, thereby becoming a national network for Iceland, covering the country's two main seismic zones. In the following years, the network was gradually expanded along the rift zones that cross Iceland from southwest to northeast (pink colour in Fig. 1a). In 1994, SIL included 18 stations and by the end of 2013 the seismic network had grown to 68 stations (Fig. 1a).

South Iceland is a rather densely populated farming area with many small towns and several critical infrastructures, such as hydropower plants and associated reservoirs, geothermal power plants,

[1] Icelandic Meteorological Office, Reykjavík, Iceland. E-mail: panzerafrancesco@hotmail.it
[2] Università di Catania, Dipartimento di Scienze Biologiche, Geologiche e Ambientali, Catania, Italy.
[3] Swiss Seismological Service (SED), ETH, Zurich, Switzerland.

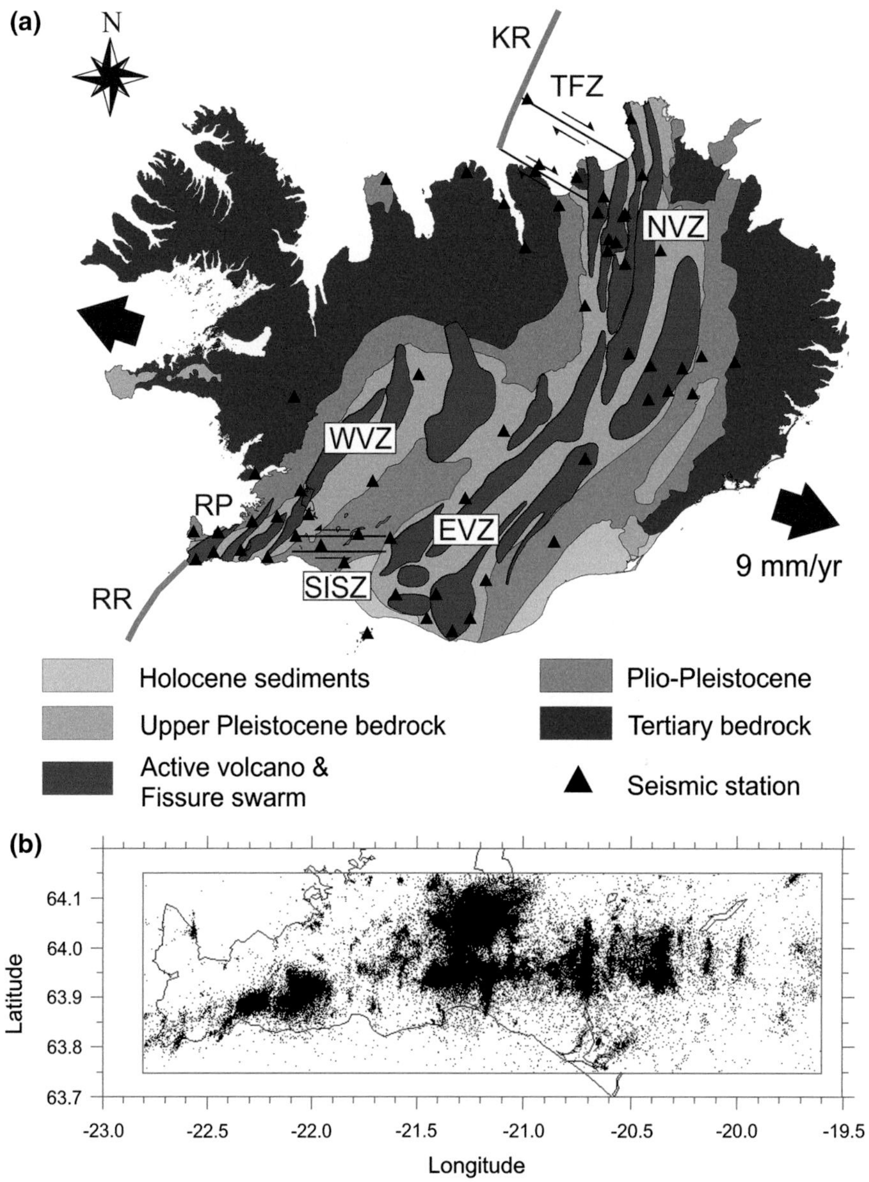

Figure 1
a Schematic representation of the plate boundary in Iceland (modified from Angelier *et al.* 2004). **b** Map of the study area. *Black dots* events in the selected catalogue. *EVZ* Eastern Volcanic Zone, *NVZ* Northern Volcanic Zone, *KR* Kolbeinsey Ridge, *RP* Reykjanes Peninsula, *RR* Reykjanes Ridge, *TFZ* Tjörnes Fracture Zone, *SISZ* South Iceland Seismic Zone, *WVZ* Western Volcanic Zone

industrial plants and transportation infrastructures. More hydropower plants and reservoirs are planned in the area in coming years. Because of the presence of human population and critical infrastructures, southern Iceland has the highest seismic risk in the country. We would like to build models to mitigate the seismic risk; almost any such model one can imagine requires a seismicity catalogue as input and, therefore, we study the SIL catalogue to obtain a first-order understanding of seismicity in this region and to understand the quality of the catalogue. Moreover, since the year 2000 a major earthquake sequence is ongoing in the study area, where three earthquakes of $M_W > 6$ have already occurred and more events of up to moment magnitude (M_W) 7.0 can be expected in the coming years to decades (Einarsson *et al.* 1981;

Decriem *et al.* 2010). The historical seismicity in this region is well known (Einarsson *et al.* 1981; Ambraseys and Sigbjörnsson 2000) and many major faults have been mapped on the surface (Einarsson *et al.* 1981; Bergerat and Angelier 2000; Clifton and Einarsson 2005) and at depth by high-precision locations (HjaltadÓttir *et al.* 2005; Hjaltadóttir 2009).

Quality check of the seismic catalogue is important because results obtained from a contaminated dataset may be misleading (Gulia *et al.* 2012). To make such a check, careful identification of man-made changes should represent the first step in any statistical analysis of seismicity. Such changes, for example, redefining the magnitude scale in a catalogue may introduce errors in statistical analyses of seismicity patterns and they mislead researchers by generating spurious apparent variations in the observed seismicity rate (Habermann 1987; Tormann *et al.* 2010). Our catalogue check emphasizes removing events with very high or even unknown location uncertainty (filtering), removing quarry blasts (cleaning), and re-estimating magnitudes (correcting). After this filtering and corrections to the reported magnitudes, we report a spatio-temporal analysis of South Iceland seismicity based on *b value* and magnitude of completeness. The revised SIL catalogue is available from us upon request, and it should be used as the starting point for future analyses and modelling of South Iceland seismicity.

2. Tectonic Setting

In this section, we give only a brief summary of the regional tectonics (for more details, see Einarsson 1991).

The complex tectonic setting of Iceland is linked to the eastward shifting of the Mid-Atlantic Ridge axis, a consequence of its westward migration away from the hotspot currently situated under the Vatnajökull glacier (Einarsson 2008). As illustrated in Fig. 1a, this movement gives rise in southern Iceland to two sub-parallel volcanic zones. The first is the continuation of the Atlantic Ridge that comes on shore on the Reykjanes Peninsula (RP) and continues as the Western Volcanic Zone (WVZ). The second, the Eastern Volcanic Zone (EVZ), which continues in northern Iceland as the Northern Volcanic Zone (NVZ), is shifted eastward along the SISZ transform zone. Finally, in northern Iceland the rift shifts back westward along the Tjörnes Fracture Zone (TFZ) to connect with the Kolbeinsey Ridge (KR).

The SISZ is a 10–20-km-wide, EW striking sinistral shear zone, but faulting occurs on sub-parallel, NS striking dextral faults (Hackman *et al.* 1990; Roth 2004), as demonstrated by the shape of the damage zones of historical earthquakes (Einarsson *et al.* 1981; Ambraseys and Sigbjörnsson 2000), their mapped surface traces (Einarsson 1991; Clifton and Einarsson 2005) and subsurface fault mapping by high-precision earthquake relocations (Hjaltadóttir 2009).

The RP is a highly oblique spreading segment of the Mid-Atlantic Ridge, oriented about 30° from the direction of absolute plate motion (Clifton and Kattenhorn 2006). There are four distinct volcanic fissure swarms on the peninsula (Jakobsson *et al.* 1978), with an average strike of 40°N. Similar to the SISZ, large earthquakes on Reykjanes peninsula occur on NS striking faults that intersect the volcanic fissure swarms (Árnadóttir *et al.* 2004; Keiding *et al.* 2008; Clifton *et al* 2003; Antonioli *et al.* 2006).

3. Data Collection and Magnitude Estimation

The initial earthquake catalogue considered for this study spans the period 1991–2013 and contains 205,016 earthquakes with depth ranging between 0 and 50 km and with $M_L \geq -2.0$. The selected area includes both the SISZ and the RP and is within the limits 63.75° to 64.15° in latitude and −22.8° to −19.6° in longitude (Fig. 1b). Originally, the incoming seismic data were pre-processed by a computer at the site of each seismic station and transients were detected and defined by onset time, amplitude, and duration. Apparent velocity, azimuth, and spectral parameters were calculated, and this information was packed into short messages and sent to the data centre at the Icelandic Meteorological Office in Reykjavík, where an automatic phase association process defined events and sent requests for waveform data to the stations (Böðvarsson *et al.* 1996). In 2013–2014, transmission of continuous waveforms was initiated for all the sites and the analysis was moved to the data centre at IMO. Automatic locations and magnitudes of earthquakes are usually available within 1–2 min of their occurrence. These are manually reviewed by analysts at the SIL data centre. This includes readjusting or adding phase arrivals where necessary, relocating events and recalculating magnitudes. This process is a routine operation and special evaluation of larger events is not part of the analysis.

The SIL system uses two methods to estimate each earthquake's magnitude. The first is based on an empirical local magnitude relationship:

$$M_{\mathrm{L}} = \log_{10}(A) + 2.1\ \log_{10}(D) + 4.8 \quad (1)$$

where A is the maximum velocity amplitude in m/s of high-pass filtered waveforms with a corner frequency (f) at 1.5 Hz and scaled to the response of Lennartz 1.0 Hz sensor and Nanometrics RD3 digitizer, and D is the epicentral distance in km (Gudmundsson *et al.* 2006).

The other magnitude scale is a "local" moment magnitude scale, M_{LW} that was originally constructed by Slunga *et al.* (1984) to agree with local magnitude scales in Sweden:

$$M_{\mathrm{LW}} = \log_{10}(m_0) - 10, \quad \text{for } M_{\mathrm{LW}} \le 2.0 \quad (2)$$

where m_0 is the seismic moment in Nm (Rögnvaldsson and Slunga 1993). For magnitudes above 2.0, the formula is slightly modified so that the slope decreases with increasing magnitude (Pétursson and Vogfjörd 2009). In particular, considering the factor:

$$m = \log_{10}(m_0) - 10 \quad (3)$$

the M_{LW} formula becomes:

$$M_{\mathrm{LW}} = m \quad \text{for } m \le 2.00 \quad (4a)$$

$$M_{\mathrm{LW}} = 2.0 + (m - 2.00) \cdot 0.9 \quad \text{for } 2.00 < m \le 3.11 \quad (4b)$$

$$M_{\mathrm{LW}} = 3.0 + (m - 0.89) \cdot 0.8 \quad \text{for } 3.11 < m \le 5.11 \quad (4c)$$

$$M_{\mathrm{LW}} = 4.6 + (m + 1.1) \cdot 0.7 \quad \text{for } 5.11 < m \le 6.25 \quad (4d)$$

$$M_{\mathrm{LW}} = 5.4 + (m + 2.25) \cdot 0.5 \quad \text{for } 6.25 < m \le 7.25 \quad (4e)$$

$$M_{\mathrm{LW}} = 5.9 + (m + 3.25) \cdot 0.4 \quad \text{for } 7.25 < m \le 9.25 \quad (4f)$$

$$M_{\mathrm{LW}} = 6.3 + (m + 5.25) \cdot 0.35 \quad \text{for } 9.25 < m \quad (4g)$$

4. *Verifying the Quality of the Data and Revising the Catalogue*

In this section, we consider the quality of the SIL catalogue and revise it in three stages: we remove events with very large or unknown location uncertainty (filtering); we identify and delete quarry blasts that were misidentified as earthquakes (cleaning); and we re-estimate magnitudes (correcting). The procedure is illustrated in Fig. 2, including the number of events affected at each stage, and the three stages are described in the following three subsections.

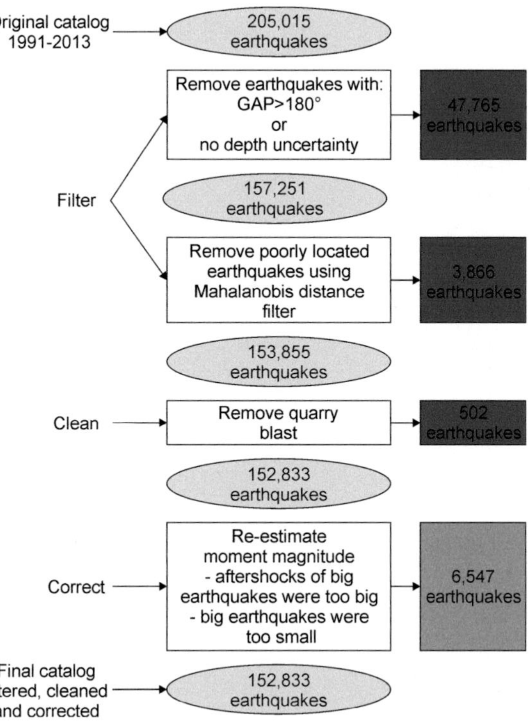

Figure 2

Flowchart that summarizes the filtering, cleaning, and correcting used to produce the revised catalogue. *Boxes in the right column* the number of events affected by each step

4.1. *Earthquake Location Precision: Filtering the Catalogue*

Despite all the checks in the SIL system, often earthquakes stored in the catalogue are not well located, have large azimuthal GAP (the largest angle between any two stations that recorded the earthquake), and large uncertainties in latitude, longitude, and depth (blue histograms in Fig. 3). Workers have suggested many methods to filter a catalogue so that those events that are not well located are removed. The most popular of these are the network criteria methods (NCM) discussed by GOMBERG *et al.* (1990), BONDAR *et al.* (2004), and HUSEN and HARDEBECK (2010). These authors proposed to keep only the events with GAP smaller than 180° and with at least 8 arrival times for *P* and *S* waves, of which at least one is an *S* wave arrival time. We applied this filter to the SIL catalogue (red histograms in Fig. 3), but the SIL network is less dense than regional networks elsewhere and, therefore, these criteria seem to be too restrictive, especially for microseismicity.

Uncertainties in earthquake locations arise from (unknown) errors in seismic arrival times, network geometry, signal-to-noise ratio, dominant frequency of the arriving phase, and the velocity model used for the location, so any catalogue should include uncertainty estimates for each earthquake's latitude, longitude, and depth. We used these estimates directly as an alternative to filter the catalogue. As in the NCM, we removed all the earthquakes with GAP higher than 180°. Noting that a histogram of the depth distribution (Fig. 4; left panel) shows spikes at exact depths 0, 1, 3, and 5 km, we also removed all earthquakes without any estimate of depth error, i.e. $D_E = 0$. These represent earthquake locations for which the location program did not converge to a solution, and thus the SIL analyst fixed the depth at 0, 1.3, or 5 km.

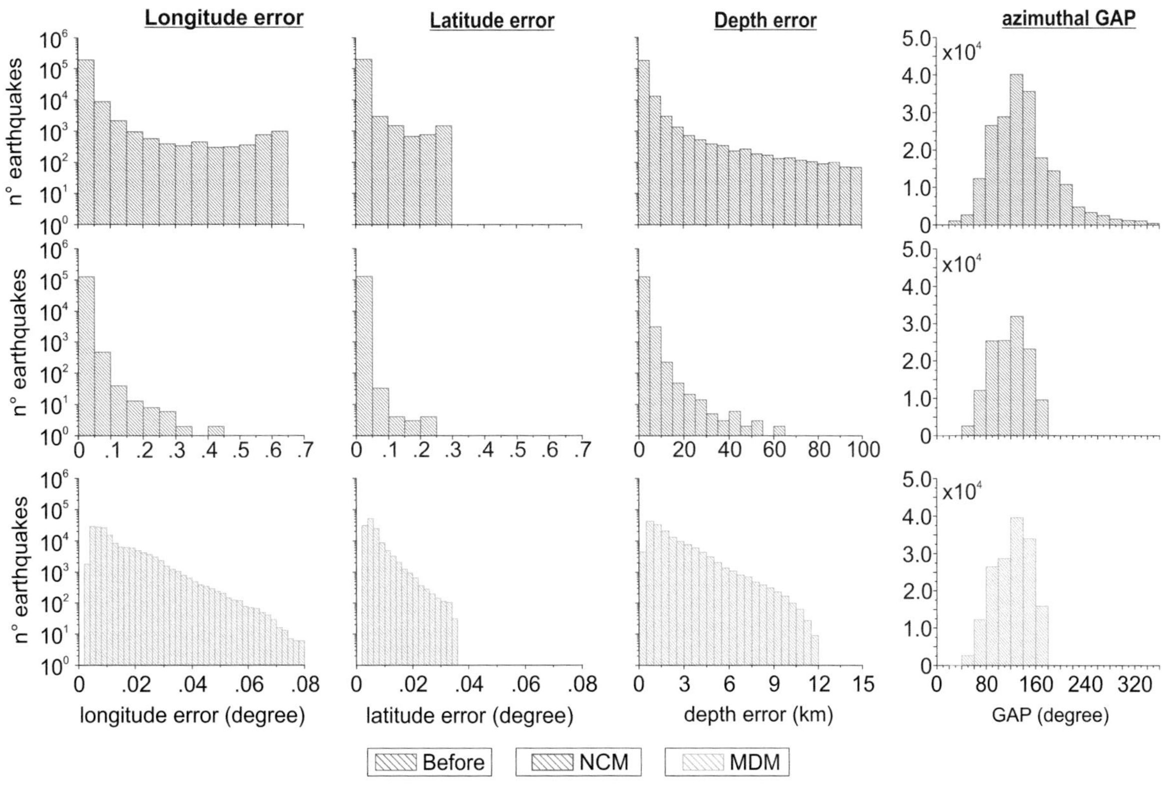

Figure 3

Histograms showing the distribution of longitude, latitude, and depth errors as well as azimuthal gap before (*blue*) and after the filtering by the network criteria method (NCM, *red*) and the Mahalanobis distance method (MDM, *green*). Note that for the *first 3 columns* the scale is semi-logarithmic, and in the *third row* a different scale for error histograms is used

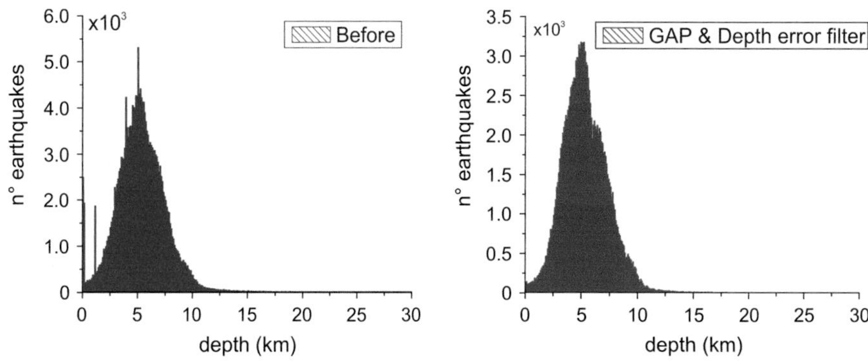

Figure 4
Distribution in depth before filtering (*blue*) and after removing GAP > 180° and $D_E = 0$ (*red*)

After removing these events with large GAP and artificially fixed depths, we applied a technique commonly used in multivariate analysis to identify outliers: filtering based on Mahalanobis distance (e.g. Everitt 2005, Section 1.4). We treat the uncertainties in latitude, longitude, and depth as multivariate data and compute the Mahalanobis distance for each earthquake; this is the distance of a point $x(x_1; x_2; x_3)$ from the population mean $\mu(\mu_1; \mu_2; \mu_3)$:

$$d_m(\vec{x}) = \sqrt{(\vec{x} - \vec{\mu})^T S^{-1} (\vec{x} - \vec{\mu})} \quad (5)$$

where S^{-1} is the inverse covariance matrix of the independent variables. Then, if the data distribution is multivariate normal, the squared Mahalanobis distances should follow a Chi-squared (χ^2) distribution with p degrees of freedom satisfying:

$$(\vec{x} - \vec{\mu})^T S^{-1} (\vec{x} - \vec{\mu}) \leq \chi_p^2(\alpha) \quad (6)$$

where p is the dimension of the data (in our case, 3) and α is the probability of an equal or larger value. We take $\alpha = 0.05$, and the corresponding χ^2 critical value is 7.82. Consequently any earthquake that has a squared Mahalanobis distances in the tail of the corresponding χ^2 distribution (i.e. with value higher than 7.82) is considered an outlier and removed. We call this the MDM (Mahalanobis Distance Method).

The NCM did reduce the mean error in the catalogue (Table 1), but many earthquakes with large uncertainties in latitude, longitude and depth remained (see the second row of Fig. 3 and note the different scale for the second and third rows). On the other hand, the MDM results in a catalogue of earthquakes with a much narrower distribution of location uncertainties. Moreover, the MDM-filtered catalogue contains 153,385 earthquakes compared to the 118,823 using NCM. Because it contains more, and more precise, information, we proceed with the MDM-filtered catalogue. The MDM method is appropriate for an area like South Iceland, where the network is optimized to record microearthquakes. These events are well located despite not being recorded by several seismic stations (e.g. 4–5 seismic stations). The MDM filter could be applied also in areas that are not optimized for microearthquakes detection because it filters based on the uncertainties directly, whereas the NCM filter uses network criteria as a proxy.

4.2. *Analysis of Explosion Contamination–Cleaning the Catalogue*

For analysis of seismicity, the SIL catalogue should only contain earthquakes, but sometimes quarry blasts are erroneously identified as earthquakes. These explosion events have low magnitudes and produce an enrichment in the number of small earthquakes; such an increase of microseismicity may be misinterpreted as a change in the natural phenomena (Gulia *et al.* 2012).

A histogram of the number of seismic events during each hour of the day is one way to identify the presence of blasts. In a contaminated catalogue, the plot is characterized by a peak during daytime hours (Wiemer and Baer 2000). As noted by Gulia *et al.* (2012), this behaviour is due to the fact that during

Table 1

Range of GAP, and mean and standard deviation of the latitude, longitude, and depth errors during the filtering process

	Before filtering	After removing GAP > 180° and $D_E = 0$	NCM	MDM
GAP (degree)	37°–360°	37°–179.9°	37°–179.9°	37°–179.9°
Longitude (degree)	0.027° ± 0.070°	0.014° ± 0.024°	0.009 ± 0.008	0.012° ± 0.006°
Latitude(degree)	0.013° ± 0.031°	0.007° ± 0.011°	0.005° ± 0.003°	0.005° ± 0.004°
Depth (km)	2.956 ± 6.680	2.247 ± 3.509	1.506 ± 1.404	1.904 ± 1.531
Eq. number	205,015	157,251	118,823	153,385

daytime ambient noise interferes with the detection of earthquakes and a decrease in the number of detected events is generally observed, whereas the quarry-rich areas show the opposite trend. The SIL catalogue, after filtering by Mahalanobis distance, seems to be only slightly contaminated by the presence of explosions. A small peak observed in the activity around noon is likely due to less human noise during lunch hour, when stations are not affected by traffic and industrial activity (Fig. 5a). Moreover, contamination by explosions is found when areas around the cities are analysed in more detail.

To identify blasts, in Fig. 5b we present a map of the ratio of the number of daytime events to the number of nighttime events using the Wiemer and Baer (2000) algorithm. This normalized ratio, Rq, is defined as:

$$Rq = \frac{NdLn}{NnLd} \tag{7}$$

where Nd is the total number of events in the daytime, Nn in the nighttime, Ld is the number of hours in the daytime period and Ln in the nighttime period. The ratio is calculated using a regularly spaced grid covering the studied area. We used a spatial grid of 0.03° in latitude and 0.07° in longitude that gives approximately equal distance in km, and sampled the 100 nearest events to each node. The quarry contamination map is then obtained using the nearest neighbour gridding method. It is important to highlight that interpolation algorithms may show edge effects, especially when the data are not homogenously distributed. To reduce edge effects, we only identify anomalous behaviour where earthquakes have occurred (see Fig. 1b for earthquakes location).

In the obtained map, as suggested by Gulia *et al.* (2012), areas with $Rq \geq 1.5$ are identified as areas contaminated by explosions. In particular, using interactively selected polygons all the areas with Rq higher than 1.5 were selected and removed. The resulting map shows that the study area is only contaminated by explosions along the Reykjanes peninsula coast, at Helguvík, Hafjarfjördur, east Reykjavík, and in the south near Thorlákshöfn. This result is also supported by the hourly distribution of events for such areas (Fig. 5b).

This analysis led to removal of 502 events that were recognized as explosions, resulting in a catalogue of 152,883 earthquakes.

4.3. Problems with Magnitudes: Correcting the Catalogue

In the left panel of Fig. 6 are the M_L versus M_{LW} magnitude values from the filtered and cleaned catalogue. Many events fall above the 1:1 line, indicating that M_{LW} gives magnitude values higher than those obtained using M_L. The difference between the two magnitude scales seem to increase for larger earthquakes. This behaviour is probably due to the fact that M_L is a better estimate for small earthquakes, but for large earthquakes it underestimates the magnitude because short-period seismometers and high-pass filtering ($f > 1.5$ Hz) are used. We also note that the theoretical M_{LW} values derived from Slunga *et al.* (1984) for the magnitude range $1.8 < M_{LW} < 6.6$, are larger than global M_W, while outside this range M_{LW} is smaller (Fig. 6; right panel).

In addition to these systematic discrepancies—i.e. the M_L underestimation of large events and the non-linear behaviour of the M_{LW} scale—there are other problems in the SIL magnitude estimates. In particular the two main concerns are: (1) the underestimation of the m_0 of large earthquakes in the routine processing performed to obtain M_{LW} in the SIL catalogue (Rögnvaldsson and Slunga 1993)

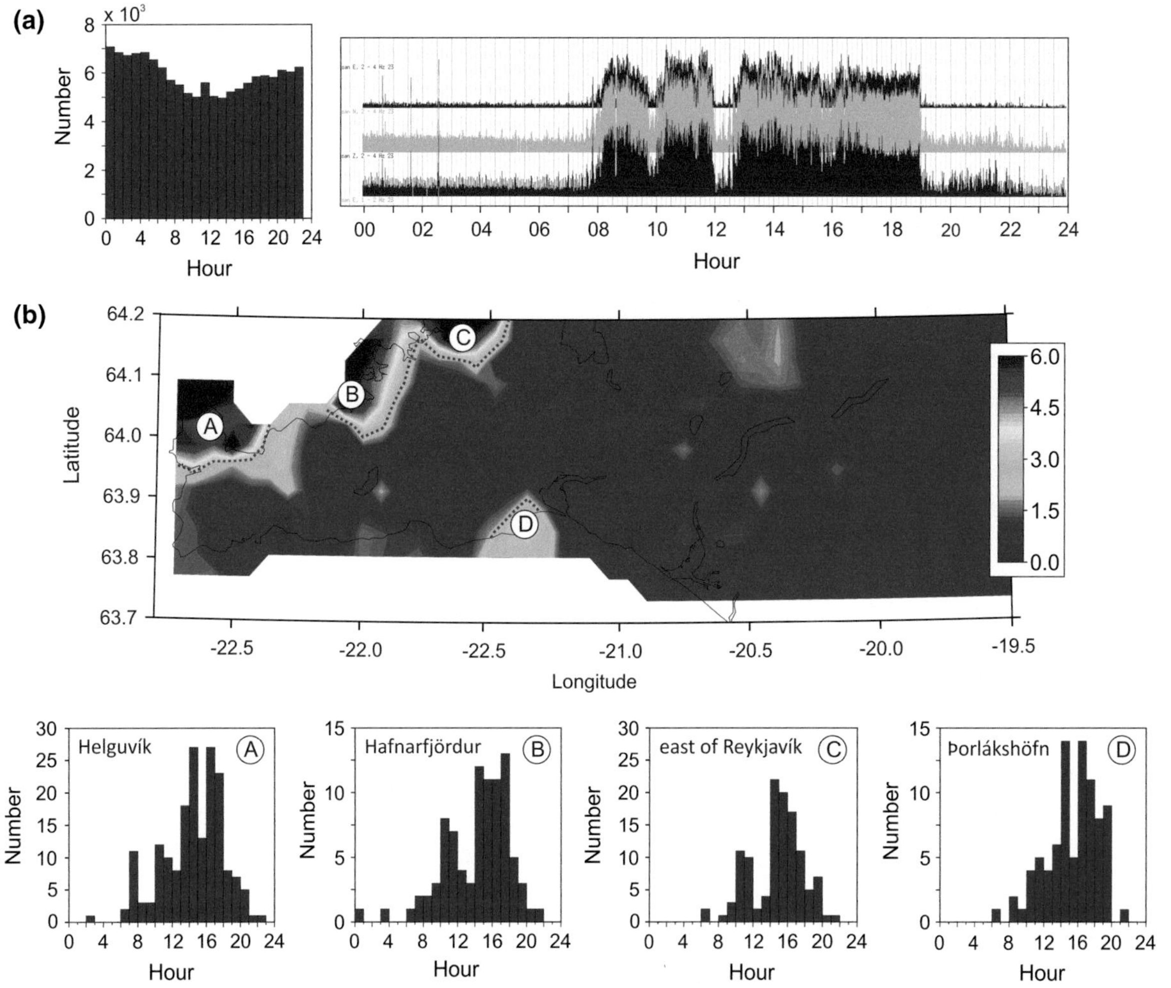

Figure 5
a *Histogram* of the number of events per hour of the day for the catalogue after the application of the Mahalanobis distance filter and example of noise amplitude versus time at the station *san*, which is located near a quarry. **b** Map of the ratio between daily and nightly events and histograms of the number of events per hour of the day for suspected areas

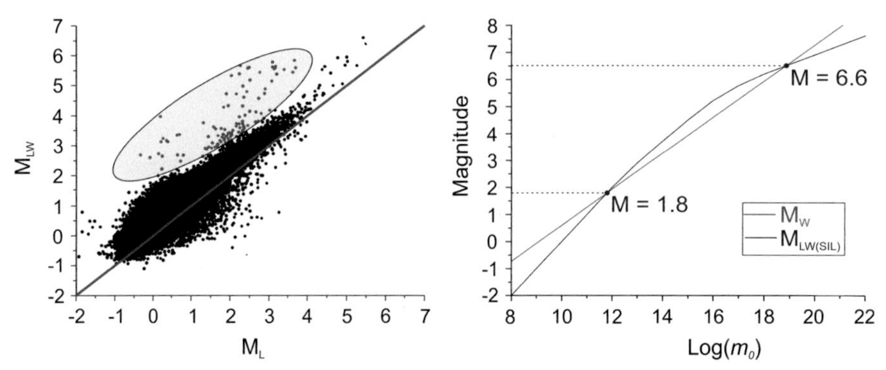

Figure 6
Left panel M_L versus M_{LW} scatter plot in which the *red line* is a line with slope one and *yellow ellipse* contains outliers due to erroneous magnitude estimates. *Right panel* the comparison between the M_{LW} and M_W magnitudes

and (2) the overestimation of magnitude assigned to aftershocks of big earthquakes, due to contamination by the mainshocks.

The main reason for the underestimated m_O is that for large earthquakes, the routine analysis window around the P and S waves is too short to contain all of the source time function, and the frequency band within which the corner frequency of the source spectrum is searched does not extend to low enough frequencies. Additionally, most of the SIL seismic stations have short-period seismometers (0.2 and 1 Hz) that do not extend to sufficiently low frequencies to allow for undisturbed calculation of seismic moment for the largest earthquakes. The underestimation of m_O becomes evident for M_{LW} greater than 3.0, which corresponds to an M_L of about 2.5.

Table 2 shows that the moment magnitude of larger earthquakes obtained using m_O in the SIL catalogue ($M_{W\text{-}SIL}$) is severely underestimated compared to the magnitudes ($M_{W\text{-}GCMT}$) reported in the Global Centroid Moment Tensor catalogue (GCMT; Dziewonski *et al.* 1981; Ekström *et al.* 2012). Consequently it is not possible to use the seismic moment (m_O), which is used for estimating M_{LW}, to directly estimate M_W. Another evidence of the problem concerning m_O estimation is shown in the left panel of Fig. 6 (yellow ellipse), where we highlight outliers characterized by surprisingly large values of M_{LW} given their M_L. These are erroneous M_{LW} values assigned to aftershocks of big earthquakes, and the errors are caused by contamination of the waveform from the mainshock. Usually, the overestimation of aftershock magnitudes occurs in the first few minutes after the mainshock. We suspect that the low-frequency component of the mainshock coda waves make the corner frequency appear to be lower than it should be.

For all these reasons the magnitude values in the catalogue must be corrected before any statistical analysis. Although the M_L estimate is influenced by the high-pass filtering, we prefer to use these estimates to find a method for magnitude correction, because they are more stable than the M_{LW} estimates. Moreover, there are 902 earthquakes in the catalogue with M_L higher than 2.5 for which it is necessary to make the correction, and it would be very time-consuming to re-compute the seismic moment for all of these earthquakes.

To correct magnitudes, we used a technique based on peak ground velocity magnitude (M_{PGV}) estimates using an attenuation relationship for South Iceland (Pétursson and Vogfjörd 2009). This relationship uses an empirical form for the decay of peak ground velocity and acceleration (PGV and PGA) with distance based on waveforms from 45 events in southern Iceland in the magnitude range $3.0 \leq M_W \leq 6.5$ and for distances in the range 5–300 km. The relations were scaled with M_W and include a non-linear, near-source term. A log-linear approximation to the relation, valid outside the near-source region, is a robust real-time measure of magnitude for events of $M_w \geq 3.0$. The log-linear part of the equation for velocity is:

$$\log_{10}(\mathrm{PGV}) = -1.63\log_{10}(r) + M - 4.88 \qquad (8)$$

where r is the epicentral distance in km and M is the moment magnitude. For our study we used Eq. (8) to estimate M_{PGV} through the reverse formula:

Table 2

List of earthquakes in South Iceland for which the GCMT reports a moment magnitude. In the table $M_{W\text{-}SIL}$ are the underestimated moment magnitude values obtained by using the seismic moment in the SIL catalogue, named m_0(SIL), whereas $M_{W\text{-}GCMT}$ and m_0(GCMT) are corresponding values in Global Centroid Moment Tensor catalogue

Dd/mm/yy	Origin time	$M_{W\text{-}SIL}$	$M_{W\text{-}GCMT}$	M_0(SIL) (dyne·cm)	M_0(GCMT) (dyne·cm)
04/06/98	21:36:53.81	4.4	5.4	4.96×10^{22}	1.56×10^{24}
13/11/98	10:38:34.42	4.5	5.1	5.93×10^{22}	5.43×10^{23}
17/06/00	15:40:40.98	6.2	6.5	2.36×10^{23}	7.05×10^{25}
21/06/00	00:51:46.99	6.6	6.4	1.03×10^{26}	5.44×10^{25}
23/08/03	02:00:11.79	4.4	5.1	4.90×10^{22}	5.96×10^{23}
29/05/08	15:45:58.91	5.1	6.3	6.16×10^{23}	3.38×10^{25}
29/05/08	21:33:49.11	4.3	4.8	3.70×10^{22}	2.21×10^{23}

$$M_{PGV} = \log_{10}(PGV) + 1.63\log_{10}(r) + 4.88 \quad (9)$$

This equation is also used in the Icelandic real-time ShakeMap system (Vogfjörd *et al.* 2012; Icelandic Meteorological Office website: http://hraun.vedur.is/ja/alert/shake/). This type of approach was used with good results by several authors in others part of the world (e.g. Lin and Wu 2010; Eshaghi *et al.* 2013).

To calibrate a conversion between M_L with M_{PGV}, we used the 45 events used by Pétursson and Vogfjörd (2009) and additional 59 events taken from the Icelandic ShakeMap website (all events are listed in Table 3).

We fit a generalized orthogonal regression (GOR) model to the data in Table 3 to describe the relationship between M_{PGV} and M_L (Fig. 7, left panel). We used GOR rather than standard least-squares because the two magnitude scales have similar uncertainties (Castellaro *et al.* 2006). The resulting GOR relationship is:

$$M_{PGV} = (1.215 \pm 0.031)M_L + (-0.373 \pm 0.110), \quad M_L \geq 2.5 \quad (10)$$

We used Eq. (10) to correct estimates of $M_{PGV} \approx M_W$ for all the earthquakes with M_L higher than 2.5, since those are events for which the seismic moment is thought to be underestimated with respect to the GCMT catalogue. For earthquakes with M_L lower than 2.5 we used the SIL seismic moment to estimate M_W. After these corrections, we performed comparisons among the derived M_{PGV}, $M_{W\text{-}SIL}$, $M_{W\text{-}GCMT}$ and body wave magnitude (m_b) from the International Seismological Centre (ISC) catalogue (Storchak *et al.* 2013). In Fig. 8a (left panel) we present M_{PGV} versus $M_{W\text{-}GCMT}$ for 14 earthquakes in South Iceland (Table 4) reported in the GCMT catalogue. We noted that the relationship is nearly 1:1 between M_{PGV} and $M_{W\text{-}GCMT}$. Since 14 events are too few and the M_W magnitude range is short, for a better comparison we used also m_b for 38 events from the ISC catalogue with magnitudes in the range 3.5–6.0 (Table 4). We fitted a GOR model using M_{PGV} and m_b (Fig. 8a, right panel) and then we compared the obtained model with that proposed by Wason *et al.* (2012). Although the number of events used and the range of magnitudes considered by Wason *et al.* (2012) are different from those of our analysis, there seems to be a good agreement between the two models (Fig. 8a, right panel). On the contrary, as already discussed concerning the problems with the SIL magnitude estimates, Fig. 8b highlights a poor correlation among the $M_{W\text{-}SIL}$, $M_{W\text{-}GCMT}$ and m_b listed in Table 4.

This procedure allowed us to overcome the problem of underestimated seismic moment for events with $M_L \geq 2.5$. However, below M_L 2.5 there are several earthquakes with poorly estimated local moment magnitude (yellow ellipse in the left panel of Fig. 6), and we also seek to correct these. To correct the erroneous moment magnitudes of earthquakes with $M_L < 2.5$, we divided the earthquakes by M_L into bins of $\pm$ 0.1 (see Table 5) and for each bin we calculated the mean value and two standard deviations (2σ) of M_W (see Table 5), obtained from the SIL seismic moment. For each bin any earthquake having an M_W more than 2σ from the mean were considered outliers. For these outliers, we used a linear relationship between the M_L central values and M_W mean values in Table 5 to estimate a corrected M_W:

$$M_W = 0.7654 + 0.6944\,M_L, \quad \sigma = 0.0326, \quad 0.0 \leq M_L < 2.5 \quad (11)$$

Following this procedure, we corrected the moment magnitude estimates of 5645 earthquakes.

After these magnitude re-estimations, we have a revised catalogue that has been filtered, cleaned, and corrected. Again, we refer you to Fig. 2 for a graphical summary. In the next section, we describe an analysis of the revised catalogue, which contains 152,833 earthquakes.

5. *Analysis of Seismicity*

5.1. *Magnitude of Completeness and b value*

Estimating the magnitude of completeness, M_C, is important for analysing earthquake rate changes, mapping seismicity parameters, forecasting seismicity, and assessing seismic hazard. It is defined as the magnitude above which all events in a given space

Table 3

List of M_L and M_{PGV} used to obtain the conversion law in Eq. (10). M_{PGV} of 45 events was taken from Pétursson and Vogfjörd (2009), while the additional 59 events are computed using Eq. (8)

Dd/mm/yyyy	Hh:mm:ss	Lat	Lon	Depth	M_L	M_{PGV}
20/11/1992	10:28:33	63.93	−21.98	6.8	3.8	4.3
27/12/1992	12:23:22	64.02	−21.18	0.8	4.2	4.6
19/09/1993	10:00:31	63.89	−22.26	5.3	3.0	3.6
17/08/1994	06:29:30	64.06	−21.19	3.0	3.6	4.0
19/08/1994	19:18:42	64.03	−21.25	1.5	3.9	4.3
20/08/1994	16:40:26	64.04	−21.24	1.7	3.9	4.4
30/04/1995	00:57:59	64.07	−21.16	3.4	3.7	4.2
23/07/1995	09:28:55	64.06	−21.32	5.2	3.4	3.6
20/08/1995	16:57:04	64.07	−21.22	2.1	3.1	3.9
27/12/1995	04:26:07	64.07	−21.39	0.0	3.1	3.7
14/03/1996	05:34:57	64.04	−21.21	4.1	3.6	4.0
14/10/1996	20:59:58	64.05	−21.05	4.2	3.8	4.4
23/02/1997	00:35:48	63.93	−22.08	4.6	3.6	4.0
23/02/1997	08:45:03	63.94	−22.08	4.3	3.6	4.1
12/04/1997	23:04:44	64.07	−21.24	3.7	3.8	4.3
24/08/1997	03:20:02	64.05	−21.26	4.6	3.7	4.0
24/08/1997	03:04:22	64.03	−21.26	5.4	4.4	5.0
29/12/1997	10:37:31	64.02	−21.18	5.2	3.4	3.8
03/06/1998	06:47:42	64.06	−21.26	4.1	3.6	4.0
03/06/1998	18:46:09	64.07	−21.21	3.5	3.0	3.6
03/06/1998	23:23:49	64.07	−21.17	4.3	3.4	3.9
04/06/1998	12:23:27	64.04	−21.31	4.2	3.5	3.9
04/06/1998	19:04:45	64.07	−21.30	4.0	4.0	4.6
04/06/1998	21:36:54	64.04	−21.29	5.9	4.7	5.5
04/06/1998	22:04:40	64.05	−21.29	4.8	3.3	3.9
04/06/1998	22:59:57	63.99	−21.31	3.1	4.0	4.8
13/11/1998	10:38:34	63.96	−21.35	5.0	4.6	5.2
13/11/1998	10:46:31	63.96	−21.38	9.5	4.1	4.3
14/11/1998	04:36:40	63.94	−21.41	5.3	3.6	4.2
14/11/1998	04:21:14	63.94	−21.39	4.2	4.0	4.2
14/11/1998	14:24:07	63.96	−21.24	4.4	4.4	5.0
30/11/1998	10:41:16	63.93	−22.00	5.6	3.6	4.1
25/05/1999	13:19:40	64.06	−21.15	5.3	4.0	4.3
25/05/1999	18:03:05	64.05	−21.18	5.6	3.2	3.5
20/07/1999	06:04:02	63.90	−22.03	5.4	2.7	3.1
27/09/1999	16:01:15	63.97	−20.79	6.0	4.2	4.5
28/09/1999	21:50:20	63.98	−20.79	4.9	3.8	4.1
18/04/2000	19:49:08	64.06	−21.32	3.9	3.0	3.4
17/06/2000	15:40:41	63.97	−20.37	6.4	5.5	6.4
17/06/2000	15:42:51	63.94	−20.46	6.0	5.0	5.7
17/06/2000	16:24:04	64.06	−21.31	4.1	3.6	4.1
21/06/2000	00:51:47	63.97	−20.71	5.0	5.4	6.5
23/08/2003	02:00:12	63.91	−22.09	3.7	4.3	5.0
07/01/2004	23:25:25	64.02	−21.22	6.2	3.7	4.0
06/03/2006	14:31:55	63.92	−21.92	8.1	4.2	4.5
20/11/2007	18:48:54	63.95	−20.99	1.8	3.1	3.5
29/05/2008	15:46:06	63.97	−21.06	5.1	5.3	6.3
03/06/2008	19:49:00	63.92	−21.18	4.2	3.1	3.4
04/11/2008	17:47:27	63.86	−22.44	5.0	3.6	3.8
06/12/2008	14:16:34	63.97	−21.42	7.2	3.4	3.9
29/04/2009	02:57:56	63.95	−21.26	7.8	3.8	3.9
08/05/2009	19:27:11	63.94	−21.41	7.6	2.7	3.2
29/05/2009	21:33:50	63.89	−22.34	6.4	4.2	4.7
30/05/2009	02:00:04	63.88	−22.33	5.3	3.0	3.5

Table 3 *continued*

Dd/mm/yyyy	Hh:mm:ss	Lat	Lon	Depth	M_L	M_{PGV}
30/05/2009	02:05:07	64.02	−22.36	6.2	2.8	3.0
30/05/2009	07:47:43	63.92	−22.27	7.7	2.8	3.0
30/05/2009	08:33:08	63.92	−22.27	6.1	2.9	3.2
30/05/2009	13:35:23	63.91	−22.27	6.6	4.0	4.3
30/05/2009	15:13:14	63.94	−22.25	6.7	2.8	3.0
30/05/2009	16:11:16	63.91	−22.32	5.4	2.8	3.1
30/05/2009	17:05:39	63.91	−22.25	7.5	3.7	3.9
31/05/2009	07:45:14	63.90	−22.32	7.4	2.8	3.0
19/06/2009	18:13:20	63.88	−22.09	4.5	3.9	4.2
19/06/2009	20:37:13	63.89	−22.09	4.9	3.8	4.1
25/06/2009	17:20:17	63.91	−22.01	5.6	3.7	4.0
25/06/2009	19:20:33	63.92	−21.99	6.5	3.1	3.3
31/07/2009	23:46:22	63.92	−22.07	5.0	2.7	3.0
19/08/2009	13:42:16	63.91	−21.99	6.2	3.0	3.2
08/02/2010	22:08:25	63.86	−22.76	10.9	2.8	3.0
10/02/2010	16:48:06	63.72	−22.98	9.5	2.8	3.0
21/05/2010	18:56:56	64.07	−20.55	6.1	3.0	3.0
31/05/2010	07:33:25	63.95	−21.39	7.4	3.0	3.1
13/12/2010	12:58:09	63.89	−22.04	5.0	2.7	3.1
25/12/2010	17:03:54	63.93	−22.03	5.6	3.0	3.1
27/02/2011	05:20:33	63.94	−22.04	4.4	2.7	3.1
27/02/2011	05:46:03	63.92	−22.03	4.1	3.0	3.3
27/02/2011	09:05:59	63.92	−22.02	4.9	3.6	4.0
27/02/2011	09:11:13	63.90	−22.04	4.7	2.8	3.1
27/02/2011	09:16:11	63.89	−22.02	4.4	2.8	3.1
27/02/2011	09:48:42	63.93	−22.04	4.6	3.0	3.0
27/02/2011	17:27:36	63.91	−22.03	4.7	3.8	4.2
02/03/2011	08:32:18	63.93	−22.05	4.4	2.8	3.1
02/03/2011	17:56:53	63.89	−22.04	4.1	3.5	3.5
12/08/2011	04:03:17	63.93	−21.98	9.6	3.1	3.0
16/08/2011	22:14:18	63.85	−22.44	2.4	3.3	3.6
17/08/2011	01:34:23	63.85	−22.38	5.4	3.1	3.1
23/09/2011	15:10:16	64.06	−21.38	5.1	2.6	3.1
23/09/2011	15:20:04	64.06	−21.39	4.9	2.8	3.3
23/09/2011	15:22:46	64.05	−21.38	3.9	3.0	3.4
07/10/2011	11:00:31	64.05	−21.41	4.1	2.7	3.0
08/10/2011	23:04:14	64.06	−21.40	2.8	3.1	3.3
15/10/2011	09:00:46	64.07	−21.41	2.4	2.8	3.1
15/10/2011	09:03:08	64.05	−21.42	3.3	3.7	4.0
15/10/2011	09:46:04	64.07	−21.40	2.0	3.9	4.0
27/11/2011	10:17:06	63.91	−22.14	5.8	2.6	3.0
03/01/2012	21:12:42	63.89	−22.09	4.8	3.8	4.0
01/03/2012	00:29:16	64.00	−21.82	4.2	3.2	3.6
01/03/2012	01:03:07	63.99	−21.81	4.3	4.0	4.2
21/04/2012	22:36:40	64.05	−21.43	4.0	2.7	3.2
30/08/2012	11:59:02	64.00	−21.59	5.5	3.9	4.6
01/09/2012	16:33:47	63.89	−22.25	7.3	2.8	3.0
13/10/2013	09:38:52	63.83	−22.64	5.8	3.3	3.1
07/01/2014	12:11:49	64.00	−21.84	0.2	3.2	3.1

and time are expected to be detected by a seismic network (Woessner and Wiemer 2005; Schorlemmer and Woessner 2008; Mignan and Woessner 2012).

Most methods for assessing the magnitude of completeness rely on the Gutenberg–Richter (GR; Gutenberg and Richter 1944) relation that describes the frequency of earthquake magnitudes:

$$\log_{10} N(\geq M) = a - bM, \qquad (12)$$

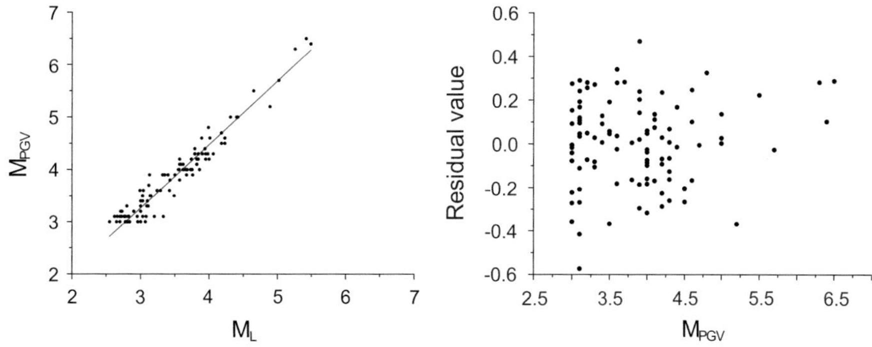

Figure 7
Left panel M_L versus M_{PGV}, *red line* the best fit obtained using generalized orthogonal regression. *Right panel* residual values of predicted M_{PGV} using Eq. (10) for the earthquakes listed in Table 2

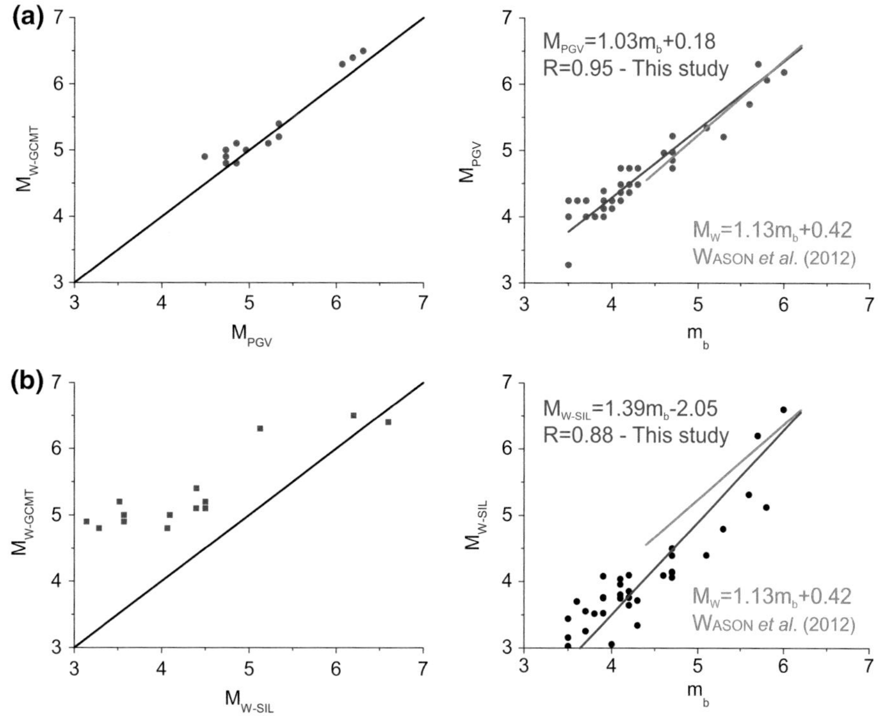

Figure 8
Comparisons of the derived M_{PGV}, $M_{W\text{-}SIL}$, $M_{W\text{-}GCMT}$ and m_b for the earthquakes listed in Table 4. **a** M_{PGV} versus $M_{W\text{-}GCMT}$ *left panel* and m_b vs M_{PGV} *right panel*, respectively. **b** $M_{W\text{-}SIL}$ versus $M_{W\text{-}GCMT}$ *left panel* and m_b vs $M_{W\text{-}SIL}$ *right panel*, respectively. The *black lines* in the figure are 1:1 lines showing how closely the values match. *Blue lines* the best fit obtained using GOR, whereas *green lines* the model proposed by Wason *et al.* (2012)

where $N(\geq M)$ is the number of earthquakes with magnitude at least as large as M. Consequently, M_C can be identified from data as the minimum magnitude at which the cumulative frequency magnitude distribution departs from the exponential decay (e.g. Zuniga and Wyss 1995). In this study, we used the maximum curvature method (Wiemer and Wyss 2000) to estimate M_C, and we estimated the *b value* using maximum likelihood (Utsu 1965). Figure 9a shows the results before and after the quality check of the catalogue. There are no strong differences in M_C, *b value* and *a* values using the two considered catalogues, but the linear part of the GR relation appears to fit the corrected catalogue better (right

Table 4

List of earthquakes plotted in Fig. 8

Dd/mm/yyyy	Hh:mm:ss	M_L	$M_{W\text{-}SIL}$	$M_{W\text{-}GCMT}$	M_{PGV}	M_b (ISC)
27/12/1992	12:23:20	4.2	3.7	–	4.7	4.3
20/08/1994	16:40:22	3.9	3.6	–	4.4	4.2
30/04/1995	00:57:57	3.7	3.5	–	4.1	3.9
14/10/1996	20:59:58	3.8	3.8	–	4.2	4.1
23/02/1997	00:35:47	3.6	3.6	–	4.0	3.7
23/02/1997	08:45:02	3.6	3.5	–	4.0	3.8
12/04/1997	23:04:44	3.8	3.8	–	4.2	3.9
24/08/1997	03:04:22	4.4	4.1	–	5.0	4.7
03/06/1998	06:47:42	3.6	3.8	–	4.0	–
04/06/1998	19:04:46	4.0	3.8	–	4.5	4.2
04/06/1998	21:36:54	4.7	4.4	5.4	5.3	5.1
04/06/1998	22:59:58	4.0	3.9	–	4.5	4.2
13/11/1998	10:38:34	4.6	4.5	5.1	5.2	4.7
14/11/1998	14:24:07	4.4	4.1	–	5.0	4.7
25/05/1999	13:19:40	4.0	3.7	–	4.5	4.1
27/09/1999	16:01:14	4.2	4.0	–	4.7	4.1
17/06/2000	15:40:41	5.5	6.2	6.5	6.3	5.7
17/06/2000	15:42:51	5.0	5.3	–	5.7	5.6
17/06/2000	15:45:27	4.6	4.8	–	5.2	5.3
17/06/2000	17:09:26	3.9	4.1	–	4.4	3.9
17/06/2000	17:40:17	4.4	4.1	5.0	5.0	4.6
21/06/2000	00:51:47	5.4	6.6	6.4	6.2	6.0
23/08/2003	02:00:11	4.3	4.4	5.1	4.9	4.7
06/03/2006	14:31:55	4.2	4.1	–	4.7	4.2
27/02/2007	05:24:23	3.9	3.6	4.9	4.4	–
27/02/2007	05:51:51	4.0	3.6	5.0	4.4	–
29/05/2008	15:46:00	5.3	5.1	6.3	6.1	5.8
29/05/2008	17:07:31	4.1	3.7	–	4.6	3.6
29/05/2008	17:09:42	3.8	3.4	–	4.2	3.5
29/05/2009	21:33:51	4.2	4.1	4.8	4.7	4.7
30/05/2009	13:35:21	4.0	3.3	–	4.5	4.3
19/06/2009	18:13:20	3.8	2.8	–	4.2	4.2
19/06/2009	20:37:13	3.8	2.9	–	4.2	4.0
25/06/2009	17:20:17	3.7	3.1	–	4.1	4.0
23/10/2010	20:34:31	4.2	3.1	4.9	4.7	–
23/10/2010	21:59:00	4.7	3.5	5.2	5.3	–
27/02/2011	09:06:00	3.6	3.0	–	4.0	3.5
27/02/2011	09:49:04	3.1	3.2	–	3.4	3.5
27/02/2011	17:27:37	3.8	3.3	–	4.2	3.7
03/01/2012	21:12:42	3.8	3.0	–	4.2	3.5
30/08/2012	11:59:04	3.9	4.0	–	4.4	4.1
09/05/2013	19:20:40	3.7	3.3	4.8	4.2	–
13/10/2013	07:34:06	4.7	4.5	5.2	5.3	–

panel in Fig. 9a). For the investigated area we estimate M_C as 0.79 ± 0.05, which demonstrates the SIL system's ability to reliably detect small earthquakes; the estimated *b value* is 0.85 ± 0.03.

In case of an aftershock sequence or changes to the seismic network, one would expect to see changes in M_C and *b value*. To investigate this, we checked for temporal variations in M_C and *b value* using the maximum curvature technique, a window size of 500 events, a moving window of 50 events, and a 25 % smoothing function (Fig. 9b).

The results show that both the M_C and the *b value* are affected by large earthquake sequences, such as the June 2000 and May 2008 sequences, which both included $M_W > 6.0$ events. The mainshock and the subsequent moderate magnitude earthquakes

Table 5

M_L central value, bin amplitude, mean values of M_W and two standard deviations (σ)

M_L bin	Mean M_W	2σ
−0.1–0.1	0.79	0.37
0.1–0.3	0.92	0.40
0.3–0.5	1.06	0.43
0.5–0.7	1.19	0.42
0.7–0.9	1.31	0.40
0.9–1.1	1.44	0.39
1.1–1.3	1.56	0.39
1.3–1.7	1.70	0.42
1.5–1.7	1.85	0.44
1.7–1.9	2.02	0.46
1.9–2.1	2.17	0.53
2.1–2.3	2.37	0.64
2.3–2.5	2.41	0.72

($M_W > 4.0$) in the following days, indeed, introduce the largest variation in M_C when it reaches values of about 1.4–1.5 (Fig. 9b, upper panel). In the case of *b value*, strange variations could be seen as for instance starting from 2000 until 2005 for the occurrence of a long aftershocks sequence following the 17 and 21 June 2000 earthquakes (Fig. 9b, lower panel). In particular, in this case the *b value* increased steadily reaching 1.4 after those events. We also note that the *b value* exceeds in many periods the value estimated from the GR relationship of all events during the considered time period (Fig. 9b). These results could be related to the behaviour of the GR curve (Fig. 9a). The GR curve has a high slope in the magnitude range 0.79–2.6, whereas the slope is low at magnitudes greater than 2.6. A possible explanation could be an interaction between tectonic and volcanic activity, which in Iceland are tightly linked. A similar result was obtained in Guagua Pichincha volcano (Ecuador) by Legrand *et al.* (2004). The authors suggested that a non-linear Frequency Magnitude Distribution (FMD) of GR may be understood as the superposition of various processes, such as classic elastic rupture (*b value* $\approx$ 1.00) and hydraulic fracturing (*b value* up to 2.0). Therefore, deviations from linearity in the FMD may be related to temporal and spatial variations in *b value* as observed in our results.

The spatial variations of the M_C and *b values* for South Iceland were achieved using the common catalogue-based mapping approach of Wyss *et al.* (1999) and Wiemer and Wyss (2000). The method consists of computing M_C and *b value* through FMD, using a combination of a fixed number of events, N in each node of a spatial grid and the constant radius method, which consists of selecting all earthquakes within a certain distance. This means that if a node in the grid does not meet the requirement of N earthquakes, then those events within distance R from the node are included, to fulfil the requirement. In particular, we used a spatial grid of 0.03° in latitude and 0.07 in longitude and $R = 15$ km. The fixed number of earthquakes, N to compute the GR was 300 and the minimum number of earthquakes required to estimate M_C was 30. Finally the maps were smoothed by using a Gaussian kernel filter with sigma equal to 3.

In Fig. 10a we show the results for the original catalogue; these results seem to be strongly influenced by the explosions near Reykjavík and Reykjanesbaer with very high *b values* and M_C (>2.0). The *b value* for the revised catalogue is in the range 0.8–1.0 except for the eastern zone, which is characterized by high *b value*, probably due to a small number of earthquakes (Fig. 10b). The central area of the South Iceland Seismic Zone is characterized by low values of M_C ($\approx$0.6–0.8) whereas the eastern and western areas show higher values (>1.0). In particular, the detection capability of the seismic network decreases near the Reykjanes peninsula coast, probably due to the proximity to the city, and in the eastern SISZ where the seismic network configuration is more spread out.

5.2. Seismic Rate Analysis

In Fig. 11a, we plot the number of earthquakes as a function of time to explore possible changes in seismicity rate, considering only the earthquakes with magnitude greater than M_C (0.79). The trend of the revised catalogue is not markedly different from the original catalogue, although of course the cumulative number of earthquakes is different. Three prominent steps in the curve's slope are identified corresponding to the April 1994, June 2000, and May 2008 earthquake sequences in the investigated area (dashed line in Fig. 11a). There is a change in slope (i.e.

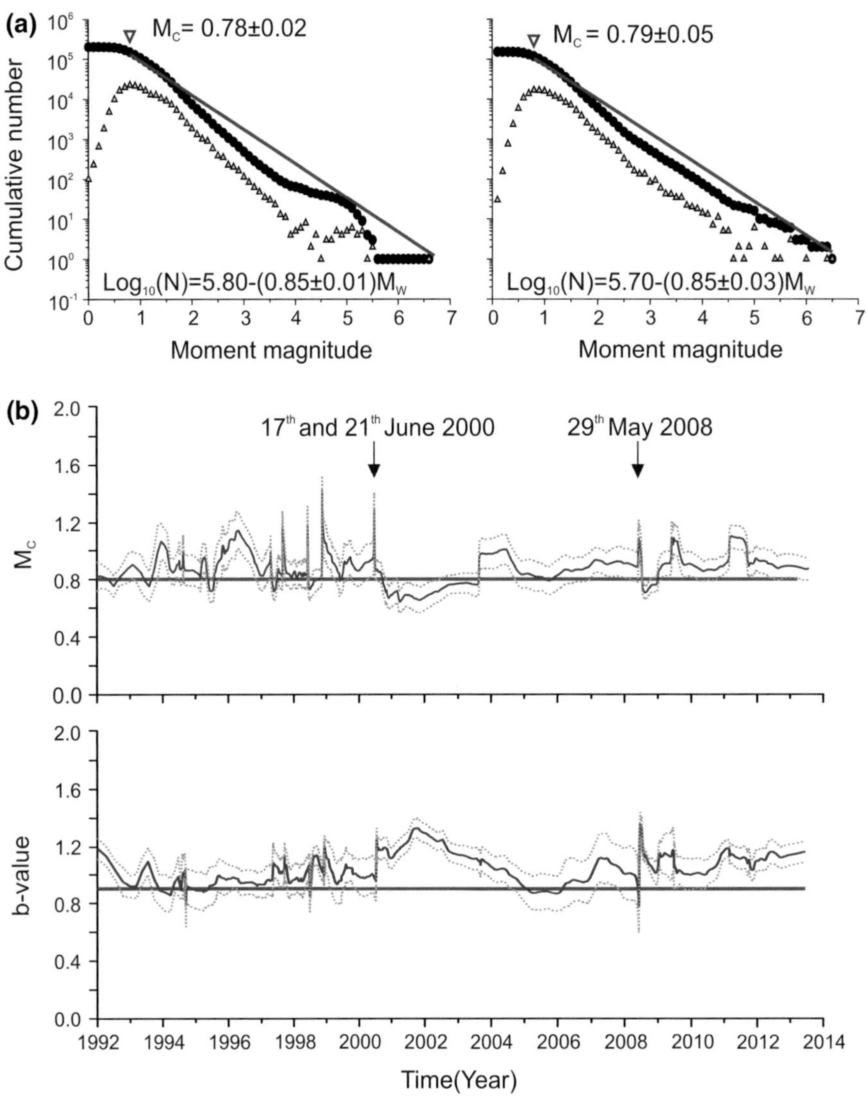

Figure 9

a Magnitude distribution of the original catalogue (*left panel*) and of the revised catalogue (*right panel*). **b** M_C (*top panel*) and *b value* (*lower panel*) as a function of time. *Red line* the M_C and *b value* estimated from the GR relationship of all events during the considered time period and the *black arrows* the onset of the June 2000 and May 2008 earthquake sequences

increased average seismicity rate) between mid-1994 and 1999, which is due to an intense volcano-tectonic interaction episode in the Hengill region (Feigl *et al.* 2000). The effect on the overall trend, caused by the Hengill swarm, is evident by removing all the seismicity around this region (black curve in Fig. 11a) and verifying that the slope becomes fairly regular for the studied period (1991–2013). It is also interesting to observe that the seismicity filter demonstrates more clearly another step in the cumulative curve in addition to those related to the occurrence of strong earthquakes (June 2000 and May 2008). In particular, the step linked to the two earthquakes of November 1998, previously hidden in the Hengill swarm, appears evident (Fig. 11a).

There is also a noticeable change around 2011. This change in rate may be explained by the fact that the seismic network was improved in the eastern part of the investigated area in late 2010. Distribution of earthquakes in the studied area was also investigated

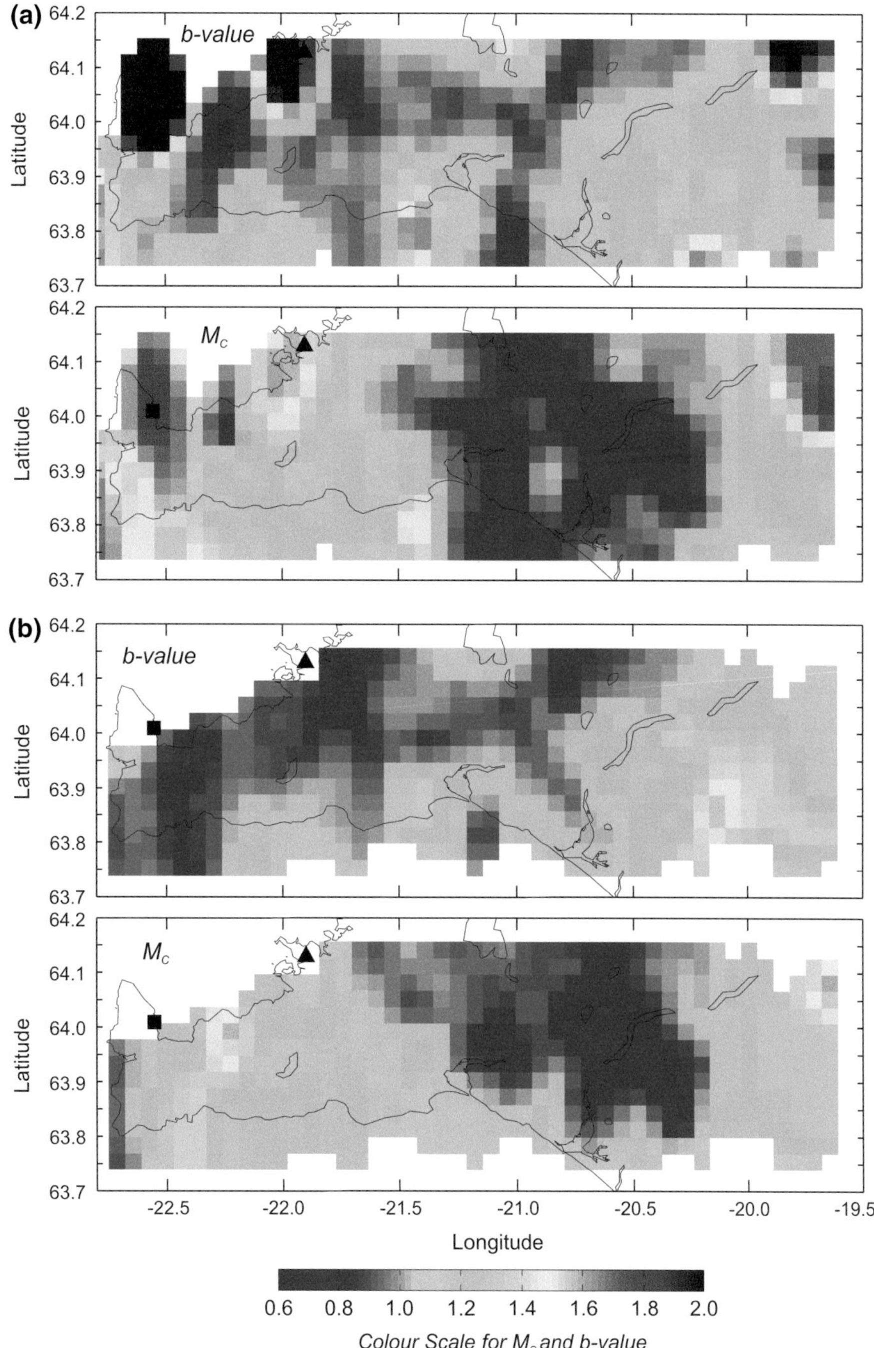

Figure 10
b value and M_C maps for the **a** the original catalogue and **b** revised catalogue. *Black triangle* and *rectangle* the position of Reykjavík and Reykjanesbaer, respectively

by drawing an earthquake density map (Fig. 11b), as the logarithm of earthquakes per square km, obtained using a spatial grid of 0.03° in latitude and 0.07 in longitude. From this map we can identify the zones with major occurrence of earthquakes. The seismicity is not uniformly distributed, but rather it is

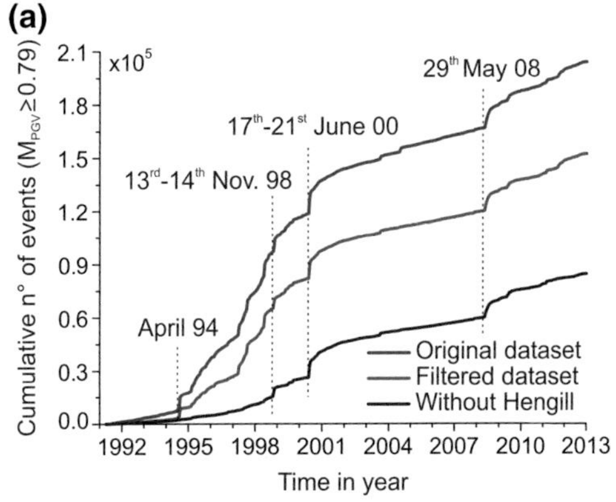

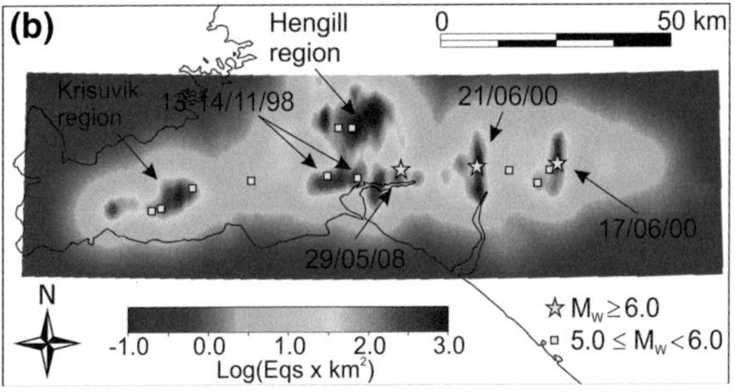

Figure 11
a Number of earthquakes versus time before and after revising the catalogue. **b** Earthquake density in which *stars* indicate the areas with high earthquake activities

concentrated around the major active faults, geothermal centres in the Krísuvík region and the geothermal/volcanic centre in Hengill (Fig. 11b).

6. Concluding Remarks

Twenty years after the installation of the SIL system, we have analysed the South Iceland earthquake catalogue to understand its strengths and weaknesses. In a sense, this is necessary preliminary work for any future study on seismicity forecasting or regional seismic hazard assessment. The revised catalogue is available upon request to the authors.

The main results of this study can be summarized as follows:

- We explored the consequences of using the network criteria method to filter the SIL catalogue in Iceland. In this region, where the emphasis is on recording microseismicity, but the network is not as dense as in many other parts of the world, we found the filter to be too strict: almost all microearthquakes were filtered out because they did not have enough arrival time picks. As an alternative, we suggest a filter based on the multivariate distribution of hypocentral uncertainties. This filter, based on Mahalanobis distance, results in a catalogue with smaller uncertainties and more events than the network criteria method (75 % of events pass the filter).
- We searched the catalogue for quarry blasts that were incorrectly identified as earthquakes. We found and removed 502 such events ($\sim$0.4 % of earthquakes in the filtered catalogue).
- We noted and corrected several problems with the local magnitude and so-called local moment magnitude estimates reported in the SIL catalogue. In particular, the magnitude of the largest events is often underestimated, and the magnitude of smaller events following large events was often overestimated.
- We compared estimates of the magnitude of completeness, b value, and seismicity rate based on the original catalogue and the filtered, corrected catalogue and identified possible explanations for the differences.

This work emphasizes the importance of carefully assessing an earthquake catalogue before proceeding with further analysis, and as such we hope that our revised catalogue will be used for future studies and models of seismicity in South Iceland.

Acknowledgments

This study was supported with funds provided by the European project REAKT Strategies and tools for Real-Time EArthquake RisK ReducTion. All the maps and graphics were obtained through open-source and freely available matlab code MapSeis, that can be downloaded from http://www.corssa.org/mapseis. The authors wish to thank the Guest Editor Prof. Jiancang Zhuang, Dr. Marie Keiding and an anonymous reviewer for constructive comments which contributed to improve the quality of the paper. The authors are also grateful to the Icelandic Meteorological Office for providing access to the SIL catalogue data.

References

Ambraseys, N.N., and Sigbjörnsson, R. (2000), *Re-Appraisal of the Seismicity of Iceland*, Polytechnica-Engineering Seismology, No. 3, Earthquake Engineering Research Centre, University of Iceland.

Angelier, J., Slunga, R.F., Bergerat, F., Stefansson, R., and Homberg, C. (2004), *Perturbation of stress and oceanic rift extension across transform faults shown by earthquake focal mechanisms in Iceland*, E.P.S.L., *219*, 271–284.

Antonioli, A., Belardinelli, M. E., Bizzari, A., and Vogfjörd K. S. (2006). *Evidences of instantaneous triggering during the seismic sequence of year 2000 in south Iceland.* J. Geophys. Res., *111*. doi:10.1029/2005JB003935.

Árnadóttir, T.H., Geirsson, H., and Einarsson P. (2004). *Co-seismic stress changes and crustal deformation on the Reykjanes Peninsula due to triggered earthquakes on June 17, 2000.* J. Geophys. Res., *109*, B09307. doi:10.1029/2004JB003130.

Bergerat, F., and Angelier, J. (2000), *The South Iceland Seismic Zone: tectonic and seismotectonic analyses revealing the evolution from rifting to transform motion*, Journal of Geodynamics, *29*, 211–231.

Böðvarsson, R., Rögnvaldsson, S.T., Jakobsdóttir, S.S., Slunga, R., and Stefánsson, R., (1996), *The SIL data acquisition and monitoring system*, Seism. Res. Lett., *67*, 35–46.

Böðvarsson, R., Rögnvaldsson, S.T., Slunga, R., and Kjartansson, E. (1999), *The SIL data acquisition system-at present an beyond year 2000*, Phys. Earth Planet. Inter., *113*, 89–101.

Bondar, I., Myers, S.C., Engdahl, E.R., and Bergman, E.A. (2004), *Epicentre accuracy based on seismic network criteria*, Geophys. J. Int., *156*, 483–496.

Castellaro, S., Mulargia, F., and Kagan, Y.Y. (2006), *Regression problems for magnitudes*, Geophys. J. Int., *165*, 913–930. doi:10.1111/j.1365-246X.2006.02955.x.

Clifton, A. E., Pagli, C., Jónsdóttir, J. F., Eythórsdóttir, K., and Vogfjörd, K. (2003). *Surface effects of triggered fault slip on Reykjanes Peninsula, SW Iceland.* Tectonophysics, *369*, 145–154.

Clifton, A., and Einarsson P. (2005). *Styles of surface rupture accompanying the June 17 and 21, 2000 earthquakes in the South Iceland Seismic Zone.* Tectonophysics, *396*, 141–159.

Clifton, A.E., and Kattenhorn, S.A. (2006), *Structural architecture of a highly oblique divergent plate boundary segment*, Tectonophysics, *419*, 27–40.

Decriem, J., Árnadóttir, T., Hooper, A., Geirsson, H., Sigmundsson, F., Keiding, M., Ófeigsson, B.G., Hreinsdóttir, S., Einarsson, P., Lafemina, P., and Bennett, R.A. (2010). *The 2008 May 29 earthquake doublet in SW Iceland*, Geophys. J. Int. (2010) *181*, 1128–1146. doi: 10.1111/j.1365-246X.2010.04565.x.

Dziewonski, A.M., Chou, T.A., and Woodhouse, J.H. (1981), *Determination of earthquake source parameters from waveform data for studies of global and regional seismicity*, J. Geophys. Res., *86*, 2825–2852. doi:10.1029/JB086iB04p02825.

Einarsson, P. (1991), *Earthquakes and present-day tectonism in Iceland*, Tectonophysics, *189*, 261–279.

Einarsson, P. (2008), *Plate boundaries, rifts and transforms in Iceland*, JÖKULL, *58*, 35–58.

Einarsson, P., Björnsson, S., Foulger, G., Stefánsson, R. and Skaftadóttir, T. (1981) *Seismicity Pattern in the South Iceland Seismic Zone*, in Earthquake Prediction (eds D. W. Simpson and P. G. Richards), American Geophysical Union, Washington, D. C. Maurice Ewing Series *4*, 141–151, doi:10.1029/ME004p0141.

Ekström, G., Nettles, M., and Dziewonski, A.M. (2012), *The global CMT project 2004-2010: Centroid-moment tensors for 13,017 earthquakes*, Phys. Earth Planet. Inter., 200–201, 1–9. doi:10.1016/j.pepi.2012.04.002.

Eshaghi, A., Tiampo, K.F., Ghofrani, H., and Atkinson, G.M. (2013), *Using Borehole Records to Estimate Magnitude for Earthquake and Tsunami Early-Warning Systems*, Bull. Seismol. Soc. Am., *103*(4), 2216–2226. doi:10.1785/0120120319.

Everitt, B.S. (2005), *An R and S-PLUS companion to multivariate analysis*, Springer: London, p. 221.

Feigl, K.L., Gasperi, J., Sigmundsson, F., and RIGO, A. (2000). *Crustal deformation near Hengill volcano, Iceland 1993–1998: Coupling between magmatic activity and faulting inferred from elastic modelling of satellite radar interferograms.* J. Geophys. Res., *105* (B11), 25655–25670.

Gomberg, J.S., Shedlock, K.M., and Roecker, S.W. (1990), *The effect of S-Wave arrival times on the accuracy of hypocenter estimation*, Bull. Seism. Soc. Am., *80*, 1605–1628.

Gudmundsson, G.B., Vogfjörd, K.S. and Thorbjarnardóttir, B.S. (2006). *SIL data status report*, in: *PREPARED – third periodic report, February 1, 2005 – July 31, 2005.* R. Stefánsson et al. (Eds.). Icelandic Meteorological Office report, no 06008, VI-ES-05, Appendix 3, pp 127–131.

Gulia, L., Wiemer, S., and Wyss, M. (2012), *Catalog artifacts and quality controls*, Community Online Resource for Statistical Seismicity Analysis. doi:10.5078/corssa-93722864. http://www.corssa.org.

Gutenberg, R., and Richter, C.F. (1944), *Frequency of earthquakes in California*, Bulletin of the Seismological Society of America, *34*, 185–188.

Habermann, R.E. (1987), *Man-made changes of seismicity rates*, Bull. Seism. Soc. Am., *77* (1), 141–159.

Hackman, M.C., King, G.C.P., and Bilham, R. (1990), *The mechanics of the South Iceland Seismic Zone*, Journal of Geophysical Research. doi:10.1029/90JB01043.

Hjaltadóttir, S. (2009) *Use of relatively located microearthquakes to map fault patterns and estimate the thickness of the brittle crust in Southwest Iceland. Sub-surface fault mapping in Southwest Iceland*, Master's thesis, Faculty of Earth Sciences, University of Iceland, pp. 104. ISBN 978-9979-9914-6-5. http://hdl.handle.net/1946/3990.

Hjaltadottir, S., Vogfjord, K. S., and Slunga, R. (2005), *Mapping subsurface faults in southwest iceland using relatively located microearthquakes*, Geophysical research abstracts, *7*, 06664.

Husen, S., and Hardebeck, J.L. (2010), *Earthquake location accuracy*, Community Online Resource for Statistical Seismicity Analysis. doi:10.5078/corssa-55815573. http://www.corssa.org.

Jakobsson, S.P., Honsson, J., and Shido, F. (1978), *Petrology of the western Reykjanes Peninsula, Iceland*, Journal of Petrology, *19*, 669–705.

Keiding, M., Árnadóttir, TH., Sturkell, E., Geirsson, H., and Lund, B. (2008). *Strain accumulation along an oblique plate boundary: The Reykjanes Peninsula, southwest Iceland.* Geophys. J. Int., *172*(1):861–872. doi:10.1111/j.1365-246X.2007.03655.x.

Legrand, D., Villagómez, D., Yepes, H., and Calahorrano, A. (2004) *Multifractal dimension and b value analysis of the 1998–1999 Quito swarm related to Guagua Pichincha volcano activity,*

Ecuador, J. Geophys. Res., *109*, B01307. doi:10.1029/2003 jb002572.

Lin, T., and Wu, Y.M. (2010), *Magnitude determination using strong ground motion attenuation in earthquake early warning*, Geophys. Res. Lett., *7*, L07304. doi:10.1029/2010GL042502.

Madigan, D., Bartlett, P., Bühlmann, P., Carroll, R., Murphy, S., Roberts, G., Scott, M., Távare, S., Triggs, C., Wang, J-L., Wasserstein, R., and Zuma, K. (2014), *Statistics and science: a report of the London Workshop on the Future of the Statistical Sciences*. http://bit.ly/londonreport. last accessed 31 July 2014.

Mignan, A., Woessner, J. (2012), *Estimating the magnitude of completeness for earthquake catalogs*, Community Online Resource for Statistical Seismicity Analysis. doi:10.5078/corssa-00180805. http://www.corssa.org.

Pétursson, G.G. and Vogfjörd, K.S. (2009), *Attenuation relations for near- and far field peak ground motion (PGV, PGA)and new magnitude estimates for large earthquakes in SW-Iceland*, Icelandic Meteorological Report no VI 2009-012, pp 43. ISSN 1670-8261.

Roth, F. (2004), *Stress Changes Modelled for the Sequence of Strong Earthquakes in the South Iceland Seismic Zone Since 1706*. Pure & Applied Geophysics, *161*, 1305–1327.

Rögnvaldsson, S.T. and SLUNGA, R. (1993), Routine fault plane solutions for local networks: A test with synthetic data, Bull. Seismol. Soc. Am., *83*(4), 1232–1247.

Schorlemmer, D., and Woessner, J. (2008), *Probability of detecting an earthquake*, Bull. Seismol. Soc. Am., *98*(5), 2103–2117. doi:10.1785/0120070105.

Slunga, R., Norrman, P., and Glans, A. (1984), *Seismicity of Southern Sweden – Stockholm: Försvarets Forskningsanstalt*, FOA Report, C2 C20543-T1, 106 pp.

Stefánsson, R., Böðvarsson, R., Slunga, R., Einarsson, P., Jakobsdóttir, S.S., Bungum, H., Gregersen, S., Havskov, J., Hjelme, J., and Korhonen, H. (1993), *Earthquake prediction research in the South Iceland seismic zone and the SIL project*, Bull. Seism. Soc. Am., *83*, 696–716.

Storchak, D.A., Di Giacomo, D., Bondár, I., Engdahl, E.R., Harris, J., Lee, W.H.K., Villaseñor, A., and Bormann, P. (2013). *Public Release of the ISC-GEM Global Instrumental Earthquake Catalogue (1900–2009)*, Seism. Res. Lett., *84*(5), 810–815. doi:10.1785/0220130034.

Tormann, T., Wiemer, S., Hauksson, E. (2010). *Changes in reporting Rates in the Southern California Earthquake Catalog, Introduced by a New Definition of Ml*, Bull. Seism. Soc. Am., *100*(4), 1733–1742.

Utsu, T. (1965), *A method for determining the b-value in a formula logN = a−bm showing magnitude frequency relation for earthquakes*, Hokkaido Univ. Geophys. Bull., *13*, 99–103, in Japanese with English abstract.

Vogfjörd, K.S., Kjartansson, E., Slunga, R., Halldórsson, P., Hjaltadóttir, S., Gudmundsson, G.B., Sveonbjörnsson, H., Ármannsdóttir, S., Thorbjarnardóttir B., and Jakobsdóttir, S. (2012). *Development and implementation of seismic early warning processes in South-west Iceland*, Icelandic Meteorological Report n° VI 2010-012, pp 83. ISSN 1670-8261.

Wason, H.R., Das, R., and Sharma M.L. (2012), *Magnitude conversion problem using general orthogonal regression*, Geophys. J. Int., *190*(2), 1091–1096.

Wiemer, S., and Baer, M. (2000), *Mapping and removing quarry blast events from seismicity catalogs*, Bull. Seism. Soc. Am., *90*, 525–530.

Wiemer, S., and Wyss, M. (2000), *Minimum magnitude of completeness in earthquake catalogs: examples from Alaska, the Western United States, and Japan*, Bull. Seism. Soc. Am., *90*, 859–869.

Woessner, J., and Wiemer, S. (2005), *Assessing the quality of earthquake catalogues: Estimating the magnitude of completeness and its uncertainty*, Bull. Seismol. Soc. Am., *95*. doi:10. 1785/012040007.

Wyss, M., Hasegawa, A., Wiemer, S. and Umino, N. (1999), *Quantitative mapping of precursory seismic quiescence before the 1989, M 7.1, off-Sanriku earthquake, Japan*, Annali di Geophysica, *42*, 851–86.

Zuniga, F.R., and Wyss, M. (1995), *Inadvertent changes in magnitude reported in earthquake catalogs: their evaluation through b-value estimates*, Bull. Seism. Soc. Am., *85*, 1858–1866.

(Received September 7, 2014, revised March 27, 2015, accepted May 29, 2015, Published online June 22, 2015)

Pure Appl. Geophys. 173 (2016), 117–124

DOI 10.1007/s00024-015-1133-7

Pure and Applied Geophysics

The Spatial Scale of Detected Seismicity

A. Mignan[1] and C.-C. Chen[2]

Abstract—An experimental method for the spatial resolution analysis of the earthquake frequency-magnitude distribution is introduced in order to identify the intrinsic spatial scale of the detected seismicity phenomenon. We consider the unbounded magnitude range $m \in (-\infty, +\infty)$, which includes incomplete data below the completeness magnitude m_c. By analyzing a relocated earthquake catalog of Taiwan, we find that the detected seismicity phenomenon is scale-variant for $m \in (-\infty, +\infty)$ with its spatial grain a function of the configuration of the seismic network, while seismicity is known to be scale invariant for $m \in [m_c, +\infty)$. Correction for data incompleteness for $m < m_c$ based on the knowledge of the spatial scale of the process allows extending the analysis of the Gutenberg–Richter law and of the fractal dimension to lower magnitudes. This shall allow verifying the continuity of universality of these parameters over a wider magnitude range. Our results also suggest that the commonly accepted Gaussian model of earthquake detection might be an artifact of observation.

Key words: Spatial scale, Earthquake magnitude, Earthquake detection, Completeness magnitude.

1. Introduction

Earthquakes represent a complicated geographical phenomenon, which is shown to be self-similar in a wide range of scales (Bak *et al.* 2002). The universal Gutenberg–Richter scaling relation (Gutenberg and Richter 1944) states that earthquake magnitudes m are distributed according to the exponential law

$$\lambda_0(m) = a10^{-bm} \tag{1}$$

where λ_0 is the number of expected earthquakes, a the earthquake productivity and b the magnitude scaling parameter. Assuming self-similarity in space, Eq. (1) becomes

$$\lambda_0(m, L) = a10^{-bm}L^D \tag{2}$$

with L the size of the spatial cell and D the spatial scaling parameter or fractal dimension of earthquake epicenters (Kosobokov and Mazhkenov 1994; Molchan and Kronrod 2005). Equations (1) and (2) hold only for $m \geq m_c$ with m_c the completeness magnitude. It has been shown that this parameter is ambiguous with the $\widehat{m}_c$ estimate depending at a same location on the computation method and on the spatial scale considered (Mignan and Chouliaras 2014). The behavior of seismicity at $m < m_c$ is also ambiguous due to the fact that the intrinsic spatial scale of the detected seismicity phenomenon has not been considered so far. In other words, there is an implicit assumption that the detection process is also self-similar, which has recently been shown to be incorrect (i.e., the frequency-magnitude distribution does not display the same statistical properties or same shape at different scales; see Fig. 16 in Mignan 2012).

In this article, we investigate the shape of the earthquake frequency-magnitude distribution over the range $m \in (-\infty, +\infty)$ to determine the fundamental spatial unit of the geographical phenomenon that is detected seismicity (Sect. 2; Pereira 2001). Knowing the scale of the detected seismicity phenomenon (Sect. 3) allows us (i) to study the behavior of the scaling parameters b and D on a wider magnitude range by including smaller events, and (ii) to gain more insight into the earthquake detection process (Sect. 4).

For this analysis, we use the earthquake catalog of Taiwan relocated by Wu *et al.* (2008). We consider the period 1994–2005 and volume 119°E $\leq x \leq$ 123°E, 21°N $\leq y \leq$ 26°N and $z \leq 35$ km. Year 2005 corresponds to the last year analyzed in Wu *et al.* (2008).

[1] Institute of Geophysics, Swiss Federal Institute of Technology Zürich, Zurich, Switzerland. E-mail: arnaud.mignan@sed.ethz.ch
[2] Department of Earth Sciences, National Central University, Zhongli, Taiwan.

2. *Definition of the Spatial Scale of Detected Seismicity*

We show in this section that the spatial scale of detected seismicity can be inferred from the analysis of the shape of the earthquake frequency-magnitude distribution (FMD). This was first suggested by MIGNAN (2012) who observed that the FMD shape differs between local and regional datasets. This is conceptualized below and verified in an earthquake catalog in the next section.

The FMD is described over the range $m \in (-\infty, +\infty)$ by

$$\lambda(m) = \lambda_0(m)q(m) \tag{3}$$

where $q(m) = \Pr(\text{Detected} \mid m)$ is the earthquake detection function. MIGNAN (2012) coined the term *elemental FMD* to describe the fundamental FMD, whose shape does not depend on m_c variations. This elemental FMD takes the form

$$\lambda(m|k,b,m_c) = \begin{cases} a10^{(k-b)(m-m_c)}, & m < m_c \\ a10^{-b(m-m_c)}, & m \geq m_c \end{cases} \tag{4}$$

with $k > b$ a detection parameter. The elemental FMD has an angular shape with q $(m < m_c)$ an exponential law and q $(m \geq m_c) = 1$. Since $m_c = f(x,y)$, the elemental FMD only holds in an elemental spatial cell of width L_e. The FMD observed in larger cells is the sum of elemental FMDs, leading to a rounded shape of that composite FMD (MIGNAN 2012). This shape can be approximated by defining q (m) by the cumulative normal distribution ϕ (RINGDAL 1975; OGATA and KATSURA 1993; 2006; MIGNAN 2012). Figure 1 shows an FMD at the regional scale ($L = 5°$, full catalog) and an FMD at a local scale ($L = 0.1°$, $x = 120.95°$, $y = 23.15°$). They are rounded and angular, respectively, indicating that the regional FMD is a composite FMD (trivial) and that the local FMD is an elemental FMD, which means $L \leq L_e$ at (x, y).

The concept of elemental cell brings us to the scale of the process. Scale is defined by two components: the grain ($\sim$grid cell) and the extent ($\sim$entire grid). Those components can be artifacts of observation or attributes of phenomena (PEREIRA 2001). While this distinction is common practice in ecological studies, it is not the case in seismology where all processes are assumed scale invariant (BAK *et al.* 2002; OGATA and KATSURA 1993). Figure 2 represents maps with a same observational extent (the Taiwanese region). Figure 2a shows the spatial distribution of seismicity and seismic stations used in this study. Figure 2b, c shows maps of the estimator $\widehat{m}_c = m(\max(\lambda))$ using a same display resolution (0.1° pixel) but a different observational resolution with a constant coarse grain in Fig. 2b (with radius $r = L/2 = 50$ km) and a variable grain in Fig. 2c (with $r = L/2 = f(d_n)$, d_n being the distance to the nth seismic station). Using a variable grain function of the seismic network geometry was proposed by MIGNAN *et al.* (2011) to minimize spatial heterogeneities in m_c, m_c varying faster in the denser parts of a seismic network. We develop upon this idea by defining the fundamental spatial unit (or intrinsic grain) of the detected seismicity by the area within which the elemental FMD is observed. Visually, the elemental FMD represents the highest possible resolution (sharp angular shape, Fig. 1b) while the composite FMD represents a lower resolution (blurred rounded shape, Fig. 1a). Note that the intrinsic extent of the earthquake detection process corresponds to the area comprising all the seismicity declared from a given seismic network (here the same as the observational extent, Fig. 2a).

3. *Spatial Resolution Analysis*

We investigate the shape of FMDs located in a partitioned space using Voronoi tessellation (VORONOI 1908). We test $N_V = 1{,}000$ realizations with the number of generators n_V randomly drawn from the $\log_{10}$ space in the range $[10, 10^3]$. Each generator, or point, is defined from random geographical coordinates (x_V, y_V) located in the region of interest. For each point, there is a corresponding region, or Voronoi cell, consisting of all points closer to that point than to any other. We choose this type of tessellation to objectively produce areas of various shapes and sizes (e.g., Fig. 2a) with no a priori knowledge on the spatial scale of the detected seismicity phenomenon (see also KAMER and HIEMER 2015). Each Voronoi cell

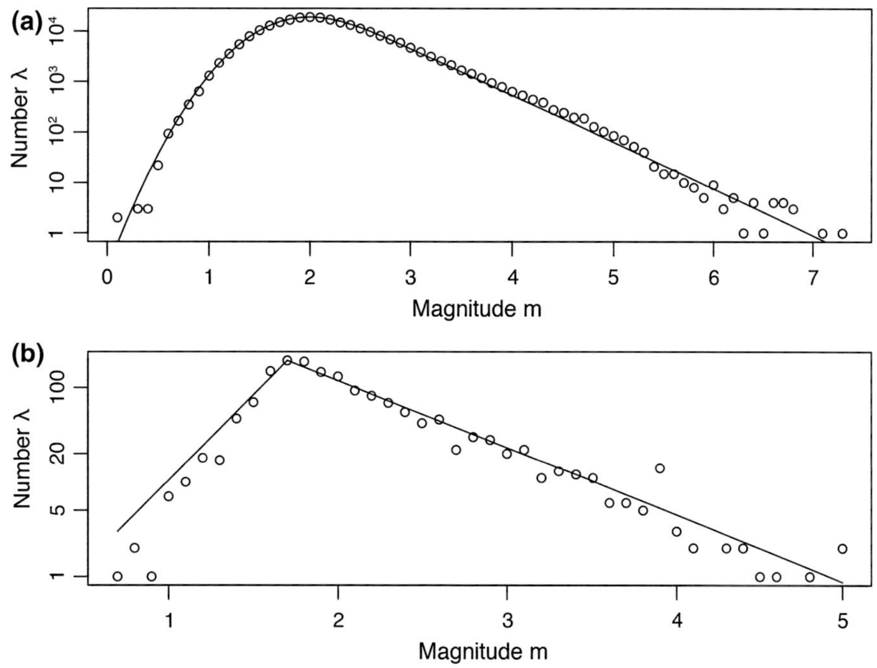

Figure 1
Shape of the earthquake frequency-magnitude distribution (FMD) at different observational scales over the range $m \in (-\infty, +\infty)$: **a** Regional FMD in a 5° cell (full catalogue), rounded and best fitted by $\lambda_0(m)\ \Phi(m)$ with $b = 0.9$, $\mu = 2.0$ and $\sigma = 0.4$ (i.e. composite FMD); **b** local FMD located in a 0.1° cell centered at ($x = 120.95°$, $y = 23.15°$), angular and best fitted by Eq. (4) with $b = 0.7$, $k = 1.8$ and $m_c = 1.7$ (i.e. elemental FMD)

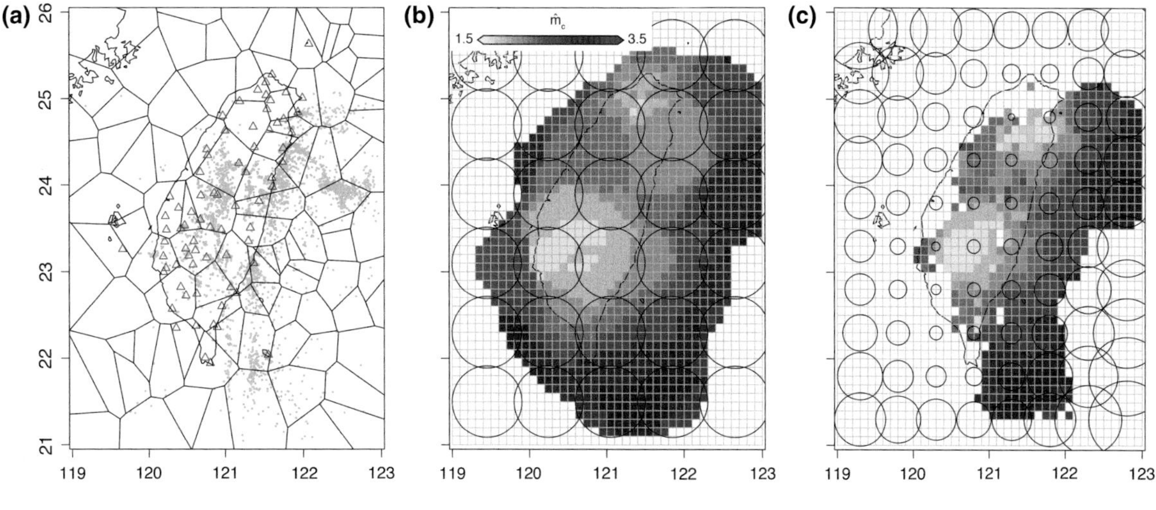

Figure 2
Observational spatial scale: **a** Spatial distribution of seismicity for magnitudes $m \geq 4.0$ (*gray dots*), spatial distribution of seismic stations (*red triangles*) and example of Voronoi tessellation with $n_V = 100$ (*black polygons*); **b** $\hat{m}_c = m(\max(\lambda))$ estimated for a constant coarse grain of radius $r = L/2 = 50$ km (*black circles*); **c** $\hat{m}_c = m(\max(\lambda))$ for a variable grain of radius $r = L/2 = f(d_5)$, d_5 being the distance to the 5th nearest seismic station (Mignan *et al.* 2011). The same observational spatial extent is used in (**a–c**) and the same display resolution (0.1° pixel) is used in (**b–c**) (*gray squares*)

is characterized by an area A, a distance $d_n = \sqrt{(x_V - x_n)^2 + (y_V - y_n)^2}$ between the generator and the nth nearest seismic station of coordinates (x_n, y_n), and an FMD defined from all earthquakes located in the cell. Let us note that the distance d_n represents a proxy to the spatial density of seismic stations—and therefore to the degree of m_c variations—and that the choice of n (commonly between 3 and 5) is not critical (Mignan *et al.* 2011, 2013; Kraft *et al.* 2013; Vorobieva *et al.* 2013; Mignan and Chouliaras 2014; Tormann *et al.* 2014).

Two FMD models are tested and compared using the Bayesian Information Criterion $\mathrm{BIC} = -2\log\hat{L} + K\log N$ with $\hat{L}$ the maximized value of the likelihood function f of the model, K the number of free parameters and N the number of data points (Schwarz 1978). The elemental FMD model (Fig. 1b) is described by the following likelihood function

$$f(m|\kappa,\beta,m_c) = \frac{1}{\frac{1}{\kappa-\beta}+\frac{1}{\beta}}\begin{cases}\exp((\kappa-\beta)(m-m_c)), & m<m_c\\ \exp(-\beta(m-m_c)), & m\geq m_c\end{cases} \tag{5}$$

where $\kappa = k\log(10)$ and $\beta = b\log(10)$ (Mignan 2012). In practice, we use a faster method introduced by Kamer (2014), which takes advantage of the piecewise structure of Eq. (4), uses the Aki maximum likelihood method (Aki 1965) and shows that $K = 2$ instead of 3 (i.e., $b/(k - b) = N_i/N_c$ with N_c and N_i the number of events in the complete and incomplete parts of the FMD, respectively). The composite FMD model (Fig. 1a) is described by

$$f(m|\beta,\mu,\sigma) = \beta\exp\left(-\beta(m-\mu) - \beta^2\frac{\sigma^2}{2}\right)\Phi(m|\mu,\sigma) \tag{6}$$

where $\Phi(m \mid \mu, \sigma)$ is the cumulative normal distribution and $K = 3$ (Ogata and Katsura 1993; 2006). The two models represent homogeneous data (elemental area with constant m_c) and heterogeneous data (composite area with variable m_c), respectively, which allows us to differentiate intrinsic grains and extrinsic grains relative to the spatial scale of detected seismicity.

For each Voronoi cell of all N_V realizations (yielding a total of ~60,000 cells), we calculate the distance d_5 to the fifth nearest seismic station, the width of the cell $L = 2\sqrt{A/\pi}$ and the FMD model choice C

$$\begin{cases}C = 1, & \text{if BIC (elemental FMD)} \leq \text{BIC (composite FMD)}\\ C = 0, & \text{otherwise}\end{cases} \tag{7}$$

Event location in the Taiwanese national catalog requires at least three stations and at least five phases P or S and it is therefore equally reasonable to use $n = 3, 4$ or 5 in d_n. Mignan *et al.* (2011) showed that the models $m_c = f(d_n)$ with $n = \{3, 4, 5\}$ give similar results in Taiwan. We then compute the ratio $\sum C_b/n_b$ with n_b the number of choices C_b, which are all the choices C located in $\Delta d_5 \times \Delta L$ bins defined in the (d_5, L) space. We fix $\Delta d_5 = \Delta L = 10$ km ~0.1° (Fig. 3). This ratio links the observational spatial resolution (or grain) L to the intrinsic spatial resolution L_e. Figure 3 shows that a constant observational spatial grain L (i.e., any horizontal line) is a poor estimator of the detected seismicity phenomenon since it includes heterogeneities in the FMD shape potentially leading to artifacts of observation at $m \leq m_c$. Considering $10\text{ km} < L < 150\text{ km}$ only (removal of earthquake location errors and of

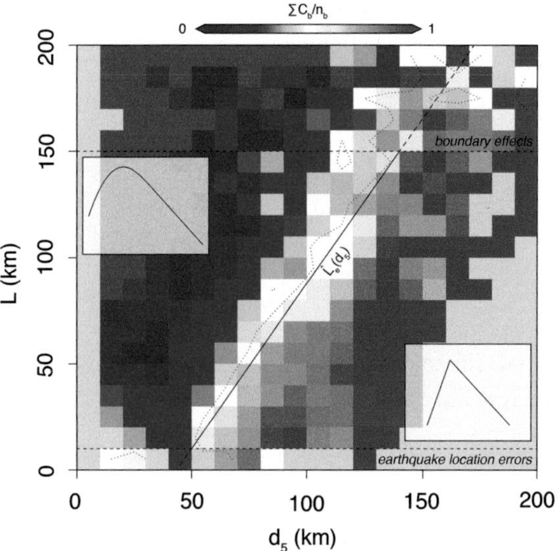

Figure 3
Spatial scale of detected seismicity as a function of distance to the fifth nearest seismic station d_5 and of cell size L. The ratio $\sum C_b/n_b$ with n_b the number of choices C_b in $\Delta d_5 \times \Delta L$ bins (Eq. 7) links the observational spatial resolution (or grain) L to the intrinsic spatial resolution L_e (*black line*, Eq. 8)

boundary effects linked to the spatial extent of the region), we find

$$\widehat{L} = 1.56d_5 - 67.8 \qquad (8)$$

the boundary (black line) between observational grains L which contain heterogeneities (in purple, $L > \widehat{L}_e$) and which are homogeneous (in red, $L \leq \widehat{L}_e$). This relationship verifies that m_c varies faster in the denser parts of a seismic network, as independently observed by Mignan *et al.* (2011), and describes the intrinsic spatial scale of the detected seismicity phenomenon. This phenomenon is thus shown to be scale-variant for $m \in (-\infty, +\infty)$ with its spatial grain a function of the configuration of the seismic network (i.e., $\widehat{L}_e = f(d_n)$), which is opposed to the scale invariance of seismicity known for $m \in [m_c, +\infty)$ for which the Gutenberg–Richter law holds and the network configuration is irrelevant.

4. Discussion

Does a better understanding of the spatial scale of the detected seismicity phenomenon allow in turn a better understanding of the behavior of seismicity below completeness and of the process of earthquake detection itself? In the aim of initiating a debate on such challenging issues, we develop upon these two important aspects with some basic illustrations.

4.1. Implications for the Universal Scaling Parameters b and D

We investigate if our results may help extending the analysis of the Gutenberg–Richter law and of the fractal dimension to lower magnitudes $m < m_c$. We first test Eq. (2) in Taiwan by measuring the number of events

$$N(m, L) = \frac{\sum_i N_i(m, L)^2}{N(m, L_0)} \qquad (9)$$

with N_i the number of events of magnitude m in cell i of size $L = L_0/2^h$, $L_0 = 4°$ and $h = \{0, \ldots, 5\}$ the level of hierarchy (Kosobokov and Mazhkenov 1994). Figure 4a shows that a data collapse is obtained on the Gutenberg–Richter law of slope $b = 0.9$ (solid line) when normalizing to $N(m, L)/L^D$ with $D = 1.2$, which is in agreement with results in other regions for earthquake epicenters (Bak *et al.* 2002). For illustration purposes, we here assume that Eq. (2) is verified if $N(m, L)/L^D$ is within a factor 2 of the theoretical curve (dotted lines). The law is verified in the range $2.1 \leq m \leq 5.0$ for $h \in [0, 5]$ while the data deviate from the law at $m < 2.1$ for all levels of hierarchy h (i.e. $m_c = 2.1$).

We then consider the corrected case $N_{ic}(m) =$ - $N_i(m)/q_i(m)$ with q_i the detection function (see Eq. 3). Since this approach assumes that the Gutenberg–Richter law holds over the unbounded magnitude range, any deviation from the law observed after correction for incompleteness would suggest that a different physical process is in play at low magnitudes. Using a fixed L in Eq. (9) means that the q model depends on cell i (Fig. 3; Eq. 8). We use the cumulative normal distribution model for $L > \widehat{L}_e$ (heterogeneous composite model; Fig. 1a) and the exponential model for $L \leq \widehat{L}_e$ (homogeneous elemental model; Fig. 1b). Results are shown in Fig. 4b and indicate that the data collapse continues in the range $0.4 \leq m < 2.1$ for at least some h when the seismicity data are corrected for incompleteness. Scattering remains high and relates to under-sampling of the original data at low m values, similarly to the under-sampling observed at high m values. The scattering at low magnitudes is, however, lower here than in the case where only one detection model q would be used. These results suggest that the behavior of events at $m < m_c$ might be informative about b and D once the scale variance of the detected seismicity phenomenon is considered. This has yet to be confirmed by robust sensitivity analyses in various applications (e.g., in b-value mapping; Tormann *et al.* 2014; Kamer and Hiemer 2015).

4.2. Implications for the Earthquake Detection Process

The Gaussian detection model $q(m) = \Phi(m \mid \mu, \sigma)$ was proposed long ago based on the assumptions that both the true magnitude m_A and the threshold magnitude m_T are normal variables—an event being detected if $m_A > m_T$ (Ringdal 1975). This would reflect a lognormal distribution of the seismic noise

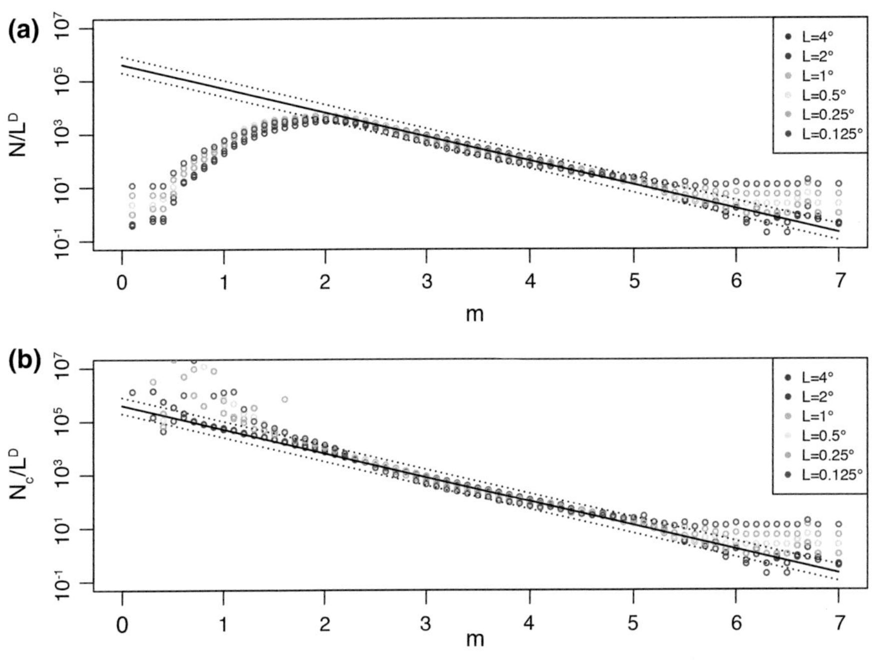

Figure 4

Data collapse for universal scaling parameters $b = 0.9$ and $D = 1.2$ in Taiwan with $N\,(m, L)$ the number of events, m the magnitude and L the cell size: **a** original data incomplete below magnitude $m = 2.1$; **b** data corrected for incompleteness, extending universality to lower magnitudes

amplitude (Richter 1935). Noise amplitude, defined as

$$A_{\mathrm{norm}} = \left| \frac{A(t) - \overline{A(t)}}{\sqrt{\overline{A(t)^2} - \overline{A(t)}^2}} \right| \tag{10}$$

with $A\ (t)$ the raw amplitude time series, has been approximated as lognormal in studies of station detection capability (Freedman 1967; Zheng *et al.* 2012).

An exponential detection model of magnitudes (Eq. 4) means a linear detection model of amplitudes and therefore suggests a triangular distribution of the noise amplitude $A\ (t)$. This is illustrated in Fig. 5a where 100 s of waveform data from the WSF station (east–west direction) of the Taiwanese seismic network is shown for comparison (use of other stations, directions and time periods does not significantly change the results). The model $q(A) = \Pr(\text{Detected} \mid A) \propto A$ for $A < A_c$, with A_c the completeness amplitude, is sketched in gray. Figure 5b shows the same empirical distribution in terms of A_{norm}, which shows some skewness in the $\log_{10}$ space. Here, we simulate 1000 $A\ (t)$ samples from the triangular distribution shown in Fig. 5a and compute the A_{norm} distribution (0.05 and 0.95 quantiles as dotted gray curves in Fig. 5b). While a systematic analysis of waveform data would be needed to prove or disprove a triangular distribution of the seismic noise, our aim is not here to validate the empirical model of Mignan (2012) but to very briefly consider the earthquake detection process in a novel way, by looking at the possible relationship between noise amplitude distribution and FMD shape (Fig. 5). Such a link has never been made, to the best of our knowledge.

Our results, however, emphasize the fallacy of conclusions on the earthquake detection process inferred from regional FMD analyses. Use of the cumulative normal distribution $\Phi(m \mid \mu, \sigma)$ is often justified based on the curved shape of regional FMDs (Ringdal 1975; Ogata and Katsura 1993) although its extrapolation to higher resolutions has already been shown to be incorrect (Mignan 2012). Therefore, the parameters μ and σ of Φ have no physical meaning and only represent

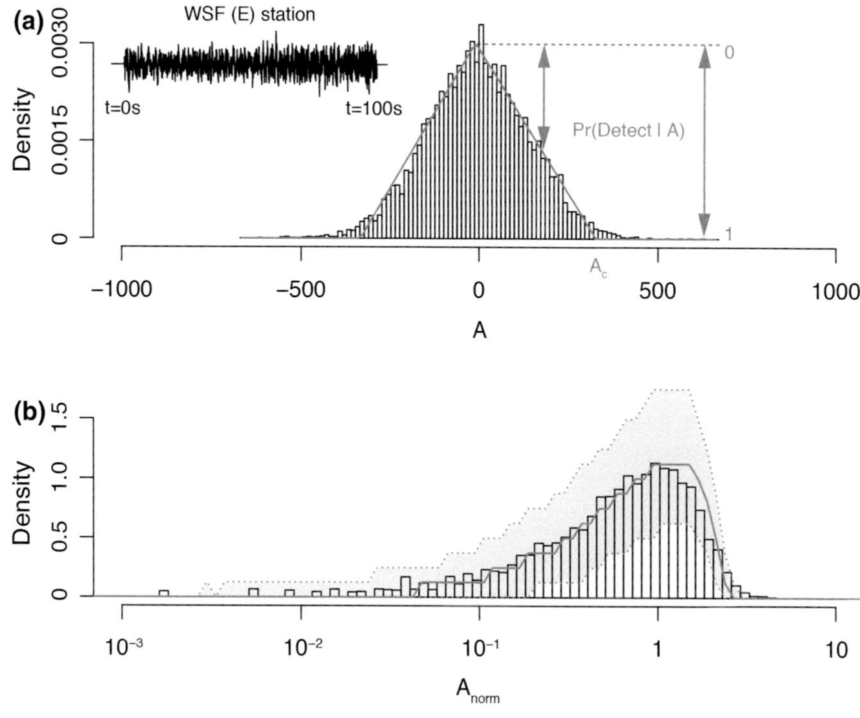

Figure 5
Seismic noise amplitude distribution, empirical (*histograms*) and inferred from the elemental FMD model (*gray curves*): **a** Raw amplitude A. The *inset* shows an example of 100 s of waveform data from the WSF station (east–west direction). Its distribution in amplitude may be approximated by a *triangular distribution*, which would explain Eq. (4) or in amplitude space $q(A) = \Pr(\text{Detected} \mid A) \propto A$ for $A < A_c$, with A_c the completeness amplitude; **b** Normalized amplitude A_{norm} (Eq. 10). The *gray curves* (*solid and dotted*) represent, respectively, the 0.50, 0.05 and 0.95 quantiles obtained for 1000 $A(t)$ simulations sampled from the *triangular distribution* shown in (**a**)

an artifact of observation due to the difference between observational scale and the intrinsic scale of the detected seismicity phenomenon previously described. In this view, the physical origin of parameter k of Eq. (4) has yet to be understood and interpreted in terms of amplitude measurement uncertainty.

5. Conclusions

Earthquake detectability issues have always hampered the study of seismicity at the lower magnitudes. From rule of thumb, using only $m \geq m_c$ means that up to about half of all the data are systematically discarded. In this article, we have identified the intrinsic spatial scale of the detected seismicity over $m \in (-\infty, +\infty)$ (Fig. 3; Eq. 8), which allows us to better understand changes in the shape of the FMD, and therefore to reduce ambiguity on m_c and to correct for data incompleteness.

As noted by Pereira (2001), not considering the intrinsic scale of the process leads to artifacts of observation, which may explain why data below m_c had previously been considered unstable (Kagan 2002). While the usefulness of the data below m_c has yet to be confirmed in future studies, there would be a clear advantage in being able to exploit this information. Statistical analyses would gain using larger datasets, for instance in the investigation of the prognostic value of earthquake precursors (see meta-analysis by Mignan 2014), which remains one of the main challenges in the field of statistical seismology.

Acknowledgements

We thank three anonymous reviewers for their useful comments.

References

Aki, K. (1965), *Maximum Likelihood Estimate of b in the Formula log N = a – bM and its Confidence Limits*, Bull. Earthquake Res. Inst. Univ. Tokyo, *43*, 237–239.

Bak, P., K. Christensen, L. Danon and T. Scanlon (2002), *Unified Scaling Law for Earthquakes*, Phys. Rev. Lett., *88*, 178501.

Freedman, H. W. (1967), *Estimating earthquake magnitude*, Bull. Seismol. Soc. Am., *57*, 747–760.

Gutenberg, B. and C. F. Richter (1944), *Frequency of earthquakes in California*, Bull. Seismol. Soc. Am., *34*, 185–188.

Kagan, Y. Y. (2002), *Seismic moment distribution revisited: I. Statistical results*, Geophys. J. Int., *148*, 520–541.

Kamer, Y. (2014), *Minimum sample size for detection of Gutenberg-Richter's b-value*, arXiv:1410.1815v1.

Kamer, Y. and S. Hiemer (2015), *Data-driven spatial b-value estimation with applications to California seismicity: To b or not to b*, J. Geophys. Res. Solid Earth, in press, doi:10.1002/2014JB011510.

Kosobokov, V. G. and S. A. Mazhkenov (1994), *On similarity in the spatial distribution of seismicity*, Computational Seismol. Geodyn., *1*, 6–15.

Kraft, T., A. Mignan and D. Giardini (2013), *Optimization of a large-scale microseismic monitoring network in northern Switzerland*, Geophys. J. Int., *195*, 474–490, doi:10.1093/gji/ggt225.

Mignan, A., M. J. Werner, W. Wiemer, C.-C. Chen and Y.-M. Wu (2011), *Bayesian Estimation of the Spatially Varying Completeness Magnitude of Earthquake Catalogs*, Bull. Seismol. Soc. Am., *101*, 1371–1385, doi:10.1785/0120100223.

Mignan, A. (2012), *Functional shape of the earthquake frequency-magnitude distribution and completeness magnitude*, J. Geophys. Res., *117*, B08302, doi:10.1029/2012JB009347.

Mignan, A., C. Jiang, J. D. Zechar, S. Wiemer, Z. Wu and Z. Huang (2013), *Completeness of the Mainland China Earthquake Catalog and Implications for the Setup of the China Earthquake Forecast Testing Center*, Bull. Seismol. Soc. Am., *103*, 845–859, doi:10.1785/0120120052.

Mignan, A. (2014), *The debate on the prognostic value of earthquake foreshocks: A meta-analysis*, Sci. Rep., *4*, 4099, doi:10.1038/srep04099.

Mignan, A. and G. Chouliaras (2014), Fifty Years of Seismic Network Performance in Greece (1964–2013): *Spatiotemporal Evolution of the Completeness Magnitude*, Seismol. Res. Lett., *85*, 657–667, doi:10.1785/0220130209.

Molchan, G. and T. Kronrod (2005), *On the spatial scaling of seismicity rate*, Geophys. J. Int., *162*, 899–909, doi:10.1111/j.1365-246X.2005.02693.x.

Ogata, Y. and K. Katsura (1993), *Analysis of temporal and spatial heterogeneity of magnitude frequency distribution inferred from earthquake catalogues*, Geophys. J. Int., *113*, 727–738.

Ogata, Y. and K. Katsura (2006), *Immediate and updated forecasting of aftershock hazard*, Geophys. Res. Lett., *33*, L10305, doi:10.1029/2006GL025888.

Pereira, G. M. (2001), *A Typology of Spatial and Temporal Scale Relations*, Geographical Analysis, *34*, 21–33.

Richter, C. F (1935), *An instrumental earthquake magnitude scale*, Bull. Seismol. Soc. Am., *25*, 1–32.

Ringdal, F. (1975), *On the estimation of seismic detection thresholds*, Bull. Seismol. Soc. Am., *65*, 1631–1642.

Schwarz, G. (1978), *Estimating the dimension of a model*, Ann. Stat., *6*, 461–464.

Tormann, T., S. Wiemer and A. Mignan (2014), *Systematic survey of high-resolution b value imaging along Californian faults: inference on asperities*, J. Geophys. Res. Solid Earth, *119*, 2029–2054, doi:10.1002/2013JB010867.

Vorobieva, I. C. Narteau, P. Shebalin, F. Beauducel, A. Nercessian, V. Clouard and M.-P. Bouin (2013), *Multiscale Mapping of Completeness Magnitude of Earthquake Catalogs*, Bull. Seismol. Soc. Am., *103*, 2188–2202, doi:10.1785/0120120132.

Voronoi, G. F. (1908), *Nouvelles applications des paramètres continus à la théorie de forms quadratiques*, J. für die Reine und Angewandte Mathematik, *134*, 198–287.

Wu, Y.-M., C.-H. Chang, L. Zhao, T.-L. Teng and M. Nakamura (2008), *A Comprehensive Relocation of Earthquakes in Taiwan from 1991 to 2005*, Bull. Seismol. Soc. Am., *98*, 1471–1481, doi:10.1785/0120070166.

Zheng, Z., T. Takaishi, N. Sakurai, X. Zhang and K. Yamasaki (2012), *Statistical Regularities of Seismic Noise*, Progr. Theoretical Phys. Suppl., *194*, 193–201.

(Received October 24, 2014, revised June 23, 2015, accepted June 27, 2015, Published online July 10, 2015)

Pure Appl. Geophys. 173 (2016), 125–132
© 2015 Springer Basel
DOI 10.1007/s00024-015-1034-9

Pure and Applied Geophysics

Visibility Graph Analysis of the 2003–2012 Earthquake Sequence in the Kachchh Region of Western India

LUCIANO TELESCA,[1] MICHELE LOVALLO,[2] S. K. AGGARWAL,[3] P. K. KHAN,[4] and B. K. RASTOGI[3]

Abstract—A visibility graph (VG) is a rather novel statistical method in earthquake sequence analysis; it maps a time series into networks or graphs, converting dynamical properties of the time series into topological properties of networks. By using the VG approach, we defined the parameter window mean interval connectivity time $<T_c>$, that informs about the mean linkage time between earthquakes. We analysed the time variation of $<T_c>$ in the aftershock-depleted catalogue of Kachchh Gujarat (Western India) seismicity from 2003 to 2012, and we found that $<T_c>$: i) changes through time, indicating that the topological properties of the earthquake network are not stationary; and, ii) appeared to significantly decrease before the largest shock (M5.7) that occurred on March 7, 2006 near the Gedi fault, an active fault in the Kachchh region.

Key words: Visibility graph, seismicity, Kachchh.

1. Introduction

Recently, the visibility graph (VG) method has represented a novel approach in describing statistical characteristics of earthquakes by focusing on an earthquake's topological properties. The VG method was developed by LACASA *et al.* (2008), who proposed a way to map a time series into networks or graphs. They showed that such mapping converts dynamical properties of a time series into topological properties of networks and vice versa, and the features of networks can be informative of the characteristics of the associated time series. Specifically, periodic time series are mapped into regular networks, random time series are mapped into random graphs, and fractal time series are mapped into scale-free networks (LACASA *et al.* 2008; DONNER *et al.* 2011; CAMPANHARO *et al.* 2011).

In the VG approach, a segment connects any two values of a series visible to each other, meaning that such segment is not broken by any other intermediate value. In terms of graph theory, each time series value represents a node, and two nodes are connected if visibility exists between them. The mathematical definition of the visibility criterion (LACASA *et al.* 2008) is the following: two arbitrary data values (t_a, y_a) and (t_b, y_b) are visible to each other if any other data (t_c, y_c) placed between them fulfills the following constraint:

$$y_c < y_b + (y_a - y_b)\frac{t_b - t_c}{t_b - t_a}. \quad (1)$$

Let's indicate the connectivity degree with k_i, which is the number of connections of each node i. The following properties always hold (LACASA *et al.* 2008): (1) Connection: each node is visible at least by its nearest neighbors (left and right); (2) Absence of directionality: no direction is defined in the links; (3) Invariance under affine transformations (rescaling of both axes and horizontal and vertical translations) of the time series.

The VG method was mainly applied to investigate the dynamical properties of continuous signals. Using the VG method, PIERINI *et al.* (2012) demonstrated that the time series of hourly means of wind speed recorded at two wind stations in central Argentina (one inland and the other coastal), finding that the topological properties of the two series are similar and do not depend on the characteristics of the two sites. Analysis of the v–k (sample value–connectivity

[1] National Research Council, Institute of Methodologies for Environmental Analysis, C.da S.Loja, 85050 Tito, Italy. E-mail: luciano.telesca@imaa.cnr.it
[2] ARPAB, 85100 Potenza, Italy.
[3] Institute of Seismological Research, Gandhinagar, Gujarat, India.
[4] Indian School of Mines, Dhanbad, Jharkhand, India.

degree) plots also suggested that higher series values are not necessarily "hubs," that is, values characterized by large connectivity. Telesca *et al.* (2012) analysed ocean tide records in central Argentina using the VG method and discriminated local effects (linked with the coastal conditions) from global effects (linked to more general and common atmospheric forcing and ocean current conditions) by only analysing the properties of the connectivity degree distribution curve.

The VG method was recently also applied to point processes (a sequence of events randomly occurring over time; see Fig. 1 for a visual sketch). Telesca and Lovallo (2012a, b) applied the VG method to the seismicity of whole country of Italy, finding power-law behavior in the distribution of the connectivity degree, independent of the time-clustering structure and independent of an increase in the magnitude threshold. Telesca *et al.* (2013) performed the VG analysis over the sequences of earthquake occurred in the subduction zone of Mexico and found that the *k*-*M* plots (the relationship between the magnitude *M* of each event and its connectivity degree *k*) were characterized by an increasing *k* trend associated with an increasing *M*, thus revealing the property of a hub as typical of the higher magnitude events.

In the present paper, we investigated the earthquake series that occurred in the Kachchh region of Gujarat (Western India) between the years 2003 and 2012. This area has generated great interest amongst the seismological community because it was struck by the strong Bhuj earthquake (Mw 7.7) on the 26th of January 2001, causing more than 160,000 injuries and more than 20,000 deaths, thus instigating a challenging search for precursory earthquake signatures. After the Bhuj earthquake, moderate seismicity still occurs, making the region a natural seismic laboratory for seismologists in terms of moderate earthquakes (magnitude up to 5.7). The area of interest extends to several nearby faults, such as the Allah band fault (ABF), the Island belt fault (IBF), the Gedi fault (GF), the North Wagad fault (NWF), the South Wagad fault (SWF), the Kachchh Mainland fault (KMF), and the Katrol Hill fault (KHF) (Fig. 2).

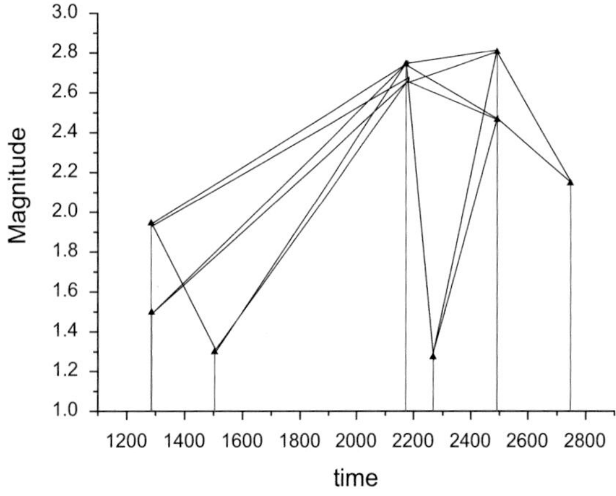

Figure 1
Sketch of the VG method applied to a synthetic seismic sequence

2. Seismotectonic Settings

According to Rastogi *et al.* (2012), the Kachchh region is considered one of the most seismically active intraplate regions worldwide. It is known to have an elevated earthquake hazard due to the occurrence of several large earthquakes. including the 2001 Bhuj earthquake that occurred along the south-dipping NWF, a blind dip-slip fault system in Kachchh, which slips in reverse to the motion in response to the prevailing north–south compressional forces attributed to the northward motion of the Indian plate (Mandal *et al.* 2004a; Bodin and Horton 2004). This event continues to produce aftershocks within the same fault and triggers seismic activity within several other nearby faults. Further, the presence of inter-connected, rupture-nucleated trends suggest a possible link between these faults (Mishra and Zhao 2003; Talwani 2001).

The Kachchh region is an example of a tectonically controlled landscape whose physiographic features are the manifestation of plate movement along the tectonic lineament of the pre-Mesozoic basinal configuration that was produced by a primordial fault pattern in the pre-Cambrian basement (Biswas and Deshpande 1970; Biswas 1987). The lithosphere of the epicentral region of the 2001 earthquake is inferred to be hot and thin (only 70 km as compared to a normal 100 km) and its crustal thickness is also small (34 km as compared to 40 km in the surrounding region), caused by rifting approximately 184 million years ago. The restructuring of

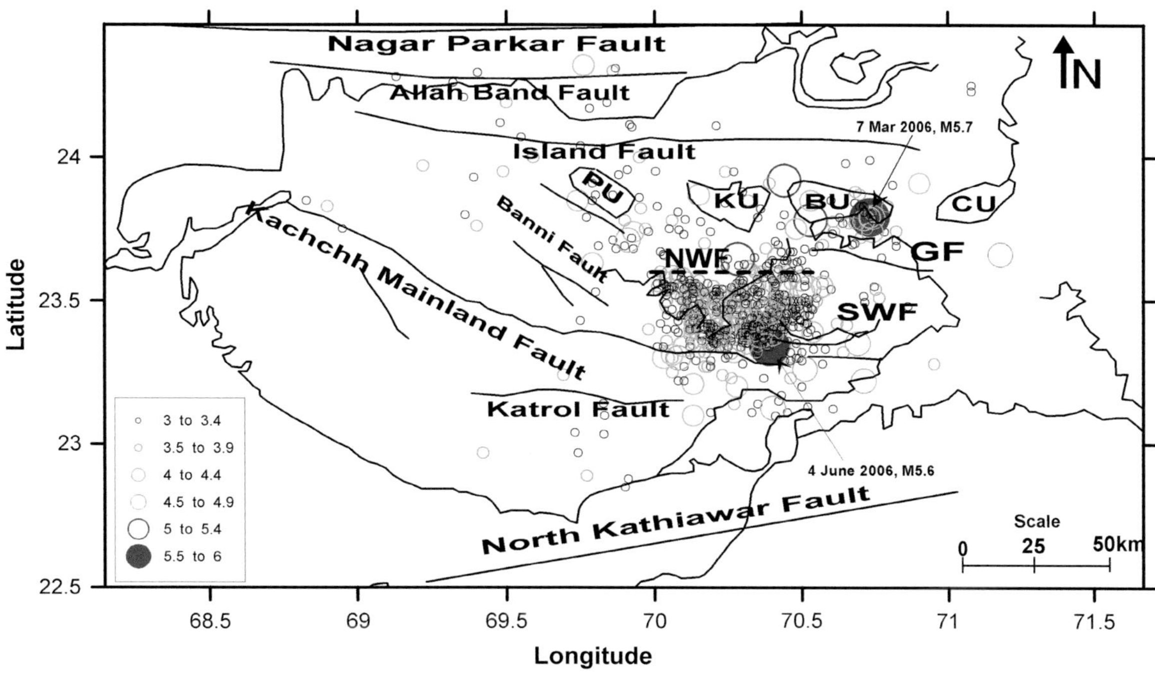

Figure 2
Spatial distribution of de-clustered Kachaah seismicity from 2003 to 2012

this warm and thin lithosphere might have occurred due to a thermal plume 65 million years ago (Mandal and Pandey 2010). The KMF is the major fault along the rift axis and has become the active principle fault; the fault presently shows a right lateral strike slip movement as an overstep by the SWF in the eastern part of the basin (Biswas 2005). The overstepped zone between the two wrench faults is a convergent transfer zone that has undergone transpressional stress in the strained eastern part of the basin.

3. Results

We analysed the seismicity of the Kachchh area for the period from 2003 to 2012. The Indian Seismometric Network has undergone a continuous upgrade, leading to a progressive decrease in the completeness magnitude of the seismic catalogue with time. Figure 3 shows the time variation of the comleteness magnitude calculated by using the maximum curvature method (Woessner and Wiemer 2005). As it can be seen, the completeness magnitude strongly changes with time. In our study, we

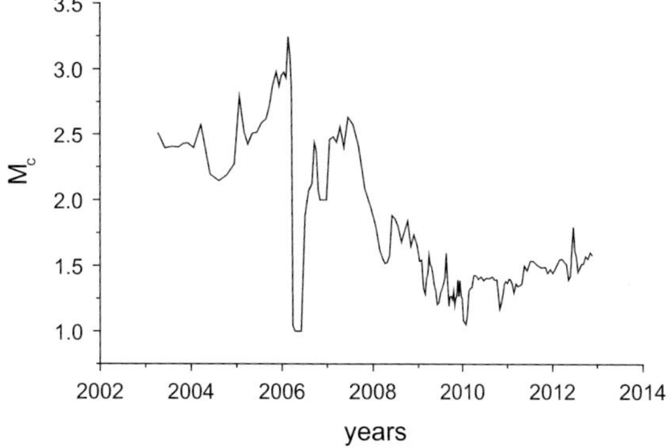

Figure 3
Time variation of the completeness magnitude calculated by using the maximum curvature method (Woessner and Wiemer 2005)

preferred to be very conservative and considered the sequence of events with magnitudes equal or larger than 3.0. Before analysing the seismic catalog by means of the VG method, we removed the aftershocks by using the Gardner and Knopoff (1974) technique jointly with the Uhrhammer space–time window (van Stiphout *et al.* 2012). For each earthquake in the catalog with magnitude *M*, the successive events are identified as aftershocks if they occur within a specified time interval and within a distance interval defined by the Uhrhammer relations (van Stiphout *et al.* 2012) to, respectively, the occurrence time and epicenter of the first earthquake. The obtained de-clustered catalogue contains 865 events with magnitudes larger or equal to 3.0, the epicentres of which are shown in Fig. 2. We applied the VG method to the de-clustered seismic catalogue. In particular, the time variation of the VG window mean interval connectivity time $<T_c>$, defined below, was analysed. Considering a sliding window of N events with a shift of one event between two successive windows, for each event in the window, we computed the interval connectivity time T_c as the time interval between two visible events (events that satisfy the visibility condition of Eq. 1). For each event, we calculated the mean interval connectivity time $<T_c>$, the average of all the intervals T_c intervals corresponding to that event. Now, for each window we calculated the average of all the $<T_c>$, obtaining the window mean interval connectivity time $<T_c>$. Each calculated value of $<T_c>$ was associated with the time of occurrence of the last event in the sliding window. Figure 4 shows the time variation of $<T_c>$ (red curve) along with the largest earthquake that occurred on the 7th of March 2006 with a magnitude 5.7 within the same time span of the variability of $<T_c>$. We used a window length of $N = 100$ events. We can observe that before the occurrence of the largest shock, $<T_c>$ decreases almost suddenly, while afterwards, it starts to increase; the rate of increase after the earthquake is lower than that of the decrease before that event. After the increasing phase, the $<T_c>$ recovers to its initial conditions, evolving in a more or less stable behavior. In order to check that the behavior shown by $<T_c>$ is not due to chance, we applied the same analysis on the 100 shuffled seismic sequences. Each shuffle is obtained, maintaining the order of the magnitudes as fixed, but shuffling the occurrence times; in this way, we obtain a random earthquake sequence that shares with the original sequence the same probability density function of the inter-event times and the same magnitude distribution. For each shuffle, we calculated the time variation of the parameter $<T_c>_{\text{shuf}}$ considering a sliding window of 100 events, similar to the original sequence. Therefore, at each time, 100 values of $<T_c>_{\text{shuf}}$ are obtained. The 95 % confidence band was estimated as the 2.5th and the 97.5th quantiles of the distribution of the $<T_c>_{\text{shuf}}$ at that time. Figure 4 shows that the decrease of $<T_c>$ before the occurrence of the M5.7 earthquake is outside the 95 % confidence band of $<T_c>_{\text{shuf}}$ (black dotted lines), indicating that this decrease is significant. In order to check the robustness of the obtained results with respect to the length of the sliding window, we repeated the analysis with a moving window of 50 and 150 events. The results, shown in Figs. 5 and 6, are similar to those obtained

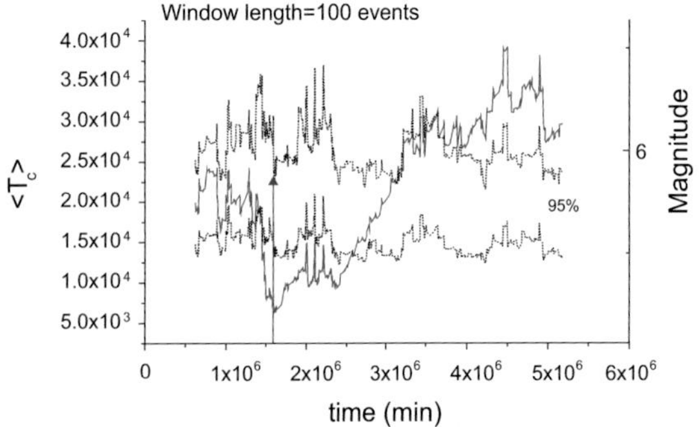

Figure 4
Time variation of $<T_c>$ (*red*) and the 95 % confidence band (*black dotted*), using a window length of 100 events. The largest earthquakes occurred on the 7th of March 2006 ($M = 5.7$), indicated by the *blue vertical arrow*

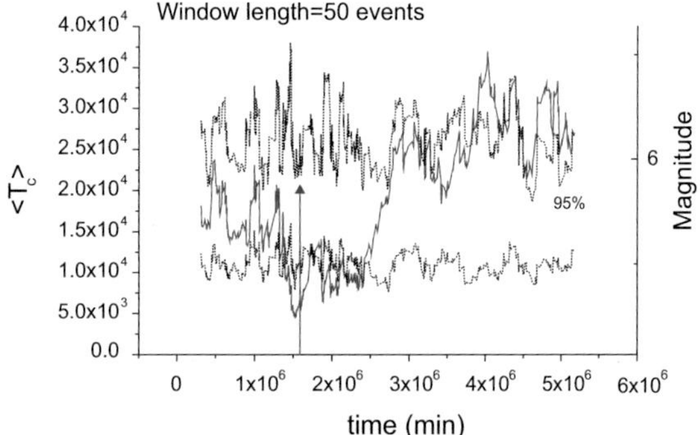

Figure 5
Time variation of $<T_c>$ (*red*) and the 95 % confidence band (*black dotted*) using a window length of 50 events. The largest earthquakes occurred on the 7th of March 2006 ($M = 5.7$), indicated by the *blue vertical arrow*

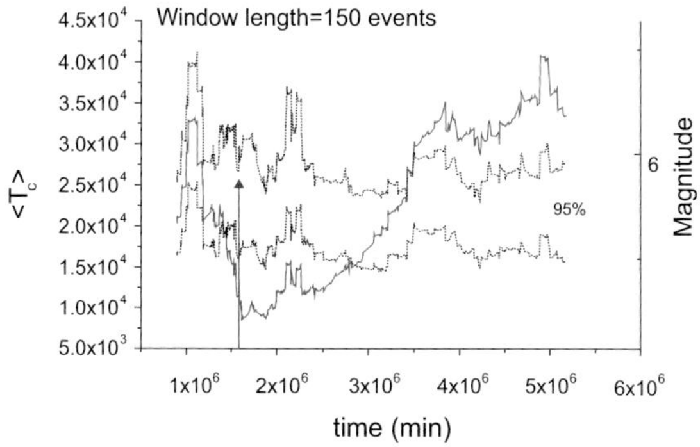

Figure 6
Time variation of $<T_c>$ (*red*) and the 95 % confidence band (*black dotted*) using a window length of 150 events. The largest earthquakes occurred on the 7th of March 2006 ($M = 5.7$), indicated by the *blue vertical arrow*

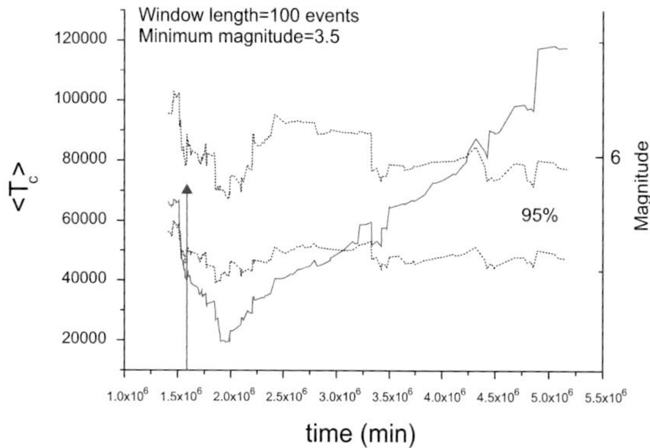

Figure 8
Time variation of $<T_c>$ (*red*) and the 95 % confidence band (*black dotted*) using a window length of 100 events and minimum magnitude of 3.5. The largest earthquakes occurred on the 7th of March 2006 ($M = 5.7$), indicated by the *blue vertical arrow*

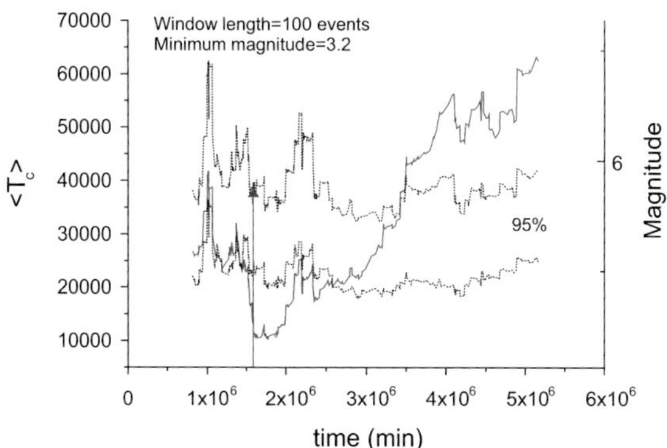

Figure 7
Time variation of $<T_c>$ (*red*) and the 95 % confidence band (*black dotted*) using a window length of 100 events and minimum magnitude of 3.2. The largest earthquakes occurred on the 7th of March 2006 ($M = 5.7$), indicated by the *blue vertical arrow*

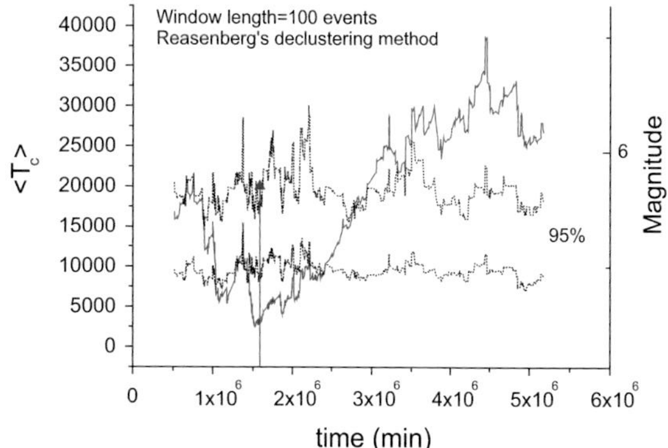

Figure 9
Time variation of $<T_c>$ (*red*) and the 95 % confidence band (*black dotted*) using a window length of 100 events and minimum magnitude of 3.0. The catalogue was de-clustyered by using Reasenberg's method. The largest earthquakes occurred on the 7th of March 2006 ($M = 5.7$), indicated by the *blue vertical arrow*

with a 100-event window, and this reinforces the significance of the decrease of $<T_c>$ before the occurrence of the largest shock of March 2006.

In order to check the robustness of the results with different threshold magnitudes and de-clustering techniques, we applied the VG method for sliding windows of 100 events to the Uhmhammer-based de-clustered catalogue with cut-off magnitudes of 3.2 (Fig. 7) and 3.5 (Fig. 8), and to the Reasenberg-based (Reasenberg 1985) de-clustered catalog with a cut-off magnitude of 3.0 (Fig. 9). The results show that the method could reliably identify anomalous decreasing behaviour of the parameter $<T_c>$ before the occurrence of the largest shock.

4. Discussion

It is well known that earthquakes are correlated with time. Several studies applying time-correlation methods have revealed such (Telesca *et al.* 2003; 2007). However, earthquakes are magnitude-marked point processes (e.g., Daley and Vere-Jones 2003, 2008), and most of the methods used to investigate the time properties of seismicity do not involve directly with magnitude (a method focused on the consecutively emitted energies of earthquakes) and, hence, magnitude, is a natural time analysis (Varotsos *et al.* 2006, 2011; Sarlis *et al.* 2009; Sarlis 2011;

Sarlis and Christopoulos 2012). The VG approach could be viewed as a topological method able to capture correlations in earthquake sequences involving both the magnitude and the occurrence time. Linking the earthquakes with each other based on the visibility criterion (Eq. 1) that takes into account both the occurrence time and the magnitude of each event converts the earthquake sequence into a clustered network. In this network, the smaller and more frequent events are linked with the nearest larger events that, having a better visibility contact with more neighbors than other events, play the role of hubs of the network. Furthermore, the larger events, more infrequent, also tend to inter-connect with each other, linking clusters with clusters, leading to an interdependence in the occurrence of major earthquakes (Romanowicz 1993; Mega *et al.* 2003).

The VG procedure seems not so different from the idea behind the single link cluster (SLC) method (Frohlich and Davis 1990), on the basis of which the analysis of spatial correlation length was proposed as a precursor of large earthquakes (Zoeller *et al.* 2001, 2002); the difference relies in the domain; that is, in the SLC approach, the link between two earthquakes is a space-type link, and in the VG method it is a time-type link.

What does the VG parameter $<T_c>$ mean? On the basis of its definition, the parameter $<T_c>$ relies, substantially, on the mean inter-event time between one earthquake and all those that are "visible" to it in the VG domain. Two earthquakes that are "visible" to each other could also be not consecutive; in fact it's the magnitude of the events that governs the "visibility" and, thus, the linkage between the earthquakes. Therefore, $<T_c>$ averages all these means over a group of earthquakes (those included in the moving window). If $<T_c>$ is high, this means that, on average, more time has to pass so that two earthquakes will be linked to each other, while if it is low, this means that less time has to pass so that two earthquakes will be visibly connected.

In our case, we observe a decrease of the VG parameter $<T_c>$ before the occurrence of the largest shock of the analysed sequence; this indicates that before its occurrence, the average linkage time between "visible" earthquakes lowers. This behaviour could also suggest that before the largest shock, time-clustering of the events occurring before is enhancing, and could be considered consistent with what was observed by Tyupkin and Di Giovambattista (2005). Even though these authors analysed the decrease of the spatial correlation length before a large earthquake, indicating the approach to a critical point behaviour, they have reported a relationship between the characteristic time interval between events involved in the correlation length estimation and the linear dimension of the domain where the spatial correlation length is evaluated. From this relationship it is possible, then, to deduce that the characteristic time should be low such that the decrease in the spatial correlation length could be observed. This lowering of the VG parameter $<T_c>$ before the largest event could be thought as related to such characteristic time and its decrease would be considered consistent with the decrease in the correlation length before some large earthquakes observed by Tyupkin and Di Giovambattista (2005).

The decrease of $<Tc>$ is also compatible with the recent observation that the fluctuations of the seismicity order parameter in natural time exhibits remarkable minima before the strongest earthquakes (Varotsos *et al.* 2011, 2013; Sarlis *et al.* 2013)

5. Conclusions

The VG method represents a novel approach in investigating the time dynamics of seismicity. Of course, it should not be considered as being better than the standard statistical methods that are well known and well assessed in getting information about the time dynamics of earthquake sequences. However, the VG method could be viewed as an alternative method that allows one to investigate an earthquake sequence focusing on their topological properties and interconnectivity. Application of the VG method to the seismicity of the Kachchh area has revealed that a decrease in the connectivity time between (also non-consecutive) earthquakes appears before the largest shock of the sequence; and such decrease is significant at 95 % and also robust with respect to the length of the event window. This result strengthens the reliability of the VG approach in detecting precursory signatures in earthquake sequences.

Acknowledgments

The present study was supported by the ISR project "Statistical Seismic Data Analysis of Gujarat Region, Western India." The authors from The Institute of Seismological Research are thankful to the Department of Science and Technology, Gujarat for providing all necessary facilities to carry out this research.

References

Biswas, S. K., and Deshpande, S. V., 1970. Geological and Tectonic maps of Kutch. In: Bull. Oil. Nat. Gas Commn., *7*, 115–116.

Biswas, S. K., 1987. *Regional tectonic framework structure and evolution of western marginal basin of India.* Tectonophysics, *135*, 307–327.

Biswas, S.K., 2005. *A review of structure and tectonics of Kutch basin, western India, with special reference to earthquakes.* Current Science, *88*, No. 10, pp. 1592–1600.

Bodin, P. and Horton, S., 2004. *Source parameter and tectonic implication of aftershocks of the Mw7.6 Bhuj earthquake of January 26, 2001.* Bulletin of the Seismological Society of America, *94*, 818–827.

Campanharo A. S. L. O., Sirer M. I., Malmgren R. D., Ramos F. M., Amaral L. A. N., 2011, *Duality between Time Series and Networks.* PLoS ONE, *6*, e23378.

Daley, D.J., Vere-Jones, D., 2003. An Introduction to the Theory of Point Processes Volume I: Elementary Theory and Methods. Springer, New York, 471p.

Daley, D.J., Vere-Jones, D., 2008. An Introduction to the Theory of Point Processes Volume II: General Theory and Structure. Springer, New York, 573p.

Donner R. V., Donges J. F., 2011, *Visibility graph analysis of geophysical time series: Potentials and possible pitfalls.* Acta Geophysica, *60*, 589–623.

Frohlich, C., Davis, S. D., 1990, *Single-link cluster analysis as a method to evaluate spatial and temporal properties of earthquake catalogues.* Geophysical Journal International, *100*, 19–32.

Gardner, J. K., Knopoff, L., 1974, *Is the sequence of earthquakes in Southern California, with aftershocks removed, Poissonian?* Bulletin of the Seismological Society of America, *64*, 1363–1367.

Lacasa L., Luque B., Ballesteros F., Luque J., Nuño J. C., 2008, *From time series to complex networks: The visibility graph.* Proceedings of the National Academy of Sciences, *105*, 4972–4975.

Mishra, O. P., and Zhao, D., 2003. *Crack density, saturation rate and porosity at the 2001 Bhuj, India earthquake hypocenter: a fluid-driven earthquake*? Earth Planet Science Letters, *212*, 393–405.

Mandal, P., Rastogi, B. K., Satyanarayana, H. V. S., Kousaliya, M., Vijayraghavan, C., Satyamurthy, C., Raju, I. P., Sharma, A. N. S., and Kumar, N. 2004a. *Characterization of the causative fault system for the 2001 Bhuj earthquake of Mw7.7.* Tectonophysics, *378*, 105-121.

Mandal, P., and Pandey, O. P., 2010. *Relocation of aftershocks of the 2001 Bhuj earthquake: a new insight into seismotectonics of the Kachchh seismic zone, Gujarat, India.* Journal of Geodynamics, *49*, 254–260.

Mega, M. S., Allegroni, P., Grignolini, P., Latora, V., Palatela, L., Rapisarda, A., Vinciguerra, S., 2003, *Power-law time distribution of large earthquakes.* Physical Review Letters, *90*, 188501.

Pierini, J. O., Lovallo, M., Telesca, L., 2012. *Visibility graph analysis of wind speed records measured in central Argentina.* Physica A, *391*, 5041–5048.

Rastogi, B. K., Kumar, S., Aggrawal, S. K., 2012. *Seismicity of Gujarat.* Natural Hazards, doi:10.1007/s11069-011-0077-1.

Reasenberg, P.A., 1985, *Second-order moment of Central California Seismicity.* Journal of Geophysical Research, *90*, 5479–5495.

Romanowicz, B., 1993, *Spatiotemporal patterns in the energy-release of great earthquakes.* Science, *260*, 1923–1926.

Talwani, P. and Gangopadhyay A., 2001, *Tectonic framework of the Kachchh earthquake of 26 January 2001.* Seismological Research Letters, *72*, 336-345.

Telesca L., Lovallo M., 2012, *Analysis of seismic sequences by using the method of visibility graph.* Europhysics Letters, *97*, 50002.

Telesca L., Lovallo M., Ramirez-Rojas A., Flores-Marquez L., 2013, *Investigating the time dynamics of seismicity by using the visibility graph approach: Application to seismicity of Mexican subduction zone.* Physica A *392*, 6571–6577.

Telesca, L., Lapenna, V., Macchiato, M., 2003, *Spatial variability of time-correlated behaviour in Italian seismicity.* Earth and Planetary Science Letters, *212*, 279–290.

Telesca, L., Lovallo, M., Lapenna, V., Macchiato, M., 2007, *Long-range correlations in 2-dimensional spatio-temporal seismic fluctuations.* Physica A, *377*, 279–284.

Telesca, L., Lovallo, M., 2012. *Analysis of seismic sequences by using the method of visibility graph.* Europhysics Letters, *97*, 50002.

Tyupkin, Y. S., Di Giovambattista, R., 2005. *Correlation length a san indicator of critical point bhevior prior to a large earthquake.* Earth and Planetary Science Letters, *230*, 85–96.

Varotsos, P., Sarlis, N., Skordas, E., Lazaridou, M., 2006. *Attempt to distinguish long-range temporal correlations from the statistics of the increments by natural time analysis.* Physical Review E, *74*, 021123.

Varotsos, P. A., Sarlis, N. V., Skordas, E. S., 2011. *Scale-specific order parameter fluctuations of seismicity in natural time before mainshocks.* Europhysics Letters, *96*, 59002.

Varotsos, P. A., Sarlis, N. V., Skordas, E. S., Lazaridou, M. S., 2013. *Seismic Electric Signals: An additional fact showing their physical interconnection with seismicity.* Tectonophysics, *589*, 116–125.

Sarlis, N.V., Skordas, E. S., Varotsos, P. A., Nagao, T., Kamogawa, M., Tanaka, H., Uyeda, S., 2013. *Minimum of the order parameter fluctuations of seismicity before major earthquakes in Japan.* Proceedings of the National Academy of Sciences, *110*, 13734–13738.

Sarlis, N., Skordas, E., Varotsos, P., 2009. *Multiplicative cascades and seismicity in natural time.* Physical Review E, *80*, 022102.

Sarlis, N. V., 2011. *Magnitude correlations in global seismicity.* Physical Review E, *84*, 022101.

Sarlis, N.V., Christopoulos, S.-R.G., 2012. *Natural time analysis of the Centennial Earthquake Catalog.* CHAOS, *22*, 023123.

van Stiphout, T., Zhuang, J., Marsan, D., 2012. Seismicity declustering, Community Online Resource for Statistical Seismicity Analysis, doi:10.5078/corssa-52382934. Available at http://www.corssa.org.

Woessner, J., Wiemer, S., 2005. *Assessing the quality of earthquake catalogues: estimating the magnitude of completeness and its uncertainty*. Bulletin of the Seismological Society of America, *95*, 684–698, doi:10.1785/0120040007.

Zoeller G., Hainzl S., 2002. *A systematic spatiotemporal test of the critical point hypothesis for large earthquakes*. Geophysical Research Letters, *29*, 53/1–4.

Zoeller G., Hainzl S., Kurths J., 2001. *Observation of growing correlation length as an indicator for critical point behavior prior to large earthquakes*. Journal of Geophysical Research, *106*, 2167–2176.

(Received July 31, 2014, revised December 27, 2014, accepted January 8, 2015, Published online February 1, 2015)

Pure Appl. Geophys. 173 (2016), 133–152

DOI 10.1007/s00024-015-1103-0

Pure and Applied Geophysics

Analysis of Foreshock Sequences in California and Implications for Earthquake Triggering

Xiaowei Chen[1] and Peter M. Shearer[2]

Abstract—We analyze foreshock activity in California and compare observations with simulated catalogs based on a branching aftershock-triggering model. We first examine foreshock occurrence patterns for isolated $M \geq 5$ earthquakes in southern California from 1981 to 2011 and in northern California from 1984 to 2009. Among the 64 $M \geq 5$ mainshocks, excluding 3 swarms and 3 doubles, 53 % of the rest are preceded by at least one foreshock within 30 days and 5 km. Foreshock occurrence appears correlated with mainshock faulting type and depth. Foreshock area is correlated with the magnitude of the largest foreshock and the number of foreshocks, however, it is not correlated with mainshock magnitude. We then examine the occurrence pattern of all seismicity clusters without a minimum magnitude requirement, and the possibility that they are "foreshocks" of larger mainshocks. Only about 30 % of the small clusters lead to a larger cluster. About 66 % of the larger clusters have foreshock activities, and the spatial distribution pattern is similar to $M \geq 5$ mainshocks, with lower occurrence rates in the Transverse Range and central California and higher occurrence rates in the Eastern California Shear Zone and the Bay Area. These results suggest that foreshock occurrence is largely controlled by the regional tectonic stress field and fault zone properties. In special cases, foreshock occurrence may be useful for short-term forecasting; however, foreshock properties are not reliably predictive of the magnitude of the eventual "mainshock". Comparison with simulated catalogs suggest that the "swarmy" features and foreshock occurrence rate in the observed catalogs are not well reproduced from common statistical models of earthquake triggering.

1. Introduction

Mogi (1963) distinguished three main types of earthquake sequences: (1) mainshocks with both foreshocks and aftershocks; (2) mainshocks and aftershocks but no foreshocks; and (3) earthquake swarms that lack clear mainshocks. There have been studies of the triggering process involved in each category. Aftershocks are usually assumed triggered by dynamic or static stress changes imposed by the mainshocks (e.g., Toda *et al.* 2012), and earthquake swarms are thought to result from underlying crustal transient processes (e.g., Vidale and Shearer 2006; Chen *et al.* 2012). Foreshocks are of great interest because of their possible triggering role and predictive value, but their relationship to mainshocks is still poorly understood. The successful evacuation prior to the 1975 M7.3 Haicheng earthquake is a promising example for earthquake prediction; however, many mainshocks occur abruptly without foreshocks (e.g., the 2004 Parkfield earthquake), or the foreshocks are only recognized retrospectively (e.g., the 1992 Landers earthquake) Mignan (2014).

Two models have been suggested to explain foreshock occurrence: (1) "rupture model", where foreshocks and aftershocks can be explained with a common triggering model, as indicated by statistical tests of California seismicity Felzer *et al.* (2004), therefore, the mainshock is just an accidentally larger aftershock; (2) "pre-slip model", where foreshocks are triggered by quasi-static slip occurring within the mainshock nucleation zone, and foreshock properties are possibly predictive of mainshock magnitude (Dodge *et al.* 1996). Dodge *et al.* (1996) reported scaling of foreshock area with mainshock magnitude, which is similar to the scaling relationship of the proposed nucleation phase of Ellsworth and Beroza (1995). However, Felzer *et al.* (2004) found a much stronger correlation between foreshock area and the magnitude of the largest foreshock, instead of the magnitude of the mainshock, suggesting that foreshock area is not a useful predictor of the eventual mainshock size. Recent observations have found that foreshocks may be driven by an independent slow-slip phase (not part of the nucleation process), concurrently occurring within the fault zone (Kato

[1] University of Oklahoma, Norman, OK 73019, USA. E-mail: xiaowei.chen@ou.edu

[2] University of California, San Diego, La Jolla, CA 92093, USA.

et al. 2012; Chen and Shearer 2013), or within a wide region along the plate interface for interplate earthquakes (Bouchon *et al.* 2013).

The relative location and time between foreshocks and mainshocks is of great importance in recognizing foreshock sequences. For the foreshock sequences in Abercrombie and Mori (1996), all but one continue to the last day before the mainshock within 5 km of its hypocenter. In Bouchon *et al.* (2013), 70 % of interplate earthquakes have foreshocks continuous to the last day, but within a much larger spatial extent (up to 50 km). For the three M7 mainshocks in southern California, high-resolution earthquake catalogs reveal foreshock activities concentrated within hours of the mainshock Chen and Shearer (2013) within 0.5–2 km of the mainshock hypocenter. Some oceanic transform faults have enhanced immediate foreshock activities within hours before mainshocks within 5 km McGuire (2005). These observations suggest that, if foreshocks exist, they typically continue to immediately before the mainshock; thus, if a sequence is identified as a foreshock sequence, such as the 2014 Chile earthquake (Kato and Nakagawa 2014), the location and time of the eventual mainshock may be predicted.

Studies have found foreshock occurrence is dependent on the regional stress field, e.g., normal faulting versus reverse faulting Abercrombie and Mori (1996), or intraplate earthquakes versus interplate earthquakes (Bouchon *et al.* 2013), suggesting that the occurrence of foreshocks is not purely random, but may be influenced by the regional stress field. In this regard, retrospective searches of foreshock occurrence patterns in a variety of tectonic settings will be useful for future prospective forecasts.

Recently developed high-resolution catalogs provide opportunities to review previously identified foreshock features, and further probe the possible relationships among precursory seismicity, characteristics of earthquake clusters, and mainshock properties, which may be helpful in developing or improving earthquake forecasting models. In this study, we first retrospectively search for foreshocks for mainshocks ($M \geq 5$) and compare foreshock occurrence with mainshock faulting type, location, and foreshock and mainshock magnitudes, in order to see if there are any patterns in the apparent randomness of foreshock occurrence, and if there is any relationship between foreshock properties and mainshock parameters. We then investigate the occurrence patterns of small clusters that resemble swarm-like foreshock sequences but which do not always lead to larger events, and perform comparisons with synthetic catalogs based on an ETAS-like triggering model (epidemic-type-aftershock-sequence: a branching point process where the total seismicity rate is a summation of all triggered aftershocks from prior events, Ogata 1999), in order to examine to what extent the statistical model can explain the observed seismicity patterns.

2. *Foreshock Occurrence Pattern for $M \geq 5$ Earthquakes*

We search for isolated mainshocks with $M \geq 5$ using two waveform relocated catalogs in California (with relative location accuracy typically less than 200 m): (1) the Hauksson *et al.* (2012) catalog for southern California from 1981 to 2011, excluding events north of 35.5° and south of 32.0°; (2) the double-difference catalog for northern California from 1984 to 2009, excluding events south of 35.5° and north of 39.5° (data source: http://www.ldeo.columbia.edu/~felixw/NCAeqDD/) Waldhauser and Schaff (2008). The areas excluded are beyond the coverage of the regional network recording the events and thus likely have higher detection thresholds and larger location errors. To reduce potential catalog incompleteness issues for smaller earthquakes, we use events with $M \geq 1.5$ throughout this study. We select mainshocks that are relatively isolated from other large events, i.e., events that are not part of aftershock sequences or immediate foreshocks of larger events. Specifically, we exclude: (1) smaller events within 10 days and 50 km after $M \geq 5$ events; (2) smaller events within 120 days after $M \geq 6$ events; (3) smaller events immediately before a $M \geq 5$ event within 2 days and 5 km. These requirements are not an attempt to decluster the catalog, but rather to ensure the mainshocks that we analyze are largely independent from other large events (e.g., not within direct aftershock sequences or when the catalog is temporarily influenced by the occurrence of a large event). Tests without applying such criteria resulted in several large "mainshocks" that are clear aftershocks of previous larger events (two are within the aftershock zone of a M6 earthquake and one is within a long-lasting swarm in the Long Valley volcanic region), in which their "foreshocks" cannot be distinguished from aftershocks of the earlier event. In total, 70 mainshocks in the two catalogs meet our criteria. Visual examination found five of these events are part of long-duration continuous sequences, and are excluded from the final list. The M6.6 event on the Superstition Hills fault in 1987 occurred 12 h following the M6.2 Elmore Ranch earthquake (noted with "*" in Table 1) and is excluded from the final list, because its precursory seismicity is dominated by seismicity following the first event.

Among the 64 mainshocks, 3 are within earthquake swarms in Nevada and the Salton Trough, and 3 are earthquake doublets (two events of similar magnitude occurring almost instantaneously, listed with "**" in Table 1). For the remaining 58 mainshocks, we examine the precursory activity within 50 km and 100 days before the mainchock. The scatter plot of days before the mainshock and distance to the mainshock suggests that most of the precursory activity is concentrated within 5 km of the mainshock hypocenters (Fig. 1). For most of the mainshocks that have precursory activity within 5 km, the cumulative number of foreshocks steadily grows until up to 30 days before the mainshocks, an increased rate of foreshocks occurs within approximately 20 days before, and significantly enhanced activity occurs within 2 days before the mainshocks (see Fig. 2). To check if the apparent acceleration behavior is dominated by a few larger sequences, we normalize event occurrence time by the duration of the precursory sequence (see Table 2 for calculation of duration), then calculate the cumulative density function (CDF) for each sequence, and average all sequences to get an averaged CDF for all mainshocks. Individual sequences have considerable scatter, however, the averaged CDF suggests enhanced precursory activities within the last 20 % of the total precursor duration (see Fig. 3), where the seismicity rate is significantly above the steady background rate.

Based on the broad spatial–temporal behavior of the precursory activities, we define foreshocks in this study as immediate precursory activity within 30 days and 5 km of the mainshock. Because foreshocks can only be identified in relation to their spatial and temporal proximity to mainshocks, there is no perfect method to separate foreshocks from 'background' activity, just as there is no way to uniquely discriminate very late aftershocks from background activity. We believe our 30-day and 5-km cutoff is a reasonable and practical choice (see scatter plot in Fig. 1) to ensure that the vast majority of our foreshocks are indeed foreshocks and not background activity. Using a larger spatial and/or temporal window might yield more foreshocks, but at the cost of including many background events. We use a fixed selection window regardless of mainshock magnitude to avoid biasing any comparisons between foreshock and mainshock properties.

We find that 27 of the 58 mainshocks (excluding the 3 swarms and 3 doublets) have no foreshocks within 30 days and 5 km. Among the 31 mainshocks with foreshocks, 14 mainshocks have "swarm-like" foreshocks (with more than 3 events, so we are able to estimate foreshock area in the following section, and the foreshocks do not start with the largest foreshock). Some special cases are included (noted with "*" in Table 1): (1) the 1986 Mt. Lewis sequence has a swarm with 14 events that occurred 7 days before the mainshock, which are included in the foreshocks; (2) for the 1986 Chalfant earthquake, a M5.9 event occurred 1 day before the M6.4 event, and the former is assumed to be the mainshock, with 40 foreshocks. Thus, from the 64 mainshocks examined here, excluding 3 swarms and 3 doublets, 53 % have at least one foreshock (58 % if including the swarms and doublets). A list of foreshocks is in Table 1 and a map view of the mainshock locations is shown in Fig. 4.

Our observed 53 % rate of foreshock occurrence is consistent with previous work. Abercrombie and Mori (1996) found a 44 % rate of $M \geq 2$ foreshocks prior to $M \geq 5$ events in the western United States. However, as noted by Reasenberg (1999), one expects the rate of foreshock occurrence to increase for lower foreshock magnitude cutoffs compared to the mainshock, so studies are best compared by dividing the

Table 1

List of foreshocks with $M \geq 5$ included in this study

Time	Location	Depth	Mag	N_{fore}	Focal mechanism		Fault type	$Plane_{diff}$
1981/09/04 15:50:49.62	33.650°, −119.121°	10.92 (5.00)	5.45	0	134.77°, 169°	(311.90°, 180°)	0.12 (0.00)	12.94
1986/07/08 09:20:44.06	34.001°, −116.606°	13.25 (10.00)	5.65	0	298.38°, −177°	(294.37°, 156°)	−0.03 (0.27)	3.33
1986/07/13 13:47:9.12[a]	32.988°, −117.863°[oa]	21.36 (5.00)[a]	5.45[a]	0[a]	359.72°, −167°	(126.37°, 106°)[a]	−0.14 (0.82)[a]	43.86[a]
1987/10/01 14:42:19.66	34.067°, −118.092°	13.50 (11.00)	5.90	0	262.23°, 83°	(270.31°, 98°)	0.92 (0.91)	10.71
1988/06/10 23:06:42.52	34.931°, −118.742°	9.46 (N/A)	5.37	0	162.83°, 176°	(N/A)	0.04 (N/A)	N/A
1988/12/03 11:38:26.26	34.142°, −118.138°	12.74 (N/A)	5.02	0	157.86°, 169°	(N/A)	0.12 (N/A)	N/A
1991/06/28 14:43:54.47	34.266°, −117.989°	9.64 (11.00)	5.80	0	56.25°, 74°	(93.43°, 130°)	0.82 (0.56)	35.15
1991/12/03 17:54:36.20	31.718°, −115.821°	18.31 (N/A)	5.32	0	N/A	(N/A)	0.00 (N/A)	N/A
1993/05/28 04:47:40.26	35.132°, −119.116°	23.90 (N/A)	5.19	0	114.71°, 170°	(N/A)	O.11 (N/A)	N/A
1994/01/17 12:30:54.96	34.206°, −118.549°	21.07 (18.00)	6.70	0	113.36°, 106°	(278.42°, 65°)	0.82 (0.72)	13.00
2001/12/08 23:36:10.03	32.035°, −114.963°	17.23 (10.00)	5.70	0	N/A	(141.59°, −149°)	N/A (−0.34)	N/A
2004/09/29 22:54:54.20	35.385°, −118.629°	7.30 (3.50)	5.03	0	105.82°, 173°	(293.71°, −169°)	0.08 (−0.12)	11.13
2008/02/09 07:12: 6.84	32.410°, −115.312°	18.65 (2.90)	5.10	0	147.66°, −175°	(226.79°, 3°)	−0.06 (0.03)	24.25
2008/07/29 18:42:15.28	33.946°, −117.767°	14.89 (14.70)	5.39	0	296.66°, 146°	(44.55°, 29°)	0.38 (0.32)	34.85
2008/12/06 04:18:42.29	34.812°, −116.423°	9.33 (7.30)	5.06	0	172.79°, −157°	(253.83°, 6°)	−0.26 (0.07)	12.02
1984/01/23 05:40:20.03	36.390°, −121.886°	7.73 (N/A)	5.10	0	65.85°, 10°	(N/A)	0.11 (N/A)	N/A
1984/04/24 21:15:18.75	37.310°, −121.682°	7.97 (8.00)	6.20	0	240.80°, 10°	(333.76°, 179°)	0.11 (0.01)	16.66
1988/02/20 08:39:57.49	36.798°, −121.306°	8.22 (N/A)	5.10	0	45.60°, 10°	(N/A)	0.11 (N/A)	N/A
1988/06/27 18:43:22.65	37.129°, −121.894°	11.54 (N/A)	5.30	0	35.85°, 30°	(N/A)	0.33 (N/A)	N/A
1988/09/19 02:56:31.33	38.458°, −118.344°	6.72 (N/A)	5.30	0	40.50°, −10°	(N/A)	−0.11 (N/A)	N/A
1989/08/08 08:13:27.51	37.153°, −121.926°	12.59 (N/A)	5.40	0	45.65°, 30°	(N/A)	0.33 (N/A)	N/A
1989/10/18 00:04:15.39	37.043°, −121.877°	16.41 (19.00)	7.00	0	130.75°, 130°	(235.41°, 29°)	0.56 (0.321)	46.77
1991/09/17 21:10:29.35	35.815°, −121.322°	8.01 (N/A)	5.20	0	80.55°, 50°	(N/A)	0.56 (N/A)	N/A
1993/05/17 23:20:49:15[a]	37.171°, −117.782°[oa]	2.39 (7.00)[a]	6.40[a]	0[a]	250.65°, 20°	(210.30°, −93°)[a]	0.22 (−0.97)[a]	42.81[a]
1996/11/27 20:17:23.54	36.090°, −117.628°	6.56 (1.00)	5.10	0	N/A	(244.71°, −3°)	N/A (−0.03)	N/A
2003/12/22 19:15:56.21	35.701°, −121.099°	8.05 (7.60)	6.50	0	105.35°, 80°	(296.32°, 88°)	0.89 (0.98)	9.73
2004/09/28 17:15:24.31	35.818°, −120.366°	8.20 (8.80)	6.00	0	145.85°, −170°	(321.72°, −178°)	−0.11 (−0.02)	13.02
1988/12/16 05:53:4.48	33.983°, −116.688°	11.47 (N/A)	5.03	6	292.41°, 148°	(N/A)	0.36 (N/A)	N/A
1987/02/07 03:45:14.97[a]	32.388°, −115.317°[oa]	24.44 (5.00)[a]	5.38[a]	2[a]	235.87°, 98°	(202.70°, 2°)[a]	0.91 (0.02)[a]	17.55[a]
1997/04/26 10:37:30.38	34.376°, −118.673°	13.67 (N/A)	5.07	1	358.52°, −141°	(N/A)	−0.43 (N/A)	N/A
2001/10/31 07:56:16.22	33.504°, −116.503°	16.83 (N/A)	5.02	1	301.35°, 172°	(N/A)	0.09 (N/A)	N/A
2002/02/22 19:32:41.50	32.309°, −115.315°	19.91 (7.00)	5.70	2	N/A	(190.66°, −4°)	N/A (−0.04)	N/A
2005/06/12 15:41:46.19	33.533°, −116.570°	15.48 (14.20)	5.20	1	304.58°, 172°	(305.53°, −179°)	0.09 (−0.01)	5.03
2006/05/24 04:20:26.05	32.303°, −115.223°	14.38 (N/A)	5.37	2	0.0°, 0°	(N/A)	N/A (N/A)	N/A
2009/09/19 22:55:17.64[a]	32.344°, −115.256°[oa]	19.08 (3.00)[a]	5.08[a]	1[a]	67.38°, 75°	(125.7°, −172°)[a]	0.83 (−0.09)	44.09[a]
2009/12/30 18:48:56.69[a]	32.417°, −115.149°[oa]	23.81 (9.00)[a]	5.80[a]	2[a]	205.40°, −92°	(328.82°, −178°)[a]	−0.98 (−0.02)	45.98[a]
1984/11/23 18:08:25.25	37.455°, −118.606°	11.11 (N/A)	6.10	1	65.65°, 30°	(N/A)	0.33 (N/A)	N/A
1986/01/26 19:20:51.18	36.803°, −121.284°	7.10 (7.00)	5.50	1	260.80°, −110°	(166.90°, 180°)	−0.11 (0.00)	9.97
1987/02/14 07:26:50.39[a]	36.171°, −120.339°[oa]	13.55 (13.0)[a]	5.30[a]	1[a]	150.50°, 90°	(300.46°, 38°)[a]	1.00 (0.42)[a]	20.24[a]
1988/06/13 01:45:36.38	37.395°, −121.739°	8.87 (9.00)	5.30	1	60.90°, −10°	(325.76°, −175°)	−0.11 (0.06)	13.97
1990/10/24 06:15:20.01	38.053°, −119.125°	12.38 (12.00)	5.80	1	70.55°, 10°	(56.59°, −10°)	0.11 (−0.11)	8.58
1996/01/07 14:32:52.82	35.772°, −117.622°	9.56 (N/A)	5.10	2	160.80°, −170°	(N/A)	−0.11 (N/A)	N/A
2007/10/31 03:04:54.87	37.432°, −127.777°	7.49 (10.00)	5.40	1	55.85°, 0°	(324.81°, 176°)	0.00 (0.04)	10.18
1994/09/12 12:23:42.94	38.793°, −119.702°	2.94 (14.00)	5.90	1	40.40°, −40°	(42.74°, −13°)	−0.44 (−0.14)	34.01
1987/11/24 01:54:14.21	33.082°, −115.779°	10.07 (5.00)	6.20	6*	280.86°, 171°	(305.90°, 180°)	0.10 (0.00)	3.95
1990/02/28 23:43:36.23	34.138°, −117.708°	7.28 (10.00)	5.51	4	132.89°, 167°	(307.73°, 169°)	0.14 (0.12)	16.00
1992/04/23 04:50:22.73	33.968°, −116.313°	13.71 (10.00)	6.10	6	344.85°, 171°	(81.87°, −1°)	0.10 (−0.01)	5.47
1992/06/28 11:57:33.85	34.202°, −116.435°	7.01 (5.00)	7.30	27	173.85°, −177°	(341.70°, −172°)	−0.03 (−0.09)	15.14
1992/11/27 16:00:57.39	34.337°, −116.892°	0.00 (N/A)	5.29	5	128.88°, 167°	(N/A)	0.14 (N/A)	N/A
1997/03/18 15:24:47.70	34.966°, −116.822°	4.02 (N/A)	5.26	3	154.75°, −163°	(N/A)	−0.19 (N/A)	N/A
1999/10/16 09:46:43.95	34.595°, −116.271°	9.06 (0.00)	7.10	18	5.90°, 159°	(336.80°, 174°)	0.23 (0.07)	9.95
2010/04/04 22:40:42.16	32.264°, −115.295°	16.47 (6.00)	7.20	26	264.49°, 165°	(223.84°, −2°)	0.17 (−0.02)	36.64
1985/08/04 12:01:55.85[a]	36.138°, −120.153°[oa]	10.35 (5.00)[a]	5.60[a]	6[a]	70.20°, 40°	(138.10°, 105°)[a]	0.44 (0.83)[a]	66.04[a]
1986/03/31 11:55:39.93	37.479°, −121.691°	8.39 (6.00)	5.70	15*	355.80°, −180°	(353.79°, −170°)	0.00 (−0.11)	1.06
1986/07/20 14:29:45.47	37.567°, −118.437°	6.16 (8.00)	5.90	40*	205.85°, −10°	(223.54°, −35°)	−0.11 (−0.39)	31.28
1990/04/18/ 13:53:51.62	36.931°, −121.652°	4.61 (N/A)	5.40	4	55.80°, 40°	(N/A)	0.44 (N/A)	N/A
1997/11/02 08:51:52.83	37.863°, −118.190°	1.65 (5.00)	5.30	8	N/A	(238.63°, 15°)	N/A (0.17)	N/A

Table 1

continued

Time	Location	Depth	Mag	N_{fore}	Focal mechanism		Fault type	$Plane_{diff}$
1998/08/12 14:10:25.15	36.759°, −121.452°	7.75 (8.80)	5.10	3	225.75°, −10°	(48.85°, −1°)	−0.11 (−0.01)	10.01
1981/04/26 12:09:28.26	33.088°, −115.619°	10.60 (6.00)	5.75	349	158.86°, −151°	(249.45°, −8°)	−0.32 (−0.09)	45.03
2005/09/02 01:27:19.46[a]	33.154°, −115.633°[a]	5.77 (9.80)[a]	5.11[a]	387[a]	190.62°, −106°	(335.76°, −167°)[a]	−0.82 (−0.14)[a]	18.21[a]
2008/04/26 06:40:10.76	39.522°, −119.927°	2.28 (1.40)	5.10	214	N/A	(328.86°, 180°)	N/A (0.00)	N/A
2001/07/17 12:07:26.24	36.005°, −117.871°	8.73 (N/A)	5.20	41**	80.90°, 0°	(N/A)	0.00 (N/A)	N/A
2004/09/18 23:02:17.72	38.012°, −118.691°	3.26 (5.00)	5.60	42**	65.90°, −10°	(330.76°, −171°)	−0.11 (−0.10)	13.97
2009/10/03 01:15:59.75	36.396°, −117.858°	0.29 (2.50)	5.10	52**	N/A	(214.56°, −36°)	N/A (−0.40)	N/A

CMT solutions are in parentheses

Events with "*" are special cases and events with "**" indicate earthquake doublets (see text for details)

[a] Events with a high degree of disagreement between CMT solutions and the regional moment tensor catalog

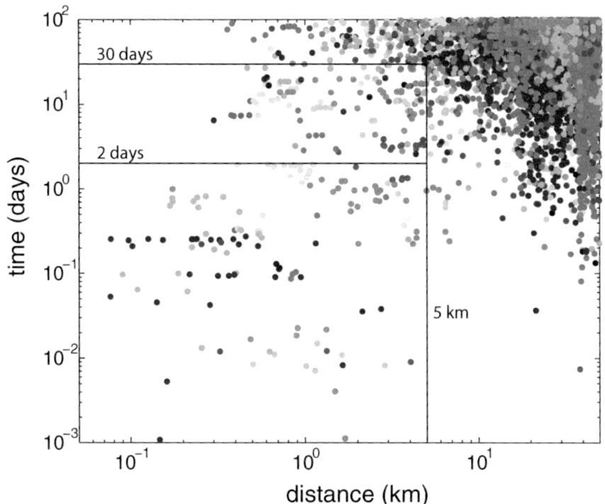

Figure 1
Scatter plot of all events within 50 km and 100 days before each mainshock (excludes the 3 swarms and 3 doublets). *Dots with the same color* belong to the same foreshock–mainshock sequence

apparent foreshock rate by the differential magnitude. In our case, we obtain a foreshock rate density of 15 % per magnitude unit. For comparison, Abercrombie and Mori (1996) obtained 15 % per magnitude unit, and Reasenberg (1999) obtained 13 % per magnitude unit from a global study of $M \geq 6$ events.

2.1. *Relationship Between Foreshock Properties and Mainshock Parameters*

Next, we examine if there is any relationship between foreshock properties and mainshock source

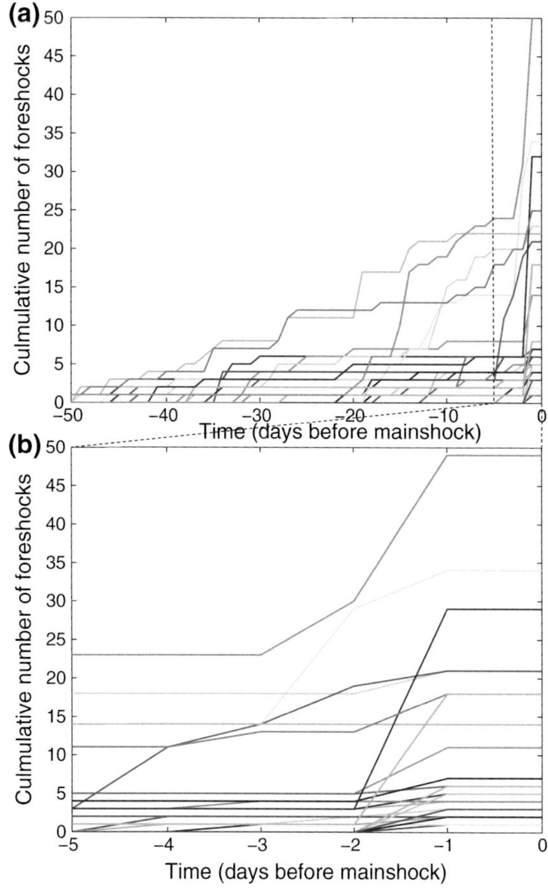

Figure 2
a Cumulative number of foreshocks within 5 km before each mainshock. *Vertical dashed line* marks 5 days before mainshock. **b** Closeup of the last 5 days before the mainshocks. The *colors* correspond to individual mainshock sequences in Fig. 1

Table 2

Definition of parameters, the superscript letter a are parameters analyzed in this study

T_i	Time of each event
T_0	Time of the first event
X_i	3D locations of each event
$M_0(i)$	$10^{1.5*M(i)+9.1}$
t_i	$(T_i - T_0)/\text{mean}(T_i - T_0, i = 1\ldots N)$
$\bar{t}$	$\sum_i^N t_i M_0(i) / \sum_i^N M_0(i)$
μ_3	$\sum_i^N (t_i - \bar{t})^3 M_0(i)$
δ	$\sqrt{\sum_i^N (t_i - \bar{t})^2 m_0(i)}$
C_i^j	Centroid location= $\text{median}(X_k, k = i\ldots j)$
r^a	Radius $= \text{median}(\lvert X_i - C_1^N \rvert, i = 1\ldots N)^a$
$t_{dura}{}^a$	Duration $= \text{median}(\lvert T_i - T_0 \rvert, i = 1\ldots N)^a$
$t_{max}{}^a$	$t_i, \{M(i) = \max(M)\}^a$
μ^a	μ_3/δ^a
$\Delta\sigma_{quasi}{}^a$	$\frac{7\sum_i^N M_0(i)_a}{16r^3}$
$d_s{}^a$	$(\lvert C_1^{N/2} - C_{N/2}^N \rvert)/r^a$
$M_{f_{max}}{}^a$	Maximum magnitude of foreshocks[a]
$N_{fore}{}^a$	Number of foreshocks[a]
$M_{max}{}^a$	Magnitude of mainshock[a]
$F_{area}{}^a$	Area of foreshocks[a]
Swarm-type[a]	$t_{max} \geq 0.2^a$
Aftershock-type[a]	$t_{max} < 0.2^a$

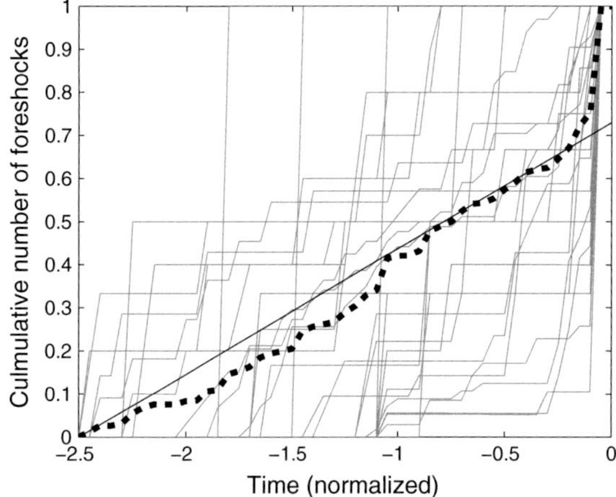

Figure 3

Cumulative density functions (CDF) with origin time normalized by duration of each precursory period, (see text and Table 2 for details). *Each grey line* corresponds to a individual mainshock, the *thin black line* is an approximation of a constant background rate. The *black dashed line* is the averaged CDF of all the *individual grey lines*, note the clear deviation from the constant rate within the last 20 % of the precursory period

parameters. We first compare foreshock occurrence with focal mechanism. We obtain focal mechanism solutions from: (1) the YHS (by Yang, Hauksson and Shearer 2012) catalog for southern California, recalculated using the HASH method (in Hardebeck and Shearer 2003) with data from the Southern California earthquake center (Yang *et al.* 2012); (2) the northern California moment tensor catalog [*Northern California Earthquake Catalog and Phase Data*]; and (3) Global Centroid-Moment-Tensor (CMT) solutions when available. We compute faulting type based on rake angle (−1 is normal fault, 0 is strike–slip fault, 1 is reverse fault):

$$f = \begin{cases} \lambda/90 & \text{if } |\lambda| \leq 90; \\ (180 - |\lambda|) * (\lambda/|\lambda|)/90 & \text{if } |\lambda| > 90. \end{cases} \quad (1)$$

For 8 mainshocks, there is a high degree of mismatch (>40° between fault plane orientations and $df = |f_{regional} - f_{cmt}| > 0.4$ between the global CMT solution and the regional network solution (see Table 1; Fig. 4). The mismatch may be due to reduced azimuthal station coverage for events outside the regional network (e.g., events off-shore and events in Mexico), or complexity in the earthquake rupture process (e.g., the rupture initiated with a sub-event with a different focal mechanism). Because the regional networks do not provide unique solutions for many events, we use CMT solutions when available.

We first examine foreshock occurrence in 10 faulting type bins from −1 to 1. From Fig. 5, there is a higher foreshock occurrence rate for mainshocks with dilational components ($f < 0$) and reverse-faulting mainshocks tend to have lower foreshock occurrence rates. Although some mainshocks have a larger degree of mismatch between the regional and CMT solutions, these events do not affect the overall trend of decreasing foreshock occurrence for reverse-faulting events. There is only one pure normal-faulting mainshock, which has no foreshock. However, with the regional focal mechanism catalog, all the normal-faulting earthquakes are preceded by at least one foreshock. Our result is most reliable for faulting types from −0.4 to 0.5, where an increase in the compressional component decreases foreshock occurrence for strike–slip faults. We also compare foreshock occurrence with mainshock depth from the two regional catalogs. In Fig. 5, for shallower events (mostly ≤5 km), the majority of mainshocks have

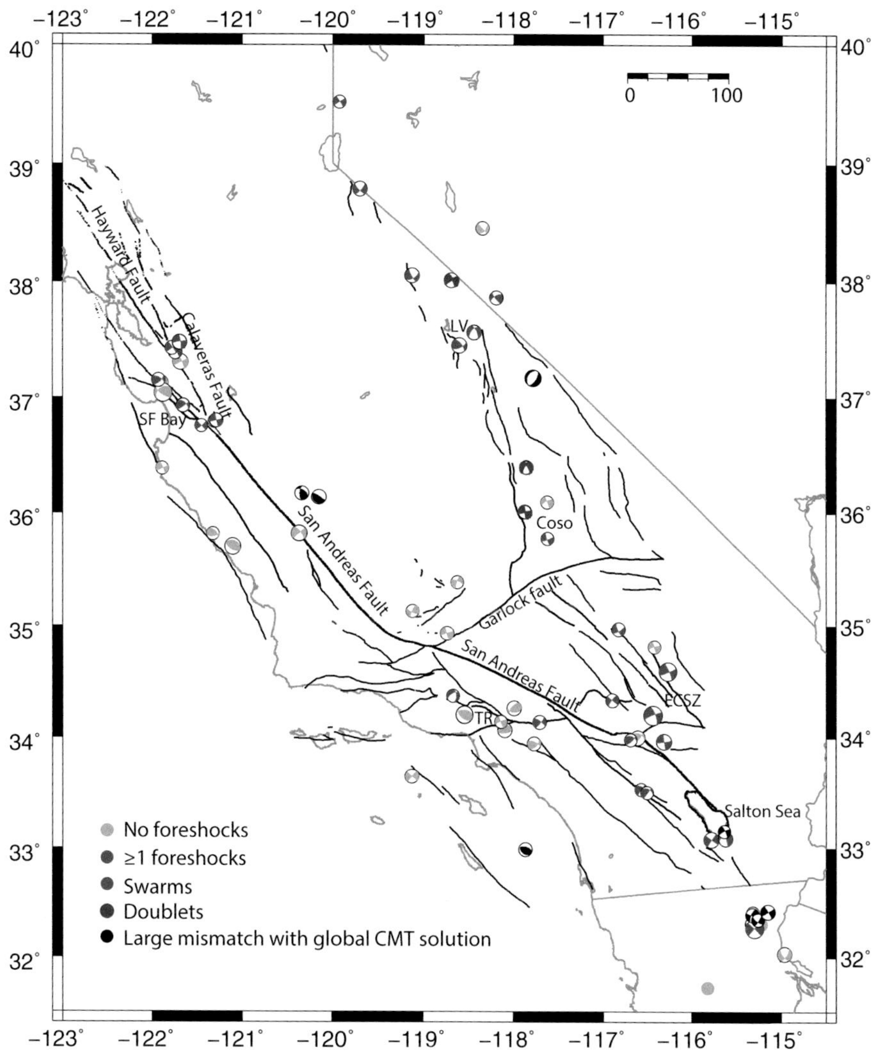

Figure 4
Map view of $M \geq 5$ mainshocks. Events with a large mismatch with CMT solutions correspond to events with red colors in Table 1). Focal mechanisms are from CMT solutions when available. Regions in the map: *TR* Transverse Range, *LV* Long Valley

foreshocks, and the occurrence rate decreases with depth. We test the statistical significance of the pattern with the Student's *t* test (Abercrombie and Mori 1996). For this, we divide foreshock occurrence rate into several groups in faulting type and depth: for faulting type, we use −0.5–0 (*f*1), 0–0.5 (*f*2), and 0.5–1 (*f*3); for depth, we use 0–5 and 5–25 km. For faulting type, group *f*1 is different from *f*2 and *f*3 at a 97 % confidence level, while *f*2 and *f*3 are similar; for depth, the shallow group is different from the deeper group at a 98 % confidence level. We also examine the relationship between faulting type and depth (see Fig. 6). The shallowest events spread evenly between $f = -0.5$ and $f = 0.5$. For events deeper than 5 km, there is no clear dependence between faulting type and depth, and a faulting type dependence of foreshock occurrence is clear. Thus, the most reliable trends are a dependence on faulting type for $-0.5 \leq f \leq 1$ and a higher foreshock occurrence rate at shallow depth.

We compare foreshock properties with mainshock magnitude by examining: (1) the radius of foreshock rupture area for the 14 swarm-like foreshock sequences; (2) the number of foreshocks; (3) the magnitude of the largest foreshock; and (4) foreshock duration (see Table 2 for definitions). For the three

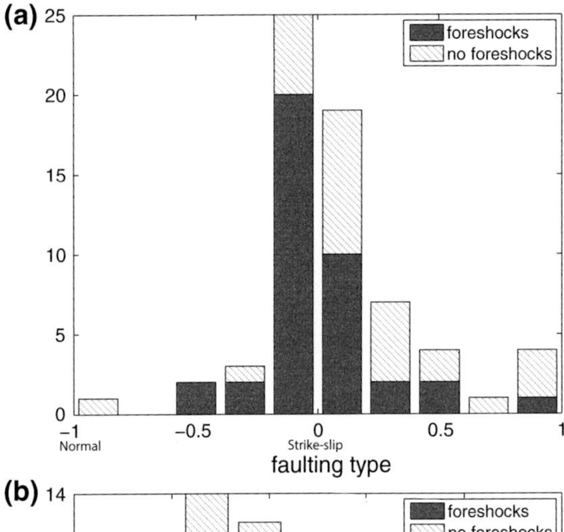

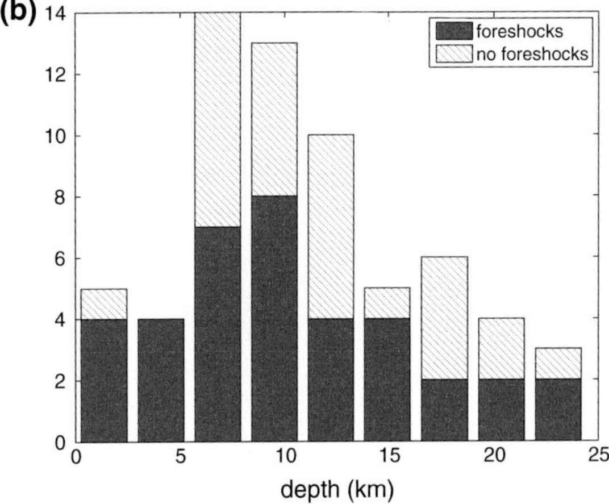

Figure 5

a Histogram of faulting types for 27 mainshocks without foreshocks and 37 mainshocks with foreshocks. Faulting types from −1 to 0 to 1 correspond to normal faulting–strike slip–reverse faulting, respectively. **b** Histogram of the depth distribution for 27 mainshocks without foreshocks and 37 mainshocks with foreshocks. Note the prevalence of foreshock occurrence at shallow depth and for transtentional faults ($f \leq 0$, for definition of f, refer to Eq. 1). We perform a Student's t test to examine the statistical significance of the pattern (Abercrombie and Mori 1996). For faulting types, the group of −0.5 to 0 is different from 0 to 1 at 97 % confident level, and the shallow group (≤ 5) is different from the rest at 98 % confident level

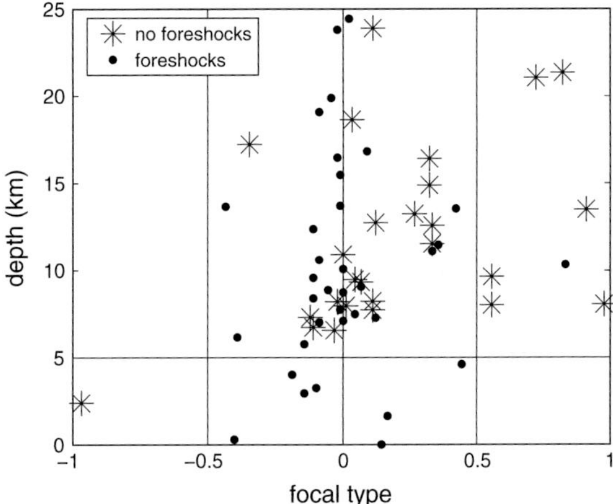

Figure 6

Scatter plot of depth versus faulting type for each mainshock. *Black dots* are mainshocks with foreshocks, *black stars* are mainshocks without foreshocks. The *vertical lines* corresponds to the three groups that we test against for faulting types. The *horizontal line* marks the depth groups that we test against. For events deeper than 5 km, there is no clear relationship between faulting type and depth, and the most prominent feature is a faulting type dependence of foreshock occurrence

$M > 7$ mainshocks (the 1992 Landers, the 1999 Hector Mine and the 2010 El Mayor-Cucapah earthquakes), we have shown that none of these properties are correlated with mainshock magnitude (Chen and Shearer 2013). For each of the 14 mainshocks, we visually check seismicity within each mainshock rupture zone using an interactive tool, to ensure a good selection of foreshocks from the automatic search process described above. We also compare two different methods to estimate foreshock area, specifically: (1) we calculate area based on the radius assuming a circular area; (2) we compute the convex polygon that includes all foreshocks. Both methods yield similar results, where mainshock magnitude is correlated with the number of foreshocks, but not foreshock area, while foreshock area is correlated with the number of foreshocks and the magnitude of the largest foreshock (see Table 3). The lack of correlation of foreshock area with mainshock magnitude, but with foreshock number and magnitude, suggests foreshock processes are controlled by interactions within foreshocks themselves, rather than being an indicator of mainshock magnitude. Most of the foreshocks do not start with their largest event, suggesting the "swarmy" nature of foreshock sequences. The fact that mainshock magnitude is correlated with the number of foreshocks suggests that swarms with many events increase the probability of large earthquakes, such as the 1975 Haicheng sequence, the 1992 Landers and the 1999 Hector Mine earthquakes.

Overall, observations of foreshock dependence on faulting type and depth are consistent with the results in Abercrombie and Mori (1996); however, our result

Table 3

Correlation between foreshock properties and mainshock parameters

$M_{max} - F_{area}$ (convex)	$cc = 0.46, p = 0.11$
$M_{max} - F_{area}$ (radius)	$cc = 0.32, p = 0.28$
$M_{max} - N_{fore}$[a]	$cc = 0.62, p = 0.02$[a]
$M_{max} - M_{fmax}$	$cc = 0.07, p = 0.86$
$M_{max} - t_{dura}$	$cc = 0.10, p = 0.75$
$M_{fmax} - F_{area}$ (convex)[a]	$cc = 0.56, p = 0.05$[a]
$M_{fmax} - F_{area}$ (radius)[a]	$cc = 0.60, p = 0.03$[a]
$N_{fore} - M_{fmax}$	$cc = 0.11, p = 0.73$
$N_{fore} - F_{area}$ (convex)[a]	$cc = 0.69, p = 0.01$[a]
$N_{fore} - F_{area}$ (radius)[a]	$cc = 0.58, p = 0.04$[a]

(cc is correlation coefficient, and p is the statistical significance of correlation, $p \leq 0.05$ is generally considered as significantly above random chance)

[a] Significant correlation

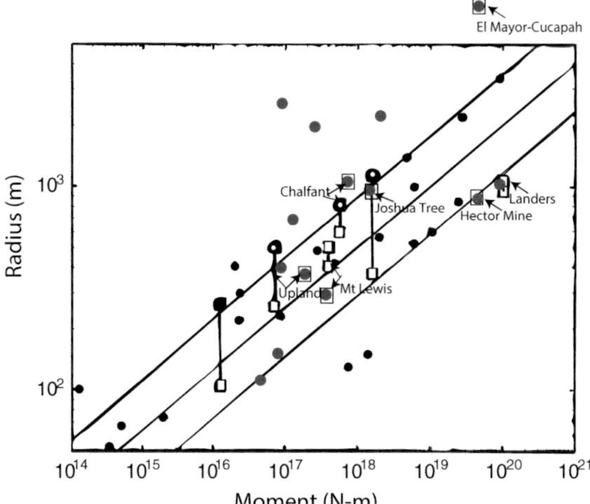

Figure 7
Foreshock radius versus mainshock moment plotted on Figure 16 from Dodge *et al.* (1996). *Red dots* are 14 mainshocks in this study, *boxes* indicate named events in the catalog. *Black dots* are source radii estimated in Beroza and Ellsworth (1996), *straight lines* are best-linear fit and 1σ boundaries. Note the scattered *red dots* above the 1σ limit. The p value of the correlation for the *red dots* is 0.28, indicating no statistical significance (see Table 3)

is more reliable for strike–slip faults and shallow depths, because we have used improved catalogs, examined more events and obtain consistent results with both regional and CMT catalogs. It is interesting to compare our foreshock radius estimates with results in Dodge *et al.* (1996) (see Fig. 7). In general, for the events in common, our radius is consistent with previous measurements; however, about half of the points are above the 1σ boundaries of nucleation radius estimated from the slow onset of mainshock waveforms (Ellsworth and Beroza 1995). There is no correlation between foreshock duration and mainshock depth ($cc = 0.15, p = 0.62$), which is inconsistent with previous observations (Jones 1984; Abercrombie and Mori 1996). There is some degree of spatial separation between different types of mainshocks: within the Transverse Range region in Los Angeles county and central California, most mainshocks do not have foreshocks; in contrast, in the San Francisco Bay area and Eastern California Shear Zone (ECSZ), at the intersection of different faults, the occurrence rate of foreshocks is relatively high (see Fig. 4).

2.2. Precursory Seismicity

The rupture area of a M5.8 earthquake is about 5 km, assuming a stress drop of 2 MPa (the average stress drop for southern California from Shearer *et al.* (2006), calculated from Shearer (2009), with $r = \left(\frac{7M_0}{16\Delta\sigma}\right)^{1/3}$. According to Shearer and Lin (2009), the radius of the "Mogi-doughnut" (enhanced precursory seismicity) roughly agrees with the expected rupture radius for target mainshocks. Due to the small number of M5 earthquakes in the catalog, such behavior is not reliably resolved for larger events in their study. For the 64 M $\geq$ 5 mainshocks, we examine the averaged precursory seismicity for: (1) 27 mainshocks without foreshocks; (2) 31 mainshocks with foreshocks; (3) 3 earthquake swarms; and (4) 3 earthquake doublets. For each group, we calculate seismicity density for each time and distance bin based on: $D = \frac{n}{N(t_2-t_1)(4/3)\pi(r_2^3-r_1^3)}$, where n is the total number of precursory events in the space/time bin, and N is the number of target events in each group. We use 100 space-time bins, evenly spaced in 10 log distance bins between 0.01 and 100 km, and in 10 log time bins from 0.001 to 1000 days. From Fig. 8, within 1 day prior to the mainshock, the low seismicity zone extends beyond the 5 km criteria, consistent with the empirical scaling of "Mogi doughnut" behavior. Due to the limited number of available mainshocks, we do not attempt to perform statistical analysis to test the reliability of the "Mogi" zone relative to smaller event bins from Shearer and Lin (2009). Foreshock activity is confined within the "Mogi" zone, and well

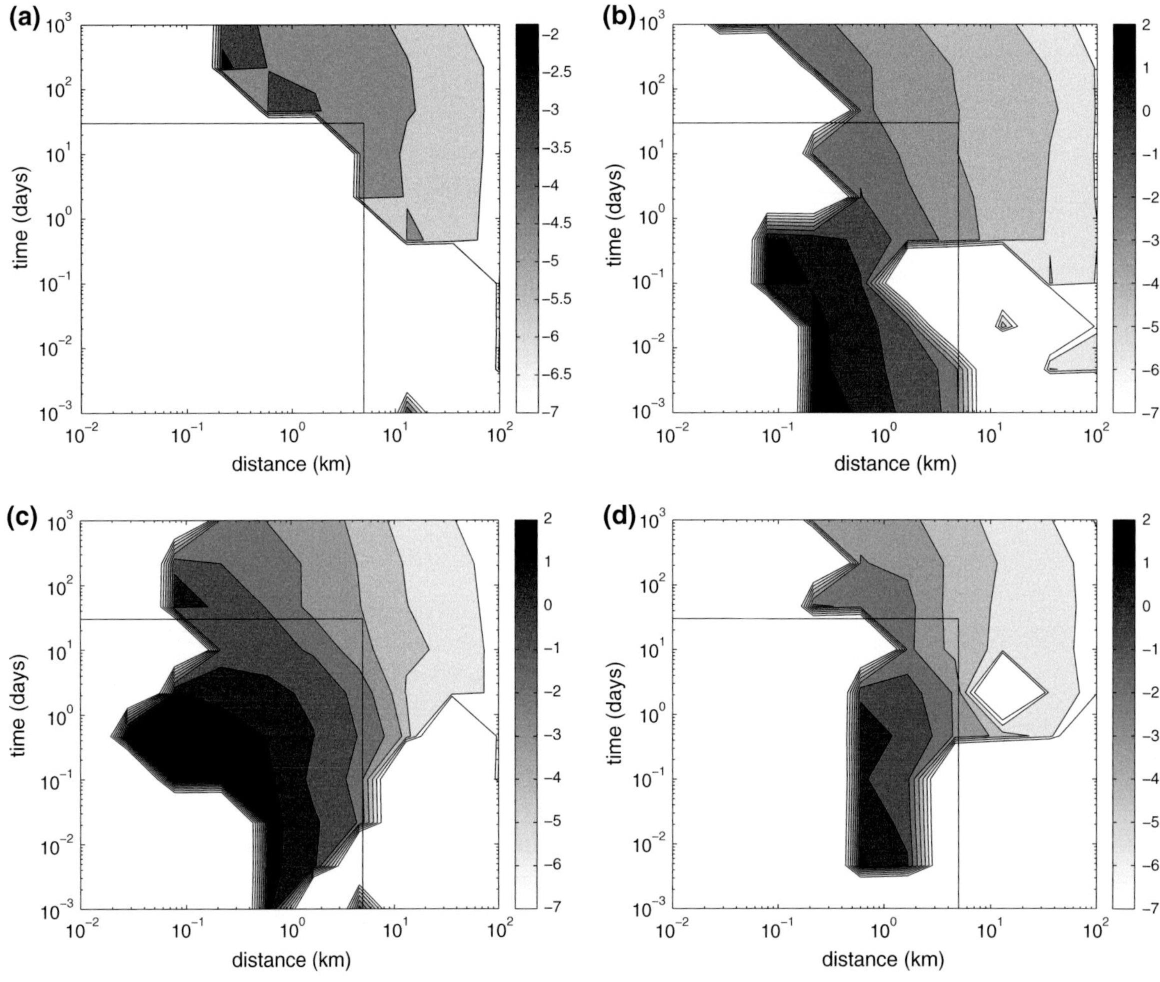

Figure 8
Precursory seismicity within 1000 days and 100 km prior to target mainshocks with $M > 5$. Four mainshock types are included: **a** mainshocks with no foreshocks; **b** mainshocks with ≥ 1 foreshocks; **c** earthquake swarms; **d** earthquake doublets. *Horizontal black lines* correspond to $T = 2$ days, *vertical black lines* correspond to $D = 5$ km. The *color scale* shows $\log_{10}$ of seismicity rate in each distance-time bin

separated from background seismicity. The high event density per time/distance bin suggests that foreshocks are highly localized within future mainshock rupture zones (also see Fig. 1).

3. *Seismicity Clusters that Resemble Foreshock Sequences*

So far, our analysis has been limited to $M \geq 5$ target events, and any foreshock activities within 5 km and 30 days. However, there are other important questions related to precursory activities, such as: (1) how are foreshock sequences different from random small clusters? (2) How often do small clusters lead to larger clusters that might include a larger event? To address this, we remove the magnitude requirement, and search for small compact clusters that resemble the observed foreshock sequences. Specifically, we search for small clusters that have at least $N \geq 10$ events within 5 km and 2 days, and fewer than 5 events in the previous 7 days within 5 km. The 2-day requirement is based on the observation that enhanced activities typically occur within 2 days before mainshocks (see Fig. 2), the number requirement is to ensure the relative independence of

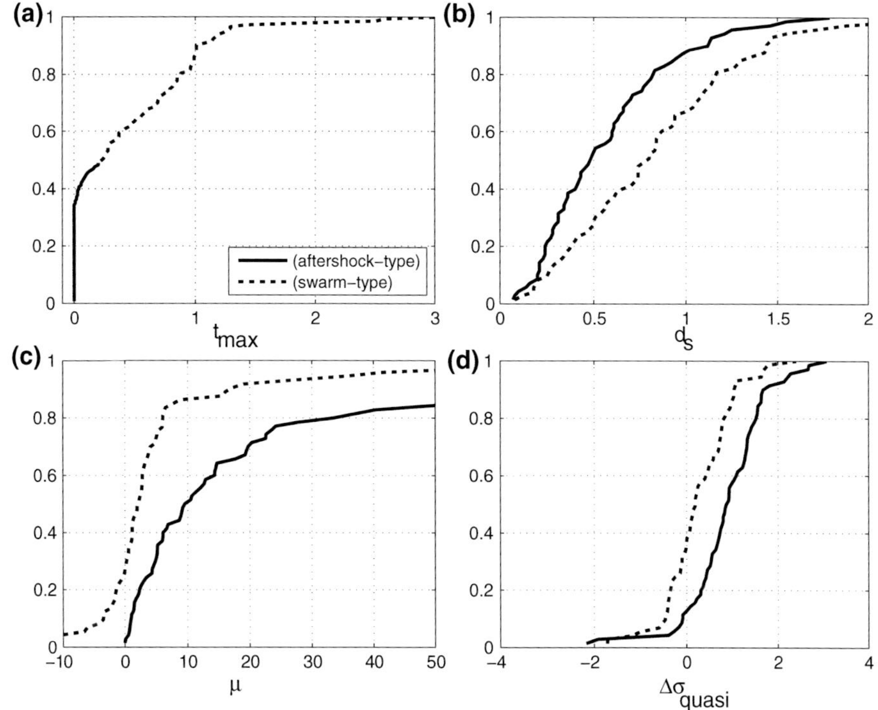

Figure 9
Distribution of different parameters for: "aftershock-type" ($t_{max} \leq 0.2$) (*solid line*) and "swarm-type" ($t_{max} \geq 0.2$) (*dashed line*) sequences from combined observations in southern and northern California. **a–d** Correspond to the four parameters discussed in the text: t_{max}, d_s, μ, $\Delta\sigma_{quasi}$

the small cluster, and minimize the number of random scattered clusters. We apply the same criteria as before to temporarily remove immediate aftershock sequences from the catalog for consistency.

Next, we check whether these clusters grow into larger clusters. To define a larger cluster, we require: (1) $N_{cluster} \geq 50$ within 28 days and 5 km following the small cluster; (2) $N_{outside} \leq 0.2 N_{cluster}$ within 28 days between 5 and 10 km. The latter requirement is to ensure the cluster is spatially isolated from background seismicity; however, it does not affect the relative distribution of different types of clusters. To characterize the cluster type, we compute four parameters as described in Chen *et al.* (2012) and listed in Table 2: (1) t_{max}, the relative timing of the largest event in the cluster with respect to the mean delay time; (2) the skew (μ) of moment release history; (3) the distance separation of the first and second half of the cluster normalized by the radius of the cluster (d_s); and (4) $\Delta\sigma_{quasi} = \frac{7\sum_n^1 M_0^i}{16r^3}$, the effective stress drop. The parameter d_s is a proxy to measure the spatial migration of seismicity clusters, and $\Delta\sigma_{quasi}$ is a parameter to measure the effectiveness of moment release compared with the rupture area. The observations in Chen *et al.* (2012) suggested clusters with $t_{max} \geq 0.2$ are more prone to spatial migration controlled by external aseismic transients. We use this empirical relationship, and define aftershock-type clusters as those with $t_{max} \leq 0.2$, and swarm-type clusters as those with $t_{max} \geq 0.2$. Foreshocks are any earthquakes occurring before the largest earthquake (mainshock) in the cluster, and all swarm-type clusters have foreshocks by definition.

For the southern California catalog, we identify 311 small clusters, of which 87 grow into larger clusters, and 27 start with their mainshock. Small clusters are less common in northern California, where we identify 184 small clusters, 56 of which grow into larger clusters, and 21 start with their mainshock. From Fig. 9, the CDF of the four parameters are consistent with Chen *et al.* (2012):

aftershock-type clusters have lower d_s, higher μ and higher $\Delta\sigma_{\text{quasi}}$.

Overall, in total 495 small clusters are identified, of which 30 % eventually grow into a larger cluster. Among the larger clusters, 66 % have precursory activities. If we only focus on aftershock-type larger clusters (see Table 2), there are 39 % with precursory activities. For southern California, we observe a similar relationship to $M \geq 5$ mainshocks between precursory occurrence and small mainshock faulting type (foreshock occurrence rate is higher for faulting-type <0), which is consistent with Chen *et al.* (2012), however, the correlation is not so clear for northern California. For smaller events, the focal mechanism is likely poorly determined compared

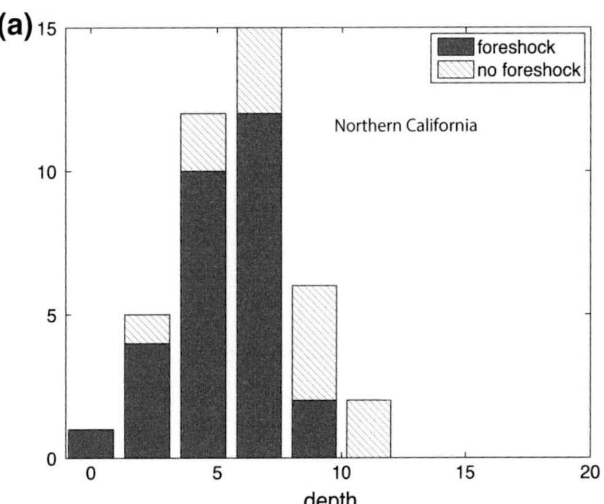

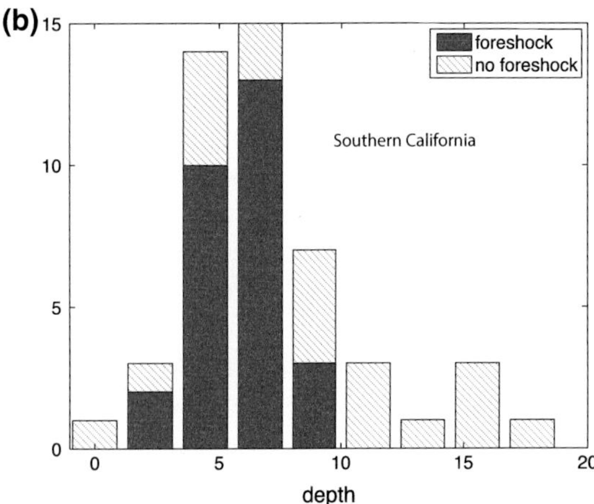

Figure 10
Depth distribution of all clusters with at least one foreshock and without foreshocks. **a** Northern California, **b** Southern California. The majority of mainshocks shallower than 8 have foreshocks, while all mainshocks deeper than 10 km have no foreshocks

with $M \geq 5$ events, and the reduced correlation may be due to uncertainties in the fault plane solutions (as shown in Table 1, there is sometimes a large mismatch for $M \geq 5$ events). The southern California YHS catalog is improved compared to the routine catalog solutions, and likely is more accurate (Yang *et al.* 2012). For both catalogs, we observe a prevalence of precursory activities at shallow depth and a lack of precursory activities at deeper depth (Fig. 10).

The overall spatial distribution is similar for $M \geq 5$ mainshocks (see Figs. 4, 11). For example, the Transverse Ranges and central California are still dominated by mainshocks without foreshocks. The Bay Area, the ECSZ, the Salton Trough and the Long Valley region are dominated by mainshocks with foreshocks. Considering the geological features of these regions, this suggests that foreshocks tend to occur within extensional step overs, high heat flow regions, and complex fault zones, while a lack of foreshocks is expected at thrust fault zones and relatively simple planar fault zones. The consistency between foreshocks for $M \geq 5$ mainshocks and general clustering types suggests that localized fault zone properties control precursory activities.

We next investigate whether foreshock properties are predictive of mainshock parameters. We find that:

1. Neither foreshock area nor the number of foreshocks is correlated with mainshock magnitude; none of the other foreshock properties correlate with mainshock parameters.
2. Foreshock area is well correlated with the number of foreshocks: correlation coefficient $(cc) = 0.8, p = 10^{-5}$ (see Table 4).
3. If we consider swarms ($t_{\max} \geq 0.2$) as foreshock–mainshock sequences, then for clusters with $M_{\max} \geq 4.5$, foreshock area is correlated with the maximum magnitude of foreshocks ($M_{f\max}$) with $cc = 0.55, p = 0.005$, however, for clusters with $2 \leq M_{\max} \leq 4.5$, there is no correlation between foreshock area and $M_{f\max}$ (see Table 4).
4. The magnitude difference (d_m) between $M_{\max}$ and $M_{f\max}$ is dependent on $M_{\max}$: for $M_{\max} \geq 4.5$, d_m is approximately uniformly distributed between 0.5 and 2, and the median value is 1.47; for $2 \leq M_{\max} \leq 4.5$, d_m is skewed towards lower values, and is mostly below 1 (Fig. 12).

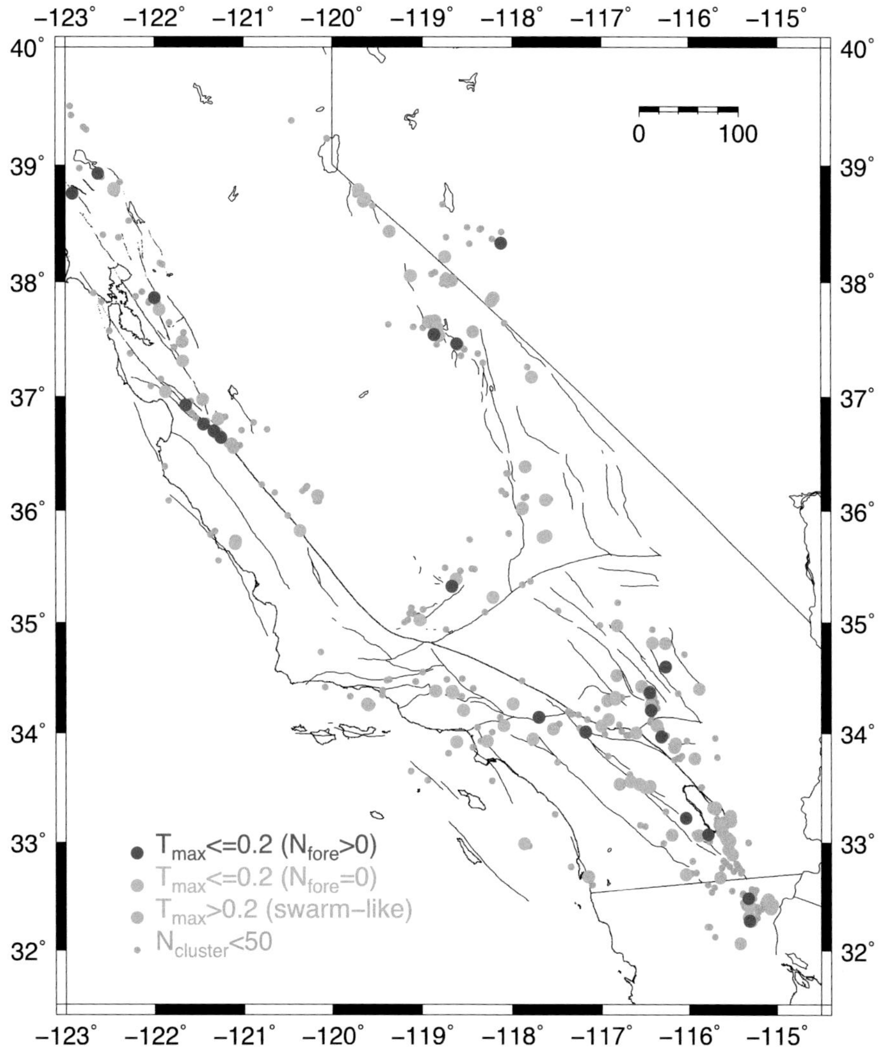

Figure 11

Map view of seismicity bursts in California. *Red dots* aftershock-type clusters with at least one foreshocks; *green dots* aftershock-type clusters without any foreshocks; *orange dots* swarm-type clusters (have foreshocks by definition); *grey dots* small random clusters that do not grow into large clusters

Table 4

Correlation between foreshock area F_{area}, mainshock magnitude M_{max}, magnitude of largest foreshock $M_{f max}$ and number of foreshocks (N_{fore}) from combined result of southern and northern California

Type	$2 \leq M_{max} \leq 4.5$	$M_{max} \geq 4.5$
F_{area} and M_{max}	$cc = 0.19, p = 0.15$	$cc = 0.17, p = 0.43$
F_{area} and $M_{f max}$	$cc = 0.23, p = 0.083$	$cc = 0.55, p = 0.005$
F_{area} and N_{fore}	$cc = 0.74, p = 2.9e - 11$	$cc = 0.81, p = 1.4e - 6$

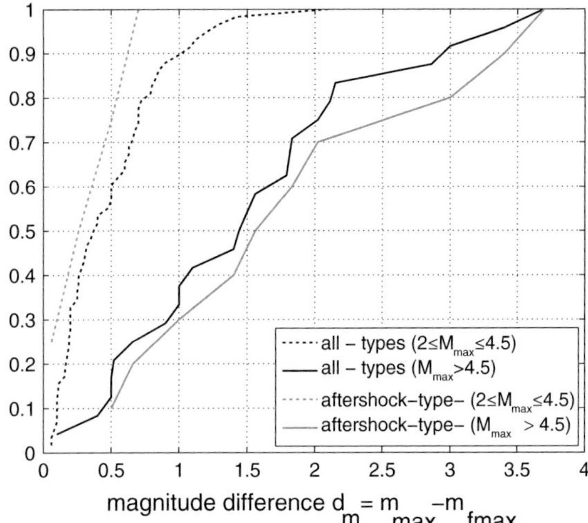

Figure 12

Magnitude difference between the mainshocks and their largest foreshock for combined observations from southern California and northern California. *Black lines* are for all cluster types (include both aftershock-type and swarm-type), *grey lines* only include aftershock-type clusters. *Dash lines* $2 \leq M_{\max} \leq 4.5$, *solid lines* $M_{\max} \geq 4.5$

Overall, results in this section confirm the relationship between foreshock parameters and $M \geq 5$ mainshocks in the previous section. However, we also find that the correlation between foreshock area and foreshock magnitude, and the magnitude difference distribution, are different for clusters with $2 \leq M_{\max} \leq 4.5$ and $M_{\max} \geq 4.5$, which may suggest different triggering processes for different sized earthquakes. Next, we examine whether or not these observations are consistent with synthetic catalogs generated based on empirical statistical relationships.

4. Comparison with Synthetic Catalogs

Earthquake-to-earthquake triggering models relate the probability of earthquakes to the past history of nearby earthquakes using Omori's law for aftershocks and other empirical relationships and have been extensively described by a number of authors (e.g., OGATA1999; HELMSTETTER *et al.* 2003, 2005; FELZER *et al.* 2004). To investigate which features of our observations are consistent with triggering models, we create synthetic catalogs based on the self-similar triggering model described by SHEARER (2012a) that is similar to the ETAS model (epidemic-type-aftershock-sequence) (OGATA 1999). In the models, each "parent" event independently triggers its own aftershock chain of "daughter" events. The number of direct aftershocks N_{asl} for an event with magnitude m follows a productivity law, and event probabliity follows a power-law decay in with time and distance:

$$\begin{aligned} N_{asl} &= Q10^{\alpha(m-m_1)} \\ N(r,t) &= N_{asl}(t+c)^{-p}r^{-q}, \end{aligned} \tag{2}$$

where m_1 is the minimum magnitude earthquake, Q is the aftershock productivity, α describes aftershock generation efficiency, $N(r, t)$ is the aftershock rate as a function of distance and delay time from the triggering event, c and p are Omori's law parameters, and q is the distance decay constant. The magnitude of background events, or triggered events, is a random variable drawn from the Gutenberg–Richter (G–R) distribution:

$$N(\geq m) = 10^{a-bm}, \tag{3}$$

where a is related to the total number of earthquakes, and b (the b value) describes the relative numbers of larger events compared with smaller events. In practice, for computer simulations, m_r (individual magnitude drawn from G–R distribution) is computed as:

$$m_r = m_1 - \log_{10} x_r, \tag{4}$$

where x_r is a random variable drawn from the uniform distribution between $10^{(m_1-m_2)}$ and 1. For the model here, we use $m_1 = 1.5$ and $m_2 = 7.0$, which correspond to the magnitude of completeness and largest earthquake in the catalog.

Recent studies suggest that $\alpha = b = 1$ (used in this study) produces self-similar behavior, for which the increased triggering caused by larger magnitude events is compensated by their decreased numbers in the G–R relation (SHEARER 2012a). For the self-similar triggering case, the branching ratio that describes the average number of first generation aftershocks to the number of background events is

$$n = Qb\ \ln(10)(m_2 - m_1) \tag{5}$$

and $n = 0.39$ for the case of $m_1 = 1.5$ and $m_2 = 7.0$ in order to satisfy Båth's law (BATH 1965), i.e., that the magnitude difference between the mainshock and the largest aftershock is, on average, 1.2.

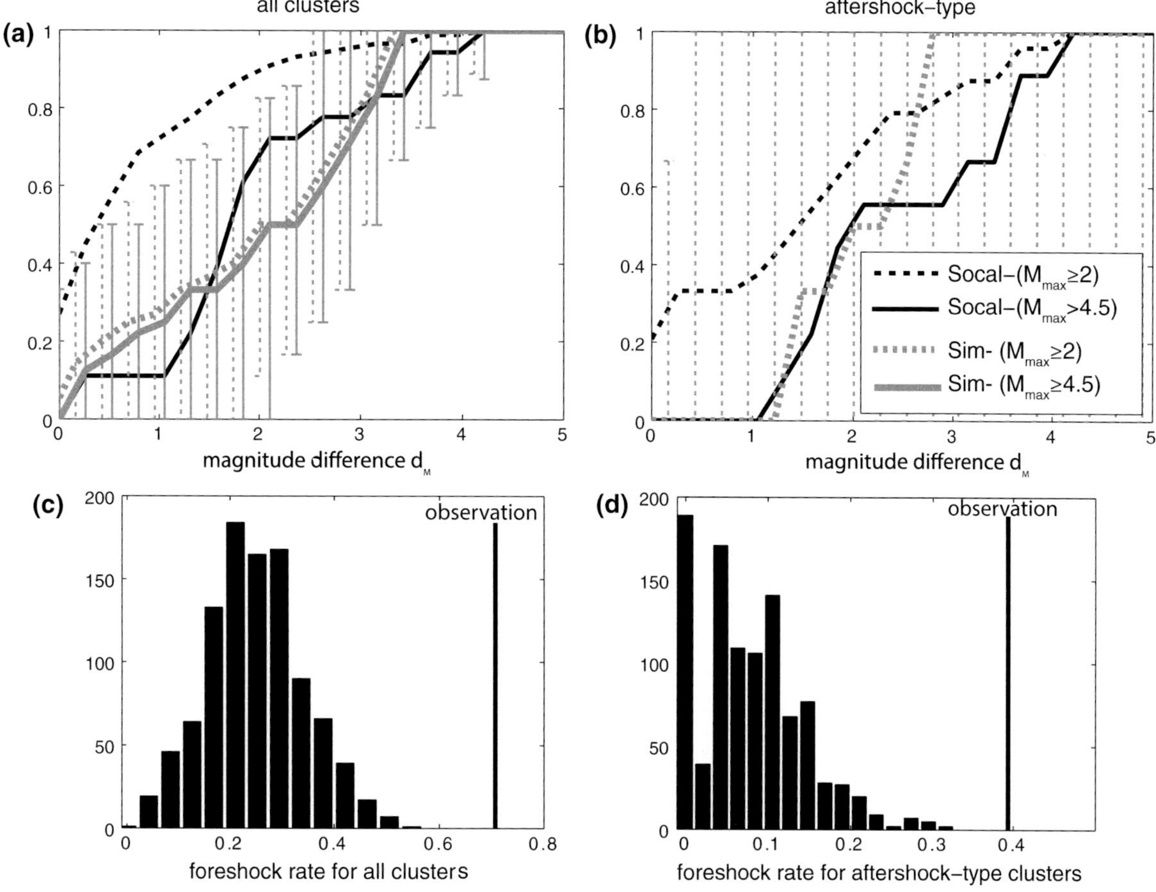

Figure 13

Top magnitude difference between mainshock and largest foreshock. **a** All cluster types (both aftershock-type and swarm-type); **b** only aftershock-type clusters. In both **a** and **b**, *solid lines* $M_{max} \geq 4.5$; *dashed lines* all M_{max} ($M_{max} \geq 2$). *Grey* synthetic catalogs, *black* observations. *Error bars* are 5 and 95 % percentiles from 1000 synthetic catalogs, line styles match corresponding data types. In **a**, the distribution that includes small mainshocks ($M_{max} \leq 4.5$) is clearly above the 95 % limit from synthetic catalogs, thus is significantly different from synthetic catalogs. In **b**, the long *vertical dashed lines* indicate that the distribution of d_m from synthetic catalogs is considerably variably due to the small number of clusters; the *grey solid line* is not applicable due to small number of clusters from synthetic catalogs. *Bottom* foreshock occurrence rate from observations and synthetic catalogs. **c** Foreshock occurrence rate for all clusters, including swarm-type clusters; **d** foreshock occurrence rate only considering aftershock-type clusters. *Vertical lines* are observations (66 % for all clusters, and 39 % for only aftershock-type clusters, these are the results from general cluster search, and differs from the 53 % for only $M \geq 5$ mainshocks), note the observed foreshock occurrence rate for with and without swarm-type clusters is significantly higher than in synthetic catalogs

For the spatial–temporal decay, we use $c = 0.001$ days, $p = 1$ and $q = 1.0$. The p and c values are consistent with Shearer (2012b), the q value is slightly lower than the 1.37 used in Shearer (2012b), but does not strongly affect the results. We generate synthetic catalogs using the declustered southern California catalog with all events $M \geq 1.5$. The declustering process follows Reasenberg (1985), where each aftershock sequence is replaced by a single equivalent event (the mainshock), and the residual catalog approximates a Poisson process. We generate 1000 synthetic catalogs using the declustered catalog as "parent" earthquakes (background seismicity), and generate triggered aftershocks using the above relationships. The synthetic catalogs are then processed the same way as we did for the real catalogs.

On average, only 20 % of the $M \geq 5$ mainshocks in the synthetic catalogs have foreshocks, and the probability of observing a 53 % foreshock occurrence rate among our 70 mainshocks is only 3 % (based on a Gaussian distribution of the occurrence rate from the 1000 synthetic catalogs). Next, we examine

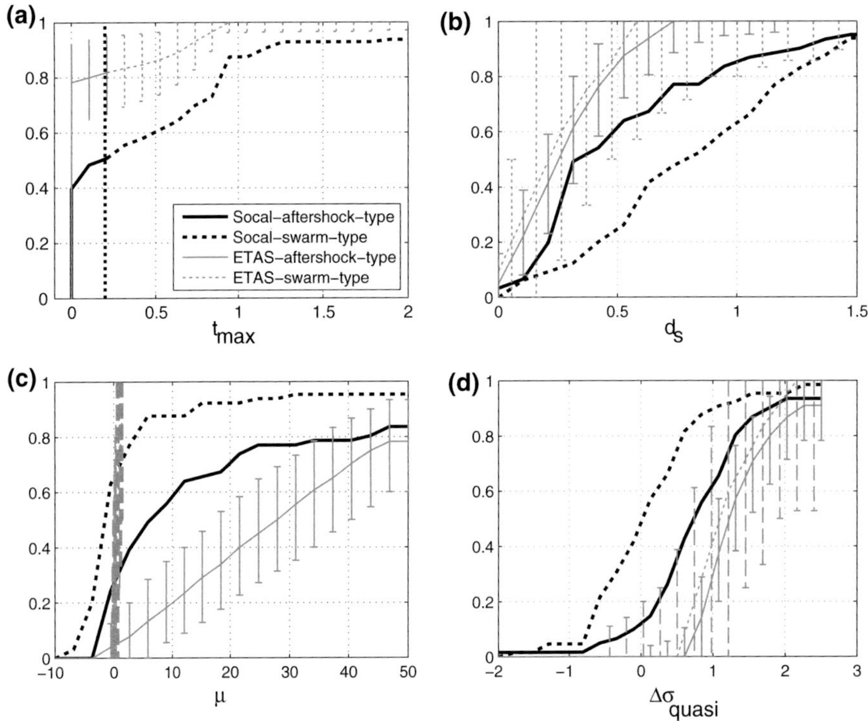

Figure 14
Comparison of distributions of different parameters between observations and synthetic catalogs for southern California. In all figures: *solid lines* aftershock-type, *dashed lines* swarm-type. *Grey* synthetic catalogs, *black* observations. *Vertical error bars* correspond to the 5 and 95 % percentiles from 1000 synthetic catalogs. **a–d** Correspond to the four parameters discussed in text: $t_{\max}$, d_s, μ, $\Delta\sigma_{\text{quasi}}$. In all cases, the *dashed line* (for swarm-type) is significantly outside the variability limit from synthetic catalogs, while the *black line* is sometimes within model limits except the parameter μ that describes the moment release time history

seismicity clusters in the synthetic catalogs, as defined and described for the data in the previous section. We compare the CDF distribution of the four parameters of Chen *et al.* (2012) with observations from the southern California YHS catalog, i.e., the timing of the largest event $t_{\max}$, the skew μ of moment release history, the normalized distance between the first and second halves of the cluster d_s, and the effective stress drop $\Delta\sigma_{\text{quasi}}$. The results are shown in Figs. 14 and 13, and may be summarized as follows:

1. Small random clusters are common in synthetic catalogs, and about 70 % of small clusters do not grow into large clusters, which is consistent with observations.
2. The synthetic catalogs are dominated by aftershock-type clusters, and the probability of getting the observed foreshock occurrence rate is less than 5 % (based on a Gaussian distribution of the foreshock occurrence rate from 1000 synthetic catalogs), suggesting that the synthetic catalogs and observations are statistically different in terms of foreshock occurrence (see Fig. 13c, d).
3. The magnitude difference (d_m) between foreshocks and mainshocks for clusters with $M_{\max} \geq 4.5$ are overall consistent with observations within the model variability (see Fig. 13a), and the distribution of aftershock-type clusters is considerably variable due to the small number of clusters. The distribution of d_m for all clusters with $M_{\max} \geq 2$ is statistically different from the synthetic catalogs, suggesting that the smaller events may not follow the same triggering processes as larger events (see Fig. 13a, b).
4. For the two parameters that describe the spatial evolution of rupture area (d_s and $\Delta\sigma_{\text{quasi}}$), the distributions for swarm-type and aftershock-type

clusters are indistinguishable in the synthetic catalog, while observations find swarm-type clusters have higher d_s and lower $\Delta\sigma_{\text{quasi}}$. This suggests that the gradual spatial expansion observed for "swarmy" clusters is not produced from triggering models with a pure power-law spatial–temporal decay (see Fig. 14b, d).

5. The skew of moment release history agrees with the temporal distribution of magnitude. The synthetic catalogs have different distributions between swarm-type and aftershock-type clusters, which is consistent with observations. The exact forms of the CDFs differ from observations, perhaps due to regional diversities of triggering parameters (see Fig. 14c).
6. 8 % of the synthetic catalogs (81 out of 1000 catalogs) have at least 3 clusters with more than 3 foreshocks (the minimum number to estimate foreshock area) and find a strong correlation between F_{area} and N_{fore}, suggesting that the observed correlation could be reproduced from simulation by random chance. The correlation between F_{area} and $M_{f\max}$ is less statistically significant.

These results depend to some extent on the parameter choices for our triggering model. One could increase the clustering and the number of foreshocks in the synthetic catalogs by increasing the aftershock productivity Q. However, as discussed in Shearer (2012a, b) this would result in catalogs that violate Bath's law, and, at larger values of Q, runaway explosions of seismicity. In any case, even if the number of clusters and foreshocks in the synthetic catalogs could be increased, the synthetic clusters would not show the spatial evolution and temporal skewness distributions of our observed clusters.

Overall, comparisons between synthetic catalogs and observations in southern California suggests that the "swarmy" features in the real catalog are not well reproduced from ETAS-like simulations. The foreshock occurrence rate is too low to be consistent with observations, and the distributions of the magnitude difference between mainshocks and foreshocks is statistically different from observations for small mainshocks. This is consistent with results documented by Shearer (2012b) for M 2.5 to 5.5 mainshocks in southern California, in which the foreshock-to-aftershock ratio is observed to be too large to be consistent with Båth's law, and suggests that the observed foreshock rate cannot be explained entirely with earthquake-to-earthquake triggering models with expected rates of aftershock productivity.

5. Discussion and Conclusions

Our analysis approach is similar to the Vidale and Shearer (2006) study of earthquake "bursts", which considers the relative independence of the clusters from other mainshocks. Although this approach may miss some events due to the selection criteria, the earthquakes selected should represent background activity that is largely free of stress changes and other transient effects from larger events. Our foreshock statistics are generally consistent with previous observations (e.g., Abercrombie and Mori 1996) and indicate that foreshock occurrence rates: (1) depend on faulting type, mainshocks in a transtension setting tend to have more foreshocks compared with mainshocks in a transpression regime; (2) depend on mainshock depth, shallow mainshocks tend to have more foreshocks. We observe this behavior for both retrospective searches of foreshocks of $M \geq 5$ earthquakes, and prospective searches of random clusters. The dependence of clustering type on focal mechanism was also noted in our study focusing on crustal "bursts" (Chen *et al.* 2012); however, the depth dependence was not explored in that study, as the majority of "bursts" occur at shallow depth.

The observed faulting type and depth dependence is consistent with theories related to stress loading during earthquake cycle, where the loading style depends on the regional stress field (Sibson 1993). For pure normal faults, a "loading-weakening" mechanism, where the shear strength reduces with stress loading, is expected; while for reverse faults, a "loading-strengthening" mechanism is expected. For strike–slip faults, depending on the actual stress values, both mechanisms are possible, in which the style changes progressively from "loading-weakening" to "loading-strengthening" as faulting type changes from transtension to transpression status. If the mean stress decreases during shear loading, it is more likely for fluid to flow into the fault zone, which may facilitate the occurrence of small events. ABERCROMBIE and MORI (1996) noted that increased normal stress is commonly expected for reverse faulting and deeper events, which may prohibit occurrence of small events.

The high foreshock occurrence rate from the "burst" approach (53 % for $M \geq 5$ mainshocks, 66 % for all clusters without a magnitude requirement) raises questions about the predictive value of foreshock activities. We investigate the relationship between various foreshock parameters and mainshock magnitude, and find that:

1. Among all the foreshock parameters (number of foreshocks, area of foreshocks, foreshock magnitude), only the number of foreshocks is weakly correlated with mainshock magnitude for carefully selected $M \geq 5$ mainshocks, and does not apply to our generalized cluster search (in Sect. 3) without a mainshock magnitude requirement.
2. For $M \geq 5$ earthquakes, comparing with DODGE *et al.* (1996), the foreshock areas for the "new" mainshocks (in this study) do not fall along the scaling relationship between nucleation zone and mainshock size (see Fig. 7), and the correlation is not statistically significant. There is no significant correlation between foreshock area and mainshock size.
3. A stronger correlation is seen between area and the magnitude of the largest foreshock; however, it also is not valid for smaller mainshocks ($M \leq 4.5$). This is consistent with results in FELZER *et al.* (2004), in which they argue that this agrees with a single-mode triggering process for foreshocks for the larger events, although this may not apply to small events (e.g., $M_{\max} \leq 4.5$) based on observations here.
4. The most consistent correlation for all sized mainshocks is between foreshock area and the number of foreshocks. About 8 % of the synthetic catalogs based on self-similar triggering produce such a correlation; however, the percentage is not statistically significant. Such a correlation confirms the "swarmy" nature of foreshock sequences, where each event may nucleate near the edge of the rupture zone of preceding events, consistent with a self-organized earthquake sequence based on a spring-block model (HAINZL and FISCHER 2002; HAINZL 2003), in which the same mechanical coupling system can produce the main statistical characteristics of both aftershock-type and swarm-type sequences given the appropriate parameter range. For example, the viscous-elastic block system reproduces earthquake swarms in the Vogtland region with higher viscous coupling between blocks (HAINZL 2003), and the same elastic strike–slip fault generates swarm-like sequences with a short critical slip distance (D_c—higher dilatancy strengthening) (YAMASHITA 1999). D_c scales with the characteristic length of small-scale heterogeneity, and within highly fractured zones, a shorter D_c is expected. The observed relationship between clustering type and faulting system is consistent with this point of view, with a higher probability of foreshock occurrence in more complex fault zones.
5. To understand what features could be explained by synthetic triggering models, we compare parameter distributions from observations with 1000 random synthetic catalogs based on a self-similar branching triggering model. The results suggest that for $M \geq 4.5$ mainshocks, the magnitude difference (d_m) is comparable with observations within the model variability. The inconsistencies are in the foreshock occurrence rate and parameters related to spatial expansion of individual clusters, most notably for the "swarm-type" clusters (see Figs. 13, 14). These results suggest that swarm-like behavior that is not well represented by a unified aftershock-triggering model. Accounts for heterogeneity in stress transfer parameters in catalog simulators may help to represent such features more accurately.

Overall, the observations and comparisons with synthetic catalogs suggest that foreshocks are not necessarily part of the "nucleation process" of the eventual mainshock, as there is no consistent correlation between foreshock properties and the mainshock. This also does not necessarily imply a single "rupture mode" for all sized events. Rather, foreshocks are likely driven by independent aseismic processes occurring within the mainshock rupture zone (as suggested for interplate earthquakes Bouchon *et al.* 2013). Regardless of the initiation of the sequence, its evolution is controlled by local heterogeneity, as a unified triggering model could not explain features observed for smaller events. Small, compact clusters are frequent in California, but only 30 % grow into larger clusters with more than 50 events. Before the eventual mainshock occurs, it is difficult to distinguish "foreshocks" from random small clusters. And even if a small cluster is recognized as containing likely "foreshocks", the eventual size of the mainshock is difficult to predict.

This imposes challenges to utilizing the predictive value of foreshocks and in creating simulated catalogs that account for the "swarmy" foreshock features. An important step is to focus more on the role of microearthquakes, as a recent literature review suggested (Mignan 2014), and incorporate spatial variations of earthquake triggering parameters. Based on the current analysis in California, the following properties may be particularly useful: (1) with a unified magnitude cutoff, the number of foreshocks correlates with mainshock magnitude for large earthquakes ($M \geq 5$); (2) the identified foreshocks for $M \geq 5$ mainshocks are localized within the "Mogi" zones prior to the mainshock, and are mostly associated with localized fault discontinuities (Jones 1984; Chen and Shearer 2013); (3) the foreshock occurrence patterns are consistent with earthquake clustering patterns. For example, in the Eastern California Shear Zone, where most mainshocks have foreshocks, there is a 20 % chance that a larger earthquake may occur following a small compact cluster (for southern California, 60 out of 311 clusters grow into larger clusters that do not start with the mainshock). The Hector Mine earthquake occurred within zones of enhanced swarm activities following the Landers earthquake, and is immediately preceded by a cluster of 18 events (Chen and Shearer 2013). Increased "swarmy" activity in a "seismic gap" area or near long fault segments may indicate a potential larger earthquake. Moreover, in several cases, foreshocks exhibit spatial migration (e.g., the Tohoku earthquake Kato *et al.* 2012 and the Chile earthquake Kato and Nakagawa 2014), and spectral differences from background seismicity (Chen and Shearer 2013). Thus, if historic incidences of foreshocks are well documented, then repetitive occurrence of similar clusters in similar tectonic settings may indicate future mainshocks.

Acknowledgments

We thank the Northern California Seismic Network, the U.S. Geological Survey, Menlo Park, and the Berkeley Seismological Laboratory, University of California, Berkeley for providing a moment tensor catalog. We thank the Global CMT Project for providing moment tensor solutions. We thank Richard Sibson for discussion on precursory behavior based on stress analysis. The maps are generated using the GMT software package.

References

Abercrombie, R. E., and J. Mori (1996), *Occurrence patterns of foreshocks to large earthquakes in the western united states*, Nature, *381*(6580), 303–307.

Bath, M. (1965), *Lateral inhomogeneities of upper mantle*, Tectonophysics, 2(6), 483.

Beroza, G. C., and W. L. Ellsworth (1996), *Properties of the seismic nucleation phase*, Tectonophysics, *261*(1–3), 209–227, doi:10.1016/0040-1951(96)00067-4.

Bouchon, M., V. Durand, D. Marsan, H. Karabulut, and J. Schmittbuhl (2013), *The long precursory phase of most large interplate earthquakes*, Nature Geosci, *6*, 299–302, doi:10.1038/ngeo1770.

Chen, X., and P. Shearer (2013), *California foreshock sequences suggest aseismic triggering process*, Geophys. Res. Lett., *40*, 2602–2607, doi:10.1002/grl.50444.

Chen, X., P. M. Shearer, and R. Abercrombie (2012), *Spatial migration of earthquakes within seismic clusters in southern California: Evidence for fluid diffusion*, J. Geophys. Res., *117*(B04301), doi:10.1029/2011JB008973.

Dodge, D. A., G. C. Beroza, and W. L. Ellsworth (1996), *Detailed observations of California foreshock sequences: Implications for the earthquake initiation process*, Journal of Geophysical Research-Solid Earth, *101*(B10), 22,371–22,392.

Ellsworth, W. L., and G. C. Beroza (1995), *Seismic evidence for an earthquake nucleation phase*, Science, *268*(5212), 851–855, doi:10.1126/science.268.5212.851.

Felzer, K. R., R. E. Abercrombie, and G. Ekstrom (2004), *A common origin for aftershocks, foreshocks, and multiplets*, Bulletin of the Seismological Society of America, *94*(1), 88–98, doi:10.1785/0120030069.

Hainzl, S., and T. Fischer (2002), *Indications for a successively triggered rupture growth underlying the 2000 earthquake swarm in vogtland/nw bohemia*, Journal of Geophysical Research-Solid Earth, *107*(B12), 2338, doi:10.1029/2002jb001865.

Hainzl, S. (2003), *Self-organization of earthquake swarms*, Journal of Geodynamics, *35*(1–2), 157–172.

Hardebeck, J. L., and P. M. Shearer (2003), *Using s/p amplitude ratios to constrain the focal mechanisms of small earthquakes*, Bulletin of the Seismological Society of America, *93*(6), 2434–2444.

Hauksson, E., W. Yang, and P. M. Shearer (2012), *Waveform relocated earthquake catalog for southern California (1981 to june 2011)*, Bulletin of the Seismological Society of America, *102*(5), 2239–2244, doi:10.1785/0120120010.

Helmstetter, A., D. Sornette, and J. R. Grasso (2003), *Mainshocks are aftershocks of conditional foreshocks: How do foreshock statistical properties emerge from aftershock laws*, Journal of Geophysical Research-Solid Earth, *108*(B1), 24, doi:10.1029/2002jb001991.

Helmstetter, A., Y. Y. Kagan, and D. D. Jackson (2005), *Importance of small earthquakes for stress transfers and earthquake triggering*, Journal of Geophysical Research-Solid Earth, *110*(B5), 13, doi:10.1029/2004jb003286.

Jones, L. M. (1984), *Foreshocks (1966-1980) in the San-Andreas system, California*, Bulletin of the Seismological Society of America, *74*(4), 1361–1380.

Kato, A., and S. Nakagawa (2014), *Multiple slow-slip events during a foreshock sequence of the 2014 Iquique, Chile mw8.1 earthquake*, Geophysical Research Letters, *41*, doi:10.1002/2014GL061138.

Kato, A., K. Obara, T. Igarashi, H. Tsuruoka, S. Nakagawa, and N. Hirata (2012), *Propagation of slow slip leading up to the 2011 m-w 9.0 tohoku-oki earthquake*, Science, *335*(6069), 705–708, doi:10.1126/science.1215141.

Mogi, K. (1963), *Some discussions on aftershocks, foreshocks and earthquake swarms - the fracture of a semi-infinite body casued by an inner stress origin and its relation to the earthquake phenomena*, Bulletin of the Earthquake Research Institute, *41*, 615–658.

McGuire, J. J., M. S. Boettcher and T. H. Jordan (2005), *Foreshock sequences and short-term earthquake predictability on East Pacific Rise transform faults*, Nature, *434*.

Mignan, Arnaud (2014), *The debate on the prognostic value of earthquake foreshocks: A meta-analysis*, Scientific reports, *4*(4099), doi:10.1038/srep04099.

Ogata, Y. (1999), *Seismicity analysis through point-process modeling: A review*, Pure and Applied Geophysics, *155*(2–4), 471–507.

Reasenberg, P. (1985), *Second-order moment of central California seismicity*, J. Geophys. Res., *90, pp. 5479–5495*.

Reasenberg, P. (1999), *Foreshock occurrence before large earthquakes*, J. Geophys. Res., *104*(B3), 4755–4768.

Shearer, P. M. (2009), *Introduction to Seismology*, second edition, Cambridge University Press.

Shearer, P. M., and G. Q. Lin (2009), *Evidence for mogi doughnut behavior in seismicity preceding small earthquakes in southern California*, Journal of Geophysical Research-Solid Earth, *114*(B01318), doi:10.1029/2008jb005982.

Shearer, P. M., G. A. Prieto, and E. Hauksson (2006), *Comprehensive analysis of earthquake source spectra in southern California*, Journal of Geophysical Research-Solid Earth, *111*(B06303), doi:10.1029/2005jb003979.

Shearer, P. M. (2012a), *Self-similar earthquake triggering, Båth's law, and foreshock/aftershock magnitudes: Simulations, theory, and results for southern California*, J. Geophys. Res., *117*(B06310), doi:10.1029/2011jb008957.

Shearer, P. M. (2012b), *Space-time clustering of seismicity in California and the distance dependence of earthquake triggering*, J. Geophys. Res., *117*(B10306), doi:10.1029/2012JB009471.

Sibson, R. H. (1993), *Load-strengthening versus load-weakening faulting*, J. Struct. Geol., *15*(2), 123–128.

Toda, S., R. S. Stein, G. C. Beroza, and D. Marsan (2012), *Aftershocks halted by static stress shadows*, Nature Geosci, *5*(410–413), doi:10.1038/ngeo1465.

Vidale, J. E., and P. M. Shearer (2006), *A survey of 71 earthquake bursts across southern California: Exploring the role of pore fluid pressure fluctuations and aseismic slip as drivers*, Journal of Geophysical Research-Solid Earth, *111*(B05312), doi:10.1029/2005jb004034.

Waldhauser, F., and D. P. Schaff (2008), *Large-scale relocation of two decades of northern California seismicity using cross-correlation and double-difference methods*, J. Geophys. Res., *113*(B08311), doi:10.1029/2007JB005479.

Yamashita, T. (1999), *Pore creation due to fault slip in a fluid-permeated fault zone and its effect on seismicity: Generation mechanism of earthquake swarm*, Pure and Applied Geophysics, *155*(2–4), 625–647.

Yang, W., E. Hauksson, and P. M. Shearer (2012), *Computing a large refined catalog of focal mechanisms for southern California (1981–2010): Temporal stability of the style of faulting*, Bulletin of the Seismological Society of America, *102*(3), pp. 1179–1194, doi:10.1785/0120110311.

(Received August 28, 2014, revised May 4, 2015, accepted May 7, 2015, Published online May 29, 2015)

Pure Appl. Geophys. 173 (2016), 153–164

DOI 10.1007/s00024-014-1030-5

Pure and Applied Geophysics

Quantitative Analysis of Seismicity Before Large Taiwanese Earthquakes Using the G-R Law

H.-C. LI,[1] C.-H. CHANG,[2] and C.-C. CHEN[1]

Abstract—Seismicity has been identified as an example of a natural, nonlinear system for which the distribution of frequency and event size follow a power law called the "Gutenberg–Richter (G-R) law." The parameters of the G-R law, namely *b*- and *a*-values, have been widely used in many studies about seismic hazards, earthquake forecasting models, and other related topics. However, the plausibility of the power law model and applicability of parameters were mainly verified by statistical error σ of the *b*-value, the effectiveness of which is still doubtful. In this research, we used a newly defined *p* value developed by CLAUSET *et al.* (*Power-Law Distributions in Empirical Data*, SIAM Rev. *51*, 661–703, 2009) instead of the statistical error σ of the *b*-value and verified its effectiveness as a plausibility index of the power-law model. Furthermore, we also verified the effectiveness of K–S statistics as a goodness-of-fit test in estimating the crucial parameter M_c of the power-law model.

1. Introduction

Since seismicity was identified as an example of a natural nonlinear system for which the distribution of frequency and event size in a sufficiently long interval follows the Gutenberg-Richter (G-R) law (i.e. a power law), the parameters of the G-R law, including completeness magnitude $M_c$and *b*- and *a*-values fitted based on M_c, have been used widely in numerous studies focusing on seismic hazards, earthquake forecasting models, and other related topics (HUANG *et al.* 2001; CHEN 2003; SCHORLEMMER *et al.* 2005; HUANG 2006; TSAI *et al.* 2006; WU and CHIAO 2006; BHATTACHARYA *et al.* 2011; RUNDLE *et al.* 2011). The ability to estimate the parameters of the power-law model objectively was definitely crucial to the reliability of these studies; therefore, several techniques were developed for this purpose. They were usually combinations of a fitting method, mainly maximum-likelihood or least-square, with a goodness-of-fit test. WOESSNER and WIEMER (2005) tested the efficiencies of several techniques systematically using natural seismic catalogs. However, the plausibility of a fitted power-law model to the observed data set was mainly evaluated by the statistical error σ of the *b*-value (MARZOCCHI and SANDRI 2003), which was ruled out as an effective index in our research.

CLAUSET *et al.* (2009) developed an elaborate method combining the maximum-likelihood method with a goodness-of-fit test based on the Kolmogorov–Smirnov (K–S) statistics to compute the *p* value to evaluate the plausibility of fitted power-law model. In their test, the earthquake catalog of California was suggested to follow a power law while a proper magnitude cut-off was used. A noticeable point that should be emphasized here is that the *p* value defined by Clauset et al. had no relationship with the widely used *p* value in Omori's law. We used their method and another widely accepted maximum-likelihood method developed by AKI (1965) to the Central Weather Bureau (CWB) catalog, which is another important catalog in the world, to investigate precursory anomalies before large earthquakes based on potential power-law behaviors. Besides using K–S statistics as a goodness-of-fit test, we also combined the "Goodness-of-Fit test" (GFT, WIEMER and WYSS 2000) with Aki's method as a comparison to investigate the effectiveness of the *p* value as an index of model plausibility. Through our tests, we verified that the effectiveness of the *p* value was better than the statistical error σ and also concluded that the application of K–S statistics was crucial in correctly

[1] Institute of Geophysics, National Central University, Jhongli, Taiwan 320 ROC. E-mail: klempervinsky@gmail.com

[2] Seismological Center, Central Weather Bureau, Taipei, Taiwan 320 ROC.

identifying power-law behaviors of seismicity rather than the fitted methods used.

2. Seismic Data

The history of instrumental observation for seismicity in Taiwan and nearby islands started in 1897, when the first seismometer was set up in Taipei by the Japanese. Since 1984 the Central Weather Bureau (CWB) has upgraded the instruments to an electromagnetic type and increased the coverage of the network by building more observation stations. After combining the telemetric seismic network of the Institute of Earth Sciences (IES) to the original seismic network of the CWB in 1991, a new real-time digital observation network was formed called the Central Weather Bureau Seismic Network (CWBSN) and is still maintained by the CWB.

In this research, we selected three large ($M \geqq 6.0$) on-land earthquakes in the Taiwanese area, including (1) the 1999 Chichi earthquake in Nantou, (2) the 2003 Chengkung earthquake in Taitung, and (3) the 2010 Jiashan earthquake in Kaohsiung, and summarize their important parameters in Table 1 (Kao and Chen 2000; Kuochen *et al.* 2007; Hsu *et al.* 2011). For each earthquake, we used 12 intervals of interval lengths 30, 60, 90,..., 360 days, respectively, to select seismicity. Each interval started from the previous day before the occurrence of the main-shock and extended back to the past, so the aftershocks were excluded in the data sets used in parameter computation. The purpose of using various interval lengths was to eliminate potential biases caused by using preferential interval lengths subjectively.

3. Methods

3.1. Maximum-Likelihood Estimate by Clauset et al. (2009)

We merely addressed major concepts about the method developed by Clauset *et al.* (2009) here; further details and derivations can be referred to in other articles (Newman 2005) if necessary. A quantity x obeys a power law if it is drawn from a probability distribution

$$p(x) \propto x^{-\alpha} \tag{1}$$

where α is a constant parameter called the "exponent" or "scaling parameter" of the distribution. In usual cases, there must be some lower bound $x_{\min}$, below which the power-law behavior no longer exists and the distribution of x belongs to other types. On the other hand, there are also some statistical fluctuations to larger values of x which are caused by the essential property of enormously low frequency of rare events. Newman (2005) derived that once $x_{\min}$ was known, the probability density of a continuous variable x was

$$p(x) = \frac{\alpha - 1}{x_{\min}} \left(\frac{x}{x_{\min}} \right)^{-\alpha} \tag{2}$$

and the exponent of power law distribution could be easily estimated by

$$\alpha = 1 + n \left[\sum_{i=1}^{n} \ln \frac{x_i}{x_{\min}} \right]^{-1}. \tag{3}$$

The quantity x_i, $i = 1, 2, \ldots, n$ gives the measured values of x. The cumulative density function (CDF) was

Table 1

Parameters of selected large on-land earthquakes in the Taiwan region

	1999 Chichi	2003 Chengkung	2010 Jiashan
Date (UT)	20 Sep. 1999	10 Dec. 2003	4 Mar. 2010
Magnitude	M_w 7.6, M_L 7.3	M_w 6.8, M_L 6.4	M_w 6.3, M_L 6.4
Epicenter	23.85°N, 120.82°E	23.06°N, 121.39°E	22.97°N, 120.71°E
Focal depth (km)	8	17.73	22.64

$$P(x) = \int_{x}^{\infty} p(x')dx' = \left(\frac{x}{x_{min}}\right)^{-\alpha+1} \quad (4)$$

which also followed the power law, but with a different exponent, $\alpha - 1$, which is the b-value of the G-R law in the field of earth science. An estimate of the expected statistical error σ in Eq. (3) is given by

$$\sigma = \sqrt{n}\left[\sum_{i=1}^{n} \ln\frac{x_i}{x_{min}}\right]^{-1} = \frac{\alpha - 1}{\sqrt{n}}. \quad (5)$$

The relation between the estimated maximum amplitude of an earthquake on seismograph and magnitude is

$$A(\Delta) \propto 10^{M_L}, \quad (6)$$

where A is the maximum amplitude of an event recorded at an epicentral distance Δ. In order to use Clauset's method for a real earthquake catalog, we had to transform all documented magnitudes in an interval of interest to the analogies of amplitude using Eq. 6 and use the transformed quantity as the variable x in Eq. (2), rather than using the values of magnitude directly.

3.2. *Maximum-Likelihood Estimate by Aki (1965)*

The widely accepted method in the field of earth science to estimate the parameters in the G-R law is the maximum-likelihood method developed by Aki (1965). In this method, the probability density function of an earthquake with a magnitude greater than M_c is assumed to obey an exponential distribution expressed by

$$f(M, b') = b'e^{-b'(M-M_c)}, \; M \geq M_c \quad (7)$$

where $b' = b/log_{10}e$. Suppose that we have a sample of n earthquakes with magnitudes M_1, M_2,..., M_n. The parameter b is estimated by

$$b = \frac{log_{10}e}{\langle M\rangle - M_c + \Delta M/2} \quad (8)$$

where $\langle M\rangle$ is the average magnitude of the used data set and ΔM is the binning width. In Aki's method, the parameter M_c is implicitly assumed to be known. However, we have to determine a suitable M_c value to select a data set and to compute a corresponding b-value in a real application. A statistical error in Eq. (8) is also estimated by Eq. (5) using b instead of $\alpha - 1$.

3.3. *Goodness-of-Fit Test*

A noticeable point is that the exponent in each method completely depends on an estimated data minimum, i.e. $x_{\min}$ and M_c in Eqs. (3) and (8) respectively. Thus, a good fitness-of-fit test which collaborates with the above-mentioned methods to determine a suitable minimum is critical. Clauset et al. used the Kolmogorov–Smirnov (K–S) statistics as the goodness-of-fit test, which is the maximum distance between the CDFs and a fitted model:

$$D = \max_{x \geq x_{\min}} |S(x) - P(x)|. \quad (9)$$

Here, $S(x)$ is the CDF of the observation data with values greater than or equal to $x_{\min}$, and $P(x)$ is the CDF for the fitted power-law model that best fits the data in the region $x \geq x_{\min}$. The $x_{\min}$ which minimizes D is the estimated lower bound of power-law behavior. We used the same combination in this research and called it C + KS. Furthermore, the K–S statistics was also used in the analysis applying Aki's method and this combination was called A + KS.

Another test used in the analysis applying Aki's method was the "Goodness-of-Fit test (GFT)", which was developed by Wiemer and Wyss (2000) to compute the difference between an observed frequency-magnitude distribution (FMD) and a synthetic distribution of a fitted model. At first we estimated the b- and a-values of the G-R law as a function of cut-off magnitude M_{co} by Eq. (8), then we generated synthetic distributions using the estimated b- and a-values for $M \geq M_{co}$. The residual between an observed and a synthetic distribution is computed by

$$R(a, b, M_{co}) = \frac{\sum_{M_{co}}^{M_{max}} |B_i - S_i|}{\sum_i B_i} \quad (10)$$

where B_i and S_i are the observed and synthetic cumulative number of earthquakes in each magnitude bin. If the residual value R is smaller, the similarity between an observed and a synthetic distribution is better. Wiemer and Wyss defined the M_c as the first M_{co} at which R was less than a fixed confidence level, 0.1 or 0.05.

We set the precision of each candidate of M_c in 0.1 rather than 0.01 because 0.01 was too small considering the measurement precision of magnitude of a small earthquake; therefore, it did not make sense to use an unrealistic value as small as 0.01.

3.4. Plausibility of the Hypothesis of the Power Law

Using the methods addressed in Sects. 3.1 and 3.2, we can estimate the parameters of hypothesized probabilistic models for each observed data set. However, it was critical for us to know whether the hypothetic power-law model was plausible for the observed data set. In this research, we adopted the approach developed by Clauset *et al.* (2009) to generate *p* values to quantify the plausibility of the hypothesis model.

The first step was to generate a synthetic data set. For the observed data set in an interval, the total number of observations was *n*, and the number of observations greater than or equal to an estimated x_{min} (i.e. M_c in this research) was n_{tail}. With the probability n_{tail}/n we generated a random number x_i using the probabilistic model with $x \geq x_{min}$ and estimated *b*-value. On the other hand, with the probability $1 - n_{tail}/n$, we selected x_i at random from the observed data values that conformed the condition $x < x_{min}$. Repeating the process for all $i = 1,\ldots,$ n, we generated a synthetic data set which followed the hypothesis model above x_{min} , but had the same distribution of the observed data below x_{min}. Based on Clauset's estimation, we generated 2,500 synthetic data sets to allow the *p* value to be accurate to about two decimal digits. Once the synthetic data sets were generated, we used the methods addressed in Sects. 3.1 or 3.2 to estimate M_c, *b*-value, and *D* or *R* of each synthetic data set.

The *p* value was defined as the fraction of synthetic data sets with *D* or *R* greater than the value calculated by real observed data. If most of the synthetic data sets had values of *D* or *R* greater than the observed data sets and, therefore, made the *p* value exceed the threshold value 0.1 (based on Clauset's definition), the hypothesis model was more suitable to describe the observed data set than most of the synthetic data sets. We concluded the hypothesis model to be plausible for the observed data set. Otherwise, if $p < 0.1$, the hypothesis model was ruled out as a plausible one for the observed data set. However, we should emphasize again that a high *p* value merely verified the plausibility of the hypothesis model to the observed data. It did not necessarily mean that the hypothesis model was the "correct" distribution for the data. It is still possible to find other hypothesis models that are plausible to the same data set.

4. Results

Tables 2, 3, and 4 list all estimated parameters of optimal fitted models, including M_c, σ, *a*-, *b*-, and *p* value, of the data set in each interval. To compare the *a*-value of each interval, we normalized all rates of earthquake to "number per 30 days" and calculated the *a*-value at magnitude 2.0.

4.1. Chichi earthquake, Nantou

By Table 2, we found that the methods C + KS and A + KS estimated the same values of M_c except the cases using 30-, 270-, and 300-day intervals. In the cases with the same M_c, the *b*-values estimated by C + KS and A + KS for the same data set had tiny differences smaller than 0.01. A similar situation also existed for several cases of the Chengkung earthquake in 2003 and the Jiashan earthquake in 2010. These tiny differences revealed an essential difference while using Eqs. 3 or 8 to compute the *b*-value. We illustrate the case using a 90-day interval in Fig. 1a–c, in which the inverted triangles indicate the observed cumulative FMD and the open circles indicate the predicted cumulative FMD of fitted power-law model using the method C + KS, A + KS, and A + G, respectively. The patterns of predicted FMD in Fig. 1a by C + KS and 1b by A + KS were highly similar by visual inspection and both were verified as plausible power-law models to the selected data set by their *p* values exceeding 0.1. Actually, the power-law model was shown to be plausible for all cases using the method C + KS or A + KS because their *p* values were all greater than 0.1.

Table 2

Parameters estimated using the data prior to the 1999 Chichi earthquake

Interval (day)	a-value			b-value			M_c			p Value			σ		
	C + KS	A + KS	A + G	C + KS	A + KS	A + G	C + KS	A + KS	A + G	C + KS	A + KS	A + G	C + KS	A + KS	A + G
30	2.7949	2.7270	2.6857	0.9872	0.9178	0.8587	2.7	2.4	2.1	0.81	0.59	0.45	0.0876	0.0607	0.0430
60	2.7216	2.7157	2.6421	0.9338	0.9238	0.8338	2.6	2.6	2.1	0.98	0.90	0.34	0.0548	0.0542	0.0310
90	2.6884	2.6840	2.6288	0.8760	0.8672	0.7942	2.5	2.5	2.1	0.64	0.77	0.07	0.0379	0.0375	0.0244
120	2.7229	2.7176	2.6476	0.8877	0.8787	0.7990	2.6	2.6	2.2	0.47	0.68	0.08	0.0357	0.0353	0.0228
150	2.7132	2.7089	2.6434	0.8699	0.8613	0.7807	2.5	2.5	2.1	0.26	0.34	0.03	0.0282	0.0280	0.0182
180	2.7091	2.7047	2.6483	0.8808	0.8719	0.8005	2.5	2.5	2.1	0.65	0.85	0.08	0.0264	0.0261	0.0170
210	2.7149	2.7104	2.6428	0.8955	0.8864	0.8019	2.5	2.5	2.1	0.83	0.53	0.04	0.0249	0.0246	0.0159
240	2.7233	2.7186	2.6597	0.9123	0.9028	0.8322	2.5	2.5	2.2	0.72	0.48	0.05	0.0237	0.0235	0.0167
270	2.7383	2.7297	2.6740	0.9211	0.9085	0.8406	2.6	2.5	2.2	0.54	0.27	0.03	0.0248	0.0221	0.0157
300	2.7557	2.7531	2.6977	0.9260	0.9187	0.8504	2.5	2.6	2.2	0.17	0.22	0.02	0.0209	0.0230	0.0146
330	2.7759	2.7699	2.7123	0.9387	0.9287	0.8580	2.6	2.6	2.2	0.31	0.33	0.02	0.0222	0.0219	0.0139
360	2.7894	2.7833	2.7242	0.9390	0.9290	0.8571	2.6	2.6	2.2	0.31	0.31	0.02	0.0209	0.0207	0.0131

Each interval started from the day before the main shock occurred and extended back to the past. The a-values in Tables 2, 3, and 4 were estimated at magnitude 2.0

C Clauset, *A* Aki, *KS* Kolmogorov–Smirnov statistics, *G* goodness-of-fit test by Woessner and Wiemer

Table 3

Parameters estimated using the data prior to the 2003 Chengkung earthquake

Interval (day)	a-Value			b-Value			M_c			p Value			σ		
	C + KS	A + KS	A + G	C + KS	A + KS	A + G	C + KS	A + KS	A + G	C + KS	A + KS	A + G	C + KS	A + KS	A + G
30	2.8014	2.8014	2.7871	0.9115	0.9021	0.8678	2.0	2.0	1.9	0.15	0.28	0.21	0.0362	0.0359	0.0317
60	2.8523	2.8513	2.8228	0.9380	0.9279	0.8646	2.1	2.1	1.9	0.70	0.41	0	0.0277	0.0274	0.0215
90	2.8559	2.8550	2.8449	0.9327	0.9228	0.9038	2.1	2.1	2.0	0.23	0.59	0.38	0.0224	0.0221	0.0197
120	2.8942	2.8522	2.8395	1.0055	0.9401	0.9162	2.7	2.1	2.0	0.86	0.02	0.06	0.0404	0.0196	0.0174
150	2.8949	2.8871	2.8421	0.9915	0.9803	0.9018	2.7	2.7	2.0	0.57	0.37	0.01	0.0352	0.0348	0.0153
180	2.8916	2.8784	2.8702	0.9831	0.9721	0.9116	3.2	3.2	2.0	0.74	0.66	0	0.0559	0.0553	0.0137
210	2.9182	2.9058	2.9085	0.9526	0.9423	0.8885	3.2	3.2	2.0	0.94	0.92	0	0.0466	0.0461	0.0118
240	2.9155	2.9027	2.9186	0.9688	0.9581	0.9273	3.2	3.2	2.1	1	0.99	0	0.0455	0.0450	0.0127
270	2.8897	2.8773	2.9145	0.9498	0.9395	0.9269	3.2	3.2	2.1	0.97	0.93	0	0.0422	0.0418	0.0120
300	2.9102	2.8974	2.8879	0.9651	0.9545	0.8896	3.2	3.2	2.0	0.91	0.84	0	0.0406	0.0402	0.0101
330	2.9132	2.9004	2.8861	0.9703	0.9596	0.8951	3.2	3.2	2.0	0.95	0.89	0	0.0391	0.0386	0.0097
360	2.9077	2.8949	2.8847	0.9701	0.9594	0.8981	3.2	3.2	2.0	0.82	0.70	0	0.0376	0.0372	0.0094

Table 4

Parameters estimated using the data prior to the 2010 Jiashan earthquake

Interval (day)	*a*-Value			*b*-Value			M_c			*p* Value			σ		
	C + KS	A + KS	A + G	C + KS	A + KS	A + G	C + KS	A + KS	A + G	C + KS	A + KS	A + G	C + KS	A + KS	A + G
30	2.5208	2.5126	2.7520	0.8487	0.8405	0.9418	3.0	3.0	2.0	0.93	0.82	0.03	0.1238	0.1226	0.0396
60	2.7675	2.7675	2.3036	0.9914	0.9802	0.7796	2.0	2.0	3.3	0	0	0.06	0.0290	0.0286	0.1248
90	2.9594	2.9544	2.9137	1.0519	1.0393	0.9817	2.4	2.4	2.2	0	0	0	0.0327	0.0323	0.0248
120	2.7360	2.7253	2.8988	0.8866	0.8776	0.9720	3.2	3.2	2.2	0.68	0.81	0	0.0647	0.0640	0.0216
150	2.8492	2.8401	2.8812	0.9445	0.9344	0.9193	2.9	2.9	2.1	0.62	0.41	0	0.0423	0.0418	0.0166
180	2.9166	2.9119	2.8652	1.0134	1.0017	0.9279	2.4	2.4	2.1	0	0	0	0.0230	0.0227	0.0156
210	2.8581	2.7516	2.8879	0.9818	0.9155	0.9753	2.9	3.3	2.2	0.28	0.58	0	0.0382	0.0573	0.0166
240	2.7878	2.9285	2.9059	0.8879	0.9631	0.9258	3.3	2.2	2.1	0.74	0	0	0.0479	0.0146	0.0128
270	2.9558	2.9538	2.9316	0.9479	0.9377	0.9018	2.2	2.2	2.1	0	0	0	0.0131	0.0129	0.0114
300	2.8306	2.8189	2.9155	0.8912	0.8821	0.9063	3.3	3.3	2.1	0.62	0.60	0	0.0411	0.0407	0.0111
330	2.9553	2.9263	2.9034	0.9867	0.9453	0.9081	2.6	2.2	2.1	0.05	0	0	0.0196	0.0122	0.0107
360	2.9171	2.9088	2.8948	0.9537	0.9434	0.8997	2.8	2.8	2.1	0.03	0.05	0	0.0231	0.0228	0.0103

On the other hand, the pattern of the model shown in Fig. 1c is quite different from Fig. 1a, b. The observed data began to deviate from the open circles slightly at magnitude 3.3 and more apparently at magnitude 3.8. A similar pattern, in which there were apparent gaps between the observed FMD and predicted FMD by model, existed generally in the cases using A + G for estimating parameters. The values of M_c estimated by A + G were much smaller than the values estimated by other methods. Almost all *p* values estimated by A + G were smaller than 0.1, and thus revealed that the statistical distribution of corresponding data sets did not obey the power law except two cases using 30- and 90-day intervals. In contrast to the performances of other methods, we suggested that the method A + G failed to determine appropriate data sets of which the statistical distributions followed the hypothetic G-R law before the Chichi earthquake.

4.2. Chengkung Earthquake, Taitung

The methods C + KS and A + KS estimated the same M_c values and, therefore, very close *b*-values in all cases except the one using the 120-day interval. But the values of M_c estimated by A + G which were in the range 1.9–2.1 were much smaller than the values by other methods for the cases using interval length longer than 120 days. The unique difference of M_c between the method C + KS and A + KS in the case using the 120-day interval was noticeable, and the results are shown in Fig. 2a–c. By visual inspection, the open circles in Fig. 2a are very close to the inverted triangles until magnitude 4.9, when the observed FMD fluctuates enormously and, therefore, clearly deviates from the predicted FMD. On the other hand, the open circles in Fig. 2b began to deviate from the observed FMD at magnitude 3.2, and the open circles in Fig. 2c also clearly deviated from the observed FMD at magnitude 3.1. The *p* values revealed that the power-law model was plausible for the data set used in Fig. 2a, but implausible for the data sets in Fig. 2b, c with smaller M_c. The method C + KS was the only one which could uncover potential power-law behaviors in the case using the 120-day interval.

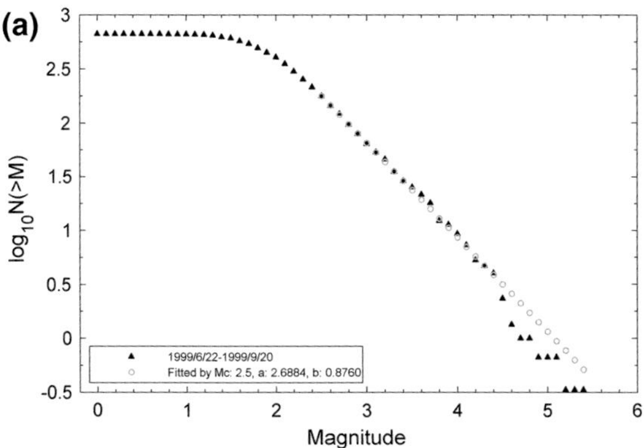

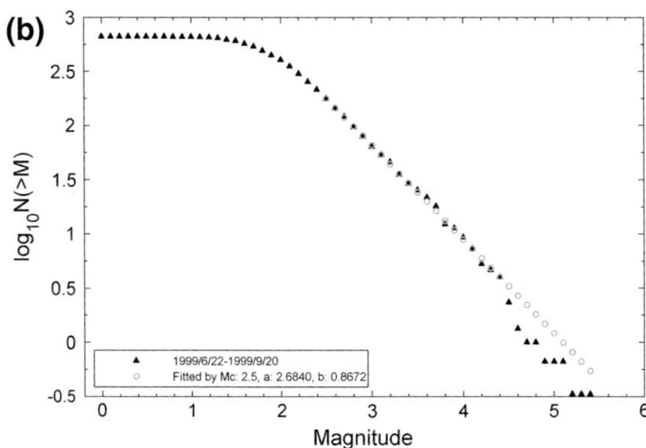

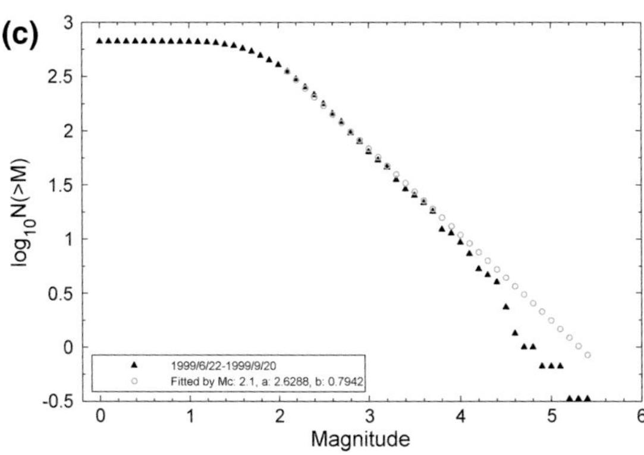

Figure 1
Observed cumulative FMD (*inverted triangle*) and corresponding fitted power-law models (*open circle*) before the 1999 Chichi earthquake. The interval length is 30 days and the method used to fit model is C + KS for (**a**), A + KS for (**b**), and A + G for (**c**)

4.3. *Jiashan Earthquake, Kaohsiung*

A noticeable property of all the observed FMDs before the 2010 Jiashan earthquake was the apparent discontinuities of the FMD curves. We illustrated this property by the case using a 150-day interval in Fig. 3a–c. The method C + KS and A + KS estimated the same M_c value, 2.9, in contrast to a smaller value, 2.1, by A + G. In Fig. 3a and b, the open circles were located very close to the inverted triangles in the magnitude ranging from 2.9 to 4.5. However, the open circles in Fig. 3c deviated clearly from the observed FMD at magnitude greater than 2.9. Based on inspection of the p value, we suggest that the power-law model is implausible for the data set used in Fig. 3c, while a smaller M_c value was estimated and, therefore, included cumbersome smaller earthquakes.

Although the method C + KS and A + KS identified appropriate data sets for which the power-law model was plausible, both methods also estimated smaller M_c values in several cases, including 60-, 90-, 180-, 270-, 330-, and 360-day intervals. Unlike the cases with larger M_c, all p values of the cases with smaller M_c were below the threshold 0.1 and, therefore, rejected the hypothesis of a power-law model. We illustrated the cases using a 240-day interval in Fig. 3d–f, in which the discontinuity of observed FMD was not so clear by visual inspection. The method C + KS estimated a larger M_c value 3.3 and p value 0.74 in contrast to the smaller M_c values and zero p values by other methods. The inverted triangles deviated from the open circles in magnitudes ranging from 3 to 4.5 in Fig. 3f and merely ranging from 3.2 to 3.4 in Fig. 3e. In fact, the deviations of the observed FMD from predicted values were not very apparent in Fig. 3e and f, but the power-law model was still rejected as plausible for the data sets by the zero p values. We suggest that these cases sufficiently illustrate the advantage in identifying the behaviors of seismicity with an objective quantitative index such as the p value rather than by visual inspection.

5. *Discussions*

We had three topics to discuss in this research, which were (1) the reliability of statistical error σ as an indicator of model plausibility, (2) the effectiveness of three fitted methods to identify potential power-law behaviors of seismicity, and (3) the implications of precursory seismic anomalies identified by appropriate fitted methods. We discussed the reliability of p value and statistical error σ first.

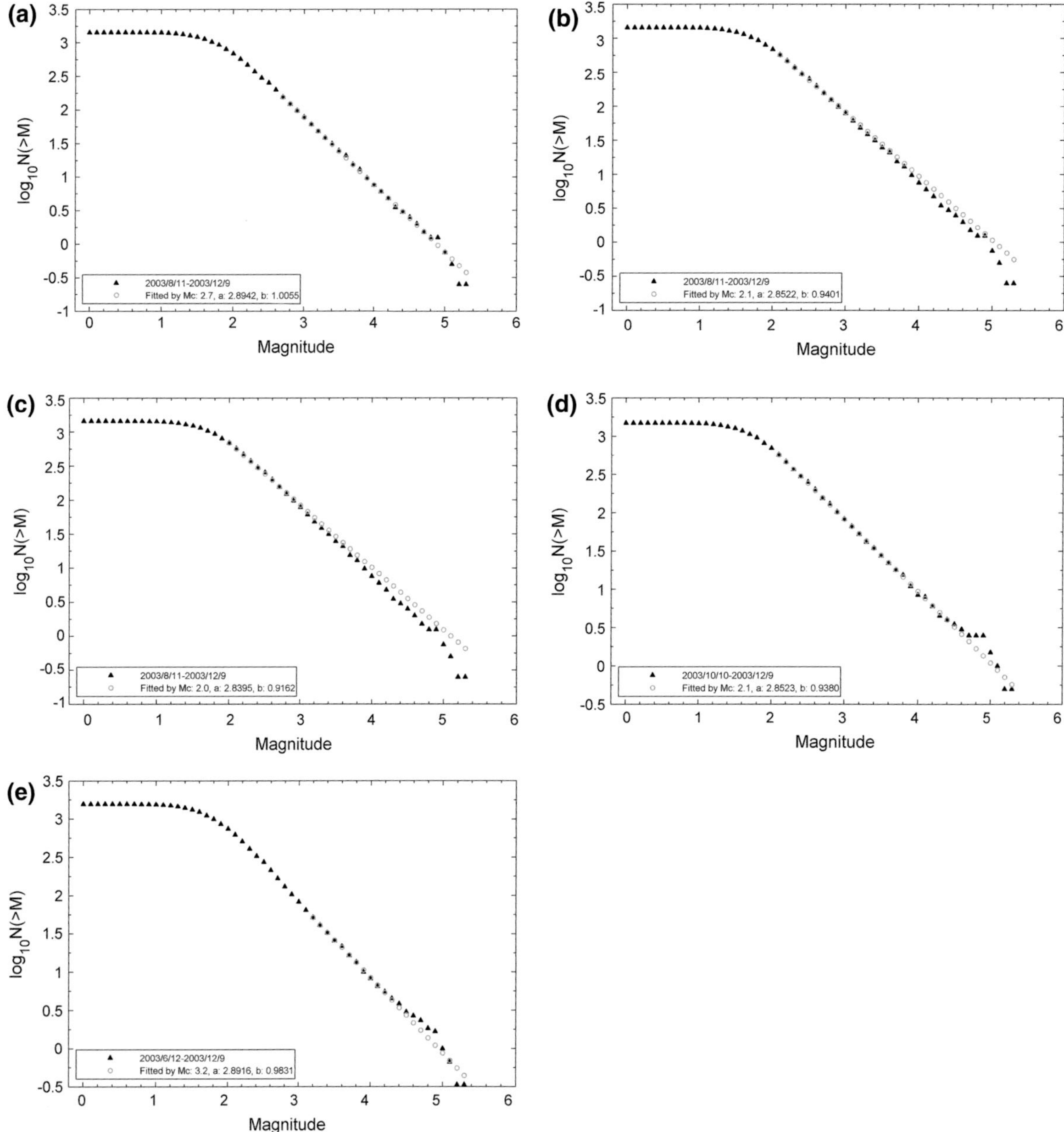

Figure 2
Observed cumulative FMD and corresponding fitted power-law models before the 2003 Chengkung earthquake. The interval length is 120 days for (**a**) to (**c**), 60 days for (**d**), and 180 days for (**e**). The fitting method is C + KS for (**a**), (**d**), and (**e**), A + KS for (**b**), and A + G for (**c**)

5.1. Reliability of σ as Indicator of Model Plausibility

An important parameter in Eq. 5 to estimate the statistical error σ is "number of earthquakes, n" in the denominator. This parameter definitely depends on the length of interval and the value of M_c which were used to select data set. If the values of $\alpha - 1$ in the nominator, i.e. b-value in the G-R law, merely fluctuated slightly between different data sets, the data sets with more data points would

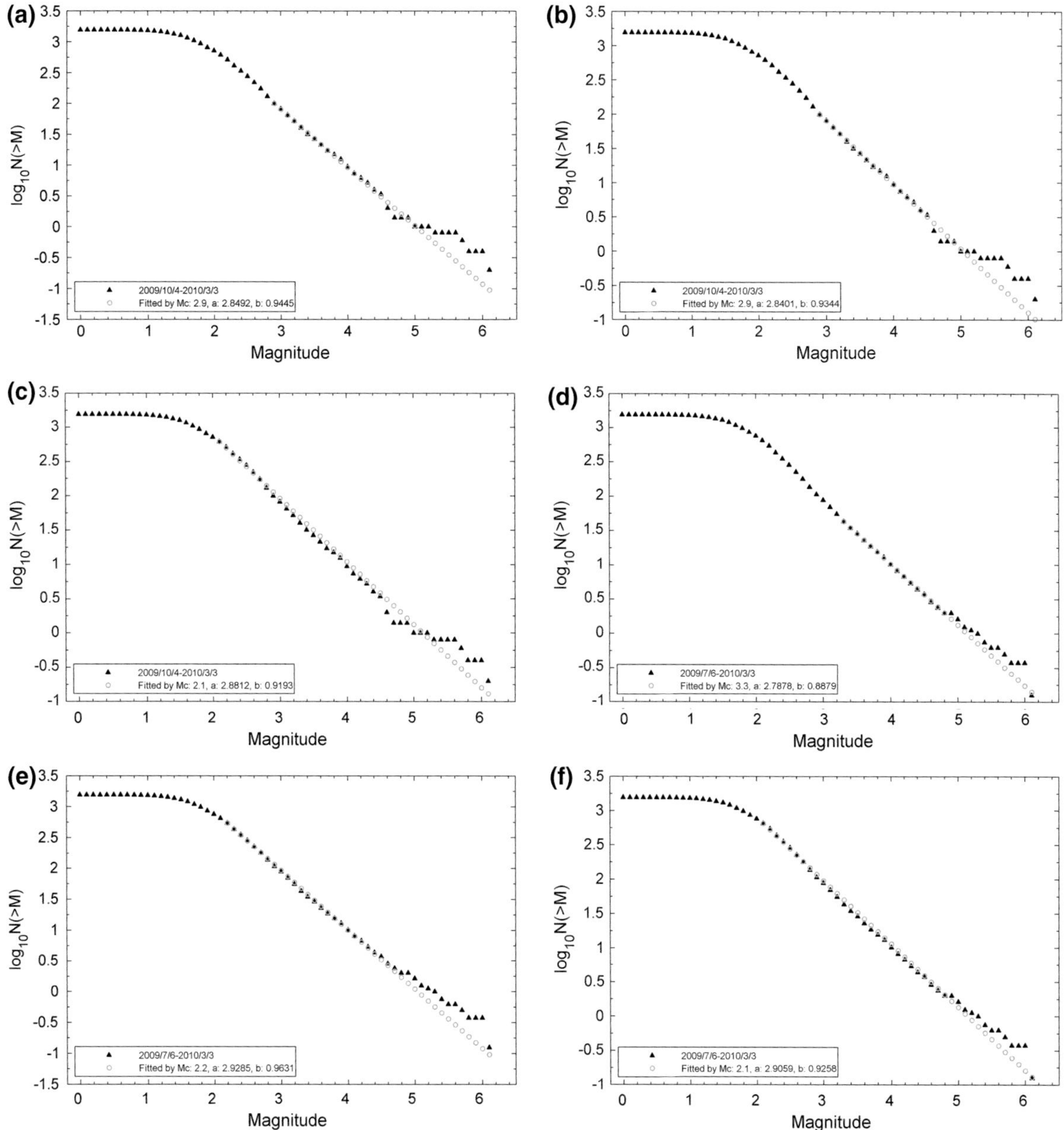

Figure 3
Observed cumulative FMD and corresponding fitted power-law models before the 2010 Jiashan earthquake. The interval length is 150 days for (**a**) to (**c**), and 240 days for (**d**) to (**f**). The fitting method is C + KS for (**a**) and (**d**), A + KS for (**b**), and (**e**), and A + G for (**c**) and (**f**)

generate smaller values of σ. This property could be easily observed in each column of σ using the method A + G in Tables 2, 3, and 4. For example, all M_c values merely fluctuated between 1.9 and 2.1, and the differences of b-value between different time intervals also merely fluctuated smaller than 0.1 in Table 3. The dominant factor affecting the number of data points is the length of time interval used, and we indeed observed that the values of σ decreased regularly with the elongation of time intervals, which also corresponded to the increase of n.

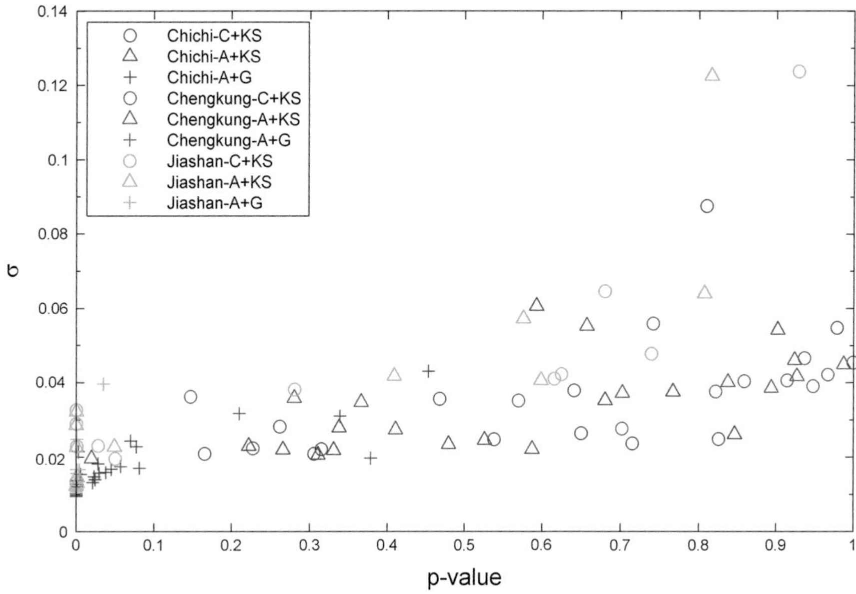

Figure 4
Plot of σ and p value. *Blue*, *red*, and *green* indexes indicate the cases of 1999 Chichi, 2003 Chengkung, and 2010 Jiashan earthquakes, respectively. Indexes marked by *circle*, *triangle*, and *cross* indicate the cases using method C + KS, A + KS, and A + G, respectively

However, smaller σ values did not guarantee the plausibility of the power-law model to a data set. We illustrated this property by considering the cases using the method A + G in Table 3. The data set selected by a 90-day interval generated a value of σ larger than other cases using longer time intervals, but its p value proved that the power-law model was plausible for this data set. Although the data sets using intervals longer 90 days had smaller σ values, the power-law models were ruled out because their p values were below the predefined threshold. A similar situation also could be observed in Tables 2 and 4.

Furthermore, we summarized the distribution between p value and σ in Fig. 4 and observed a very diverse pattern. For example, the values of σ between 0.04 and 0.05 corresponded with the p values ranging from 0.03 to 1. The value of Pearson correlation coefficient between p value and σ was 0.5833, which verified that no apparent linear correlation existed between these two parameters. An immediate consequence was that the statistical error σ could not be a reliable index to verify the plausibility of fitted models. Therefore, we strongly recommended using the p value instead of σ for the purpose of model plausibility.

5.2. *Effectiveness of Fitted Methods for Potential Power-Law Behaviors of Seismicity*

The second topic that we would like to discuss is the effectiveness of the three methods addressed in Sect. 3 in identifying potential power-law behaviors of seismicity. Our definition of effectiveness was very simple, for a case using a specific time interval, if the power-law model was proved to be plausible to either data set selected by two different methods but was ruled out in the data set selected by the third method; we would suggest that the effectiveness of the third method was not comparable to other methods. We had to emphasize that either Clauset's or Aki's maximum-likelihood method could generate a set of G-R law parameters for any input data set. The effectiveness of each method actually depended on the goodness-of-fit method which collaborated with the maximum-likelihood methods to determine the crucial parameter M_c. In this research, both the methods C + KS and A + KS excavated potential power-law behaviors in most cases. There were merely two cases, i.e. the 120-day interval of the 2003 Chengkung earthquake and the 240-day interval of the 2010 Jiashan earthquake, in which the method A + KS failed while C + KS worked. However, the hypothesis of the power-law model was ruled out in 89 % of the data sets selected using A + G and almost all of the M_c values estimated by the GFT were much smaller than the values by the K–S statistics. We suggested that this property corresponded with the observation by Woessner and Wiemer (2005) that the GFT might underestimate the value of M_c. Moreover, we found that the subjectively defined threshold value of residual R in the GFT severely affected the estimation of M_c. Any researcher who attempts to apply this method should be very careful of this affect. Therefore, we suggested that the K–S statistics was a more effective goodness-of-fit method rather than the GFT method for our purpose.

5.3. Precursory Seismic Anomalies

The final topic which we would discuss is actually based on the discussions addressed above. An objective and reliable index to evaluate the plausibility of the power-law model is the most important premise to assess precursory anomalies of seismicity based on the G-R law. With a reliable index such as the p value, it will be helpful in preventing to evaluate seismic anomalies mistakenly using improper parameters such as the b-value of G-R law. We observed both the characteristics of seismic activation and quiescence before three selected large earthquakes in this research with the aid of the p value.

In Fig. 1a and b, the observed FMD (inverted triangles) were located below the predicted FMD (open circles) at magnitude greater than 4.4. In fact, a similar property existed in all of the cases using various interval length except 30 days. We suggested that this deficiency of earthquakes with moderate magnitude lasted at least 1 year before the Chichi main-shock. Similar precursory seismic quiescence had also been reported in previous research by Wu and Chiao (2006) based on the continuously decreasing b-value using monthly seismicity. They attributed this seismic quiescence to the reduction of smaller earthquakes because the monthly numbers of larger earthquakes ($M_L > 4.0$) before the Chichi main-shock were suggested to fluctuate in a normal range consistently. An opposite observation about precursory anomalies before the 1999 Chichi earthquake was made by Chen (2003), who identified a seismic activation of earthquake with magnitude greater than 5 in the interval 1998–September 20, 1999. Chen fitted the cumulative FMD using earthquakes with moderate magnitude, but we had a lot of smaller earthquakes in the optimal data set because of M_c 2.5.

On the other hand, we observed the phenomena of seismic activation before the 2003 Chengkung and 2010 Jiashan earthquake. In Fig. 2a for the 2003 Chengkung earthquake, the observed FMD were slightly larger than the predicted FMD at the magnitude greater than 4.8. We also illustrate two cases using 60- and 180-day intervals in Fig. 2d and e of which both were estimated by the method C + KS. The observed FMD exceeded the predicted values at the magnitude greater than 4.6 in Fig. 2d and 4.3 in Fig. 2e. For the 2003 Chengkung earthquake, it was also reported that there was local activation of a moderate-size earthquake before the main-shock by Wu *et al.* (2008) based on a positive Z value surrounding the rupture region of the Chengkung earthquake. For the 2010 Jiashan earthquake, the observed FMD were obviously larger than the predicted FMD at the magnitude greater than 5.0 in Fig. 3a and 4.8 in Fig. 3d. We also observed similar characteristics in other cases that the power-law model verified to be plausible. We suggest that the 2010 Jiashan earthquakes might be another example of precursory seismic activation based on our observations, but further analyses are still necessary for more related details.

In this research, we revealed the effectiveness of the K–S statistics as a goodness-of-test method in identifying potential power-law behaviors of seismicity through the test using p value. Furthermore, either maximum-likelihood method developed by Clauset or Aki would generate almost the same values of G-R law parameters if a data set was already verified by the K–S statistics that obeyed the power-law model. The plausibility of the hypothesis model and applicability of parameters such as the b-value should always be verified first to reduce the risk of mistakenly evaluating the seismicity.

References

Aki, K. (1965), *Maximum Likelihood Estimates of b in the Formula* $\log n = a\text{-}bm$ *and Its Confidence Limits*, Bull. Earthq. Res. Inst. Univ. Tokyo *43*, 237–239.

Bhattacharya, P., Chakrabarti, B.K., and Kamal (2011), *A Fractal Model of Earthquake Occurrence: Theory, Simulations and Comparisons with the Aftershock Data*, J. Phys. Conf. Ser. *319*, 012004.

Chen, C.-C. (2003), *Accelerating Seismicity of Moderate-Size Earthquakes before the 1999 Chi-Chi, Taiwan, Earthquake: Testing Time-Prediction of the Self-Organizing Spinodal Model of Earthquakes*, Geophys. J. Int. *155*, F1–F5.

Clauset, A., Shalizi, C.R. and newman, M.E. (2009), *Power-Law Distributions in Empirical Data*, SIAM Rev. *51*, 661–703.

Hsu, Y.-J., Yu, S.-B., Kuo, L.-C., Tsai, Y.-C., and Chen, H.-Y. (2011), *Coseismic Deformation of the 2010 Jiashian, Taiwan Earthquake and Implications for Fault Activities in Southwestern Taiwan*, Tectonophysics *502*, 328–335.

Huang, Q.-H., Sobolev, G.A., and Nagao, T. (2001), *Characteristics of the seismic quiescence and activation patterns before the* $M = 7.2$ *Kobe earthquake, January 17, 1995*, Tectonophysics *337*, 99–116.

Huang, Q.-H. (2006), *Search for reliable precursors: A case study of the seismic quiescence of the 2000 western Tottori prefecture earthquake*, J. Geophys. Res *111*, B04301.

Kao, H., and Chen, W.-P. (2000), *The Chi-Chi Earthquake Sequence: Active, Out-of-Sequence Thrust Faulting in Taiwan*, Science *288*, 2346–2349.

Kuochen, H., Wu, Y.-M., Chen, Y.-G., and Chen, R.-Y. (2007), *2003 Mw6.8 Chengkung Earthquake and Its Related Seismogenic Structures*, J. Asian Earth Sci. *31*, 332–339.

Marzocchi, W., and Sandri, L. (2003), *A Review and New Insights on the Estimation of the b-value and Its Uncertainty*, Ann. Geophys. *46*, 1271–1282.

Newman, M.E. (2005), *Power Laws, Pareto Distributions and Zipf's Law*, Contemp. Phys. *46*, 323–351.

Rundle, J.B., Holliday, J.R., Yoder, M., Sachs, M.K., Donnellan, A., Turcotte, D.L., Tiampo, K.F., Klein, W., and Kellogg, L.H. (2011), *Earthquake Precursors: Activation or Quiescence?* Geophys. J. Int. *187*, 225–236.

Schorlemmer, D., Wiemmer, S., and Wyss, M. (2005), *Variations in Earthquake-Size Distribution across Different Stress Regimes*, Nature *437*, 539–542.

Tsai, Y.-B., Liu, J.-Y., Ma, K.-F., Yen, H.-Y., Chen, K.-S., Chen, Y.-I., and Lee, C.-P. (2006), *Precursory Phenomena Associated with the 1999 Chi-Chi Earthquake in Taiwan as Identified under iSTEP Program*, Phys. Chem. Earth *31*, 365–377.

Wiemer, S., and Wyss, M. (2000), *Minimum Magnitude of Completeness in Earthquake Catalogs: Examples from Alaska, the Western United States, and Japan*, Bull. Seismol. Soc. Am. *90*, 859–869.

Woessner, J., and Wiemer, S. (2005), *Assessing the Quality of Earthquake Catalogs: Estimating the Magnitude of Completeness and Its Uncertainty*, Bull. Seismol. Soc. Am. *95*, 684–698.

Wu, Y.-M., and Chiao, L.-Y. (2006), *Seismic Quiescence before the 1999 Chi-Chi, Taiwan, Mw 7.6 Earthquake*, Bull. Seismol. Soc. Am. *96*, 321–327.

Wu Y.-M., Chen, C.-C., Zhao, L., and Chang, C.-H. (2008), *Seismicity Characteristics before the 2003 Chengkung, Taiwan, earthquake*, Tectonophysics *45*, 177–182.

(Received August 31, 2014, revised December 24, 2014, accepted December 29, 2014, Published online January 24, 2015)

Pure Appl. Geophys. 173 (2016), 165–172
© 2014 Springer Basel
DOI 10.1007/s00024-014-0930-8

Pure and Applied Geophysics

Statistical Significance of Minimum of the Order Parameter Fluctuations of Seismicity Before Major Earthquakes in Japan

N. V. Sarlis,[1] E. S. Skordas,[1] S.-R. G. Christopoulos,[1] and P. A. Varotsos[1]

Abstract—In a previous publication, the seismicity of Japan from 1 January 1984 to 11 March 2011 (the time of the *M*9 Tohoku earthquake occurrence) has been analyzed in a time domain called natural time χ. The order parameter of seismicity in this time domain is the variance of χ weighted for normalized energy of each earthquake. It was found that the fluctuations of the order parameter of seismicity exhibit 15 distinct minima—deeper than a certain threshold—1 to around 3 months before the occurrence of large earthquakes that occurred in Japan during 1984–2011. Six (out of 15) of these minima were followed by all the shallow earthquakes of magnitude 7.6 or larger during the whole period studied. Here, we show that the probability to achieve the latter result by chance is of the order of 10^{-5}. This conclusion is strengthened by employing also the receiver operating characteristics technique.

Key words: Natural time analysis , Japan, receiver operating characteristics, Monte Carlo calculation, fluctuations, order parameter of seismicity.

1. Introduction

Earthquakes (EQs) exhibit complex correlations in time, space, and magnitude (Telesca *et al.* 2002; Eichner *et al.* 2007; Huang 2008, 2011; Telesca and Lovallo 2009; Telesca 2010; Lippiello *et al.* 2009, 2012; Lennartz *et al.* 2008, 2011; Sarlis 2011; Rundle *et al.* 2012; Tenenbaum *et al.* 2012; Sarlis and Christopoulos 2012). The EQ scaling laws (Turcotte 1997) point to the view (e.g., Holliday *et al.* 2006) that a mainshock occurrence may be considered an approach to a critical point. Following this view, Varotsos *et al.* (2005) (see also Sarlis *et al.* 2008; Varotsos *et al.* 2011b) suggested that the variance κ_1 of the natural time χ (see Sect. 2) may serve as an order parameter for seismicity.

The study of the fluctuations of this order parameter, denoted β (Sarlis *et al.* 2010), becomes of major importance near the critical point, i.e., near the mainshock occurrence. We assume that a few months represent the period near criticality before each mainshock (Varotsos *et al.* 2011a, 2012a, b, 2013) motivated by the following aspect: the lead time of seismic electric signals (SES) activities that are considered to be emitted when the system enters the critical stage (Varotsos and Alexopoulos 1986; Varotsos *et al.* 1993) ranges from a few weeks to a few months (Varotsos and Lazaridou 1991; Telesca *et al.* 2009, 2010; Varotsos *et al.* 2011b).

Recently, the natural time analysis of seismicity in Japan has been investigated (Sarlis *et al.* 2013) using the Japan Meteorological Agency (JMA) seismic catalog and considering all the 47,204 EQs of magnitude $M_{JMA} \geq 3.5$ in the period from 1984 to the time of the *M*9 Tohoku EQ (i.e., 11 March 2011), within the area 25°–46°N, 125°–148°E depicted in Fig. 1. It was found that the fluctuations of the order parameter of seismicity exhibit 15 distinct minima—deeper than a certain threshold—1 to around 3 months before the occurrence of large earthquakes. Six (out of 15) of these minima were followed by all the shallow earthquakes of magnitude 7.6 or larger during the whole period studied (their epicenters are shown in Fig. 1). Among the minima, the minimum before the *M*9 Tohoku EQ was the deepest [this EQ was also preceded by a seismic quiescence, as found by Huang and Ding (2012) through an improved region-time-length algorithm]. It is the scope of the present paper to investigate the statistical significance of these results obtained by Sarlis *et al.* (2013).

[1] Department of Solid State Physics and Solid Earth Physics Institute, Faculty of Physics, School of Science, National and Kapodistrian University of Athens, Panepistimiopolis Zografos, 157 84 Athens, Greece. E-mail: nsarlis@phys.uoa.gr; eskordas@phys.uoa.gr; strichr@yahoo.gr; pvaro@otenet.gr

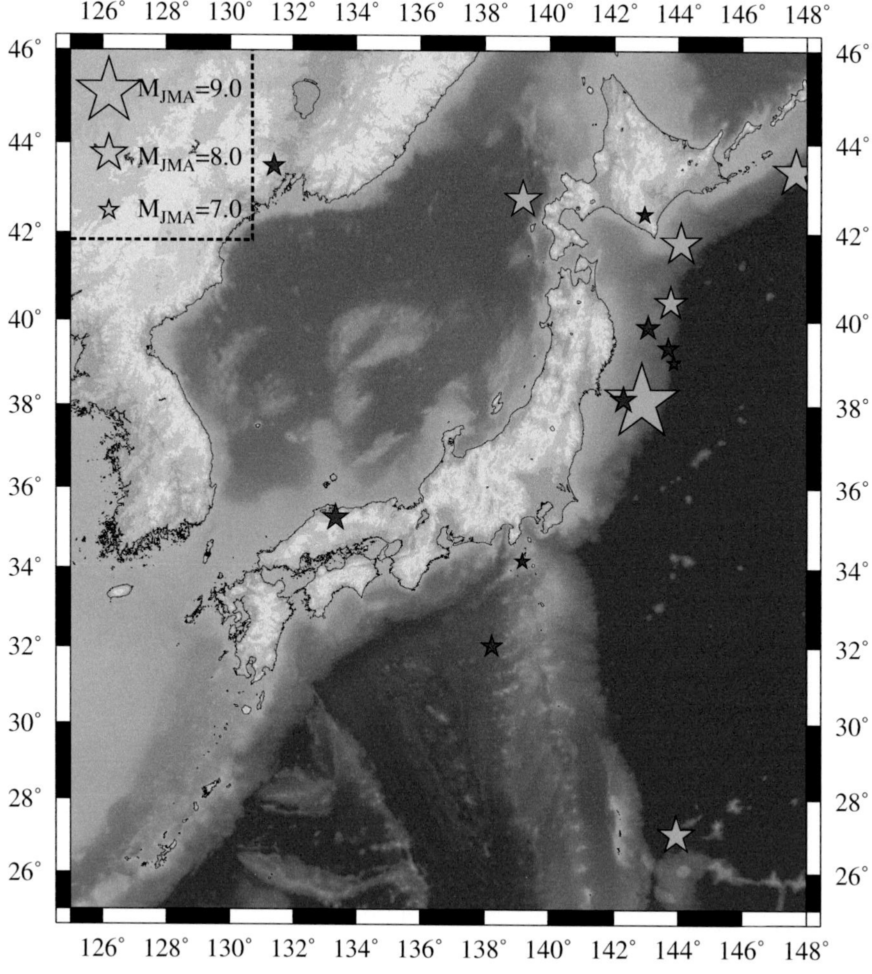

Figure 1
Epicenters (*green stars*) of all shallow EQs with magnitude 7.6 or larger (marked in *bold* in Table 1) within the depicted area N_{25}^{46} E_{125}^{148} since 1 January 1984 until the *M*9 Tohoku EQ. The *red stars* indicate the epicenters of the smaller EQs listed in Table 1

2. Methodology

Natural time analysis has been shown (Abe *et al.* 2005) to extract the maximum information possible from a given time series. For a time series comprising N events, we define the natural time for the occurrence of the kth event of energy Q_k by $\chi_k = k/N$ (Varotsos *et al.* 2001, 2002). We then study the evolution of the pair (χ_k, p_k) where

$$p_k = Q_k / \sum_{n=1}^{N} Q_n \qquad (1)$$

is the normalized energy and construct the quantity κ_1 which is the variance of χ weighted by p_k

$$\kappa_1 = \sum_{k=1}^{N} p_k \chi_k^2 - \left(\sum_{k=1}^{N} p_k \chi_k \right)^2 \equiv \langle \chi^2 \rangle - \langle \chi \rangle^2, \qquad (2)$$

where the quantity Q_k—see Eq. (1)—is estimated by means of the usual relation (Kanamori 1978)

$$Q_k \propto 10^{1.5 M_k}, \qquad (3)$$

(e.g., Varotsos *et al.* 2005, 2011a, 2012a, b, c; Sarlis *et al.* 2010; Ramírez-Rojas and Flores-Márquez 2013; Varotsos *et al.* 2013; Flores-Márquez *et al.* 2014).

The detailed procedure for the computation of β has been described by Sarlis *et al.* (2013). In short, they considered excerpts of the JMA catalog comprising W consecutive EQs and defined the quantity $\beta_W \equiv \sigma(\kappa_1)/\mu(\kappa_1)$ as the variability of the order parameter $\kappa_1 - \mu(\kappa_1)$ and $\sigma(\kappa_1)$ stand for the average value and the standard deviation of the κ_1 values for this excerpt of length W. For such an excerpt, we form its subexcerpts consisting of the nth to $(n+5)$th EQs, $(n = 1, 2, \ldots W - 5)$ and compute κ_1 for each of them by assigning $\chi_k = k/6$ and $p_k = Q_k / \sum_{n=1}^{6} Q_n$, $k = 1, 2, \ldots 6$, to the kth member of the subexcerpt (since at least $l = 6$ EQs are needed for obtaining reliable κ_1). We iterate the same process for new subexcerpts comprising $l = 7$ members, 8 members, ...and finally W members. Then, we compute the average and the standard deviation of the thus-obtained ensemble of κ_1 values [examples of the κ_1 values resulted from subexcerpts comprising $l = 6$, 40, 100, 200, and 300 members (EQs) are given in Fig. 2 for the last $\approx$10 year period before the $M9$ Tohoku EQ]. The β_W value for this excerpt W was assigned to the $(W + 1)$th EQ in the catalog, the target EQ. Hence, for the β_W value of a target EQ, only its past EQs are used in the calculation. The time evolution of the β value was then pursued by sliding the excerpt W through the EQ catalog. Since $\approx 10^2$ EQs with $M_{JMA} \geq 3.5$ occur per month on average, the values $W = 200$ and $W = 300$ were chosen, which would cover a period of a few months before each target EQ. As an example, we depict in Fig. 3 the values of β_{200} and β_{300} (red, left scale) along with all $M_{JMA} \geq 6$ EQs (black, right scale) versus the conventional time during the $\approx$10-year period before the $M9$ Tohoku EQ, i.e., since 1 January 2001 until 11 March 2011. The corresponding β values for the remaining period, i.e., since 1 January 1984 until 31 December 2010, can be visualized in Sarlis *et al.* (2013). Distinct minima of β_{200} and β_{300}—deeper than a certain threshold and having a ratio β_{300}/β_{200} close to unity (in the range of 0.95 to 1.08)—have been identified 1–3 months before all shallow EQs with magnitude 7.6 or larger in the period from 1984 to the time of the $M9$ Tohoku EQ. Especially, the minima of β_{200} precede the EQ occurrence by a lead time Δt_{200} which is at the most 96 days (cf. the entries in bold in the first and the third column of Table 1, see also Table 1 of Sarlis *et al.* 2013). Moreover, nine additional similar minima have been identified (see Table 2 of Sarlis *et al.* 2013) during the same period, which were followed by large EQs of smaller magnitude within 3 months. These 15 (=6 + 9) minima are summarized here in Table 1.

3. Statistical Evaluation by Means of Monte Carlo

The above 15 β minima have been identified during the $\approx$27 -year study period comprising

Table 1

The 15 minima of β_{200} that were found (Sarlis et al. 2013) to precede large EQs in Japan during the period 1 January 1984 to 11 March 2011

Date of β_{200} minimum	Value of β_{200} minimum	EQ date	Lat. (°N)	Long. (°E)	M	Δt_{200} (months)
13-10-1986	0.254	14-01-1987	42.45	142.93	6.6	3
08-08-1989	0.278	02-11-1989	39.86	143.05	7.1	3
05-04-1992	0.250	18-07-1992	39.37	143.67	6.9	3
23-05-1993	**0.293**	**12-07-1993**	**42.78**	**139.18**	**7.8**	**2**
1993-07-13	0.188	12-10-1993	32.03	138.24	6.9	3
30-06-1994	**0.295**	**04-10-1994**	**43.38**	**147.67**	**8.2**	**3**
15-10-1994	**0.196**	**28-12-1994**	**40.43**	**143.75**	**7.6**	**2–3**
17-02-1998	0.237	31-05-1998	39.03	143.85	6.4	3
12-04-2000	0.229	01-07-2000	34.19	139.19	6.5	3
09-07-2000	0.243	06-10-2000	35.27	133.35	7.3	3
12-05-2002	0.244	29-06-2002	43.50	131.39	7.0	2
03-07-2003	**0.289**	**26-09-2003**	**41.78**	**144.08**	**8.0**	**3**
11-06-2005	0.286	16-08-2005	38.15	142.28	7.2	2
30-11-2010	**0.232**	**22-12-2010**	**27.05**	**143.94**	**7.8**	**1**
05-01-2011	**0.157**	**11-03-2011**	**38.10**	**142.86**	**9.0**	**2**

The six cases that were followed by all the shallow EQs of magnitude 7.6 or larger are shown in bold

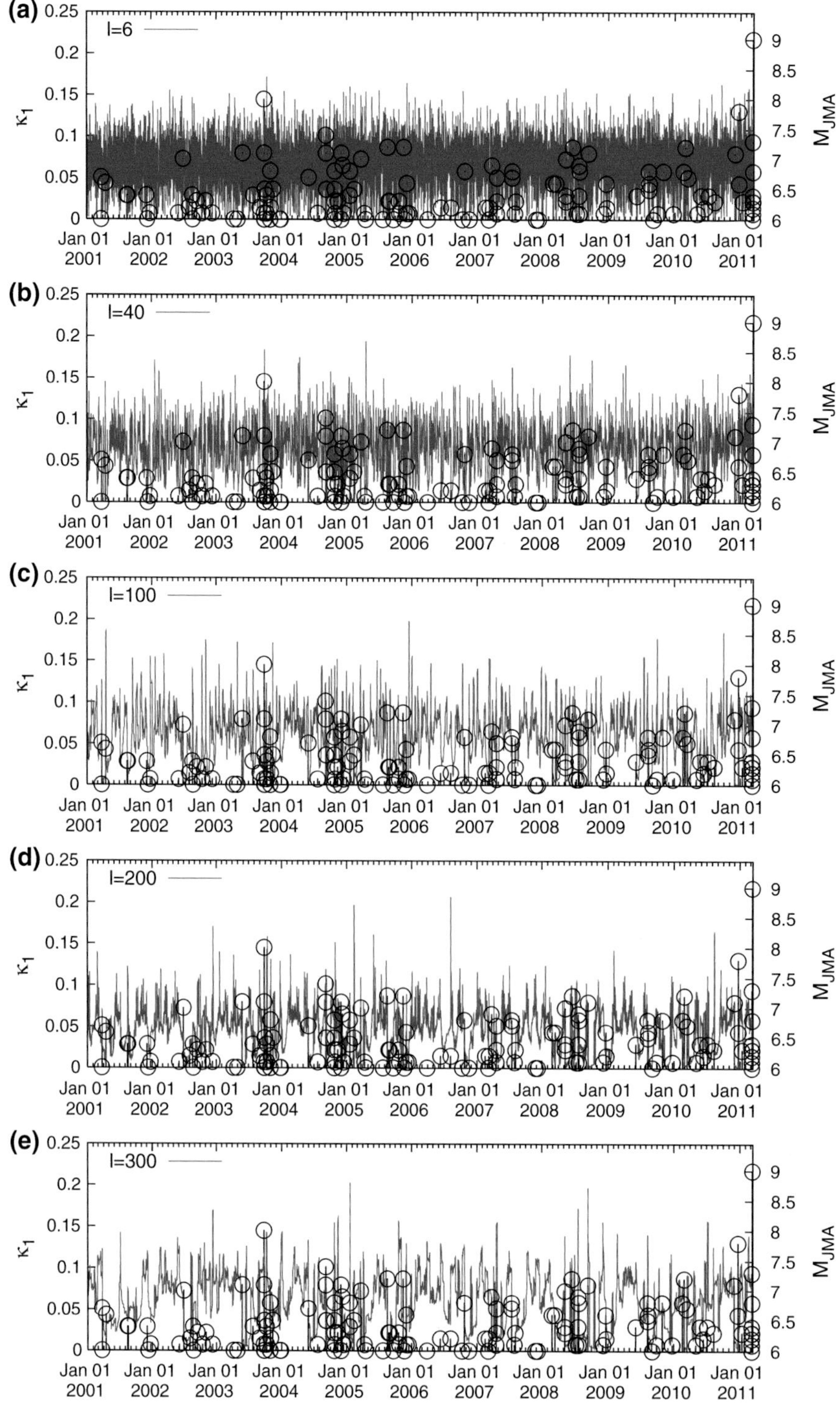

Figure 2
Plots showing how the κ_1 values (*left scale*) fluctuate during the last ≈10-year period before the *M*9 Tohoku EQ, i.e., since 1 January 2001 until 11 March 2011. Here, we depict examples of the κ_1 values computed from subexcerpts comprising $l = 6$ (**a**), 40 (**b**), 100 (**c**), 200 (**d**), and 300 (**e**) EQs. The EQs with $M_{JMA} \geq 6$ (*right scale*) are also depicted by the *vertical black lines* ending at *circles*

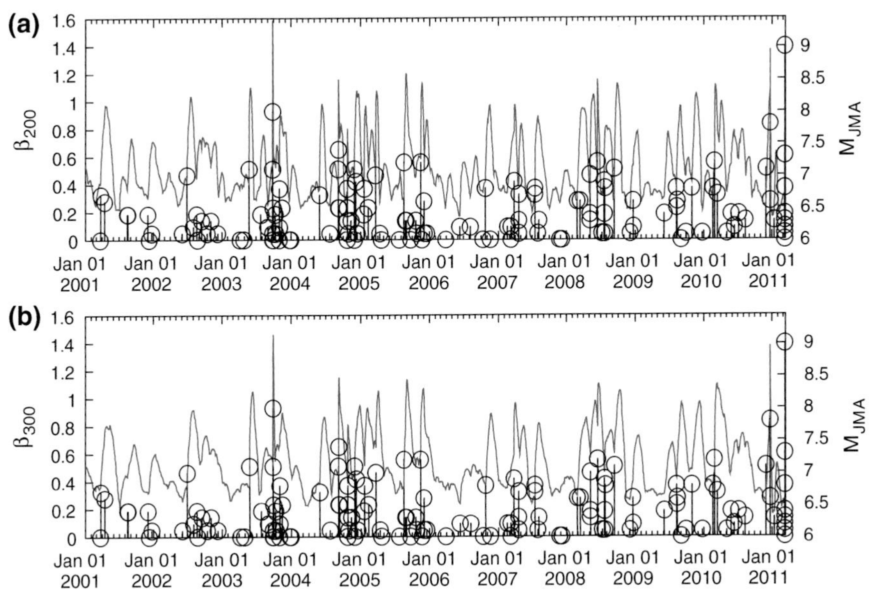

Figure 3

Plots of the β values (*left scale*) during the last $\approx$10-year period before the *M*9 Tohoku EQ, i.e., since 1 January 2001 until 11 March 2011 for $W = 200$ (**a**) and $W = 300$ (**b**) along with all $M_{JMA} \geq 6$ EQs (*right scale*)—*vertical black lines* ending at *circles*—versus the conventional time

9,931 days. The maximum lead time for $W = 200$ was found to be, as mentioned, $\Delta t_{200} = 96$ days. Since the β_W values are calculated after the occurrence of each of the 47,204 EQs, we can estimate the probability p_1 to obtain by chance a date having a lead time smaller than 97 days before an EQ of magnitude 7.6 or larger by considering the ratio of the EQs that occurred up to 96 days before an EQ of magnitude 7.6 or larger over the total number of the EQs considered. This value results in $p_1 = 4{,}768/47{,}204 \approx 10.1\,\%$, and hence the probability to obtain at least six such dates when performing 15(=6+9) attempts can be obtained by the binomial distribution which leads to $p_{bin} = 0.237\,\%$. Of course, this probability does not correspond to the probability to obtain the results of Sarlis *et al.* (2013) by chance since the six successful dates may not correspond to different EQs of magnitude 7.6 or larger.

In order to quantify the latter probability, we performed a Monte Carlo calculation in which we generated 10^6 times, 15 uniformly distributed random integers from 1 to 47,204 to select 15 EQs from the JMA catalog, the occurrence dates of which have been compared with the occurrence dates of the six shallow EQs with magnitude 7.6 or larger in order to examine whether all these six EQs have been preceded by randomly selected EQs with a maximum lead time of 96 days. This Monte Carlo calculation has been run 10^3 times and the corresponding probability is $p_{MC} = 0.00436(64)\,\%$ where the number in parenthesis denotes the standard deviation. Thus, we find that the probability to obtain by chance the results found by Sarlis *et al.* (2013) is of the order of 10^{-5}. We clarify that this probability refers only to the occurrence time of major EQs, while the relevant calculation for an EQ prediction method (e.g., the one based on SES, Varotsos and Alexopoulos 1984a, b; Varotsos *et al.* 1988) should also consider the probabilities to obtain the epicentral area and the magnitude of the impending EQs (Varotsos *et al.* 1996a, b).

4. *Statistical Evaluation by Means of Receiver Operating Characteristics*

A receiver operating characteristics (ROC) graph (Fawcett 2006) is a technique to depict the quality of binary predictions. It is a plot of the hit rate (or true positive rate) versus the false alarm rate (or false

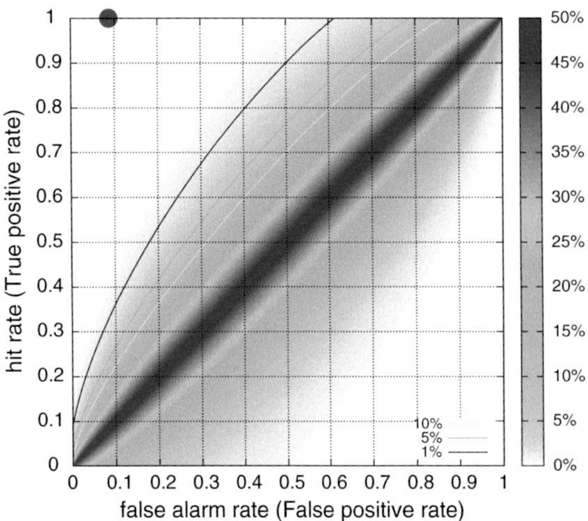

Figure 4
Receiver operating characteristics diagram for $P = 6$ and $Q = 103$ in which the *red circle* corresponds to the results obtained by Sarlis *et al.* (2013). This *circle* is far away from the *blue* diagonal that corresponds to random predictions. The colored contours present the p value to obtain by chance an ROC point based on k-ellipses (Sarlis and Christopoulos 2014); the k-ellipses with p =10, 5, and 1 % are also shown

positive rate), as a function of the total rate of alarms, which is tuned by a threshold in the predictor. The hit rate is the ratio of the cases for which the alarm was on and a significant event occurred over the total number of significant events. The false alarm rate is the ratio of the cases for which the alarm was on and no significant event occurred over the total number of non-significant events. Only if the hit rate exceeds the false alarm rate, a predictor is useful [for example, the ROC analysis has been recently used by Telesca *et al.*, (2014) to discriminate between seismograms of tsunamigenic and non-tsunamigenic EQs]. Random predictions generate equal hit and false alarm rate on average (thus, falling on the blue diagonal in Fig. 4 that will be discussed later), and the corresponding ROC curves exhibit fluctuations which depend on the positive P cases (i.e., the number of significant events) and the negative Q cases (i.e., the number of non-significant events) to be predicted. The statistical significance of an ROC curve depends (Mason and Graham 2002) on the area under the curve A in the ROC plane. Mason and Graham (2002) have shown that $A = 1 - U/(PQ)$, where U follows the Mann–Whitney U statistics (Mann and Whitney 1947). Very recently, a visualization scheme for the statistical significance of ROC curves has been proposed (Sarlis and Christopoulos 2014). It is based on k-ellipses which are the envelopes of the confidence ellipses—cf. a point lies outside a confidence ellipse with probability $\exp(-k/2)$—obtained when using a random predictor and vary the prediction threshold. These k-ellipses cover the whole ROC plane and upon using their A we can have a measure (Sarlis and Christopoulos 2014) of the probability p to obtain by chance (i.e., using a random predictor) an ROC curve passing through each point of the ROC plane.

In the present case, we divide the whole period covering 27 years and almost 3 months into 109 3-month periods (i.e., $P + Q = 109$) out of which only six included significant events ($P = 6$). These six significant events were successfully predicted by the aforementioned minima of Table 1 of Sarlis *et al.* (2013) (written here in bold in Table 1) that preceded all the shallow EQs with magnitude 7.6 or larger during the study period. Hence, the hit rate is 100 %. On the other hand, the nine minima which were followed within 3 months by smaller EQs (see Table 2 of Sarlis *et al.* 2013 which are not marked in bold in Table 1) may be considered false alarms giving rise to a false alarm rate of $9/103 \approx 8.74\,\%$. By using the FORTRAN code `VISROC.f` provided by Sarlis and Christopoulos (2014) we obtain: (a) the ROC diagram of Fig. 4 in which we depict by the red circle the operation point that corresponds to the results obtained by Sarlis *et al.* (2013) and (b) the probability p to obtain this point by chance based on k-ellipses which results in $p_{ROC} = 0.00314\,\%$. Interestingly, the value of p_{ROC} is compatible with p_{MC} estimated in the previous section strengthening the conclusion that the probability to obtain the findings of Sarlis *et al.* (2013) by chance is of the order of 10^{-5}.

5. Conclusions

Recently, the seismicity of Japan was analyzed in natural time from 1 January 1984 to 11 March 2011 using sliding natural time windows of length W comprising the number of events that would occur in a few months. Fifteen distinct minima of the variability β of the order parameter of seismicity were identified 1–3 months before large EQs. Among these minima, six were followed by the stronger EQs, namely all the six shallow EQs with $M_{JMA} \geq 7.6$ that occurred in Japan during this $\approx$27 year period. The probability to obtain the latter result by chance is of the order of 10^{-5} as shown here using Monte Carlo calculation. The same conclusion is obtained when using the ROC technique.

References

Abe S, Sarlis NV, Skordas ES, Tanaka HK, Varotsos PA (2005) *Origin of the Usefulness of the Natural-Time Representation of Complex Time Series*. Phys Rev Lett *94*:170,601.

Eichner JF, Kantelhardt JW, Bunde A, Havlin S (2007) *Statistics of return intervals in long-term correlated records*. Phys Rev E *75*:011,128.

Fawcett T (2006) *An introduction to ROC analysis*. Pattern Recogn Lett *27*(8):861–874.

Flores-Márquez E, Vargas C, Telesca L, Ramírez-Rojas A (2014) *Analysis of the distribution of the order parameter of synthetic seismicity generated by a simple spring-block system with asperities*. Physica A *393*:508–512.

Holliday JR, Rundle JB, Turcotte DL, Klein W, Tiampo KF, Donnellan A (2006) *Space-time clustering and correlations of major earthquakes*. Phys Rev Lett *97*:238501.

Huang Q (2008) *Seismicity changes prior to the Ms8.0 Wenchuan earthquake in Sichuan, China*. Geophys Res Lett *35*:L23308.

Huang Q (2011) *Retrospective investigation of geophysical data possibly associated with the Ms8.0 Wenchuan earthquake in Sichuan, China*. Journal of Asian Earth Sciences *41*(45):421–427.

Huang Q, Ding X (2012) *Spatiotemporal variations of seismic quiescence prior to the 2011 M 9.0 Tohoku earthquake revealed by an improved region-time-length algorithm*. Bull Seismol Soc Am *102*:1878–1883.

Kanamori H (1978) *Quantification of earthquakes*. Nature *271*:411–414.

Lennartz S, Livina VN, Bunde A, Havlin S (2008) *Long-term memory in earthquakes and the distribution of interoccurrence times*. EPL *81*:69,001.

Lennartz S, Bunde A, Turcotte DL (2011) *Modelling seismic catalogues by cascade models: Do we need long-term magnitude correlations?* Geophys J Int *184*:1214–1222.

Lippiello E, de Arcangelis L, Godano C (2009) *Role of static stress diffusion in the spatiotemporal organization of aftershocks*. Phys Rev Lett *103*:038501.

Lippiello E, Godano C, de Arcangelis L (2012) *The earthquake magnitude is influenced by previous seismicity*. Geophys Res Lett *39*:L05309.

Mann HB, Whitney DR (1947) *On a test of whether one of two random variables is stochastically larger than the other*. Ann Math Statist *18*:50–60.

Mason SJ, Graham NE (2002) *Areas beneath the relative operating characteristics (ROC) and relative operating levels (ROL) curves: Statistical significance and interpretation*. Quart J Roy Meteor Soc *128*:2145–2166.

Ramírez-Rojas AA, Flores-Márquez E (2013) *Order parameter analysis of seismicity of the Mexican Pacific coast*. Physica A *392*(10):2507–2512.

Rundle JB, Holliday JR, Graves WR, Turcotte DL, Tiampo KF, Klein W (2012) *Probabilities for large events in driven threshold systems*. Phys Rev E *86*:021106.

Sarlis NV (2011) *Magnitude correlations in global seismicity*. Phys Rev E *84*:022101.

Sarlis NV, Christopoulos SRG (2012) *Natural time analysis of the Centennial Earthquake Catalog*. CHAOS *22*:023123.

Sarlis NV, Christopoulos SRG (2014) *Visualization of the significance of Receiver Operating Characteristics based on confidence ellipses*. Comput Phys Commun *185*:1172–1176.

Sarlis NV, Skordas ES, Lazaridou MS, Varotsos PA (2008) *Investigation of seismicity after the initiation of a Seismic Electric Signal activity until the main shock*. Proc Japan Acad, Ser B *84*:331–343.

Sarlis NV, Skordas ES, Varotsos PA (2010) *Order parameter fluctuations of seismicity in natural time before and after mainshocks*. EPL *91*:59,001.

Sarlis NV, Skordas ES, Varotsos PA, Nagao T, Kamogawa M, Tanaka H, Uyeda S (2013) *Minimum of the order parameter fluctuations of seismicity before major earthquakes in Japan*. Proc Natl Acad Sci USA *110*(34):13,734–13,738, doi:10.1073/pnas.1312740110 .

Telesca L (2010) *Analysis of italian seismicity by using a non-extensive approach*. Tectonophysics *494*:155–162.

Telesca L, Lovallo M (2009) *Non-uniform scaling features in central italy seismicity: A non-linear approach in investigating seismic patterns and detection of possible earthquake precursors*. Geophys Res Lett *36*:L01308.

Telesca L, Lapenna V, Vallianatos F (2002) *Monofractal and multifractal approaches in investigating scaling properties in temporal patterns of the 1983–2000 seismicity in the Western Corinth Graben (Greece)*. Phys Earth Planet Int *131*:63–79.

Telesca L, Lovallo M, Ramírez-Rojas A, Angulo-Brown F (2009) *A Nonlinear Strategy to Reveal Seismic Precursory Signatures in Earthquake-related Self-potential Signals*. Physica A *388*:2036–2040.

Telesca L, Lovallo M, Carniel R (2010) *Time-dependent Fisher Information Measure of volcanic tremor before 5 April 2003 paroxysm at Stromboli volcano Italy*. J Volcanol Geoterm Res *195*:78–82.

Telesca L, Chamoli A, Lovallo M, Stabile T (2014) *Investigating the tsunamigenic potential of earthquakes from analysis of the informational and multifractal properties of seismograms*. Pure Appl Geophys. doi:10.1007/s00024-014-0862-3.

Tenenbaum JN, Havlin S, Stanley HE (2012) *Earthquake networks based on similar activity patterns*. Phys Rev E *86*:046107.

Turcotte DL (1997) Fractals and Chaos in Geology and Geophysics, 2nd edn. Cambridge University Press, Cambridge.

Varotsos P, Alexopoulos K (1984a) *Physical Properties of the variations of the electric field of the earth preceding earthquakes, I*. Tectonophysics *110*:73–98.

Varotsos P, Alexopoulos K (1984b) *Physical Properties of the variations of the electric field of the earth preceding earthquakes, II*. Tectonophysics *110*:99–125.

Varotsos P, Alexopoulos K (1986) Thermodynamics of Point Defects and their Relation with Bulk Properties. North Holland, Amsterdam.

Varotsos P, Lazaridou M (1991) *Latest aspects of earthquake prediction in Greece based on Seismic Electric Signals*. Tectonophysics *188*:321–347.

Varotsos P, Alexopoulos K, Nomicos K, Lazaridou M (1988) *Official earthquake prediction procedure in Greece*. Tectonophysics *152*:193–196.

Varotsos P, Alexopoulos K, Lazaridou M (1993) *Latest aspects of earthquake prediction in Greece based on Seismic Electric Signals, II*. Tectonophysics *224*:1–37.

Varotsos P, Eftaxias K, Lazaridou M, Antonopoulos G, Makris J, Poliyiannakis J (1996a) *Summary of the five principles suggested by Varotsos et al. [1996] and the additional questions raised in this debate*. Geophys Res Lett *23*:1449–1452.

Varotsos P, Eftaxias K, Vallianatos F, Lazaridou M (1996b) *Basic principles for evaluating an earthquake prediction method*. Geophys Res Lett *23*:1295–1298.

Varotsos P, Sarlis N, Skordas E (2011a) *Scale-specific order parameter fluctuations of seismicity in natural time before mainshocks*. EPL *96*:59,002.

Varotsos PA, Sarlis NV, Skordas ES (2011b) Natural Time Analysis: The new view of time. Precursory Seismic Electric Signals, Earthquakes and other Complex Time-Series. Springer-Verlag, Berlin Heidelberg.

Varotsos P, Sarlis N, Skordas E (2012a) *Remarkable changes in the distribution of the order parameter of seismicity before mainshocks*. EPL *100*:39,002.

Varotsos P, Sarlis N, Skordas E (2012b) *Scale-specific order parameter fluctuations of seismicity before mainshocks: Natural time and detrended fluctuation analysis*. EPL *99*:59,001.

Varotsos PA, Sarlis NV, Skordas ES (2012c) *Order parameter fluctuations in natural time and b-value variation before large earthquakes*. Natural Hazards and Earth System Science *12*:3473–3481.

Varotsos PA, Sarlis NV, Skordas ES (2001) *Spatio-temporal complexity aspects on the interrelation between seismic electric signals and seismicity*. Practica of Athens Academy *76*:294–321.

Varotsos PA, Sarlis NV, Skordas ES (2002) *Long-range correlations in the electric signals that precede rupture*. Phys Rev E *66*:011902.

Varotsos PA, Sarlis NV, Tanaka HK, Skordas ES (2005) *Similarity of fluctuations in correlated systems: The case of seismicity*. Phys Rev E *72*:041103.

Varotsos PA, Sarlis NV, Skordas ES, Lazaridou MS (2013) *Seismic electric signals: An additional fact showing their physical interconnection with seismicity*. Tectonophysics 589:116–125.

(Received June 26, 2014, revised August 13, 2014, accepted August 30, 2014, Published online September 17, 2014)

Pure Appl. Geophys. 173 (2016), 173–181

DOI 10.1007/s00024-015-1041-x

Pure and Applied Geophysics

An Explosion Aftershock Model with Application to On-Site Inspection

SEAN R. FORD[1] and PETER LABAK[2]

Abstract—An estimate of aftershock activity due to a theoretical underground nuclear explosion is produced using an aftershock rate model. The model is developed with data from the Nevada National Security Site, formerly known as the Nevada Test Site, and the Semipalatinsk Test Site, which we take to represent soft-rock and hard-rock testing environments, respectively. Estimates of expected magnitude and number of aftershocks are calculated using the models for different testing and inspection scenarios. These estimates can help inform the Seismic Aftershock Monitoring System (SAMS) deployment in a potential Comprehensive Test Ban Treaty On-Site Inspection (OSI), by giving the OSI team a probabilistic assessment of potential aftershocks in the Inspection Area (IA). The aftershock assessment, combined with an estimate of the background seismicity in the IA and an empirically derived map of threshold magnitude for the SAMS network, could aid the OSI team in reporting. We apply the hard-rock model to a M5 event and combine it with the very sensitive detection threshold for OSI sensors to show that tens of events per day are expected up to a month after an explosion measured several kilometers away.

Key words: OSI, SAMS, Passive method, Signal processing, Seismic.

1. Introduction

An explosion produces an aftershock sequence that is similar in character to a sequence from an earthquake (KITOV and KUZNETSOV 1990), so the same Omori decay in time and Gutenberg-Richter distribution in magnitude can be employed to model the explosion aftershock sequence. FORD and WALTER (2010) found that explosion aftershocks were fewer in number and lower in magnitude than earthquake aftershocks for similarly sized explosions and earthquakes. This observation prompted them to use an earthquake aftershock model calibrated from Western US seismicity to test the hypothesis that a given aftershock sequence is due to an earthquake.

The presence of explosion-induced aftershocks led to the requirement for a seismic capability in an On-Site Inspection (OSI). The search area for an OSI could be greatly reduced if explosion-induced aftershocks could be located. Therefore, it is important to model the spatial and temporal behavior of these aftershocks in order to assess the seismic capability in an OSI. The proposed aftershock rate model can predict this behavior for a given explosion size and location in an OSI-triggering event.

In order to determine the parameters of the aftershock model, we attempted to stack Nevada Test Site (NTS) explosion aftershock sequences, but there were not enough aftershocks detected by the regional networks at the time. We searched for high-resolution studies of explosion aftershock sequences that could provide data on aftershock rate and magnitude distribution. The requirement that magnitude data be available precluded many aftershock studies from use, since most employed an uncalibrated measurement of relative amplitude. This approach is fine for aftershock decay studies, but is not useful for producing an absolute, transportable aftershock model. Three studies were found to fit our requirements, and fortunately, they span a range of environments. In the work that follows, we develop models from each of these studies that we propose can be used as standards in each environment, and then we employ the models to predict aftershock seismicity in the context of OSI planning, deployment, and reporting.

2. Method

REASENBERG and JONES (1989, 1994) combined the Omori law that describes the power-law decay in the number of aftershocks N with time t,

[1] Lawrence Livermore National Laboratory, Livermore, USA. E-mail: sean@llnl.gov
[2] Comprehensive Test Ban Treaty Organization Preparatory Commission, Vienna, Austria.

$$N(t) = K(t - t_c)^{-p} , \quad (1)$$

for $t_c > 0$, where t_c is the time of the mainshock, p describes the decay, and K is related to the productivity of the sequence, with the Gutenberg-Richter description of magnitude M distribution

$$N(M) = A10^{-bM} , \quad (2)$$

where b describes the ratio of small to large events and A is related to the productivity of the sequence, to determine an aftershock rate equation given by

$$\lambda(t, \Delta M) = 10^{a+b\Delta M} t^{-p} , \quad (3)$$

where ΔM is the difference between the mainshock magnitude M_m and the aftershock magnitude M ($\Delta M = M_m - M$), and a is proportional to the seismicity rate (WIEMER 2000), which substitutes for the productivity terms, K and A, and is given by

$$a = \log_{10}(K) - b(M_m - M_{min}) , \quad (4)$$

where M_{min} is the magnitude of the smallest event recorded in the aftershock sequence. We calculate the parameters p, b, and K for each aftershock study by fitting the distribution of events in time and magnitude and M_m and M_{min} from the explosion magnitude and minimum recorded aftershock, respectively.

Assuming a nonhomogeneous Poissonian occurrence of aftershocks, the probability of N events occurring above magnitude M within an interval (0, t) is

$$p(N, t) = \frac{\left(\int_0^t \lambda(t)dt\right)^N}{N!} \exp\left(-\int_0^t \lambda(t)dt\right). \quad (5)$$

FELZER and BRODSKY (2006) showed that a power-law decay in aftershock density ρ with radial distance r from the mainshock is appropriate so that aftershock spatial density can be described as

$$\rho(r) = cr^{-n} \quad (6)$$

where n is the decay constant and c is related to the number of aftershocks. We calculate the parameter n by fitting the distribution of events in space. The probability density function is from the Pareto distribution of the form

$$p(r) = \frac{n-1}{r_{min}} \left(\frac{r}{r_{min}}\right)^{-n} \quad (7)$$

where we must assume a minimum distance cutoff r_{min} where the probability goes to zero at $r < r_{min}$. The implications of this will be discussed later.

3. Data and Models

We make use of three explosion aftershock data sets that encompass a range of geologies. The first is from the well-studied underground nuclear explosion, BENHAM, the second is from the high-explosive underground explosion named the Non-Proliferation Experiment (NPE), and the third is from an underground nuclear explosion in Shaft 1352. The first two took place at the NTS in volcanic tuffs and the third occurred at the Semipalatinsk Test Site (STS) in granite. Table 1 summarizes the explosion information and the aftershock reference.

HAMILTON (1972b) measured the magnitude distribution of BENHAM aftershocks and found a b value of 1.4, but noted this may be due to underestimation of larger aftershocks, in which case a value closer to 1.0 is more appropriate. STAUDER (1971) performed a more thorough study of BENHAM aftershocks and found a b value of 1.02. This is the value we will use to characterize the sequence. HAMILTON and HEALY (1969) plot the number of aftershocks per day as a function of days after the explosion. The decay has power-law behavior until the 15th day, when a M_L3.7 earthquake occurred nearby that caused an increase in measured aftershocks (HAMILTON 1972b). Therefore, we only calculate p and K from the first 15 days of observation (Fig. 1). BENHAM was a M_L6.3 (HAMILTON 1969) and the minimum magnitude measured in the HAMILTON (1972b) study was M_L1.3, which gives a $\Delta M = 5$. HAMILTON *et al.* (1972a) also plot the number of aftershocks per km as a function of distance from the explosion. We use this information to calculate n (Fig. 2).

JARPE *et al.* (1994) presented the magnitude distribution of the NPE, which we measured to find a b value of 1.10. Since a count of aftershocks per day was made at only a few times, JARPE (1994) combined

Table 1

Explosion aftershock data set information

Shot	Date	Site/location	Type	Yield (kt)	Depth (m)	References
BENHAM	19-Dec-68	NTS/Pahute Mesa	Welded tuff	1,150	1,402	HAMILTON (1972b)
NPE	22-Sep-93	NTS/Rainier Mesa	Tuff	1	390	JARPE (1994)
Shaft 1352	8-Jul-89	STS/Balapan	Granite	35	550	ADUSHKIN (1995)

NPE Non-Proliferation Experiment, *NTS* Nevada Test Site, *STS* Semipalatinsk Test Site

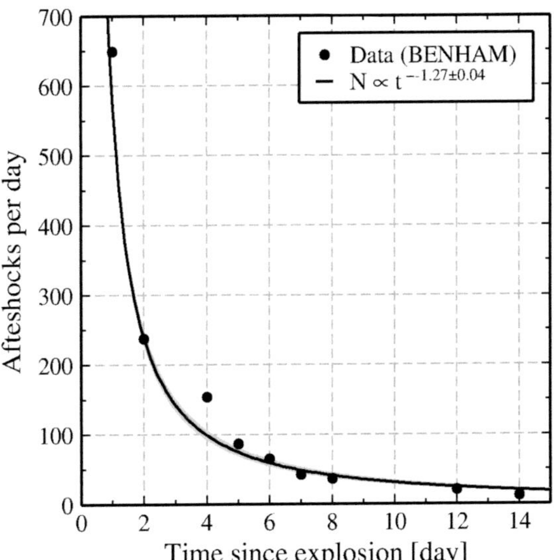

Figure 1
Aftershock rate for the explosion, BENHAM (HAMILTON 1969). The *shaded area* is the formal error in the model

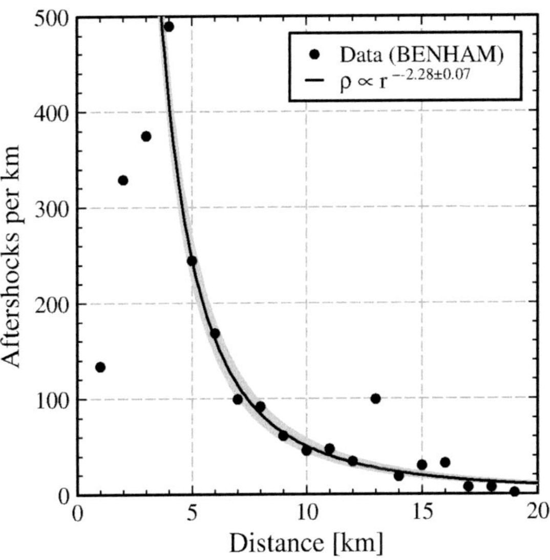

Figure 2
Aftershock linear density for the explosion, BENHAM (HAMILTON 1972b). The *shaded area* is the formal error in the model for distance > 750 m

the measurements with other explosion aftershock sequences in similar geology. We use this combined data set to measure p and K (Fig. 3); however, since there are so few measurements, the error is large. JARPE (1994) report that the NPE was a M_L4.0 and the minimum magnitude that was measured was M_L–0.5, which gives a $\Delta M = 4.5$. Explosion aftershocks were not located in any of these studies, so a determination of n cannot be made.

ADUSHKIN and SPIVAK (1995) report on several explosion aftershock studies at STS and they analyze in detail the Shaft 1352 explosion. The b value inferred from their plot of aftershock amplitude distribution is 1.00. The p and K values were extracted from the plot of aftershocks per hour, by ADUSHKIN (1995) (Fig. 4). ADUSHKIN (1995) did not report the minimum magnitude measured in the aftershock study, but rather reported peak velocities for each event in order to infer a b value. We therefore used a magnitude (MPV) equation appropriate for local-to-regional distance P waves recorded in the region, $\mathrm{MPV} = \log_{10}(A/T) + \sigma(r)$ (A. Belyashov, personal communication), where A is the displacement, T is the period of the measurement, and σ is the attenuation correction that is a function of distance r. The smallest distance for which there is a correction is 10 km, but the study at Shaft 1352 measured aftershocks at a distance of ~1 km, so we extrapolate the correction to that distance and we estimate that the aftershocks had a period of measurement of ~0.1 s. We also convert the reported smallest velocity measured in the study, 0.4 μm/s, to a displacement by dividing by 2π under the harmonic assumption. The minimum

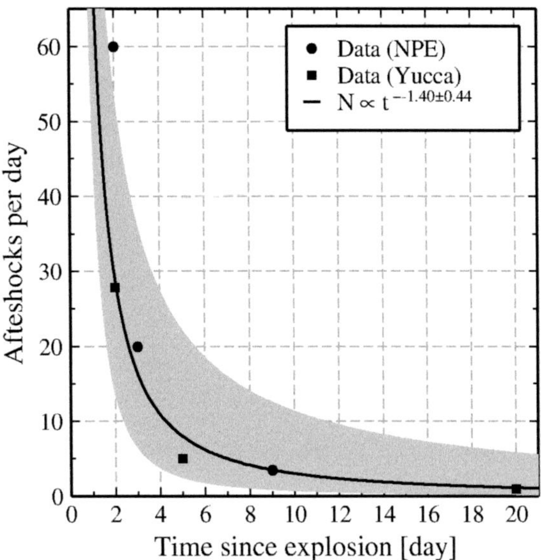

Figure 3
Aftershock rate for the explosion, NPE (*circles*), and others in similar geology (*squares*) (JARPE 1994). The *shaded area* is the formal error in the model

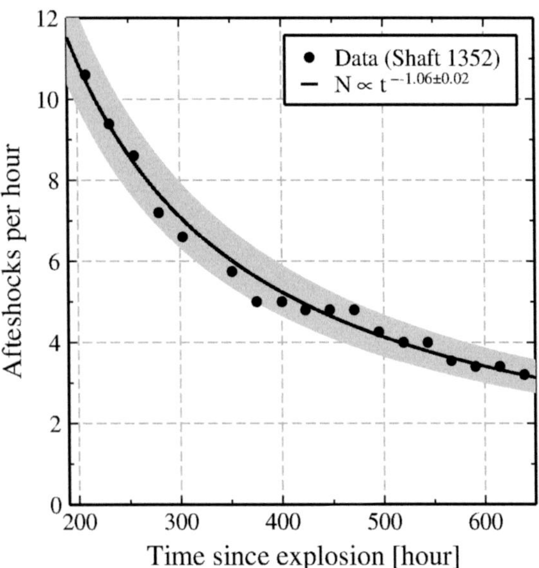

Figure 4
Aftershock rate for the Shaft 1352 explosion (ADUSHKIN 1995). The *shaded area* is the formal error in the model

magnitude in the study is M–0.3 and the explosion magnitude was 5.6 (KHALTURIN *et al.* 2001). ADUSHKIN (1995) also plot the number of aftershocks as a function of distance from the explosion. We use this information to calculate n (Fig. 5).

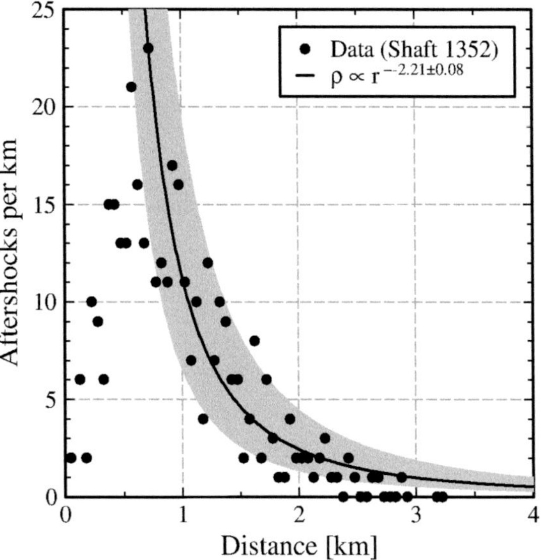

Figure 5
Aftershock linear density for the Shaft 1352 explosion (ADUSHKIN 1995). The *shaded area* is the formal error in the model

Since BENHAM occurred on Pahute Mesa in more competent, higher-strength volcanics than the NPE on Rainier Mesa, we will refer to the models from these two sequences as NTS-hard and NTS-soft, respectively. The explosion in Shaft 1352 occurred in granite at Balapan, and the model from this sequence will be named STS-hard. Table 2 summarizes the model parameters for each aftershock data set.

Using the parameters given in Table 2, we define a family of three aftershock models that span a range of geologic media, NTS-hard (BENHAM), NTS-soft (NPE), and STS-hard (Shaft 1352). Figure 6 gives the aftershock rate of events with magnitudes greater than five units less than the explosion ($\Delta M = 5$) for each of the explosion models, and for general comparison, an earthquake aftershock model used by the US Geological Survey, SoCal (GERSTENBERGER *et al.* 2004). The earthquake model has the greatest rate, followed by the explosion models ordered by decreasing medium strength or competency, STS-hard, NTS-hard, and NTS-soft. We can manipulate the event probability to estimate the probability of at least one aftershock larger than a given magnitude (1—probability of no aftershocks). Figure 7 shows these probabilities for each model as a function of time after the explosion for $\Delta M = 2$. We can also

Table 2

Aftershock model parameters

Shot	Name	M_m	M_{min}	$\log_{10}(K)$	a	p	b	n
BENHAM	NTS-hard	6.3	1.3	2.76	−2.69	1.36	1.02	2.28
NPE	NTS-soft	4.0	−0.5	2.00	−3.40	1.53	1.10	–
Shaft 1352	STS-hard	5.6	−0.3	3.38	−2.47	1.06	1.00	2.21
[a]	SoCal	–	–	–	−1.67	1.08	0.91	1.80

[a] SoCal model is from an analysis of Southern California earthquakes (Gerstenberger 2004)

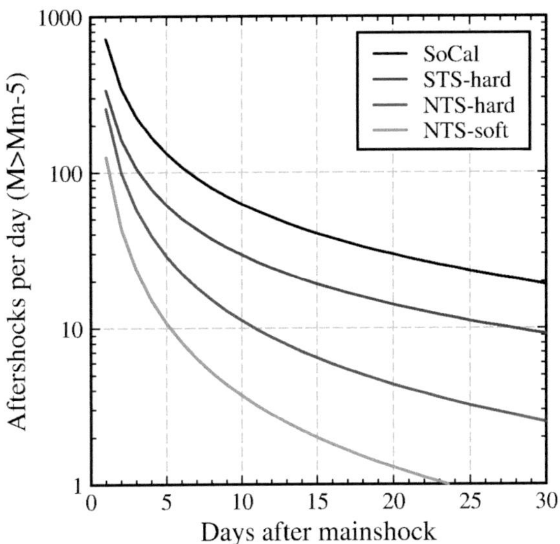

Figure 6
Aftershock rate (events/day) for the three explosion models (*color*) and an earthquake model (*black*) for events with magnitude greater than the mainshock magnitude (M_m) minus 5

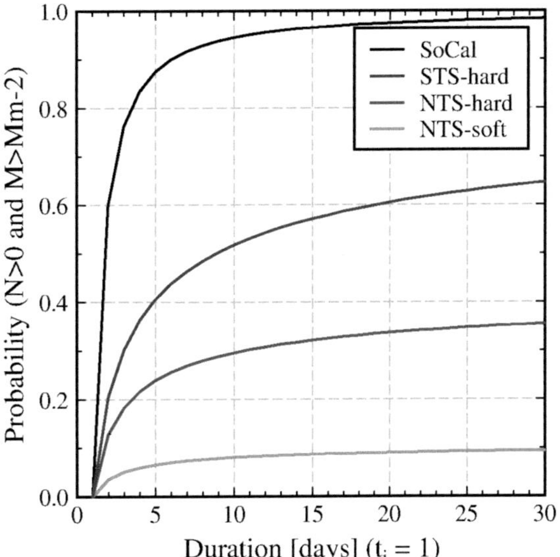

Figure 7
Probability of at least one aftershock with magnitude greater than two less than the mainshock magnitude where the duration begins (t_i) 1 day after the mainshock time (t_0) for the three explosion models (*color*) and an earthquake model (*black*)

manipulate the event probability to estimate the cumulative number of expected aftershocks given a probability and observation time after the explosion. Figure 8 shows these numbers for each model where recording starts 1 day, 1 week, and 1 month after the explosion occurs. Note that only three events are predicted in a 3-day observation period for the NTS-soft model where observation begins 1 month after the explosion (Fig. 8c).

4. Application

If we combine the explosion aftershock model with a network geometry and station detection level, we can predict, with a given probability, the observation of an explosion aftershock at a given time since the explosion. Details of the mini-arrays and processing used in a Seismic Aftershock Monitoring System (SAMS) deployment are given in Sick and Joswig (2014). Figure 9 gives the magnitude of detection (M_D) curve for the SAMS, as determined from data for a field experiment in Finland in 2009 (DE09) in competent rock, fit with a second order polynomial on log-distance. We randomly design a potential OSI deployment geometry and calculate the predicted number of aftershocks detected at each station with a probability of 0.90 for a given day after the explosion. Figure 10 shows the results for 1 week

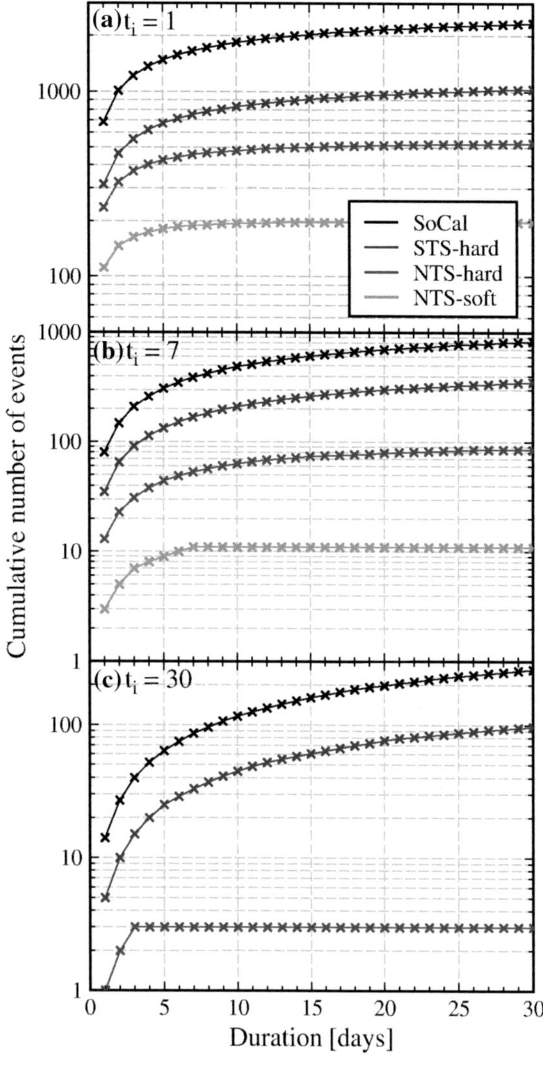

Figure 8
Expected cumulative number of events with magnitude greater than five less than the mainshock magnitude at a probability of 0.90 for the NTS-hard model where recording begins 1 day (**a**), 1 week (**b**), and 1 month (**c**) after the mainshock for the three explosion models (*color*) and an earthquake model (*black*)

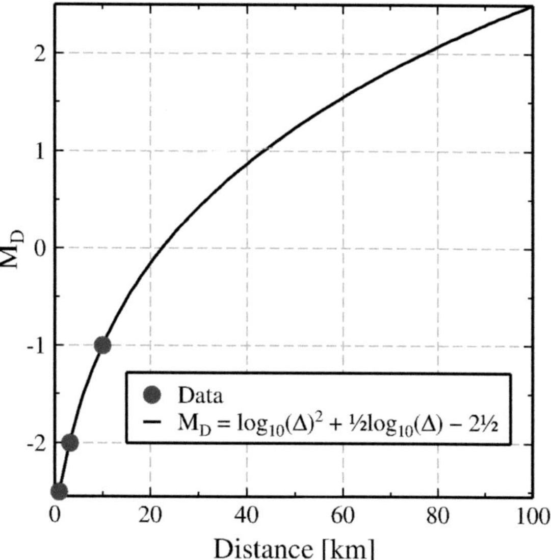

Figure 9
Magnitude of detection as a function of distance from the event. The data (*red circles*) are fit with a simple second order polynomial on log-distance (*black line*)

and 1 month after an M5 and M4 explosion-triggering event.

5. Discussion

The b value is near one for all models, which is similar to earthquake studies, though there is some suggestion that weaker source material increases the b value, but with such small studies, this is difficult to demonstrate. An apparent relationship to source medium strength is the decrease of p value and increase in a value, so that explosion aftershocks in competent rock are more abundant for a longer duration than aftershocks in weaker media. Phillips *et al.* (1999) analyzed the aftershock sequence of a controlled mine collapse and found a p value = 1.3, similar to the NTS-hard model. Kgarume *et al.* (2010) found values close to 1 for both the p and b values in gold mine tremor aftershocks, where the distance decay parameter suggested dynamic triggering. The distance decay (γ value) in explosion aftershocks is similar in both the 'hard' models and greater than some earthquake models, suggesting static stress as the dominant transfer mechanism, which is consistent with the conclusions of Parsons and Velasco (2009). Also, there seems to be an offset in distance from the detonation point, where the power-law behavior begins. This offset is greater than the cavity radius, and may be related to poor locations from simple velocity models that don't account for damage around the cavity.

The detections per station shown in Fig. 10 can be used in planning and interpretation of seismic data from an OSI for any given station configuration. These maps are very dependent on the magnitude of

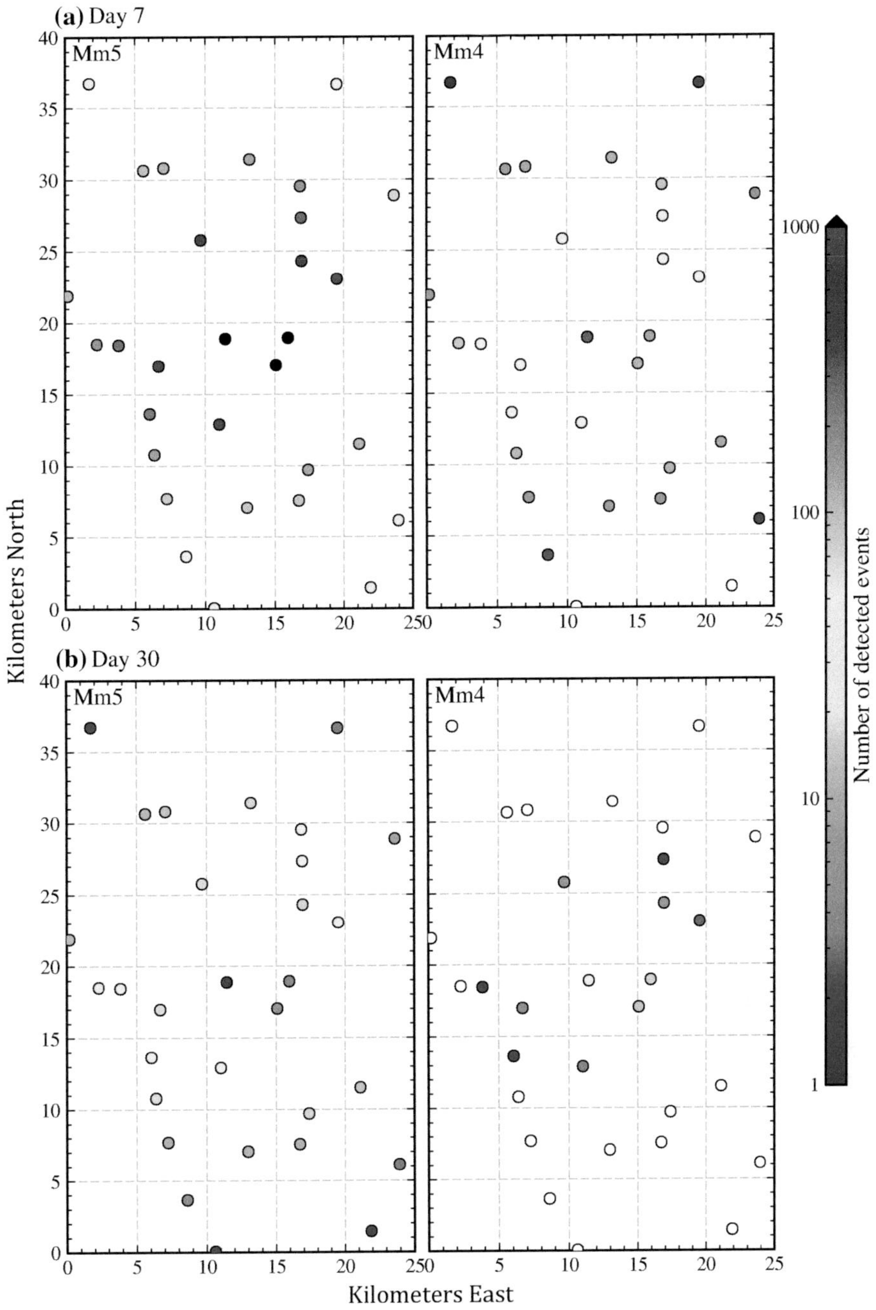

Figure 10
Number of detected events over a 25 by 40 km area at a given station (*circle*) for a explosion at the center of the deployment (12½, 20) with magnitude 5 (Mm5, *left panel*) and 4 (Mm4, *right panel*), 1 week (**a**) and 1 month (**b**) after the explosion. *White circles* are zero detections

detection relationship (as given in Fig. 9), where stations close to the explosion epicenter can detect more than a thousand events in a day even days after the large explosion. This is due to the very low detection threshold of the SAMS, so that for a explosion magnitude of 5 (left panel, Fig. 10), stations 10 km away from the epicenter detect events according to a rate equation where $\Delta M = 6$. This effect can be seen when comparing with a explosion magnitude 4 event (right panel, Fig. 10), where the number of detections is an order of magnitude less. These predictive models could be improved with more data to define the detection threshold and its error.

The predictive power of the explosion aftershock models could be used in the planning and deployment of the SAMS in a potential OSI as part of the Comprehensive Test Ban Treaty. The planned deployment could be evaluated for a predicted number of aftershocks detected, and this could guide the OSI team in its use of resources. Ideally, maps like those shown in Fig. 10 would have an additional layer of seismic noise to help in determining the true detection threshold at a site (Sick *et al.* 2013).

6. Conclusions

The spatial and temporal distribution of aftershocks due to an explosion can be described with the aftershock rate and distance decay models used for earthquake aftershocks. The parameters for the model are dependent on the source medium, and we define three parameterizations based on aftershock records from explosions in hard rock at STS and NTS, as well as less competent rock at NTS. The log-ratio of the number of large to small events (b value) for each model is approximately one, and the decay rate (p value) decreases and productivity (a value) increases with source medium strength.

The models are applied to a fictitious site inspection scenario and combined with station detection thresholds and site geometry to predict the number of events detected for a given explosion magnitude and observation day. These predictions can be used as tools to aid in the deployment and analysis of Seismic Aftershock Monitoring System data during an on-site inspection scenario.

Acknowledgments

The authors are grateful for two anonymous reviews and a review by the Associate Editor, Anton Dainty. His efforts in producing the highest quality articles regarding explosion monitoring, and seismic analyses in general have benefitted the entire community and will be greatly missed.The authors are grateful for support from the US Department of Energy, National Nuclear Security Administration, Nonproliferation and International Security and the Comprehensive Test Ban Treaty Organization Preparatory Commission, Conference on Science and Technology. This research was performed in part under the auspices of the US Department of Energy by the Lawrence Livermore National Laboratory under contract number DE-AC52-07NA27344; Information Management release number LLNL-JRNL-652465.

References

Adushkin, V., and A. Spivak (1995). Aftershock of Underground Nuclear Explosion, in Earthquakes Induced by Underground Nuclear Explosions, eds. R. Console and A. Nikolaev, Springer-Verlag, Berlin, 35–49.

Felzer, K. R. and E. E. Brodsky (2006). *Decay of aftershock density with distance indicates triggering by dynamic stresses*, Nature, *441*, 735–738.

Ford, S. R., and W. R. Walter (2010). *Aftershock Characteristics as a Means of Discriminating Explosions from Earthquakes*, Bull. Seis. Soc. Amer., *100*, 364–376, doi:10.1785/0120080349.

Gerstenberger, M. C., S. Wiemer, and L. M. Jones (2004). Real-time forecasts of tomorrow's earthquakes in California: a new mapping tool, US Geological Survey Open-File Report 2004-1390.

Hamilton, R. M. and J. H. Healy (1969). *Aftershocks of the BENHAM nuclear explosion*, Bull. Seismol. Soc. Amer., *59*, 2271–2281.

Hamilton, R. M., F. A. Mckeown, and J. H. Healy (1972a). *Seismic activity and faulting associated with a large underground nuclear explosion*, Science, *166*, 601–604.

Hamilton, R.M., B.E. Smith, F.G. Fischer, and P.J. Papanek (1972b). *Earthquakes caused by underground nuclear explosions on Pahute Mesa, Nevada Test Site*, Bull. Seismol. Soc. Amer., *62*, 1319–41.

Jarpe, S., P. Goldstein, and J. J. Zucca (1994). Comparison of the non-proliferation event aftershocks with other Nevada Test Site events, UCRL-JC-117754, in Non-proliferation Experiement Symposium, Rockville, Maryland, 19–21 April 1994.

Kgarume, T. E., S. M. Spottiswoode, R. J. Durrheim (2010). *Statistical properties of mine tremor aftershocks*, Pure Appl. Geophys., *167*, 107–117, doi:10.1007/s00024-009-0004-5.

Khalturin, V.I., T.G.Rautian, and P.G.Richards. (2001) *A study of small magnitude seismic events during 1961-1989 on and near the Semipalatinsk Test Site, Kazakhstan, Pure Appl. Geophys.*, *158*, 143–171.

Kitov, I. O., and O. P. Kuznetov. (1990) *Energy released in aftershock sequence of explosion*, Doklady Akademii Nauk SSSR, *315*, 839–842.

Parsons, T., and A. A. Velasco (2009). *On near-source earthquake triggering*, J. Geophys. Res., *114*(B10307), doi:10.1029/2008JB006277.

Phillips, W. S., D. C. Pearson, X. Yang, and B. W. Stump (1999). *Aftershocks of an explosively induced mine collapse at White Pine, Michigan*, Bull. Seis. Soc. Amer., *89*(6) 1575–1590.

Reasenberg, P. A., and L. M. Jones (1989). *Earthquake hazard after a mainshock in California*, Science, *243*, 1173–1176.

Reasenberg, P. A., and L. M. Jones (1994). *Earthquake aftershocks - Update*, Science, 265, 1251–1252.

Sick, B., N. Gestermann, and M. Joswig (2013), Seismic aftershock monitoring network optimization based on detection threshold estimation from background noise measurement, Science and Technology Conference, Vienna.

Sick, B., M. Walter, M. Joswig (2014). *Visual Event Screening of Continuous Seismic Data by Supersonograms, Recent Advances in Nuclear Explosion Monitoring Vol. 2,* Pure Appl. Geophys., *171*, 549–559, doi:10.1007/s00024-012-0618-x.

Stauder, W. (1971). *Smaller aftershock of Benham nuclear explosion*, Bull. Seismol. Soc. Amer., *61*, 417–428.

Wiemer, S. (2000). *Introducing probabilistic aftershock hazard mapping*, Geophysical Research Letters, *27*(20), 3405–3408.

(Received May 1, 2014, revised January 13, 2015, accepted January 16, 2015, Published online February 14, 2015)

Pure Appl. Geophys. 173 (2016), 183–196

DOI 10.1007/s00024-014-1019-0

Pure and Applied Geophysics

Conditional Probabilities for Large Events Estimated by Small Earthquake Rate

YI-HSUAN WU,[1] CHIEN-CHIH CHEN,[1] and HSIEN-CHI LI[1]

Abstract—We examined forecasting quiescence and activation models to obtain the conditional probability that a large earthquake will occur in a specific time period on different scales in Taiwan. The basic idea of the quiescence and activation models is to use earthquakes that have magnitudes larger than the completeness magnitude to compute the expected properties of large earthquakes. We calculated the probability time series for the whole Taiwan region and for three subareas of Taiwan—the western, eastern, and northeastern Taiwan regions—using 40 years of data from the Central Weather Bureau catalog. In the probability time series for the eastern and northeastern Taiwan regions, a high probability value is usually yielded in cluster events such as events with foreshocks and events that all occur in a short time period. In addition to the time series, we produced probability maps by calculating the conditional probability for every grid point at the time just before a large earthquake. The probability maps show that high probability values are yielded around the epicenter before a large earthquake. The receiver operating characteristic (ROC) curves of the probability maps demonstrate that the probability maps are not random forecasts, but also suggest that lowering the magnitude of a forecasted large earthquake may not improve the forecast method itself. From both the probability time series and probability maps, it can be observed that the probability obtained from the quiescence model increases before a large earthquake and the probability obtained from the activation model increases as the large earthquakes occur. The results lead us to conclude that the quiescence model has better forecast potential than the activation model.

Key words: Probabilistic forecasting, seismic quiescence, seismic activation, Gutenberg–Richter relation, Taiwan seismicity, ROC test.

1. Introduction

Research on earthquake prediction in the last few decades has increased our understanding of the earthquake process and led to several methods for producing useful estimates of seismic hazards. The earthquake prediction research here includes three different time intervals: short term from 1 day to a few months, intermediate term from a few months to a few years, and long term from several years to decades. Owing to the limited length of available observations, we have made the most significant progress in intermediate-term prediction research. Many observational studies have shown anomalous seismic behavior such as increases and decreases in the frequency of smaller events (activation/quiescence) preceding a large earthquake (WYSS *et al.* 1990, 1995; WIEMER and WYSS 1994; KOSSOBOKOV *et al.* 1999; HAINZL *et al.* 2000; BEN-ZION and LYAKHOVSKY 2002; OGATA 2004; SAMMIS *et al.* 2004; WU and CHIAO 2006; HUANG 2008; HUANG and DING 2012). These anomalous seismic behaviors can occur a few days to a few years before the occurrence of a large earthquake and continue for several months to a few years (SYKES and JAUMÉ 1990; WIEMER and WYSS 1994; WYSS *et al.* 1995; WU and CHIAO 2006). Precursors, such as activation, quiescence, and foreshocks, shed light on earthquake forecasting; we could determine the potential area that is going to have a large event by detecting and calculating precursory seismicity (TIAMPO *et al.* 2002; CONSOLE *et al.* 2007; WU *et al.* 2008, 2011). However, prediction must include either specific locations, times, and magnitudes or probabilities defined in public terms.

To arrive at a probability estimate for public use, such as public policy and insurance, various approaches have been proposed to calculate the conditional probability (VERE-JONES 1995; FERRAES 2003; GOMBERG *et al.* 2005). For earthquakes, the conditional probability $P(t|\Delta t)$ is the likelihood that a failure will occur in a time period Δt in the future, given that it has not failed before t. The likelihood is based on information regarding past earthquakes in a given area and the basic assumption that future seismic

[1] Department of Earth Sciences, National Central University, No. 300, Jhongda Rd, Jhongli, Taoyuan 32001, Taiwan (R.O.C.). E-mail: yhwu@ncu.edu.tw; maomaowyh@gmail.com

activity will follow that activity pattern observed in the past. The probability distributions of recurrence times, such as the exponential, Weibull, Gamma, and power-law distributions, are typical tools for calculating conditional probabilities (Ferraes 2003). However, various results might be yielded from the above probability distributions; for example, the exponential model might reach the maximum conditional probability for a $m \geq 6.4$ earthquake in the Tokyo area before June 2009, whereas the Weibull model might reach the maximum conditional probability for the same damaging earthquake before October 2129 (Ferraes 2003). In addition, it is difficult to describe all the earthquakes by one probability distribution; for instance, Chen *et al.* (2012) suggested the use of the Gamma distribution in modeling earthquake interevent times in Taiwan, but Wang *et al.* (2012) showed that the Gamma function is less appropriate to describe the frequency distribution of interevent times for the Taipei metropolitan area in Taiwan compared with the power-law function.

Considering the universality of probability, we use models constructed by Rundle *et al.* (2011) that are based on simple models of quiescence and activation for large earthquake probabilities. Quiescence and activation have been discovered in many cases (Wyss *et al.* 1990, 1995; Wiemer and Wyss 1994; Kossobokov *et al.* 1999; Hainzl *et al.* 2000; Ben-Zion and Lyakhovsky 2002; Ogata 2004; Sammis *et al.* 2004; Wu and Chiao 2006; Huang 2008; Huang and Ding 2012) and have also helped us develop some forecast methods (Tiampo *et al.* 2002; Huang 2004; Chen and Wu 2006; Wu *et al.* 2008, 2011; Gentili 2010). Some research has indicated that activation may be associated with the nucleation of small earthquakes that have a finite probability of growing into a large earthquake, so more small events implies a larger probability for the occurrence of a large earthquake (Langer 1967; Gunton and Droz 1983; Gunton *et al.* 1983; Rundle 1989, 1993; Klein and Unger 1983; Rundle *et al.* 1997; Shcherbakov *et al.* 2005). The physics of quiescence may be due to a mechanism such as a critical slowing down (Ma 1974; Klein and Unger 1983). When the system is driven to a critical point, fluctuations in systems with long-range interactions, such as elastic systems, tend to be suppressed prior to large nucleation events. The common point of quiescence and activation is the rate change of earthquakes; therefore, the quiescence and activation models for large earthquakes proposed by Rundle *et al.* (2011) are based on the rates of small earthquake activity. They tested these models using earthquake data from the Advanced National Seismic System (ANSS) catalog in California during the years 1985–2011 to determine which model is more consistent with the data. They found that neither the activation nor the quiescence model provides significant forecast skill from the standpoint of a reliability/attributes (R/A) test or a receiver operating characteristic (ROC) test.

We applied quiescence and activation models on earthquake data from the Central Weather Bureau (CWB) in Taiwan. The whole study area is divided into three subareas according to tectonic setting—the western, eastern, and northeastern Taiwan regions. The quiescence and activation models are tested on the whole study area and the three subareas. Furthermore, we divided the subareas into grids and calculated the conditional probability at specific times for every grid point to make two-dimensional maps of the quiescence and activation models so that an area with high probability could be observed. The ROC curves of the probability maps show that the forecast skill of the probability maps is, with few exceptions, beyond the random forecast. The results of both the probability time series and probability maps show that the quiescence model has better forecast skill for large earthquakes and that the high probability yielded by the activation model is more related to aftershocks. Based on our observation, the quiescence model could be a candidate forecast method under the conditions that an adequate interevent time can be obtained and the catalog is consistent.

2. Data

The primary seismicity dataset for Taiwan and nearby islands used in this research is the Central Weather Bureau (CWB) catalog.

The quiescence and activation models involve forecasting large earthquakes of magnitude $m \geq 6$ in this study using small earthquakes with magnitudes larger than the completeness magnitude. Considering the consistency of the catalog, the quiescence and activation models are examined using the catalog from 1994 to 2013. Nevertheless, a longer catalog from 1973 to 2013 is used for calculating the b value. Because the expected number of small earthquakes, which represents the property of the large earthquakes, is inferred from the b value, the catalog used to calculate the b value has to contain sufficient information; therefore, we adapt the catalog from 1973 to 2013 for calculating the b value. A completeness magnitude $m_c = 3$ can be obtained from both the 1994 to 2013 and the 1973 to 2013 catalogs.

Taiwan is at the complex boundary between the Eurasian Plate and the Philippine Sea Plate (TSAI *et al.* 1977; WU 1978; SHYU *et al.* 2005). Most earthquakes in Taiwan take place in the subduction zone. Because deep earthquakes usually occur offshore and cause less damage than shallow earthquakes in Taiwan, deep earthquakes are not considered in this study. The distribution of earthquake frequency with depth is shown in Fig. 1. The gray solid line shows the cumulative probability of all the earthquakes included in the CWB catalog from 1900 to 2013; the result shows that approximately 90 % of the earthquakes took place above a depth of 30 km.

In addition to the whole Taiwan region, the subareas of Taiwan shown in Fig. 2 are the study regions that we are interested in. The subareas are determined according to the seismicity distribution and tectonic setting; they are the western Taiwan region, the eastern Taiwan region, and the northeastern Taiwan region. The distributions of earthquake frequency versus depth for these three study regions are also shown in Fig. 1. The results show that approximately 90 % of the earthquakes occurred above 20 km in the western Taiwan region, approximately 90 % occurred above 30 km in the eastern Taiwan region, and approximately 90 % occurred above 60 km in the northeastern Taiwan region. To sufficiently include the shallow earthquakes and have a uniform cut-off depth for all the study regions, a depth of 40 km is regarded as the cut-off depth.

In Fig. 2, open circles represent the $m \geq 6$ earthquakes that occurred above the cut-off depth from 1900 to 2013. Significantly, there are fewer large earthquakes in the western Taiwan region, and most of them tend to occur in the same place. The major reason these large earthquakes are centered in the same place is that most of the earthquakes are aftershocks, inherently occurring in the same place as

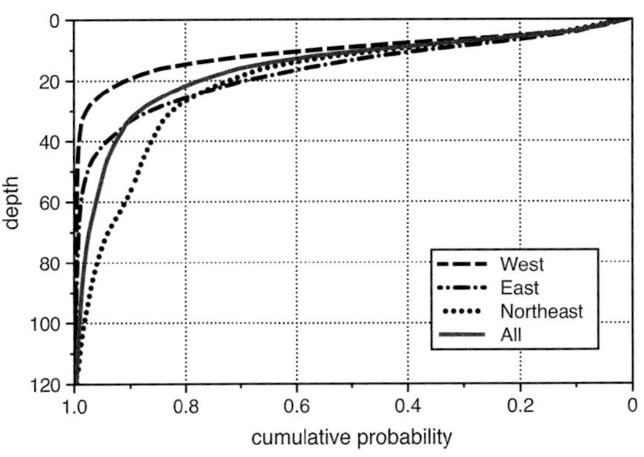

Figure 1
Distribution of earthquake frequency versus depth. The *gray solid line* shows the cumulative probability of all earthquakes from 1900 to 2013. The *black dashed/dash-dotted/dotted line* represents the cumulative probability of earthquakes in the western/eastern/northeastern Taiwan region, as shown in Fig. 2

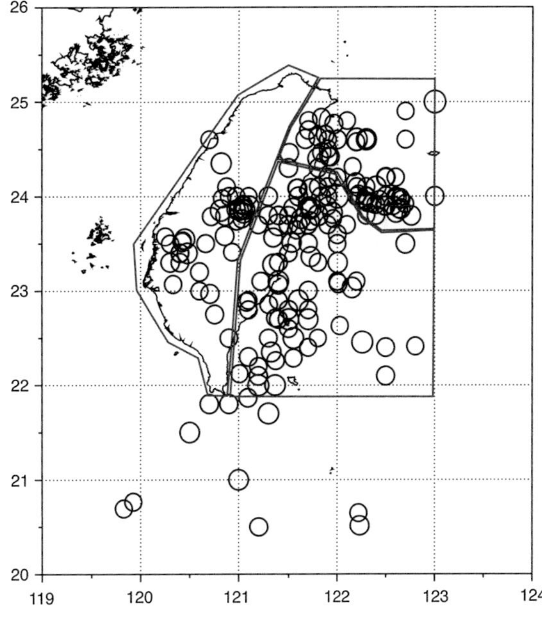

Figure 2
Distribution of $m \geq 6$ earthquakes from 1900 to 2013. *Open circles* represent $m \geq 6$ earthquakes that occurred above the cut-off depth from 1900 to 2013. The *gray lines* show three study regions of interest, which are the western, eastern, and northeastern Taiwan regions

their mainshocks. By including the aftershocks in the calculation of the interevent time, we reduce the real interevent time; however, removing aftershocks may also remove earthquakes that are not aftershocks. Therefore, we calculate the mean interevent time using all the $m \geq 6$ earthquakes that occurred from 1900 to 2013 in order to include more mainshocks.

3. Models

The model of conditional probability of failure for large earthquakes, which is based on the general formalism of reliability or hazard analysis (Ebeling 1997; NIST 2010), was developed by Rundle *et al.* (2011). The basic idea is to use the small earthquakes that have magnitudes larger than the catalog completeness magnitude m_c to evaluate the conditional probability of a subsequent large earthquake with a magnitude larger than m. The application steps for the time series of the earthquakes are the following:

1. We first calculate the expected number N_C of small earthquakes larger than the catalog completeness magnitude m_c that should be included in a cycle of large earthquakes having magnitudes larger than m by

$$N_c = 10^{b(m-m_c)}. \tag{1}$$

The parameter b is obtained by fitting the seismicity data over a long interval at least 30 years into the past, and b should be a relatively slowly varying function of time over an interval of a year or less.

2. Compute a non-declustered, nonhomogeneous, and time-varying Poisson rate of small earthquakes $v_C(t)$ by a double-averaging method:

$$v_C(t) = \frac{1}{t - t_0} \int_{t_0}^{t} \frac{\Delta n_c(t)}{T_c} \mathrm{d}t. \tag{2}$$

Here, t_0 is the onset of the time period that is going to be examined, T_C is an averaging time, and $\Delta n_C(t)$ is the number of small earthquakes over a time interval $\{t - T_C, t\}$. In Rundle *et al.* (2011), T_C is 5 years, which is approximately five times the interevent time of $m \geq 6$ earthquakes in the California region and is obtained by backtesting. Because the main purpose of our study is to obtain the conditional probability that is spontaneously reflected by the small earthquakes, we use the average interevent time of $m \geq 6$ earthquakes in the corresponding region.

3. Using the average Poisson rate of small earthquakes v_C and the expected number of small earthquakes N_C in a cycle of $m \geq 6$ earthquakes, we can determine a Poisson window:

$$T_W = \frac{N_C}{2v_C}. \tag{3}$$

The Poisson window is a moving time window for sampling the small earthquakes. The ratio of N_C to v_C represents the average time interval between large earthquakes; the sampling frequency $1/T_W$ equals twice the average recurrence frequency, and the factor of 2 in the denominator corresponds to the introduction of the Nyquist frequency.

4. Next, we compute the expected number $n_E(t)$ and observed number $n_O(t)$ of small earthquakes in the Poisson window and their ratio $R(t) = n_O(t)/n_E(t)$. If $R(t) > 1$, there are more small earthquakes observed than were expected during the previous time interval T_W, and the situation corresponds to activation. If $R(t) < 1$, there are fewer small earthquakes observed than were expected during the previous time interval T_W, and the situation corresponds to quiescence. With the definition (1), the expected number of small earthquakes during the time interval T_W is $n_E(t) = N_C/2$. To compute the observed number of small earthquakes $n_O(t)$, a window function of length T_W using tapers on the trailing edge (at time $t - T_W$) is applied over the time series of the small earthquakes. To include the triggering of large events that is sometimes observed during heightened activity at the leading edge time t, a sharp edge is used in the window function:

$$F(t) = \left[\sin\left(\frac{\tau - t + T_W/2}{T_W}\right)\pi + 1\right], \tag{4}$$

which is valid for $\tau \in \{t - T_W, t\}$. In addition, the window function is normalized so that the area under $F(t)$ is the same as the area under the boxcar window function.

5. With the above equations, two nonhomogeneous Poisson forecast models are constructed: the activation model and the quiescence model. For the activation model, we presume that the conditional probability is higher when the anomalous activity of small earthquakes increases, so the conditional probability is proportional to the ratio $R(t)$ of the observed number of small earthquakes to the expected number of small earthquakes in a Poisson window. The conditional probability of the activation model that a failure will occur in a time Δt in the future, given that it has not failed before t, is given by

$$P_{\mathrm{A}}(t|\Delta t) = 1 - \exp\{-\Delta H_{\mathrm{A}}(t, \Delta t)\}. \quad (5)$$

Here, $\Delta H_{\mathrm{A}}(t, \Delta t)$ is the cumulative conditional hazard rate function of the activation model and equals

$$\Delta H_{\mathrm{A}}(t, \Delta t) = \frac{R(t) f v_{\mathrm{C}} \Delta t}{N_{\mathrm{C}}}. \quad (6)$$

For the quiescence model, we presume that the conditional probability is higher when the anomalous activity of small earthquakes decreases, so the conditional probability is proportional to the inverse of $R(t)$. The conditional probability of the quiescence model that a failure will occur in a time Δt in the future, given that it has not failed before t, is given by

$$P_{\mathrm{Q}}(t|\Delta t) = 1 - \exp\{-\Delta H_{\mathrm{Q}}(t, \Delta t)\}. \quad (7)$$

Here, $\Delta H_{\mathrm{Q}}(t, \Delta t)$ is the cumulative conditional hazard rate function of the quiescence model and equals

$$\Delta H_{\mathrm{Q}}(t, \Delta t) = \frac{f v_{\mathrm{C}} \Delta t}{N_{\mathrm{C}} R(t)}. \quad (8)$$

The parameter f is used to optimize the forecast and make $f v_{\mathrm{c}}$ an optimal Poisson rate. The determination of f relies on standard verification tests (backtesting the forecast).

4. Results and ROC Test

For the quiescence and activation models, the interevent time of large earthquakes is a crucial parameter. We calculated the interevent times of $m \geq 6$ earthquakes that occurred from 1900 to 2013 within depths of 0 to 40 km and obtained an average interevent time of approximately 0.55 years. Considering the complexity of the tectonics and the different loading rate in Taiwan, we divided the study region into three subareas, which are western Taiwan, northeastern Taiwan, and eastern Taiwan (Fig. 2), and we grouped the earthquakes according to their locations. The average interevent times of $m \geq 6$ earthquakes at shallow depths of 0–40 km for the western, northeastern, and eastern Taiwan regions are approximately 2.65, 2.04, and 1.25 years, respectively. The b values obtained by fitting the Gutenberg–Richter scaling law to earthquakes with magnitudes larger than the cut-off magnitude m_{c} for the whole, western, northeastern, and eastern Taiwan regions are approximately 1.07, 0.95, 1.03, and 1.02, respectively.

We took 0.01 years as a time step to compute the conditional probabilities for four different study regions; Fig. 3a–d, respectively, show the results for the whole, western, eastern, and northeastern Taiwan regions. The conditional probabilities P_{Q} and P_{A} for $m \geq 6$ earthquakes that will occur within 30 days yielded from the quiescence and activation models are shown in the middle and upper panels of Fig. 3, and $m \geq 5$ earthquake sequences are shown in the bottom panels. The vertical dashed lines indicate the times of the $m \geq 6$ earthquakes. The gray lines indicate the mean probabilities obtained from the quiescence and activation models for each study region from 1994 to 2013, and the gray-colored bands show the range from the mean value to a standard deviation. The parameter f in the previous work of Rundle *et al.* (2011) is determined by optimizing the forecast using standard verification tests; thus, $f v_{\mathrm{c}}$ is considered to be an optimal Poisson rate. Instead of considering an optimal Poisson rate, we took a uniform f and considered v_{c} to be an inherently generated rate.

Comparing P_{Q} and P_{A} in Fig. 3, the probability time series obtained from the activation model shows a contrary result to the quiescence model when it approaches a large ($m \geq 6$) event and just after a large event occurs. It can be easily observed that P_{A} typically trends lower than the mean value when an $m \geq 6$ earthquake approaches, sharply increases as the event occurs, and reaches its highest value just after the earthquake occurs. In some cases, the increase of P_{A} from the mean value can be several

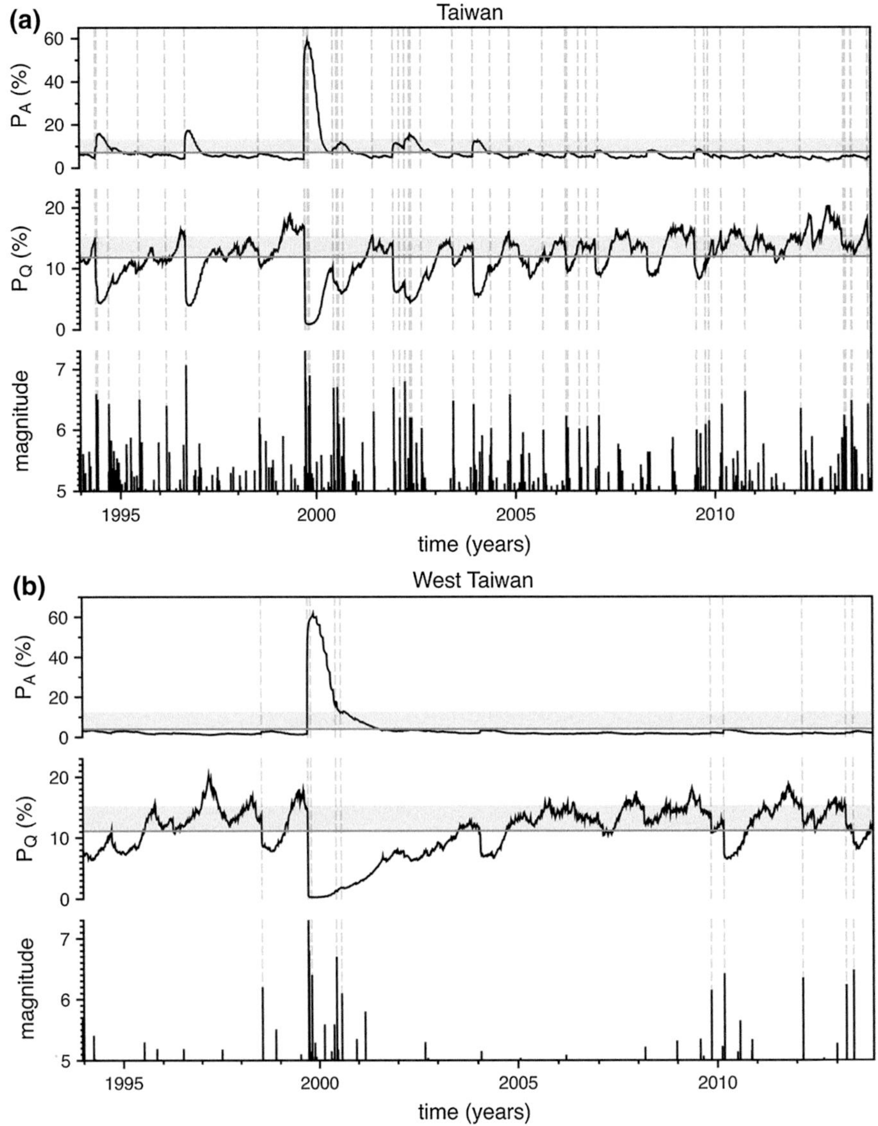

Figure 3

Time series of 30-day forecasts for **a** the whole Taiwan region, **b** the western Taiwan region, **c** the eastern Taiwan region, and **d** the northeastern Taiwan region from 1994 to 2013. The *top panel* of each figure shows the 30-day conditional probability obtained from the activation model as a function of time. The *middle panel* of each figure shows the 30-day conditional probability obtained from the quiescence model as a function of time. The *bottom panel* of each figure shows $m \geq 5$ earthquake sequences. The *dashed vertical lines* indicate the times of $m \geq 6$ earthquakes

times the standard deviation. By contrast, P_Q typically increases when an $m \geq 6$ earthquake approaches, sharply decreases as the earthquake occurs, and reaches its lowest value after the earthquake occurs. Most exceptions to this pattern occur when the $m \geq 6$ earthquakes are clustered in time.

In the whole, western, and eastern Taiwan regions (Fig. 3a–c), most $m \geq 6$ earthquakes occur after P_Q increases beyond one standard deviation (3.5, 4.2, and 1.1). However, in the northeastern Taiwan region, the increase of P_Q is usually less than one standard deviation (1.5) except for the clustered events in 1994, 2002, and 2013.

An extremely high P_A value, 60 %, which is about 8 times the mean value, was yielded after the 1999 Chi–Chi ($m = 7.3$) earthquake, as can be

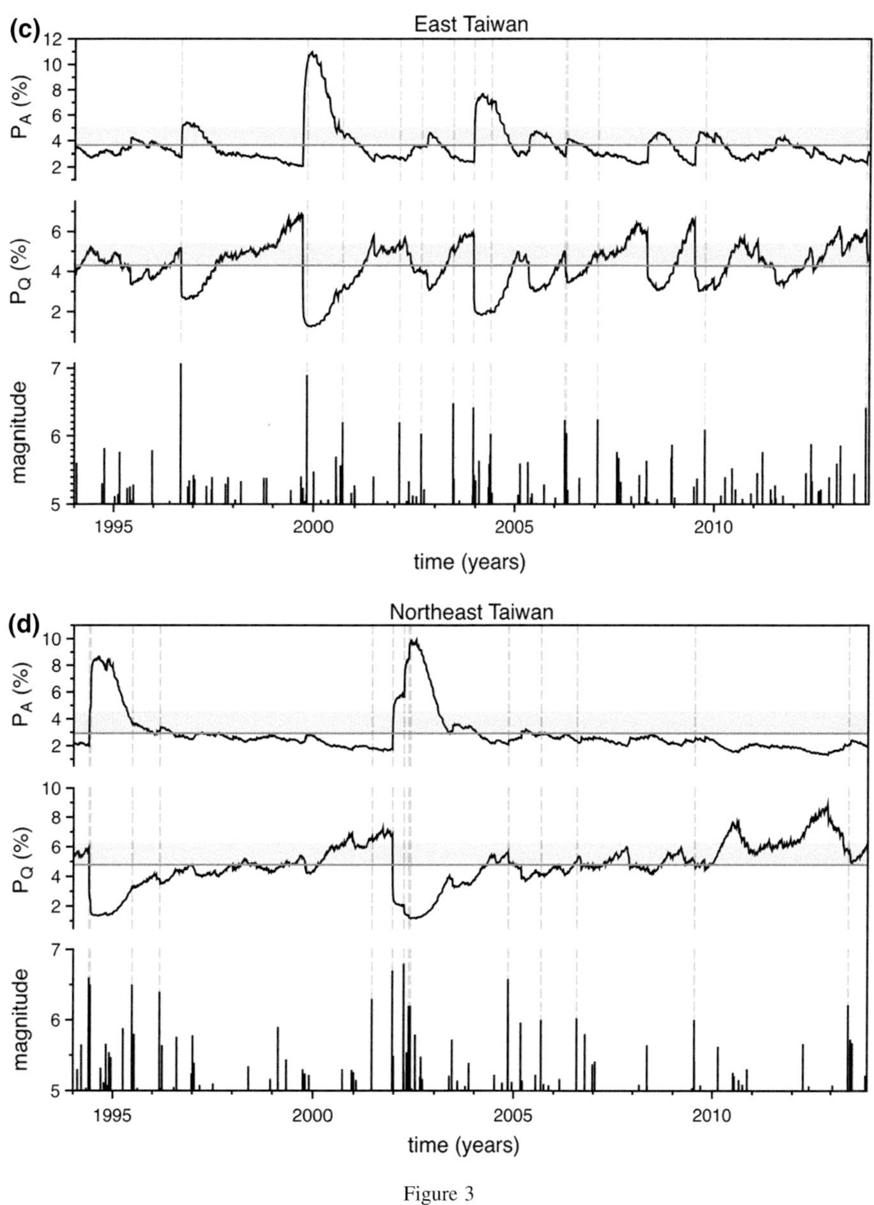

Figure 3
continued

observed in Fig. 3a. This extreme value can also be observed in Fig. 3b, c because of the Chi–Chi mainshock and its aftershocks in the western Taiwan region and some aftershocks in the eastern Taiwan region. Due to the effect of the Chi–Chi mainshock, the pattern that P_Q increases just before a large earthquake is not shown for the large earthquakes that are clustered with the Chi–Chi mainshock; in addition, an increase of P_Q before the cluster of events in 2000 is not clear as for other events.

The activation and quiescence models also produce some false alarms where P_Q increases and then sharply decreases while P_A decreases and then sharply increases but there is no corresponding $m \geq 6$ earthquake. Some cases, such as the one that occurs in mid 2008 in the whole Taiwan region (Fig. 3a), the one that occurs in early 2004 in the western Taiwan region (Fig. 3b), and the ones that occur in early 2005 and mid 2008 in the eastern Taiwan region (Fig. 3c), are even preceded by an

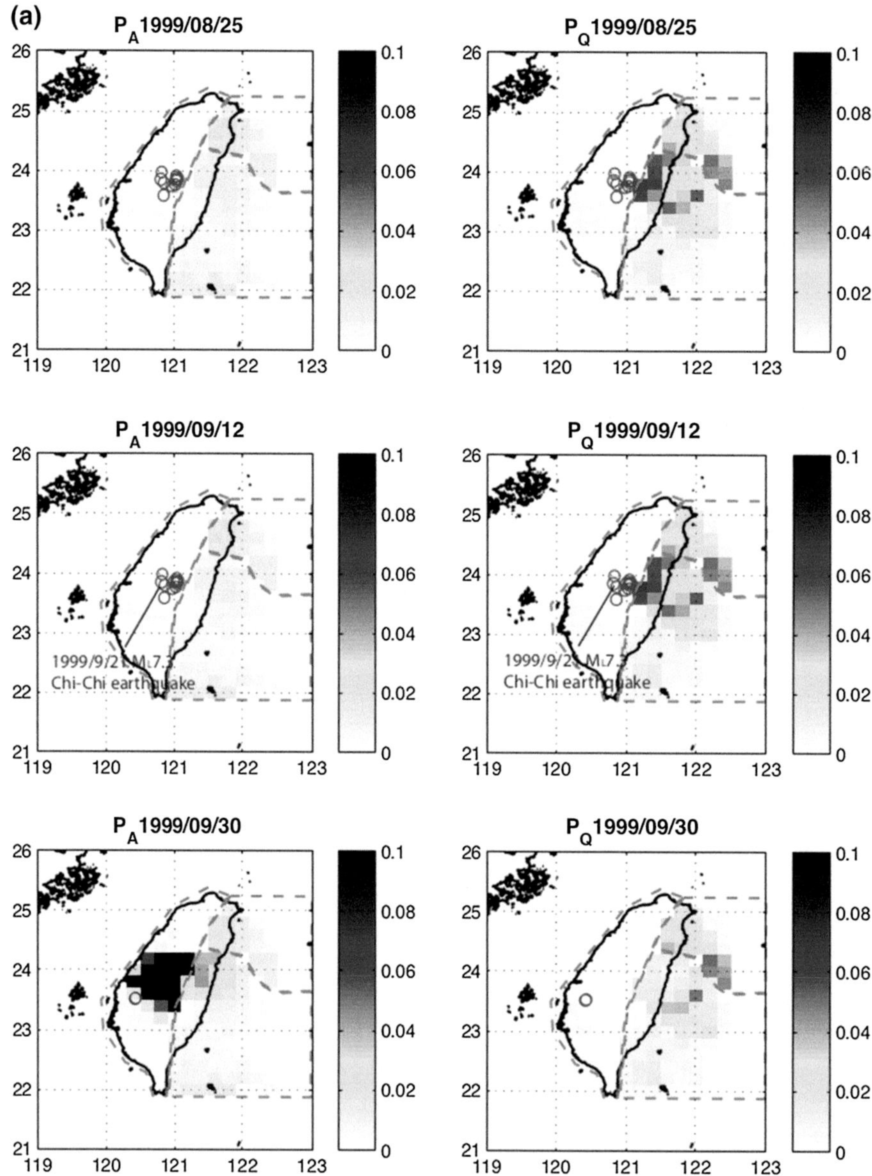

Figure 4
Maps of conditional probability for the quiescence and activation models at specific times that are close to **a** the 1999 Chi–Chi earthquake, **b** the 2003 Chengkung earthquake, and **c** the 2010 Jiashian earthquake. The *left panels* of each figure represent the conditional probabilities obtained from the activation model, and the *right panels* of each figure represent the conditional probabilities from the quiescence model for the grids in the study region at the labeled time. The *top/middle/bottom panels* of each figure show the conditional probability taken before/just before/after the large event. The *red circles* are the $m \geq 6$ earthquakes that occur within 30 days of the time labeled at the *top* of each panel. The Chi–Chi, Chengkung, and Jiashian earthquakes are indicated in the *middle panel* of each figure

increase of P_Q that is beyond one standard derivation.

The probability in Fig. 3 is a coherent property of every single region rather than of an individual location. To look into the probability change of individual locations in the study regions, we applied the probability calculation to every grid point in the study regions. We divided the research area of 119.5–122.5°E and 21.5–25.5°N into 300 nonoverlapping 0.2° × 0.2° square boxes and took the centers of the boxes in the study region as the grid points. For every grid point, the conditional

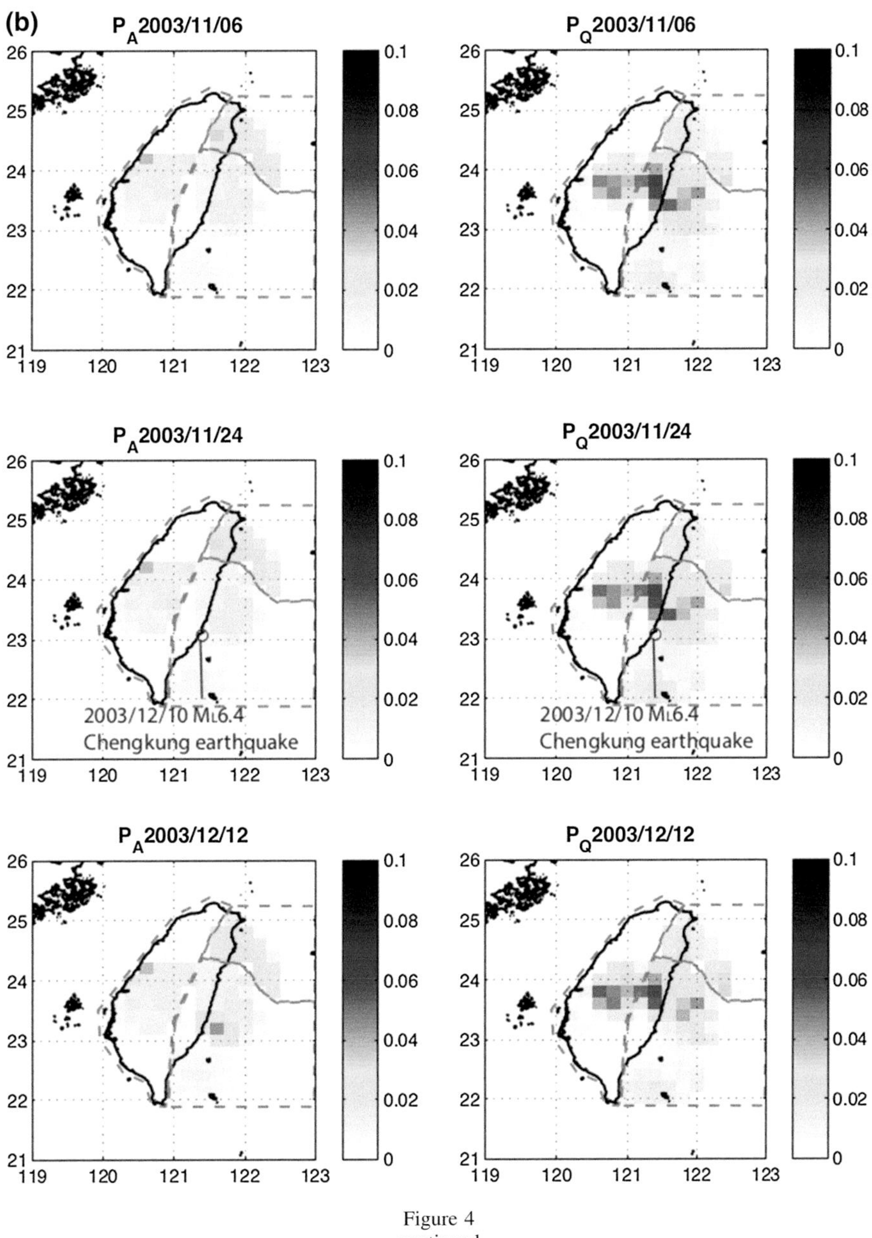

Figure 4
continued

probability was calculated using the events within a distance of 50 km from the grid point and in the subregion where the grid point lies. Considering $m \geq 6$ earthquakes to be large earthquakes, earthquakes with magnitudes larger than the catalog completeness magnitude $m_c = 3$ to be small earthquakes, and the time step to be 0.01 years, the probability time series obtained from the quiescence and activation models for every grid point are shown in Fig. 4. Given a specific time, the probabilities for the quiescence and activation models that an $m \geq 6$ earthquake occurs within 30 days after the specific time were then mapped as shown in Fig. 4.

Figure 4a–c show the probability maps before and after the 1999 Chi-Chi ($m = 7.3$) earthquake, the 2003 Chengkung ($m = 6.6$) earthquake, and the 2010 Jiashian ($m = 6.4$) earthquake. The middle panels of

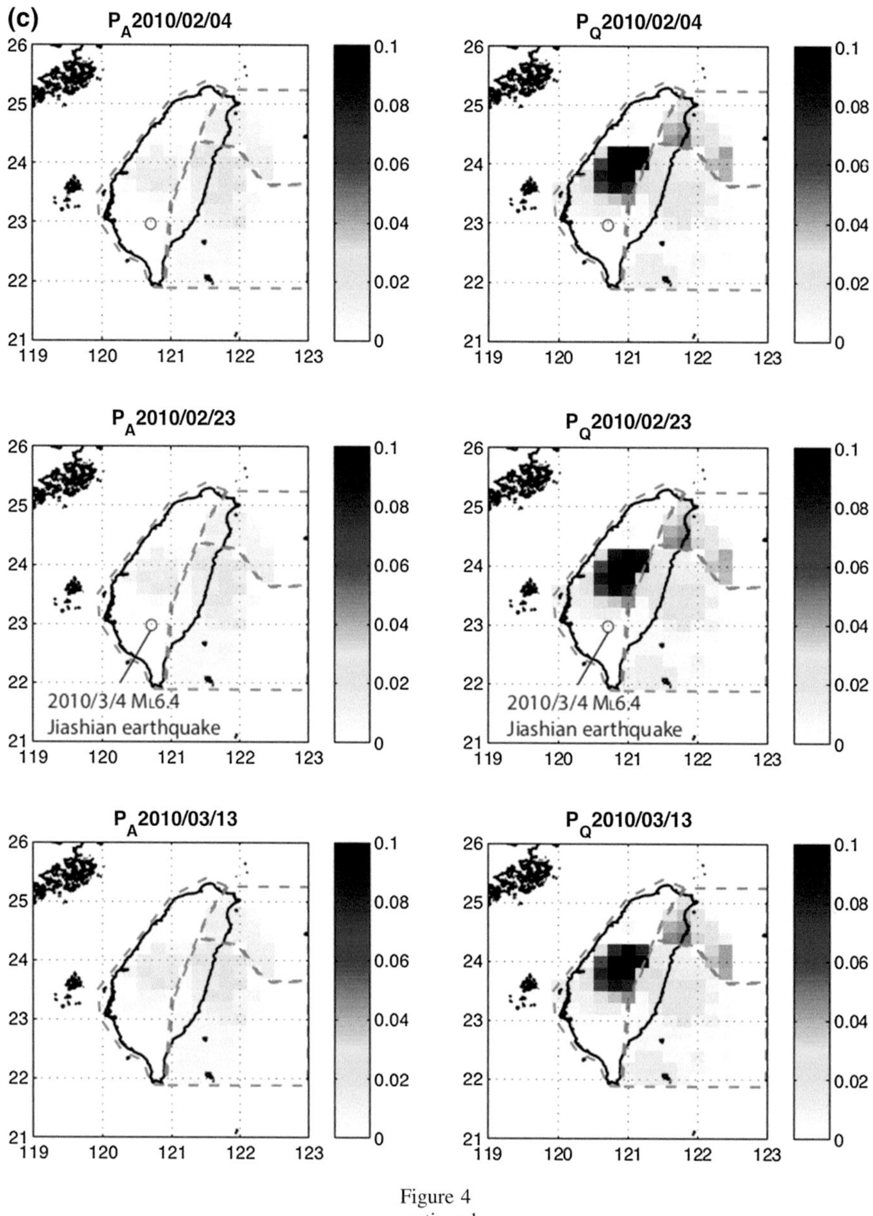

Figure 4
continued

Fig. 4a–c represent the probability maps just before the earthquakes. The locations of $m \geq 6$ earthquakes that occur within 30 days after the time labeled at the top of the figures are shown as red open circles. The probability of the quiescence and activation models is denoted by P_{Q} and P_{A} and is colored in the map; the warmer the color, the higher the probability.

In Fig. 4a, some high P_{A} and P_{Q} were yielded around the epicenter of the Chi–Chi earthquake before the event occurred; however, high P_{A} and P_{Q} were only yielded in the eastern Taiwan region, not in the western Taiwan region. After the Chi–Chi earthquake, the P_{A} around the epicenter became very high in the western Taiwan region, and the P_{Q} near the epicenter in the eastern Taiwan region decreased a lot. In Fig. 4b, high P_{Q} can be observed north of the Chengkung earthquake before the event occurred, and the P_{Q} decreased after the event. By contrast, high P_A around the epicenter was absent before the event and showed up after the event. Compared with Fig. 4a, b, Fig. 4c does not show changes in P_Q before and after the Jiashian earthquake; in addition, the distribution of high P_Q in the western Taiwan region is consistent with the distribution of the Chi–Chi earthquake and its aftershocks.

Considering that the probability maps (Fig. 4) may vary with the threshold we give in the color bar, we apply the receiver operating characteristics (ROC) test to estimate the reliability of the probability maps. A ROC curve is a graphical plot that illustrates the performance of a binary classifier system, as its discrimination threshold is varied in signal detection theory (Green and Swets 1966). A ROC curve is generally employed in medical science and social science; it is also a useful tool for evaluation of machine learning techniques (Zweig and Campbell 1993; Pepe 2003; Obuchowski 2003). The task of the ROC curve in these fields is mostly to increase the prediction of a model or to evaluate the accuracy of the default probability model.

The ROC curve has been increasingly used in verifying probability forecasts because of the binary characteristics yielded from the probability forecast, i.e., the hit rate (HR) and false alarm rate (FR). In the ROC test, a threshold R is applied to the 2D map of conditional probability to transform the forecast into a binary forecast. For a given threshold R, boxes with probability $P(t|\Delta t) \geq R$ represent the forecast locations, and boxes with large future earthquakes that occur during the forecast period represent the event locations. A forecast is successful when it makes a forecast location on an event location, and the forecast is a false alarm when it makes a forecast location on a box that is not an event location. The fraction of successful forecasts out of the total event locations is the hit rate (HR), and the fraction of false alarms out of the total boxes that are not event locations is the false alarm rate (FR). The ROC diagram is then constructed by plotting HR against FR as the threshold value R decreases.

To evaluate the performance of the activation and quiescence models in time and space at the same time, we calculated an average ROC curve over a set of conditional probability maps (Fig. 5). We sampled the time period from 2001 to 2010 by 0.05 years, and calculated a conditional probability map for the activation model at each sampled time as well as for the quiescence model, then finally averaged all the ROC curves. The black lines in Fig. 5 show the results for taking $m \geq 6$ earthquakes as target earthquakes, and the gray lines show the results for $m \geq 5$ earthquakes. The dotted lines are ROC curves of 100 bootstrap tests for taking $m \geq 6$ earthquakes as target earthquakes, and the dashed lines are the results of bootstrap tests for $m \geq 5$ earthquakes.

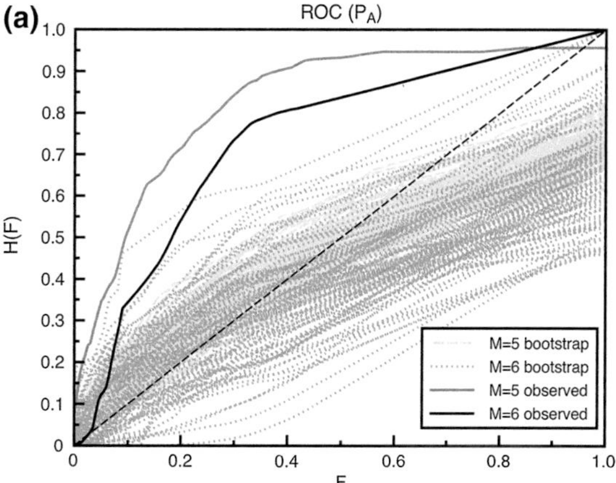

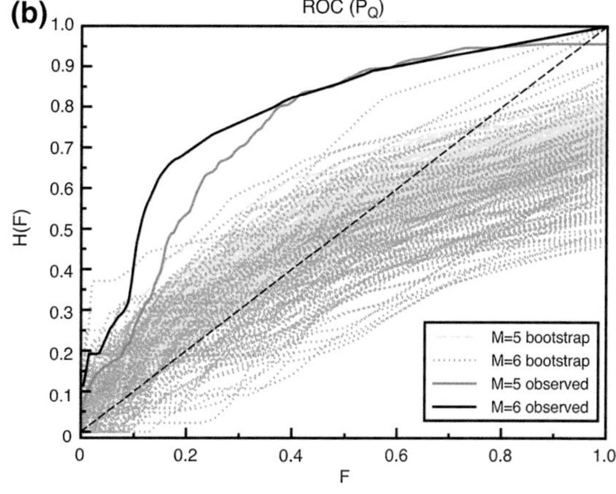

Figure 5
Receiver operating characteristic diagrams of the conditional probability map for **a** the activation model and **b** the quiescence model. The *black solid line* and *gray solid line* denote the averaged ROC curves of the conditional probability map for $m \geq 6$ and $m \geq 5$ earthquakes that occur in 30 days after the sampled times during 2001 to 2010. The *dotted* and *dashed lines* are ROC curves of the 100 bootstrap tests for the $m \geq 6$ and $m \geq 5$ earthquakes. The *diagonal line* from the *lower-left corner* (no hits or false alarms) to the *upper-left corner* (no false alarms, only successful forecasts) represents no skill (random forecast). The closer an ROC curve is to the *upper-left corner*, the higher the forecast skill

The upper-left corner of the ROC diagram represents a perfect forecast system (no false alarms, only successful forecasts). The lower-left corner (no hits or false alarms) represents a system that never warns of an event. The upper-right corner represents a system where the event is always warned. A random forecast is characterized by the condition HR = FR, which is represented on the ROC diagram by a diagonal line connecting the point at the lower-left corner to the point at the upper-right corner. A good forecast should always make more successful forecasts than false alarms; therefore, the closer any ROC curve is to the upper-left corner, the higher its forecast skill.

5. *Discussion and Conclusions*

One main purpose of this study is to look for an alternative forecast method for large earthquakes. The quiescence and activation models proposed by Rundle *et al.* (2011) were examined on the CWB catalog. In addition to the probability time series for the entire Taiwan region and the three subregions, we also made probability maps. From both the probability time series and the probability maps, it is significant that the quiescence model has better forecasting skill because of the increasing probability before the large events in the time series and the high probability distributed around the epicenter. In contrast to the quiescence model, the activation model does not have increasing probability before the event or high probability distributed around the event, but it does after the event. Because high P_A is distributed around the epicenter, the high probability of the activation model after the event should be associated with aftershocks.

The probability map enables the quiescence and activation models to identify locations with high probability; furthermore, the probability maps demonstrate that the sharply increasing P_Q or P_A in the time series is dominated by the seismicity around the epicenter of the large event before and after the event. Take the Chengkung earthquake as an example (Fig. 4b); except for the P_Q north of the epicenter, all the other P_Q remain the same before and after the event.

The properties of the seismicity associated with the large event can be observed not only in the probability maps but also in the probability time series. The probability P_Q usually drops as soon as $m \geq 6$ events occur in the eastern Taiwan region (Fig. 3c), except for two events that occurred in late 1999 and late 2009. For these two special events, some $6 > m > 5$ earthquakes occurred just before the $m \geq 6$ events, and the probability P_Q drops as the $m > 5$ earthquakes occur. Because the last $m \geq 6$ event of these two special events occurred at least 2 years earlier, the results suggest a high possibility that the $6 > m > 5$ earthquakes preceding these two special events are the foreshocks of these two special events. Another interesting characteristic reflected in the P_A and P_Q is the clustering of events in the northeastern Taiwan region. In Fig. 3d, it can be observed that P_Q only increased significantly prior to two clusters that occurred in mid 1994 and mid 2002, and P_A increased progressively to the highest value in these two clusters. Unlike most $m \geq 6$ earthquakes that took place offshore in the northeastern Taiwan region, the first two $m \geq 6$ earthquakes in 1994 took place very close to each other in time and space, and then the third $m \geq 6$ earthquake occurred at Yilan within one month. The cluster that caused high P_A and P_Q in 2002 included four $m \geq 6$ earthquakes, three of which were offshore earthquakes and the fourth at Yilan.

The ROC curves of P_A and P_Q (Fig. 5) show the significance of the activation and quiescence models. It can be observed that the probability maps of the quiescence model have better forecast ability than the activation model when taking $m \geq 6$ earthquakes as the target earthquakes. However, the probability maps of the quiescence model do not offer a better forecast than the activation model when taking $m \geq 5$ earthquakes as the target earthquakes. Because P_A usually decreases before the large events while P_Q increases and then sharply increases when the event occurs while P_Q decreases, P_A would have high value just before the $m \geq 5$ earthquakes following another large event and P_Q would have low value in this case. As a consequence, the quiescence model shows better forecasting skill as a forward forecast tool.

It is worth noting that the ROC curve in this study is not just a ROC curve of conditional probability at a single time but an averaged ROC curve of the conditional probability map over a time span. Rundle *et al.* (2011) carried out the ROC test on the time series of activation and quiescence models using the whole catalog in California and Nevada; in this study, we carried out the ROC test on the probability map on which only earthquakes with a distance smaller than 50 km from the grid point were considered. The ROC curves in California show a diagonal trend (Fig. 4 in Rundle *et al.* 2011); however, the average ROC curves in Taiwan show a trend that is close to the upper-left corner of the ROC diagram. The activation and quiescence models do not show significance in the ROC test in Rundle *et al.* (2011) because the models usually fail to forecast the events that are clustered closely in time even when these events are distant from each other in space. The events that are clustered in time but distant in space may be taken as distinct events in the probability map, therefore the significance of the activation and quiescence models can be observed in our average ROC curve.

Acknowledgments

The authors would like to express their gratitude to the Central Weather Bureau (CWB) for providing their quality earthquake catalogs and to Chien-Hsin Chang for providing the information about the catalog. The work of Y.-H.W. and C.-C.C. was supported by the National Science Council (ROC) (grant NSC-102-2811-M-008-075) and the Department of Earth Sciences, NCU (ROC).

References

Ben-Zion, Y., and Lyakhovsky, V. (2002), *Accelerated Seismic Release and Related Aspects of Seismicity Patterns on Earthquake Faults*, Pure Appl. Geophys. *159*, 2385–2412.

Chen, C.C., and Wu, Y.H. (2006), *An Improved Region-Time-Length Algorithm Applied to the 1999 Chi-Chi, Taiwan Earthquake*, Geophys. J. Int. *166*, 1144–1147.

Chen, C.H., Wang, J.P., Wu, Y.M., Chan, C.H., and, Chang, C.H. (2012), *A Study of Earthquake Inter-Occurrence Times Distribution Models in Taiwan*, Nat. Hazards. *69*, 1335–1350.

Console, R., Murru, M., Catalli, F., and Falcone, G. (2007), *Real Time Forecasts through an Earthquake Clustering Model Constrained by the Rate-and-State Constitutive Law: Comparison with a Purely Stochastic ETAS Model*, Seismol. Res. Lett. *78*, 49–56.

Ebeling, C.E., *An Introduction to Reliability and Maintainability Engineering*, (McGraw-Hill, Boston 1997).

Ferraes, S.G. (2003), *The Conditional Probability of Earthquake Occurrence and the Next Large Earthquake in Tokyo, Japan*, J. Seismol. *7*, 145–153.

Gentili, S. (2010), *Distribution of Seismicity before the Larger Earthquakes in Italy in the Time Interval 1994–2004*, Pure Appl. Geophys. *167*, 933–958.

Green, D.M., and Swets, J.A., *Signal Detection Theory and Psychophysics* (Wiley, New York 1966).

Gomberg, J., Belardinelli, M.E., Cocco, M., and Reasenberg, P. (2005), *Time-Dependent Earthquake Probabilities*, J. Geophys. Res. *110*, B05S04, doi:10.1029/2004JB003405.

Gunton, J.D., and Droz, M., *Introduction to the Theory of Metastable and Unstable States* (Springer-Verlag, Berlin 1983).

Gunton, J.D., San Miguel, M., and Sahni, P.S., The dynamics of first order phase transitions, *Phase Transitions and Critical Phenomena*, (eds. Domb, C., and Lebowitz, J.) (Academic, London 1983) pp. 269–467.

Hainzl, S., Zöller, G., Kurths, J., and Zschau, J. (2000), *Seismic Quiescence as an Indicator for Large Earthquakes in a System of Self-Organized Criticality*, Geophys. Res. Lett. *27*, 597—600.

Huang, Q.H. (2004), *Seismicity Pattern Changes Prior to Large Earthquakes-An Approach of the RTL Algorithm*, Terr. Atmos. Ocean. Sci. *15*, 469—491.

Huang, Q.H. (2008), *Seismicity Changes prior to the M_S 8.0 Wenchuan Earthquake in Sichuan, China*, Geophys. Res. Lett. *35*, L23308, doi:10.1029/2008GL036270.

Huang, Q.H., and Ding, X. (2012), *Spatiotemporal Variations of Seismic Quiescence prior to the 2011 M 9.0 Tohoku Earthquake Revealed by an Improved Region–Time–Length Algorithm*, Bull. Seismol. Soc. Am. *102*, 1878–1883.

Klein, W., and Unger, C. (1983), *Pseudospinodals, Spinodals and Nucleation*, Phys. Rev. B *28*, 445–448.

Kossobokov, V.G., Maeda, K., and Uyeda, S. (1999), *Precursory Activation of Seismicity in Advance of the Kobe, 1995, M=7.2 Earthquake*, Pure Appl. Geophys. *155*, 409–423.

Langer, J.S. (1967), *Theory of the Condensation Point*, Ann. Phys. *41*, 108–157.

Ma, S.K., *Modern Theory of Critical Phenomena* (Benjamin-Cummings, Reading, MA 1974).

NIST/SEMATECH e-Handbook of Statistical Methods, 2010. Available at: http://www.itl.nist.gov/div898/handbook/ (last accessed June 2012).

Obuchowski, N.A. (2003), *Receiver Operating Characteristic Curves and Their Use in Radiology*, Radiology *229*, 3–8.

Ogata, Y. (2004), *Seismicity Quiescence and Activation in Western Japan Associated with the 1944 and 1946 Great Earthquakes near the Nankai Trough*, J. Geophys. Res. *109*, B04305, doi:10.1029/2003JB002634.

Pepe, M.S., *The Statistical Evaluation of Medical Tests for Classification and Prediction* (Oxford, New York 2003).

Rundle, J.B. (1989), *A Physical Model for Earthquakes. 3. Thermodynamical Approach and Its Relation to Nonclassical Theories of Nucleation*, J. Geophys. Res. *94*, 2839–2855.

Rundle, J.B. (1993), *Magnitude-Frequency Relations for Earthquakes Using a Statistical-Mechanical Approach*, J. Geophys. Res. *98*, 21 943–21 949.

Rundle, J.B., Gross, S., Klein, W., and Turcotte, D.L. (1997), *The Statistical Mechanics of Earthquakes*, Tectonophysics *277*, 147–164.

RUNDLE, J.B., HOLLIDAY, J.R., YODER, M., SACHS, M.K., DONNELLAN, A., TURCOTTE, D.L., TIAMPO, K.F., KLEIN, W., and KELLOGG, L.H. (2011), *Earthquake Precursors: Activation or Quiescence?* Geophys. J. Int. *187*, 225–236.

SAMMIS, C.G., BOWMAN, D.D., and KING, G. (2004), *Anomalous Seismicity and Accelerating Moment Release Preceding the 2001 and 2002 Earthquakes in Northern Baja California, Mexico*, Pure Appl. Geophys. *161*, 2369—2378.

SHCHERBAKOV, R., YAKOVLEV, G., TURCOTTE, D.L., and RUNDLE, J.B. (2005), *Model for the Distribution of Aftershock Interoccurrence Times*, Phys. Rev. Lett. *95*, 218501, doi:10.1103/PhysRevLett.95.218501.

SHYU, J.B.H., SIEH, K., CHEN, Y.G., and LIU, C.S. (2005), *Neotectonic Architecture of Taiwan and Its Implications for Future Large Earthquakes*, J. Geophys. Res. *110*, B08402, doi:10.1029/2004JB003251.

SYKES, L.R., and JAUMÉ, S.C. (1990), *Seismic Activity on Neighbouring Faults as a Long-Term Precursor to Large Earthquakes in the San Francisco Bay Area*, Nature *348*, 595–599.

TIAMPO, K.F., RUNDLE, J.B., MCGINNIS, S.A., and KLEIN, W. (2002), *Pattern Dynamics and Forecast Methods in Seismically Active Regions*, Pure Appl. Geophys. *159*, 2429–2467.

TSAI, Y.B., TENG, T.L., CHIU, J.M., and LIU, H.L. (1977), *Tectonic Implications of the Seismicity in the Taiwan Region*, Mem. Geol. Soc. China 2, 13–41.

VERE-JONES, D. (1995), *Forecasting Earthquakes and Earthquake Risk*, Int. J. Forecasting *11*, 503–538.

WANG, J.H., CHEN, K.C., LEE, S.J., HUANG, W.G., WU, Y.H., and LEU, P.L. (2012), *The Frequency Distribution of Inter-Event Times of $M \geq 3$ Earthquakes in the Taipei Metropolitan Area: 1973–2010*, Terr. Atmos. Ocean. Sci. *23*, 269–281.

WIEMER, S., and WYSS, M. (1994), *Seismic Quiescence before the Landers ($M = 7.5$) and Big Bear ($M = 6.5$) 1992 Earthquakes*, Bull. Seismol. Soc. Am. *84*, 900–916.

WU, F.T. (1978), *Recent Tectonics of Taiwan*, J. Phys. Earth *2*, S265–S299.

WU, Y.H., CHEN, C.C., and RUNDLE, J.B. (2008), *Precursory Seismic Activation of the Pingtung (Taiwan) Offshore Doublet Earthquakes on 26 December 2006: a Pattern Informatics Analysis*, Terr. Atmos. Ocean. Sci., *19*, 743—749.

WU, Y.H., CHEN, C.C., and RUNDLE, J.B. (2011), *Precursory Small Earthquake Migration Patterns*, Terra Nova, *23*, 369–374.

WU, Y.M., and CHIAO, L.Y. (2006), *Seismic Quiescence before the 1999 Chi-Chi, Taiwan, M_W 7.6 Earthquake*, Bull. Seismol. Soc. Am. *96*, 321–327.

WYSS, M., BODIN, P., and HABERMANN, R.E. (1990), *Seismic Quiescence at Parkfield: an Independent Indication of an Imminent Earthquake*, Nature *345*, 426–428.

WYSS, M., WESTERHAUS, M., BERCKHEMER, H., and ATES, R. (1995), *Precursory Seismic Quiescence in the Mudurnu Valley, North Anatolian Fault Zone, Turkey*, Geophys. J. Int. *123*, 117–124.

ZWEIG, M.H., and CAMPBELL, G. (1993), *Receiver-Operating Characteristic (ROC) Plots: a Fundamental Evaluation Tool in Clinical Medicine*, Clin. Chem. *39*, 561–577.

(Received August 12, 2014, revised November 26, 2014, accepted December 15, 2014, Published online January 23, 2015)

Pure Appl. Geophys. 173 (2016), 197–203

DOI 10.1007/s00024-015-1056-3

Pure and Applied Geophysics

A Bayesian Assessment of Seismic Semi-Periodicity Forecasts

F. Nava,[1] C. Quinteros,[1] E. Glowacka,[1] and J. Frez[1]

Abstract—Among the schemes for earthquake forecasting, the search for semi-periodicity during large earthquakes in a given seismogenic region plays an important role. When considering earthquake forecasts based on semi-periodic sequence identification, the Bayesian formalism is a useful tool for: (1) assessing how well a given earthquake satisfies a previously made forecast; (2) re-evaluating the semi-periodic sequence probability; and (3) testing other prior estimations of the sequence probability. A comparison of Bayesian estimates with updated estimates of semi-periodic sequences that incorporate new data not used in the original estimates shows extremely good agreement, indicating that: (1) the probability that a semi-periodic sequence is not due to chance is an appropriate estimate for the prior sequence probability estimate; and (2) the Bayesian formalism does a very good job of estimating corrected semi-periodicity probabilities, using slightly less data than that used for updated estimates. The Bayesian approach is exemplified explicitly by its application to the Parkfield semi-periodic forecast, and results are given for its application to other forecasts in Japan and Venezuela.

Key words: Earthquake forecasting, Bayesian probability, semi-periodicity.

1. Introduction

An important approach to earthquake prediction is the search for statistical regularities in the time occurrence of large earthquakes. Indeed, a simplified application of the elastic rebound model (Reid 1910, as referenced in Richter 1958) with plate tectonics (e.g., Morgan 1968; Cox 1973) as the (constant rate) strain source, would lead one to expect periodic behavior (e.g., Lomnitz 1966; Rikitake 1976; and references therein).

However, seismic processes involve complex and highly non-linear systems featuring feedback, thusly depending heavily on its history. As such, this process involves self-organized criticality (SOC) (e.g., Bak *et al.* 1988; Bak and Tang 1989; Bak and Chen 1991; Turcotte 1992; Márquez 2012) with essentially random occurrences of small events and semi-periodic occurrences of large events.

The occurrence times of a semi-periodic sequence of K earthquakes are of the form:

$$t_k = t_0 + k\,\tau + \eta_k; \quad k = 1, \ldots, K, \qquad (1)$$

where τ is the period and η_k is a realization of a random variable such that $\eta \ll \tau$.

Many studies have searched for semi-periodicity, with mixed results. Of these, we will use as an example the one by Bakun and Lindh (1985) which predicted an earthquake in the region of Parkfield, California, USA, on the basis of recurrence times from a series of six earthquakes, and missed the occurrence time of the next earthquake by some 17 years.

Nava *et al.* (2014) and Quinteros *et al.* (2014), hereafter referred to as Paper I and Paper II, respectively, realized that in a given seismogenic region there may be more than one semi-periodic process, so that the observed seismicity may contain more than one semi-periodic sequence, and may also include events from long-period sequences that cannot be identified because of the limited observation time. As a result, they can be considered to occur randomly. Not all earthquakes that occur in a given seismogenic region are necessarily part of a seismic sequence. Paper I proposes a method to identify semi-periodic sequences in earthquake occurrence time series, assess their significance, and use the results for earthquake forecasting; it also illustrates the method by applying it to the Parkfield sequence. Paper II addresses some aspects of catalogue processing and presents applications to Japan and Venezuela; Quinteros and Nava (2013), hereafter referred to as Paper III, presents the application to a recent event in

[1] CICESE, Ensenada, Baja California, Mexico. E-mail: fnava@cicese.mx

Venezuela. We will use results from the abovementioned papers, where all details of the pertinent data are given, to apply and illustrate the proposed Bayesian estimation.

We will now briefly review the basic characteristics of the semi-periodicity forecasting method.

The time series of large earthquake occurrence times, considered as a point process in time, is

$$\tilde{t} = \{t_j;\ j = 1, \ldots, N\},\quad N \geq K;$$

from this series, a function is built as

$$f(t) = \sum_{j=j_1}^{j_2} \delta(t - t_j);$$

recognizing this function as a segment of an infinite series, the analytic Fourier transform (e.g., Bracewell 1965)

$$F(s) = \sum_{j=j_1}^{j_2} \mathrm{e}^{-i2\pi\, t_j s}$$

allows identification of dominant frequencies corresponding to semi-periodic sequences within the point process of earthquake occurrences in time. Figure 1 illustrates the Parkfield time series and the process of sequence identification.

Once a dominant frequency is identified in the spectrum (Fig. 1), a periodic sequence in time (referred to as a "comb") is built based on the identified spectral period τ and phase. Events of the time series possibly corresponding to comb "teeth" are identified and the rest are eliminated. The process is repeated three times using a stricter acceptance criterion during each pass (in the last pass, acceptable occurrence times t_k have to differ from that of the corresponding comb tooth t_k^c by less than $\tau/6$). Goodness of fit is measured as the root-mean-square (RMS) error of fit between sequence and comb

$$\sigma = \sqrt{\frac{\sum_{k=1}^{K} (t_k^c - t_k)}{K - 2}}. \tag{1}$$

For the Parkfield example, $\tau_p = 36.36$ year and $\sigma = 4.55$ year.

Based on the estimated comb and the error, a forecast can be made as

$$t_f = t_{0p} + K\tau \pm q\sigma, \tag{2}$$

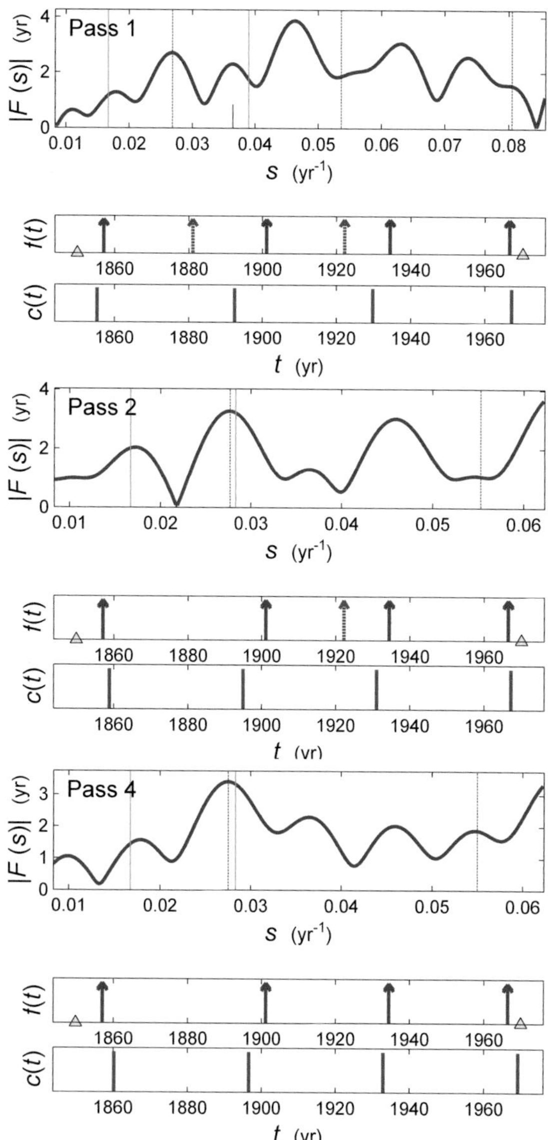

Figure 1
Example of sequence and comb determination for Parkfield (Paper I). For each pass, a section of the (absolute) spectrum is shown on top; the *vertical continuous lines* indicate the range of acceptable frequencies, and the *dotted vertical lines* indicate the chosen frequency (*left*) and some of its multiples. Below the spectra are the occurrence time series (*arrows*); *dotted arrows* show ineligible events. Below the time series the estimated comb (*vertical lines*). Pass 3 is not shown because it is exactly like pass 2

where t_f is the forecast time, t_{0p} is the time of the earliest comb tooth, determined from the phase and the period, and q is a factor that can be set to give a desired confidence interval to the forecast.

Considering $q = 2$, the forecast time of the Parkfield example is $t_f = 2005.63 \pm 9.10$ year (Paper I).

The sequence probability of non-randomness.

How significant an identified semi-periodic sequence is can be measured against the probability, P_0, of the null hypothesis that the observed sequence is a random occurrence, i.e., that earthquakes occurring with uniform probability over the observed interval $[0, T]$ could result in a sequence having K elements with RMS error $\pm q\,\sigma$, where a factor $q \geq 1$ is introduced to ensure that P_0 is not underestimated. We will now describe how this probability is estimated.

For a random occurrence with uniform probability, the distribution with the largest entropy, the probability of occurrence of n events within an interval θ is Poissonian (e.g., LOMNITZ 1994; DALEY and VERE-JONES 2002), given by

$$\Pr(n) = \frac{(\lambda\theta)^n \, e^{-\lambda\theta}}{n!},$$

where $\lambda = N/T$ is the occurrence ratio of earthquakes in the region.

In order to have exactly K elements over interval $[0, T]$, a sequence may have a period between $T/(K-1)$ and $T/K + \varepsilon$, where ε is a very small quantity introduced to ensure that no more than K elements fit within time T. We use the worst case that results in the largest random probability by considering the shortest period for which at least one event should occur, taking into account the uncertainty, within an interval of length $\Theta = T/K - \varepsilon + q\sigma$; at least one event should occur in each of $K-1$ intervals of length $\theta = 2q\sigma$. Thus, the worst-case random occurrence probability is

$$P_0 = \left(1 - e^{-\lambda\Theta}\right)\left(1 - e^{-\lambda\theta}\right)^{K-1}. \tag{3}$$

Hence, the probability that the sequence did not occur randomly is

$$P_c = 1 - P_0. \tag{4}$$

It should be mentioned that the probabilities presented here, calculated according to (3) and (4), differ slightly from those in our previous papers because, in them, P_0 was estimated approximately by a Monte-Carlo scheme.

Considering that the actual occurrence times should be distributed about the forecast time as some pdf $p(t - t_f)$ with unit area, the probability of occurrence is given by $P_c\, p(t - t_f)$, and the forecast can be represented as in Fig. 2. In this forecast, we have assumed a normal distribution $p(t) = N(t_f, \sigma)$ (see Paper I for discussion).

Actually, Fig. 2 shows an aftcast, i.e., a forecast for an event that has already occurred, based on information previous to it. The first four arrows show the identified sequence, and the curve is the forecast probability density function, while the fifth arrow indicates the occurrence of the forecast earthquake. It is evident that the actual occurrence agrees very well with the forecast, which had $P_c = 0.858$, but how exactly does the occurrence (or non-occurrence) of earthquakes after the forecast was made support or contradict the forecast and, hence, the hypothesis of semi-periodicity?

2. Bayesian Probability of Semi-Periodicity

We will now apply the Bayesian formalism to derive quantitative information from the occurrence or non-occurrence of the forecast earthquake, as well as from the accuracy of the forecast, i.e., the difference between forecast and occurrence time Δt.

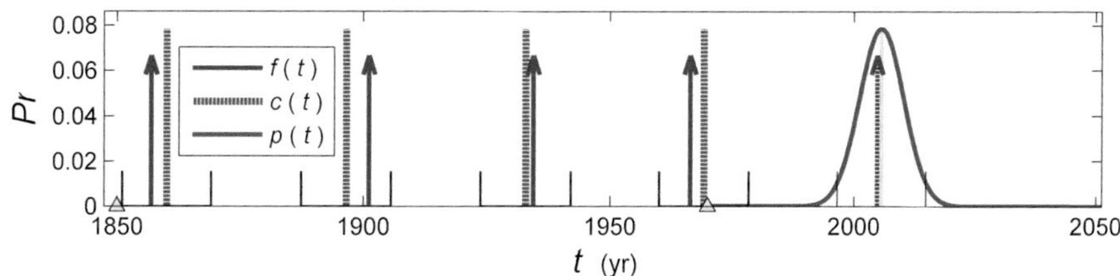

Figure 2

Forecast for the Parkfield data (Paper I). *Arrows* indicate best fitting earthquake occurrences, and *dashed vertical lines* are the comb teeth series; the curve is the forecast pdf $p(t)$. The *latest arrow* (*dotted*) marks the forecast event occurrence on $t_o = 2004.742$

2.1. Estimation of the Bayesian Probability

Let A be a semi-periodic earthquake sequence in the study region (evidence of a semi-periodic process), and let $\Pr(A)$ be the prior estimation of the probability of A.

Let B be an earthquake that occurs after the forecast has been made, at time t_0. Event B can be used to revise the probability of A by applying Bayes' formula:

$$\Pr(A|B) = \frac{\Pr(B|A)\Pr(A)}{\Pr(B|A)\Pr(A) + \Pr(B|\bar{A})\Pr(\bar{A})}, \qquad (5)$$

(e.g., Parzen 1960; Winkler 2003). We will now calculate the various probabilities needed to apply (5).

The total forecast probability, P_c, is distributed in time according to some pdf $p(t)$; thus, in order to have a finite probability for the occurrence of B at a given time, P_w, it is necessary to consider the probability of occurrence within a window of finite length w centered on the occurrence time t_0:

$$P_w = P_c \int_{t_o - w/2}^{t_o + w/2} p(t)\,\mathrm{d}t. \qquad (6)$$

The effect of the length of the time window will be discussed below. For the case $p(t) = N(t_f, \sigma)$, illustrated in Fig. 3,

$$\begin{aligned} P_w &= \frac{P_c}{\sqrt{2\pi}} \int_{t_1}^{t_2} \mathrm{e}^{-x^2/2}\mathrm{d}x \\ &= P_c\left\{\mathrm{sgn}(t_2)\,\mathrm{erf}\left(|t_2|/\sqrt{2}\right) - \mathrm{sgn}(t_1)\,\mathrm{erf}\left(|t_1|/\sqrt{2}\right)\right\} \end{aligned} \qquad (7)$$

where $t_1 = (t_f - t_o - w/2)/\sigma$ and $t_2 = (t_f - t_o + w/2)/\sigma$.

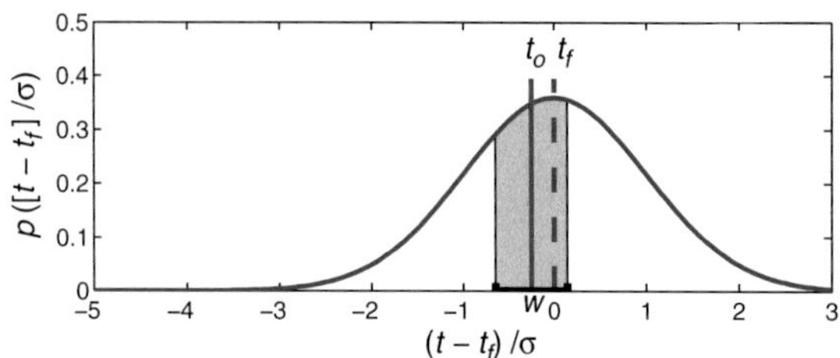

Figure 3
Forecast probability pdf $p(t)$ for normalized time

Hence, the probability of B, given A, the occurrence of at least one earthquake within window w, is the abovementioned probability of having an earthquake from the sequence, plus the Poissonian probability of having at least one earthquake that does not belong to it $\pi^*_{1+} = 1 - \mathrm{e}^{-w\,(N-K)/T}$:

$$\Pr(B|A) = P_w + \pi^*_{1+} - P_w \pi^*_{1+}, \qquad (8)$$

since $(N-K)/T$ is the occurrence ratio of earthquakes not belonging to the given sequence.

If $\bar{A}$, then the probability of B is strictly Poissonian:

$$\Pr(B|\bar{A}) = 1 - \mathrm{e}^{-wN/T} \qquad (9)$$

and, obviously,

$$\Pr(\bar{A}) = 1 - \Pr(A). \qquad (10)$$

It only remains to assign a value to $\Pr(A)$, the prior probability of there being a semi-periodic sequence. Since P_0 is the (Poissonian) probability of the observed semi-periodic sequence being due to chance, and the only alternative to this is that the semi-periodic sequence is not due to chance, it follows that

$$\Pr(A) = 1 - P_0 = P_c. \qquad (11)$$

Assuming $w = \sigma/40$ (see discussion about w below), the quantities estimated from (6) to (11) can be applied in (5).

The Parkfield earthquake series used for the semi-periodicity analysis in Paper I consists of six earthquakes with magnitudes between 6.0 and 7.9, among which a sequence of four events is identified (Fig. 2). The parameter values for this example and the results of the Bayesian appraisal are shown in the first row of Table 1; for $\Pr(A) = P_c = 0.858$ and $t_f = 2005.63$, the occurrence of the forecast event on $t_o = 2004.742$, so that $\Delta t = |t_f - t_o| = 0.888$ year yields $\Pr(A|B) = 0.952$, i.e., a probability gain (Daley and Vere-Jones 2003) $\Pr(A|B) = 0.952$.

Table 1 also shows values and results for aftcasts from Papers II and III. The aftcast of the last event in the J1 sequence in Japan is based on a sequence of $K = 3$ events among a series of $N = 8$ events with magnitudes ranging from 8.0 to 8.7; that of the last event in the R2 $K = 4$ sequence in Venezuela involves a series of $N = 13$ with magnitudes in the

Table 1

Quantities used in the Bayesian appraisal and results of the same for all examples mentioned in the text

Example	T (year)	N	K	σ (year)	t_f	P_c	t_o	Pr $(A\|B)$ P_G			P_c^U
								Pr $(A) = P_c$	Pr $(A) = 0.5$	Pr $(A) = 0.1$	
Parkfield	120	6	4	4.55	2005.630	0.858	2004.742	0.952 1.110	0.767 1.534	0.268 2.676	0.952
Japan J1	90	8	3	0.228	2003.090	0.994	2003.734	0.996 1.002	0.575 1.150	0.131 1.306	0.997
Venezuela R2	196	13	4	1.41	2010.230	0.967	2009.951	0.996 1.031	0.893 1.785	0.480 4.800	0.994
Venezuela 2013	91	20	6	1.525	2014.670	0.785	2013.778	0.892 1.137	0.694 1.389	0.202 2.015	0.861

The updated comb probability P_c^U is also shown for comparison.

5.6–7.4 range. Finally, the aftcast of the recent 2013 Venezuela $M = 6.5$ earthquake is based on a $K = 6$ event sequence among an $N = 20$ event series with magnitudes ranging from 5.9 to 6.9.

Bayes formalism allows testing other prior estimates or suppositions about possible semi-periodicity. Someone who does not like condition (11) and says that semi-periodicity may or may not occur would use Pr $(A) = 0.5$, and, for our Parkfield example, would obtain Pr $(A|B) = 0.767$ ($P_G = 1.534$). Someone who is skeptical and believes there is only a small probability of there being semi-periodicity could use, say, Pr $(A) = 0.1$, and would obtain Pr $(A|B) = 0.268$ ($P_G = 2.676$), a probability not large enough to be conclusive in favor of semi-periodicity, but certainly suggestive of it. Results for these hypothetical choices for the other aftcasts are shown in Table 1; note that, for our examples, the smaller prior probabilities result in larger probability gains. Of course Bayesian reasoning is useless for someone who firmly believes that semi-periodicity cannot exist, so that Pr $(A) = 0$, because the question becomes a matter of faith.

2.2. *Bayesian Probability and the Length of w*

Since forecasts are given as probability distribution functions in time, in order to work with finite probabilities, it is necessary to consider probabilities over some finite time interval, which we have called w. Probabilities (6–11) all increase with w, but since both the numerator and the denominator increase in (5), for $w \leq \sigma/4$, the results change only in the fourth decimal place, and tend to a limit for small w. Thus, for practical purposes, results for $w \leq \sigma/10$ can be considered independent of w. We used $w = \sigma/40$ in the results presented here.

2.3. *Bayesian Probability and Forecast Accuracy*

Among the probabilities used to calculate Pr $(A|B)$, only P_w depends on the time difference between the forecast time t_f and the actual occurrence time t_0, and attains its maximum when both times coincide, i.e., when the forecast is exact.

All aftcasts we have made so far are good enough such that the Bayesian estimates give enhanced semi-periodicity probabilities. However, this might not have been the case had the forecast earthquake occurred at a time very different from that of the forecast.

Figure 4 illustrates the behavior of Pr $(A|B)$ as a function of $\Delta t = |t_f - t_o|$ for the Parkfield example with Pr $(A) = P_c$ (top) and Pr $(A) = 0.5$ (bottom). In both cases, the Bayesian probability is maximal for $\Delta t = 0$, but there also is a Δt above which the Bayesian probability is smaller than the prior one, and clearly tells that, above this value, event B is evidence against semi-periodicity. Note that the smaller the prior estimate, the better the coincidence between actual and forecast times has to be in order to be convincing.

A common problem in evaluating forecasting performance is that it is sometimes difficult to decide whether a given event fulfills a forecast or not, particularly if it occurs with relatively low

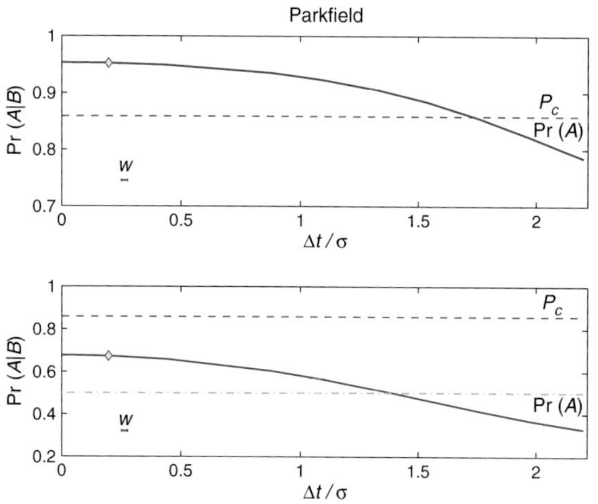

Figure 4
Dependence of the Bayesian probability on the difference between forecast and occurrence times Δt, for $\Pr(A) = P_c$ (*top*) and $\Pr(A) = 0.5$ (*bottom*). The *diamond indicates* the (normalized) Δt for the Parkfield example. The *bar* labeled *w* shows the window length used to compute probabilities

probability. Everyone knows of cases where forecasters claim that some event remotely resembling their forecast is an actual fulfillment. This problem need not arise for semi-periodicity forecasts when viewing each new occurrence in the light of Bayesian estimates, because any event resulting in $P_G < 1$ is clearly not fulfilling the forecast, so that P_G is a good estimator of how well an event fulfills a forecast.

2.4. *Bayesian vs. Updated Probabilities*

After a forecast (or aftcast) has been made, the occurrence of an event allows reevaluation of the prior probability, but it also results in a new (or updated) sequence of $K + 1$ earthquakes. Conditions are not the same for the new sequence: the number of earthquakes in the sequence has increased by one, but other "unrelated" events may have also occurred, i.e., N may be larger by more than one, and the total time T has increased by about one sequence period. Based on the new conditions, a new, updated, sequence probability P_c^U can be evaluated (the probability that would be used for a new forecast, and it is interesting to compare the updated probabilities for our examples with the corresponding Bayesian probabilities.

Table 1 shows that, for our aftcast examples, the Bayesian $\Pr(A|B)$ estimates for $\Pr(A) = P_c$ agree extremely well with the corresponding updated P_c^U values. We believe that two conclusions are derivable from this agreement: first, that the choice of P_c as the prior probability is correct; and second, since longer sequences involving larger numbers of earthquakes that occurred semi-periodically are, naturally, better and more convincing evidence of semi-periodic behavior; the Bayesian formalism does a very good job of estimating corrected semi-periodicity probabilities, using slightly less data than that used for updating.

3. *Discussion and Conclusions*

The real measure of the goodness of a forecast is whether an earthquake occurs around the forecast time and, given the occurrence, how small the absolute difference is between the forecast and the actual occurrence times (judged in terms of the period and the standard deviation of the semi-periodic sequence). Hence, the Bayesian estimation of the probability of a sequence being semi-periodic, given the occurrence of a given earthquake, is a measure of both whether the earthquake may be considered to fulfill the forecast and, if so, of the forecast goodness.

For the examples shown here, the agreement of the Bayesian estimates based on the non-randomness probability of the sequence used for the forecast, P_c, with the updated non-randomness probability of the new, longer sequence suggests that P_c is a good estimator of the probability of existence of a semi-periodic sequence.

The Bayesian estimates, together with the updated non-randomness probabilities, are a good basis upon which to support or reject the existence of semi-periodic sequences in earthquake occurrence time series.

Acknowledgments

Our sincere thanks to two anonymous reviewers and to Pageoph guest editor Matthew Gerstenberger. This project was partially funded by CONACYT scholarship 242919 (C. Quinteros) and CONACYT grant 222795, and was partially carried out within project CGL2011-29474-C01-01.

References

Bak, P., Tang, C., and Wiesenfeld, K. (1988) *Self-organized criticality*. Phys. Rev. A *38*(1), 364–374.

Bak, P., and Tang, C. (1989) *Earthquakes as a self-organized critical phenomenon*. J. Geophys. Res. *94*(B1), 15635–15637.

Bak, P., and Chen, K. (1991) *Self-organized criticality*. Sci. Am. 26–33.

Bakun, W., and Lindh A. (1985) *The Parkfield, California earthquake prediction experiment*. Science *229*, 4714, 619–624.

Bracewell, R., The Fourier transform and its applications. (McGraw-Hill Book Co., USA, 381 pp., 1965).

Cox, A. (1973) Plate tectonics and geomagnetic reversals. (W. H. Freeman and Co.).

Daley, D., Vere-Jones, D. (2002) An introduction to the theory of point processes. Vol. I. Springer, USA, 469 pp.

Lomnitz, C. (1966) *Statistical prediction of earthquakes*. Reviews of Geophysics *4*, 377–393.

Lomnitz, C., Fundamentals of earthquake prediction. (JohnWiley & 673 Sons Inc., USA, 326 pp. 1994).

Márquez, V. (2012) Multifractal analysis of the spatial distribution of seismicity and its possible prmonitory application. Exploration of a possible mechanism for fractality through semi-stochastic modeling. D.Sc. Thesis, CICESE, 31/08/2012 (in Spanish).

Morgan, W. (1968). *Rises, trenches, great faults, and crustal blocks*. J. Geophys. Res. *73*(6), 1959–1982.

Nava, F., Quinteros, C., Glowacka, E., Frez, J. (2014) *Semi-periodic sequences and extraneous events in earthquake forecasting: I. Theory and method, Parkfield*. Pageoph *171*(7), 1355–1366, doi:10.1007/s00024-013-0679-5.

Parzen, E. (1960) *Modern probability theory and its applications*. John Wiley & Sons, Inc., Japan, 464 pp.

Quinteros, C., Nava, F., Glowacka, E., Frez, J. (2014) *Semi-periodic sequences and extraneous events in earthquake forecasting: II Application, forecasts for Japan and Venezuela*. Pageoph *171*(7), 1367–1383, doi:10.1007/s00024-013-0678-6.

Quinteros, C., Nava, F. (2013) *Postnóstico (pronóstico hecho a posteriori) del sismo del 11 de octubre de 2013 en Venezuela, mediante análisis de semiperiodicidad*. GEOS ***33*** (2), 350-355..

Reid, H. (1910) The mechanics of the earthquake, the California earthquake of April 18, 1906. Report of the State Investigation Commission, Vol. 2, Carnegie Institution of Washington, Washington, D.C.

Richter, C., Elementary Seismology, (W. H. Freeman and Co., USA, 768 pp. 1958).

Turcotte, D., Fractals and Chaos in Geology and Geophysics. (Cambridge University Press, Second edition, New York. 221 pp, 1992).

Winkler, R. An introduction to Bayesian inference and decision. 2nd Edition, (Probabilistic Publishing, USA, 452 pp. 2003).

(Received July 29, 2014, revised January 15, 2015, accepted February 11, 2015, Published online February 25, 2015)

Pure Appl. Geophys. 173 (2016), 205–220
© 2015 Springer Basel
DOI 10.1007/s00024-015-1078-x

Pure and Applied Geophysics

The Determination of Earthquake Hazard Parameters Deduced from Bayesian Approach for Different Seismic Source Regions of Western Anatolia

YUSUF BAYRAK[1] and TUĞBA TÜRKER[2]

Abstract—The Bayesian method is used to evaluate earthquake hazard parameters of maximum regional magnitude ($M_{\max}$), β value, and seismic activity rate or intensity (λ) and their uncertainties for the 15 different source regions in Western Anatolia. A compiled earthquake catalog that is homogenous for $M_s \geq 4$ was completed during the period from 1900 to 2013. The computed $M_{\max}$ values are between 6.00 and 8.06. Low values are found in the northern part of Western Anatolia, whereas high values are observed in the southern part of Western Anatolia, related to the Aegean subduction zone. The largest value is computed in region 10, comprising the Aegean Islands. The quantiles of functions of distributions of true and apparent magnitude on a given time interval [0,*T*] are evaluated. The quantiles of functions of distributions of apparent and true magnitudes for future time intervals of 5, 10, 20, 50, and 100 years are calculated in all seismogenic source regions for confidence limits of probability levels of 50, 70, and 90 %. According to the computed earthquake hazard parameters, the requirement leads to the earthquake estimation of the parameters referred to as the most seismically active regions of Western Anatolia. The Aegean Islands, which have the highest earthquake magnitude (7.65) in the next 100 years with a 90 % probability level, is the most dangerous region compared to other regions. The results found in this study can be used in probabilistic seismic hazard studies of Western Anatolia.

Key words: Bayesian method, earthquake hazard parameters, quantiles, Western Anatolia.

1. Introduction

The purpose of a seismic hazard study, using available data related to earthquake events, is to determine the specific probability values for seismic activity in a region in the future and combines geological, seismological, and statistical data with other information. Seismic hazard studies are undertaken to obtain long-term predictions of the occurrences of seismic events in a particular region. Most often, the prediction is expressed in the form of probabilities of a specified earthquake magnitude over a period of time, *t*, or as the expected number of such events. Thus, the number of events over time [0,*T*] and *M* define the size of the events; then, that probability is expressed as $P\{M \geq m, (0, T)\}$ (ANAGNOS and KIREMIDJIAN 1988).

Reliable estimation of earthquake hazard in a region requires the prediction of the size and magnitude of future earthquake events and their locations. An incomplete understanding of earthquake phenomena, however, has led to the development of primarily long-term hazard assessment tools relying on the statistical averages of earthquake occurrences without considerations of specific models. One of the most important earthquake hazard parameters is the maximum regional magnitude ($M_{\max}$) and its uncertainty. In addition to $M_{\max}$, two other important earthquake hazard parameters are β value and seismic activity rate or intensity (λ). The "apparent" magnitude (TINTI and MULARGIA 1985; KIJKO and SELLEVOLL 1992), which represents the observed magnitude ($M_{\max}^{obs}$), is equal to the "true" magnitude *M*, plus an uncertainty, ε. The probability distribution of this uncertainty can be modeled by various distribution functions.

A number of statistical techniques and probabilistic models have been already used to estimate earthquake hazard parameters by various researchers. (WELLS and COPPERSMITH 1994; PISARENKO *et al.* 1996; KIJKO 2004; WHEELER 2009; MUELLER 2010). BAĞCI (1996) investigated seismic risk in Western Anatolia between 36° and 41°N and 25° and 31°E using the probabilistic model for earthquake data (1930–1990). ALTINOK

[1] Agri Ibrahim Cecen University, Agri, Turkey. E-mail: bayrak@ktu.edu.tr; ybayrak@agri.edu.tr
[2] Department of Geophysics, Karadeniz Technical University, Trabzon, Turkey. E-mail: tturker@ktu.edu.tr

(1991) evaluated the seismic risk of Western Anatolia by applying a probabilistic model. Bayrak and Bayrak (2012, 2013) investigated earthquake hazard potential using different methods for different regions in Western Anatolia.

Earthquakes have posed a persistent threat to life and property in many regions of the world. In Turkey's Anatolia region, records of devastating earthquakes can be found. A geodynamic complexity and a diversity of faulting regimes can be seen around the Aegean. The western part of the Anatolian plate is one of the most seismically active regions of Turkey. Western Anatolia has seen numerous earthquakes during past years. The consequences of large earthquakes across the globe are a primary motivation for understanding seismic hazard. Particular consideration is given to the appraisal of seismic hazard in the context of Aegean seismotectonics.

The Bayesian method has a special interest that comes from its ability to take into consideration the uncertainty of parameters in fitted probabilistic laws and a priority given to information (Morgat and Shah 1979; Campell 1982, 1983). The advantages of the method used are in its simplicity; it does not require such intermediate steps of investigation as earthquake scenarios, estimates of bimodal recurrence model of magnitude distribution, and bootstrap procedures (Lamarre *et al.* 1992). Rather, the method is straightforward and needs only a seismic catalog and seismological information.

We applied a procedure developed by Pisarenko *et al.* (1996) in order to examine earthquake hazard for the 15 different regions of Western Anatolia. For this purpose, earthquake hazard parameters ($M_{\max}$, β value, and activity rate or intensity λ) and their uncertainties are computed. In addition, the quantiles of $M_{\max}$ probabilistic distribution in future time intervals of 5, 10, 20, 50, and 100 years are estimated.

2. *The Tectonics of Western Anatolia*

The Aegean Arc and Western Anatolian Extension Zone play important roles in the geodynamic evolution of the Aegean region and Western Anatolia. Although the North Anatolian Fault is the largest fault system outside of the system, the Aegean Region is observed to commonly experience earthquake movement and is one of the regions with the most rapidly changing shapes in the world (Kahraman *et al.* 2007). The tectonics of the Aegean region and Western Anatolia have been investigated by a number of researchers (Le pichon and Angelier 1979; Şengör 1987; Barka and Reilinger 1997; Seyzitoğlu and Scott 1992, 1996; Koçyiğit *et al.* 1999; Şalk and Sari 2000; Kahraman *et al.* 2007). A number of tectonic and seismotectonic models have been investigated to determine the seismogenic structure of Western Anatolia (Dewey 1988; Seyitoğlu and Scott 1992, 1996; Koçyiğit *et al.* 1999), and researchers have found that the region has a complex structure (Blumenthal 1962; Brunn *et al.* 1971, Poisson 1984, 1990; Marcoux 1987; Kissel *et al.* 1993; Frizon *et al.* 1995). The structures in Western Anatolia have developed in the directions of NW–SE, NE–SW, N–S, and E–W, and they are oriented in the form of four separate block faults; these structures are called "cross-graben" formations (Şengör *et al.* 1985; Şengör 1987). The area is currently experiencing an approximately N–S continental extension at a rate of 30–40 mm/year (Oral *et al.* 1995; Le pichon *et al.* 1995). The Anatolian plate rotates counterclockwise with an average velocity of 24 mm/year (Mcclusky *et al.* 2000).

Western Anatolia has developed several graben trending E–W and WSW–ESE, depending on the N–S directional extension tectonics (Dewey and Şengör 1979; Jackson and Mckenzie 1984; Şengör 1982; Şengör *et al.* 1984). The Aegean Graben System (for example, Küçük Menderes, Büyük Menderes, Gediz, Bakırçay, Simav, Gökova, Kütahya, and Edremit Grabens) generally occurred on E–W trending normal faults and is located trending E–W on a number of graben blocks (Bozkurt and Sözbilir 2004; Yilmaz *et al.* 2000; Dewey and Şengör 1979; Seyitoğlu *et al.* 1992).

The eastern part of the region studied includes the NW–SE trending Beyşehir, Dinar, and Akşehir-Afyon Grabens and the NE–SW trending Burdur, Acıgöl, Sandıklı Çivril, and Dombayova Grabens and their bounding faults (e.g., Bozkurt 2001). The Büyük Menderes Graben is located between the Aegean and Denizli and is approximately 200-km long. The eastern end of the graben intersects Pamukkale around the Gediz graben (Ambraseys and Finkel 1995). Western Anatolia corresponds with the normal strike-slip component of NE–SW lines, for example, the Fethiye-Burdur Fault Zone, the Tuzla Fault, and the Bergama Foça Fault. Normal NW–SE faults are located in Southwestern Anatolia. The normal component of the Fethiye–Burdur Fault Zone is a left-lateral strike-slip fault. This fault system is a process of the northeastern Pliny-Strabo system forming the eastern flank of the Aegean Arc (Dumont *et al.* 1979; Şaroğlu *et al.* 1987; Price and Scott 1994). The E–W trending Gediz, Büyük Menderes, and Küçük Menderes Faults are located in the central region of Western Anatolia. The Simav, Kütahya, and Eskişehir Faults north of these faults show similar features. The Eskişehir Fault is a WNW–ESE trending fault and is found in the east between Bursa and Afyon. The normal component has a right lateral movement (Şaroğlu *et al.* 1987). The NE–SW basin is located among S–W and WN–ESE trending normal faults. NE–SW basins are located south of the Izmir Graben, and these trending faults are active. In addition, several NNE–SSW-trending strike-slip faults and N–S-striking active normal faults such as the Bergama-Zeytindağ Fault Zone and the Orhanlı Fault Zone are located in the region (Sözbilir 2002). The Orhanlı Fault Zone is the most continuously traceable fault. Other potentially active faults are the İzmir Fault trending in an E–W direction and the Manisa Fault near Manisa city (Bozkurt and Sözbilir 2006). The Gökova Fault must be traced on a line trending in an E–W direction along the northern coast of Gökova Bay in the southern part of the Western Anatolian zone. The Karaburun-Gulbahce Fault occurs in the Karaburun Peninsula and is believed to be predominantly a strike-slip fault (Şaroğlu *et al.* 1992; Ocakoğlu *et al.* 2004, 2005; Aktuğ and Kiliçoğlu 2006).

3. *Seismogenic Source Regions and Data*

The database used in this work was compiled from several different sources and catalogs such as IRIS (2013), the Incorporated Research Institutions for Seismology (TURKNET 2013), the International Seismological Centre (ISC), and the Scientific and Technological Research Council of Turkey (TUBİTAK); data are provided in different magnitude scales. The catalogs include different magnitude scales (M_b body wave magnitude, M_s surface wave magnitude, M_L local magnitude, M_D duration magnitude, and M_W moment magnitude), origin times, epicenters, and depth information of earthquakes.

An earthquake data set used in seismicity or seismic hazard studies must certainly be homogenous; in other words, it is necessary to use the same magnitude scale. However, the earthquake data obtained from different catalogs have been reported in different magnitude scales. Therefore, all earthquakes must be redefined in the same magnitude scale. Bayrak *et al.* (2009) developed several relationships among different magnitude scales (M_b body wave magnitude, M_s surface wave magnitude, M_L local magnitude, M_D duration magnitude, and M_W moment magnitude) in order to prepare a homogenous earthquake catalog from different data sets. The size of earthquakes that occurred before 1970 are given M_s scale in the catalogs compiled in this study. Only, the magnitudes of earthquakes that occurred after 1970 are converted to $M_{s.}$ Finally, we prepared a homogenous earthquake data catalog for M_s magnitude using relationships, and we have considered only the instrumental part of the earthquake catalog (1900–2013) for the Bayesian method.

In order to evaluate earthquake risk and/or hazard of a region, foreshocks and aftershocks should be extracted from earthquake catalogs. In other words, it is necessary to decluster the catalogs. In this study, we used the Reasenberg (1985) algorithm which uses interaction zones in space in time to link earthquakes into clusters to decluster the homogenous catalog.

The method is applied in the Western Anatolian region where a vast variation of seismicity and tectonics is observed throughout the region. In this study, we used the regions defined by Bayrak and Bayrak (2012). They divided Western Anatolia into 15 seismic regions on the basis of seismicity, tectonics, and the focal mechanism of earthquakes in order to develop a detailed analysis of seismic hazard in the region with an updated and more reliable earthquake catalog. The regions shown in Fig. 1 are as follows:

1. Region: Aliağa Fault.
2. Region: Akhisar Fault.
3. Region: Eskişehir, İnönü Dodurga Fault Zones.

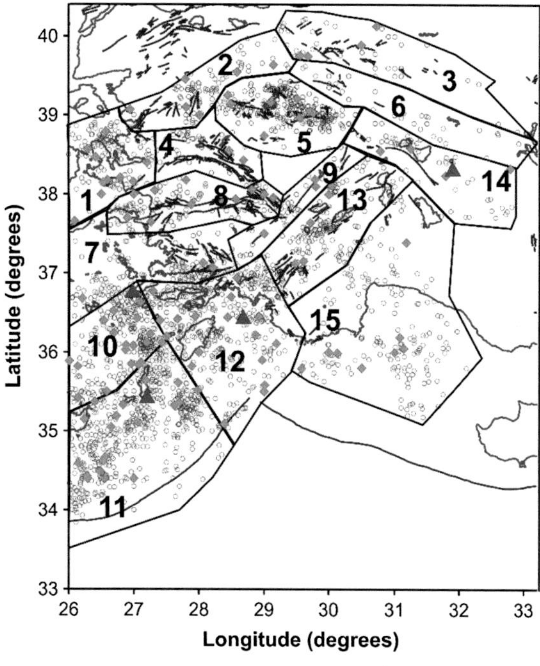

Figure 1
Delineation of the 15 different source regions of Western Anatolian the basis of seismicity, tectonics, and focal mechanism of earthquakes. The epicentral distribution of earthquakes of $M_s \geq 4$ occured during the period 1900–2013 is also shown with the *different symbols*

4. Region: Gediz Graben.
5. Region: Simav, Gediz-Dumlupınar Faults.
6. Region: Kütahya Fault Zone.
7. Region: Karova-Milas, Muğla-Yatağan Faults.
8. Region: Büyük Menderes Graben.
9. Region: Dozkırı-Çardak, Sandıklı Faults.
10. Region: Aegean Islands.
11. Region: Aegean Arc.
12. Region: Aegean Arc, Marmaris, Köyceğiz, Fethiye Faults.
13. Region: Gölhisar-Çameli, Acıgöl, Tatarlı Kumdanlı Faults, Dinar Graben.
14. Region: Sultandağı Fault.
15. Region: Kaş and Beyşehirgölü Faults.

4. Method

The technique that was used is described in detail in papers about the method (Pisarenko *et al.* 1996; Pisarenko and Lyubushin 1999; Tsapanos *et al.* 2001, 2002; Lyubushin *et al.* 2002; Tsapanos and Christova 2003; Tsapanos 2003, Lyubushin and Parvez 2010; Yadav *et al.* 2012). A brief description of the method is given below.

Let R be the value of magnitude (M), which is a measure of the size of earthquakes that occurred in a sequence on a past-time interval $(-\tau,o)$:

$$\vec{R}^{(n)} = (R_1,\ldots,R_n), \quad R_i \geq R_0, R_t = \max(R_1,\ldots,R_n), \quad 1 \leq i \leq n, \tag{1}$$

where $i = 1, 2, \ldots, n$ and R_0 is the minimum cutoff value of magnitudes (M), i.e., defined by possibilities of registration system, or it may be a minimum value from which the value written in Eq. (1) is statistically representative.

Two main assumptions for Eq. (1) were proposed. The first assumption is that Eq. (1) follows the G–R law of distribution:

$$\text{Prob}\{R<r\} = F\left(\frac{x}{R_o}, \rho, \beta\right) = \frac{e^{-\beta R_o} - e^{-\beta_x}}{e^{-\beta R_o} - e^{-\beta_\rho}}, \quad R_o \leq x \leq \rho. \tag{2}$$

Here, ρ is the unknown parameter that represents the maximum possible value of R, for instance, 'maximum regional magnitudes (M)' in a given seismogenic region. The unknown parameter b is the 'slope' of the Gutenberg–Richter law of magnitude–frequency relationship at small values of x when the dependence (Eq. 2) is plotted on double logarithmic axes.

The second assumption is that λ is an unknown parameter and a Poisson process with some activity rate or intensity λ in the sequence (Eq. 1). If three unknown parameters (ρ, β, and λ) can be written, the full vector is

$$\theta = (\rho, \beta, \lambda). \tag{3}$$

Apparent magnitude is a magnitude that is observed, i.e., those values that are presented in seismic catalogs. True magnitude is a hidden value and is unknown; it is defined by the formula

$$\bar{R} = R + \varepsilon. \tag{4}$$

Let $(x|\delta)$ be a density of probabilistic distribution of error ε where δ is a given scale parameter of the density and epsilon (ε) value is the error between the true magnitude (R) and the apparent magnitude ($\bar{R}$). We can estimate values of true magnitude taking into account different hypotheses about the probability distribution of epsilon (for example, uniform) and about parameters of this distribution. Below, we shall use the following uniform distribution density:

$$n(x|\delta) = \frac{1}{2\delta}, |x| \leq \delta \qquad n(x|\delta) = 0, |x| > \delta. \quad (5)$$

Let Π be a priori uncertainty domain of values of parameters θ

$$\Pi = \{\lambda_{\min} < \lambda \leq \lambda_{\max}, \beta_{\min} \leq \beta \leq \beta_{\max}, \rho_{\min} \leq \rho \leq \rho_{\max}\}. \quad (6)$$

We should consider the a priori density of the vector θ to be uniform in the domain Π..

According to the definition of conditional probability, α-posteriori density of distribution of vector of parameters θ is equal to

$$f\left(\theta|\vec{R}^{(n)}, \delta\right) = \frac{f\left(\theta, \vec{R}^{(n)}|\delta\right)}{f\left(\vec{R}^{(n)}|\delta\right)}. \quad (7)$$

But $f\left(\theta|\vec{R}^{(n)}, \delta\right) = f\left(\vec{R}^{(n)}|\theta, \delta\right) \times f^{a}(\theta)$, where $f^{a}(\theta)$ is the α priori density of the distribution of vector θ in the domain π. As $f^{a}(\theta) = \text{const}$ according to our assumption and taking into consideration that

$$f\left(\vec{R}^{(n)}|\delta\right) = \int_{\pi} f\left(\overset{\rightarrow(n)}{R}\middle|\theta, \delta\right) d\theta. \quad (8)$$

Then, we will obtain using a Bayesian formula (Rao 1965). The Bayesian formula is as follows:

$$F\left(\theta|\vec{R}^{(n)}, \delta\right) = \frac{f\left(\theta|\vec{R}^{(n)}, \delta\right)}{\int_{\pi} f\left(\overset{\rightarrow(n)}{R}\middle|V, \delta\right) dV}. \quad (9)$$

An expression for the function $\left(f\left(\overset{\rightarrow(n)}{R}\middle|\theta, \delta\right)\right)$ should be used in Eq. (9).

In order to use Eq. (9), we must have an expression for the function $f\left(\overset{\rightarrow(n)}{R}\middle|\theta, \delta\right)$. With the assumption of Poissonian character sequence in Eq. (1), and independent of its members, should give us

$$f\left(\vec{R}^{(n)}|\theta, \delta\right) = \bar{f}(R_1|\theta, \delta) \ldots \bar{f}(R_n|\theta, \delta) \times \frac{\exp(-\lambda(\theta, \delta)\tau) \times \left(-\bar{\lambda}(\theta, \delta)\tau\right)^{n}}{n!}. \quad (10)$$

Now, we can compute a Bayesian estimate of vector θ:

$$\theta\left(\vec{R}^{(n)}|\delta\right) = \int_{\pi} Vf\left(V|\overset{\rightarrow(n)}{R}, \delta\right) dV. \quad (11)$$

An estimate of maximum value, ρ, is one of the computations of (Eq. 11). We must obtain Bayesian estimates of any of the functions to use a formula analogous to Eq. (11).

One of the computations in (Eq. 11) contains an estimate of maximum value of ρ. Using a formula analogous to Eq. (11), we must obtain Bayesian estimates for any of the functions. The most important are estimates of quantiles of distribution functions of true and apparent values on a given future time interval [0,T], for instance for α quantiles of apparent values

$$\hat{\bar{Y}}\left(\alpha|\vec{R}^{(n)}, \delta\right) = \int_{\pi} \bar{Y}_T(\alpha|V, \delta) \times f\left(V|\overset{\rightarrow(n)}{R}, \delta\right) dV, \quad (12)$$

$\hat{Y}_T\left(\delta|\overset{\rightarrow(n)}{R}, \delta\right)$, for α quantiles for true values is written analogously to Eq. (12). We must estimate variances of Bayesian estimates (Eqs. 11, 12) using averaging over the density (Eqs. 9, 10). For example:

$$\mathrm{Var}\left\{\hat{\bar{Y}}_T\left(\alpha|\vec{R}^{(n)}, \delta\right)\right\} = \int_{\pi} \left(\bar{Y}_T(\alpha|V, \delta) - \hat{\bar{Y}}_T\left(\alpha|\vec{R}^{(n)}, \delta\right)\right)^2 \times f\left(V|\overset{\rightarrow(n)}{R}, \delta\right) dV. \quad (13)$$

First of all, we will set $\rho_{\min} = R_{\tau} - \delta$. As for the values of $\rho_{\max}$, they depend on the specific data in the series (Eq. 1) and are produced by the user of the method. Boundary values for the slope β are estimated by the formula

$$\beta_{\min} = (\beta_0 \cdot (1 - \gamma)), \beta_{\max} = \beta_0 \cdot (1 + \gamma), \quad 0 < \gamma < 1. \quad (14)$$

β_0 is the "central" value and is obtained as the maximum likelihood estimate of the slope for the Gutenberg–Richter law

$$\sum_{i-1}^{n} \ln\left\{\frac{\beta e^{-\beta R_i}}{e^{-\beta R_0} - e^{-\beta R_\tau}}\right\} \to \text{Max}; \beta, \beta\varepsilon(0, \beta_s), \quad (15)$$

where β_s is a rather large value.

For setting boundary values for the t activity rate or intensity (λ) in Eq. (6), we used the following rationale. As a consequence of normal approximation for a Poisson process for a rather large n (Cox and Lewis 1966), the standard deviation of the value $\lambda\tau$ has the approximation value $\sqrt{n} = \sqrt{\lambda\tau}$. Thus, taking boundaries at $\pm 3\sigma$, we will obtain

$$\lambda_{\min} = \lambda_o\left(1 - \frac{3}{\sqrt{\lambda_o\tau}}\right), \lambda_{\max} = \lambda_0\left(1 + \frac{3}{\sqrt{\lambda_0\tau}}\right)$$
$$\lambda_0 = \frac{\bar{\lambda}_0}{c_f(\beta_0, \delta)}, \bar{\lambda}_0 = \frac{n}{\tau}. \quad (16)$$

5. *Results and Discussion*

Earthquake hazard parameters (maximum regional magnitude $M_{\max}$, β value, and activity rate λ) have been estimated in the examined area using Bayesian statistics provided by Pisarenko *et al.* (1996) and a homogenous and complete seismic catalog of $M_s \geq 4$ during the period 1900–2013. The reliability of the estimation of hazard parameters (β value and activity rate or intensity λ) depends upon the time period covered by the instrumental catalog. The Bayesian method requires a priori distribution of unknown parameters, but the a priori distribution is negligible for a large sample. There is an advantage of this method, in that it considers magnitude uncertainties as well in the computation of hazard parameters. There is no priori advantage in using normal or Gaussian distributions, such as Kijko and Sellevoll (1992) did for the estimation of error in magnitudes as also observed by Pisarenko and Lyubushin (1997) and Tsapanos *et al.* (2001). Therefore, uniform distribution is applied in this analysis. We used the software compiled by Pisarenko and Lyubushin (1997).

The Bayesian approach is a more time consuming method (Pisarenko *et al.* 1996) but provides more stable results than unbiased approaches. With this purpose, we have also tabulated the maximum observed magnitude ($M_{\max}^{\text{obs}}$) in Table 1 with other parameters. In this study, the estimated maximum regional magnitudes are in quite good agreement with the maximum observed magnitudes and their differences. The maximum regional magnitude ($M_{\max}$) estimated by this method is comparable and more reliable than the other estimates obtained by different approaches. The close agreement between estimated $M_{\max}$ and $M_{\max}^{\text{obs}}$ validates the high quality of data used and appropriateness of the adopted cutoff magnitude.

Table 1

The estimates of the Bayesian analysis for the 15 different source regions of Western Anatolia

Region	Region adı	N	$M_{\max} \pm \sigma_{M\max}$	$M_{\max}^{\text{obs}}$	$\beta \pm \sigma_\beta$	$\lambda \pm \sigma_\lambda$
1	Aliağa Fault	129	7.29 ± 0.58	6.6	1.84 ± 0.17	0.31 ± 0.27
2	Akhisar Fault	51	7.46 ± 0.57	6.6	2.16 ± 0.30	0.12 ± 0.17
3	Eskişehir, İnönü Dodurga Fault Zones	48	7.15 ± 0.64	6.4	1.70 ± 0.27	0.11 ± 0.16
4	Gediz Graben	38	7.04 ± 0.79	5.9	2.35 ± 0.38	0.11 ± 0.18
5	Simav, Gediz-Dumlupınar Faults	331	7.01 ± 0.72	6.2	2.66 ± 0.15	0.12 ± 0.66
6	Kütahya Fault Zone	29	6.00 ± 0.88	5.3	1.81 ± 0.37	0.85 ± 0.15
7	Karova-Milas, Muğla-Yatağan Faults	172	7.33 ± 0.61	6.5	2.13 ± 0.16	0.42 ± 0.31
8	Büyük Menderes Graben	95	7.53 ± 0.52	6.8	2.01 ± 0.21	0.22 ± 0.22
9	Dozkırı-Çardak, Sandıklı Faults	52	7.17 ± 0.67	6.3	1.74 ± 0.26	0.15 ± 0.20
10	Aegean Islands	292	8.06 ± 0.25	7.7	1.99 ± 0.11	0.82 ± 0.47
11	Aegean Arc	530	7.69 ± 0.43	7.1	2.11 ± 0.93	0.13 ± 0.58
12	Aegean Arc, Marmaris, Köyceğiz, Fethiye Faults	413	7.68 ± 0.43	7.1	2.02 ± 0.10	0.11 ± 0.55
13	Gölhisar-Çameli, Acıgöl, Tatarlı Kumdanlı Faults, Dinar Graben	123	7.66 ± 0.48	6.9	3.08 ± 0.27	0.32 ± 0.28
14	Sultandağı Faults	46	7.71 ± 0.45	7.0	2.25 ± 0.33	0.11 ± 0.16
15	Kaş ve Beyşehirgölü Faults	187	7.50 ± 0.52	6.8	1.95 ± 0.14	0.49 ± 0.35

The reliability of the estimation of hazard parameters (β value and activity rate or intensity λ) depends upon the time period covered by the instrumental catalog.

The $M_{\max}$ values computed using the Bayesian method are listed in Table 1 and the map showing them is in Fig. 2 for the 15 different regions of Western Anatolia. $M_{\max}$ values vary between 6.00 and 8.06. The lowest $M_{\max}$ value ($M_{\max}^{\text{Bayes}} = 6.00$) is estimated for the Kütahya Fault Zone region. The second group of $M_{\max}$ values varying between 6.5 and 7.5 and is estimated in the regions of 1, 2, 3, 4, 5, 7, and 9, related to the northern part of Western Anatolia. The third group of $M_{\max}$ values varies between 7.5 and 8.0 and is observed in the regions of 8, 11, 12, 13, 14, and 15, related to the southern part of Western Anatolia. Earthquakes larger than 6.8 (Table 1) have occurred in these regions, including the Büyük Menderes Graben, the Aegean Arc, Marmaris, the Köyceğiz-Fethiye Faults, the Tatarlı-Kumdanlı Faults, Dinar Graben, the Sultandağı Fault, and the Kaş-Beyşehirgölü Faults. The largest earthquake in these regions was observed in regions 11 and 12 as 7.1. The highest $M_{\max}$ values (8.06) close to the size of the earthquake that occurred in 1926 (Table 1) are computed in region 10, which comprises the Aegean Islands.

Bayrak and Bayrak (2013) estimated the values of the upper bound w using the Gumbel III method (GIII) for the 15 different seismogenic source zones used in this study in Western Anatolia. Using this method, w values are considered as $M_{\max}$ values for any region. We compared the results of $M_{\max}$ values computed from the Bayesian approach in this study with the results found by Bayrak and Bayrak (2013). The distribution of $M_{\max}$ (GIII) and $M_{\max}$ (Bayes) values is shown in Fig. 3 for the different regions of Western Anatolia. The numerals on the graph represent the region numbers. Using the least squares (LS) method, we developed a relationship between $M_{\max}$ values computed by two different methods, as shown in Fig. 3:

$$M_{\max}(\text{Bayes}) = 0.85 \times M_{\max}(\text{GIII}) + 1.60. \quad (17)$$

The correlation coefficient, r, is approximately 0.90 for Eq. (17). This means that there is a strong

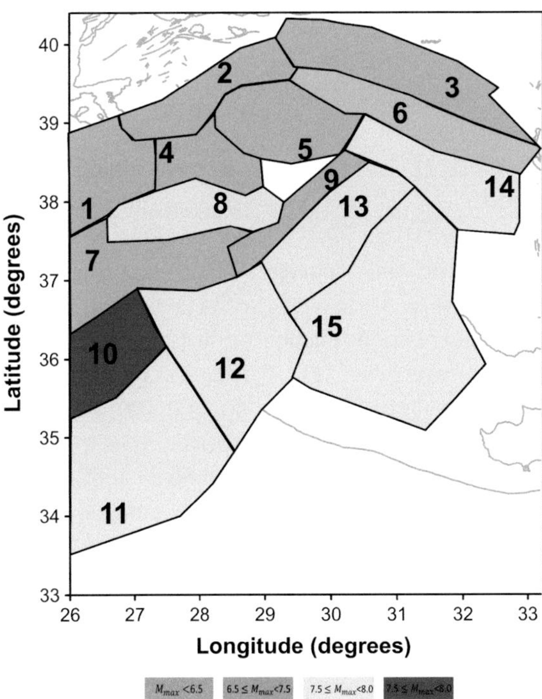

Figure 2
The map of distribution of the $M_{\max}$ values calculated by Bayesian method for the 15 different source regions of Western Anatolia

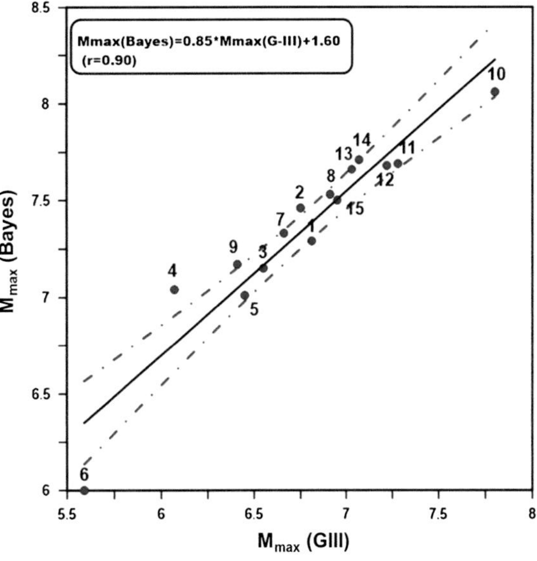

Figure 3
The relationship between $M_{\max}$(GIII) and $M_{\max}$(Bayes) values for the 15 different source regions of Western Anatolia. The regions are showed the numbers from 1 to 15 on the graph. *Straight line* is the linear regression and *dashed lines* are 95 % confidence limits and r is the correlation coefficient

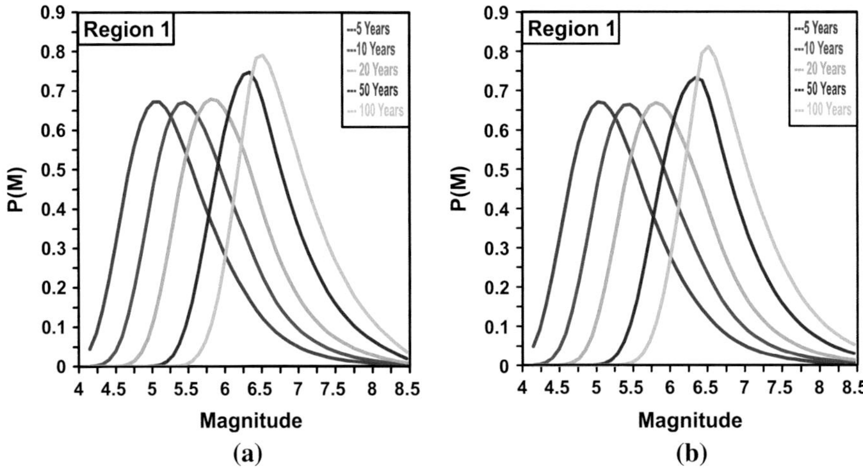

Figure 4
A posteriori probability densities of $M_{max}(T)$ for **a** 'apparent magnitude' and **b** 'true magnitude' showing statistical characteristics of seismic hazard parameters for *region 1* (Aliağa Fault) in next $T = 5$, 10, 20, 50, and 100 years

relationship between M_{max} values found by the two methods. The straight line in Fig. 3 is the linear regression, and the dashed lines are 95 % confidence limits of the linear regression. Except for region 4 and region 6, the other regions remain in the confidence interval limits. M_{max} values computed from GIII for 15 different seismogenic source regions are larger than those derived using the Bayesian method. Since the observed maximum magnitudes and the level of seismicity in the regions of 4 and 6 (Table 1) are lower than that of other regions, the computed values for these regions are outside confidence limits.

The estimated earthquake hazard parameters (β value and activity rate or intensity λ with events per day) are listed in Table 1. The method provides the mean "apparent" β and λ values as well as the 'true' values, which are listed in Table 1. As an example, for the Aliağa Fault region (region 1), we estimated the 'apparent' β value as 1.83, while the "true" mean β value was estimated to be 1.84. The mean intensity or activity rate λ is 0.31 (events/day) for "apparent" as well as "true" values. The computed β values vary between 1.70 and 3.08. The highest value is observed in region 13 (the Gölhisar-Çameli, Acıgöl, and Tatarlı Kumdanlı Faults and the Dinar Graben), while the lowest value is observed in region 3 (the Eskişehir and İnönü-Dodurga Fault Zones). Different numbers of earthquakes in different parts of the magnitude–frequency relationship are considered for the estimation of slope β value in the Bayesian method. Therefore, a significant number of earthquakes are used to estimate it for lower magnitude and fewer at larger magnitudes.

The useful probabilistic tools for earthquake hazard evaluation are estimated and demonstrated for 15 seismogenic source regions in Western Anatolia. A posteriori probability density and a posteriori density function for both apparent and true magnitudes $M_{max}(T)$ that will occur in future time intervals of 5, 10, 20, 50, and 100 years are estimated for region 1 (the Aliağa Fault). The a posteriori probability density for the apparent and true magnitudes $M_{max}(T)$ (Fig. 4) as well as the a posteriori probability distribution function for the apparent and true $M_{max}(T)$ magnitudes (Fig. 5), that will occur in future time intervals of 5, 10, 20, 50, and 100 years is illustrated for region 1 seismogenic source region. These figures are useful probabilistic tools in the earthquake hazard analysis in the examined region. We have also calculated 'tail' probabilities $P(M_{max}(T) > M)$ of the apparent and true magnitudes for all source regions, but this is shown in Fig. 6 only for region 1 for the future time intervals of 5, 10, 20, 50, and 100 years. The other important quantiles that can be considered for hazard estimation are the 'tail' probabilities $P(\mathrm{M}_{max}(T) > M)$ for the apparent as

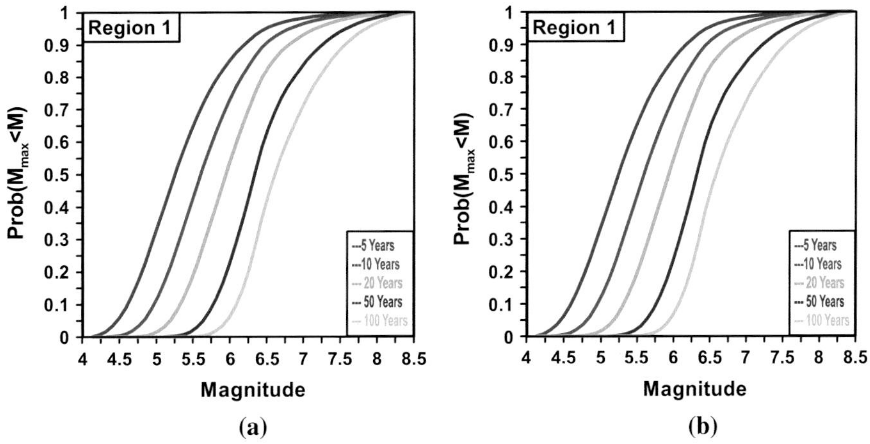

Figure 5
A posteriori probability functions of $M_{max}(T)$ for **a** 'apparent magnitude' and **b** 'true magnitude' showing statistical characteristics of seismic hazard parameters for *region 1* (Aliağa Fault) in next $T = 5$, 10, 20, 50, and 100 years

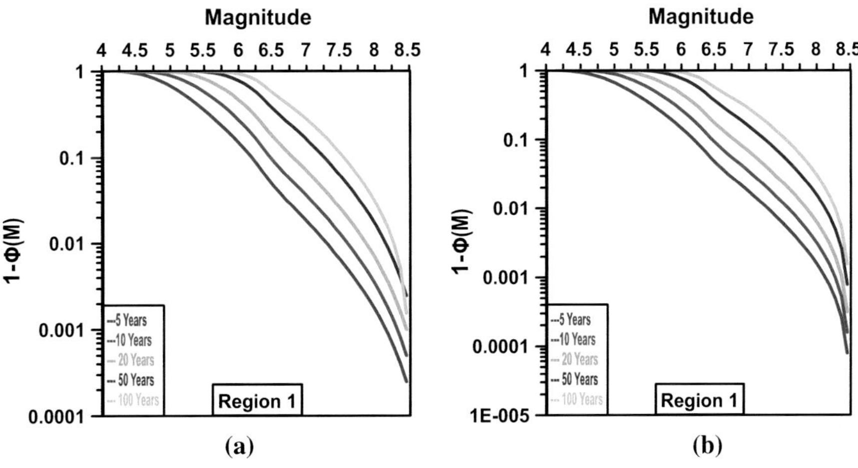

Figure 6
'Tail' probabilities $1 - \phi(M) = \text{Prob}(M_{max}(T) \geq M)$ for **a** 'apparent magnitude' and **b** 'true magnitude' showing statistical characteristics of seismic hazard parameters for *region 1* (Aliağa Fault) in next $T = 5$, 10, 20, 50, and 100 years

well as for the true magnitudes. Lastly, we have estimated a posteriori *M-quantiles* for the 15 source regions in the examined region and for probabilities 0.50, 0.70, and 0.90 in future time intervals of 5, 10, 20, 50, and 100 years. The seismogenic source regions and graphs of their distribution are illustrated (Figs. 7, 8). In these figures, the quantiles of the levels of probability ($\alpha = 0.50$, 0.70, and 0.90) are shown for each region. We also computed their confidence limits. The quantiles for both apparent and true magnitudes for probabilities of 0.50, 0.70, and 0.90 are estimated and tabulated. It can be observed that the differences between apparent and true magnitude quantiles are very low, and this is due to the good quality of the data used. The time periods $T = 50$ and 100 years are considered as appropriate time intervals for the estimation of seismic hazard, but anyone interested in shorter periods may obtain the appropriate estimate of *M*-quantiles. It has been observed that the shorter the time interval, the more appropriate are the results obtained.

We have used 110 years of seismic catalogue to calculate earthquake hazard parameters in this study which reveals that the estimates of M-quantiles for next

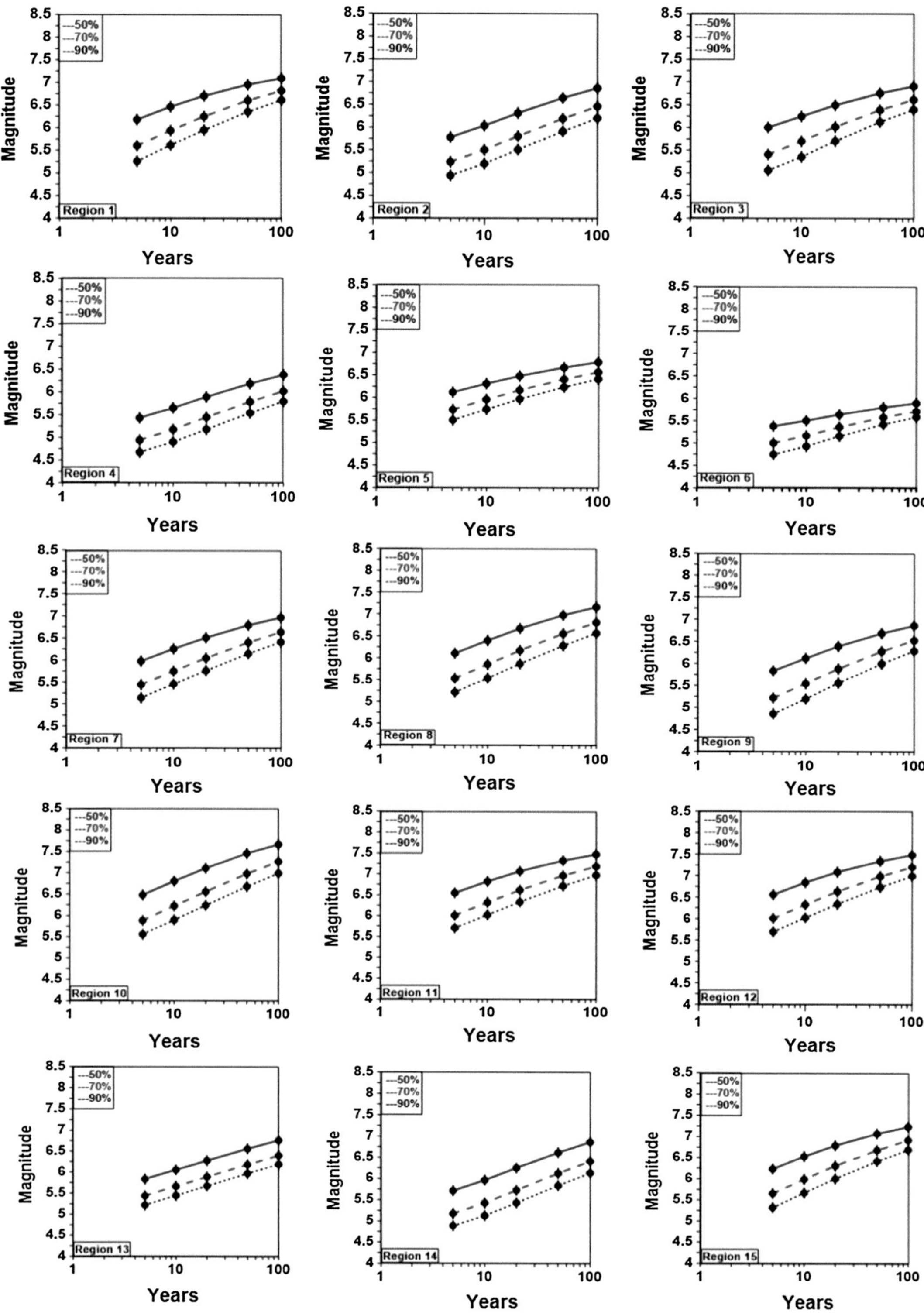

Figure 7

Quantiles for 'apparent magnitudes' (of 50, 70, and 90 %) of function of distribution of maximum values of M_{max} for a given length T of future time interval for the 15 different source regions of Western Anatolia

Table 2

The quantiles of the 'apparent magnitudes' $M_{max}(T)$ estimated for the levels of probability (1) $\alpha = 0.50$, (2) $\alpha = 0.70$, and (3) $\alpha = 0.90$ for the 15 different source regions of Western Anatolia future T = 5, 10, 20, 50, and 100 years

Region	Region name	Future years				
		5	10	20	50	100
(1) Quantiles of probability level 0.5						
1	Aliağa Fault	5.25 ± 0.10	5.61 ± 0.13	5.95 ± 0.14	6.35 ± 0.18	6.61 ± 0.22
2	Akhisar Fault	4.93 ± 0.09	5.19 ± 0.13	5.50 ± 0.17	5.91 ± 0.21	6.20 ± 0.23
3	Eskişehir, İnönü Dodurga Fault Zones	5.05 ± 0.11	5.35 ± 0.15	5.70 ± 0.18	6.12 ± 0.21	6.40 ± 0.25
4	Gediz Graben	4.67 ± 0.10	4.90 ± 0.13	5.18 ± 0.17	5.54 ± 0.21	5.79 ± 0.24
5	Simav, Gediz-Dumlupınar Faults	5.50 ± 0.07	5.74 ± 0.08	5.96 ± 0.10	6.23 ± 0.15	6.41 ± 0.21
6	Kütahya Fault Zone	4.74 ± 0.07	4.92 ± 0.10	5.15 ± 0.13	5.42 ± 0.22	5.58 ± 0.30
7	Karova-Milas, Muğla-Yatağan Faults	5.13 ± 0.09	5.45 ± 0.11	5.76 ± 0.12	6.14 ± 0.15	6.40 ± 0.18
8	Büyük Menderes Graben	5.20 ± 0.10	5.53 ± 0.13	5.86 ± 0.15	6.28 ± 0.18	6.56 ± 0.20
9	Dozkırı-Çardak, Sandıklı Faults	4.85 ± 0.13	5.19 ± 0.17	5.55 ± 0.20	5.99 ± 0.23	6.28 ± 0.26
10	Aegean Islands	5.55 ± 0.09	5.89 ± 0.11	6.24 ± 0.12	6.67 ± 0.14	6.99 ± 0.15
11	Aegean Arc	5.70 ± 0.07	6.02 ± 0.08	6.33 ± 0.09	6.71 ± 0.11	6.97 ± 0.13
12	Ege Yayı, Marmaris, Köyceğiz, Fethiye Faults	5.68 ± 0.08	6.02 ± 0.09	6.33 ± 0.10	6.73 ± 0.12	6.99 ± 0.15
13	Gölhisar-Çameli, Acıgöl, Tatarlı Kumdanlı Faults, Dinar Graben	5.21 ± 0.07	5.44 ± 0.08	5.67 ± 0.10	5.96 ± 0.13	6.18 ± 0.14
14	Sultandağı Fault	4.88 ± 0.09	5.12 ± 0.13	5.42 ± 0.17	5.83 ± 0.22	6.13 ± 0.25
15	Kaş ve Beyşehirgölü Faults	5.31 ± 0.10	5.66 ± 0.12	5.99 ± 0.13	6.40 ± 0.16	6.68 ± 0.18
(2) Quantiles of probability level 0.7						
1	Aliağa Fault	5.59 ± 0.13	5.93 ± 0.14	6.24 ± 0.16	6.60 ± 0.21	6.81 ± 0.27
2	Akhisar Fault	5.23 ± 0.13	5.50 ± 0.17	5.80 ± 0.20	6.19 ± 0.23	6.45 ± 0.25
3	Eskişehir, İnönü Dodurga Fa Zones	5.40 ± 0.15	5.70 ± 0.18	6.02 ± 0.20	6.39 ± 0.25	6.61 ± 0.30
4	Gediz Graben	4.94 ± 0.13	5.17 ± 0.17	5.44 ± 0.20	5.78 ± 0.24	6.02 ± 0.27
5	Simav, Gediz-Dumlupınar Faults	5.73 ± 0.08	5.96 ± 0.10	6.16 ± 0.13	6.40 ± 0.21	6.56 ± 0.27
6	Kütahya Fault Zone	4.99 ± 0.10	5.16 ± 0.13	5.35 ± 0.19	5.57 ± 0.30	5.71 ± 0.38
7	Karova-Milas, Muğla-Yatağan Faults	5.44 ± 0.11	5.74 ± 0.12	6.04 ± 0.14	6.39 ± 0.17	6.63 ± 0.22
8	Büyük Menderes Graben	5.53 ± 0.13	5.85 ± 0.15	6.17 ± 0.17	6.55 ± 0.20	6.80 ± 0.22
9	Dozkırı-Çardak, Sandıklı Faults	5.21 ± 0.16	5.54 ± 0.20	5.88 ± 0.22	6.27 ± 0.26	6.52 ± 0.31
10	Aegean Islands	5.88 ± 0.11	6.22 ± 0.12	6.55 ± 0.14	6.97 ± 0.15	7.26 ± 0.15
11	Aegean Arc	6.01 ± 0.08	6.32 ± 0.09	6.61 ± 0.10	6.96 ± 0.13	7.18 ± 0.17
12	Ege Yayı, Marmaris, Köyceğiz, Fethiye Faults	6.00 ± 0.09	6.32 ± 0.10	6.62 ± 0.12	6.98 ± 0.15	7.20 ± 0.19
13	Gölhisar-Çameli, Acıgöl, Tatarlı Kumdanlı Faults, Dinar Graben	5.43 ± 0.08	5.66 ± 0.10	5.88 ± 0.12	6.17 ± 0.14	6.39 ± 0.16
14	Sultandağı Fault	5.17 ± 0.13	5.42 ± 0.17	5.72 ± 0.21	6.12 ± 0.25	6.40 ± 0.27
15	Kaş ve Beyşehirgölü Faults	5.65 ± 0.12	5.98 ± 0.13	6.30 ± 0.15	6.67 ± 0.18	6.91 ± 0.22
(3) Quantiles of probability level 0.9						
1	Aliağa Fault	6.17 ± 0.16	6.46 ± 0.19	6.70 ± 0.24	6.95 ± 0.32	7.09 ± 0.38
2	Akhisar Fault	5.77 ± 0.19	6.03 ± 0.22	6.31 ± 0.24	6.64 ± 0.27	6.85 ± 0.30
3	Eskişehir, İnönü Dodurga Fault Zones	6.00 ± 0.20	6.24 ± 0.22	6.49 ± 0.27	6.75 ± 0.34	6.91 ± 0.40
4	Gediz Graben	5.43 ± 0.19	5.65 ± 0.22	5.89 ± 0.25	6.18 ± 0.30	6.38 ± 0.35
5	Simav, Gediz-Dumlupınar Faults	6.11 ± 0.12	6.31 ± 0.17	6.47 ± 0.23	6.66 ± 0.32	6.78 ± 0.39
6	Kütahya Fault Zone	5.37 ± 0.20	5.50 ± 0.25	5.64 ± 0.33	5.80 ± 0.45	5.89 ± 0.53
7	Karova-Milas, Muğla-Yatağan Faults	5.97 ± 0.14	6.25 ± 0.16	6.50 ± 0.19	6.78 ± 0.26	6.96 ± 0.32
8	Büyük Menderes Graben	6.10 ± 0.17	6.39 ± 0.19	6.67 ± 0.21	6.97 ± 0.25	7.16 ± 0.29
9	Dozkırı-Çardak, Sandıklı Faults	5.83 ± 0.21	6.11 ± 0.24	6.38 ± 0.28	6.68 ± 0.35	6.85 ± 0.41
10	Aegean Islands	6.48 ± 0.13	6.80 ± 0.14	7.10 ± 0.15	7.45 ± 0.15	7.65 ± 0.16
11	Aegean Arc	6.54 ± 0.10	6.82 ± 0.12	7.06 ± 0.15	7.32 ± 0.21	7.47 ± 0.26
12	Ege Yayı, Marmaris, Köyceğiz, Fethiye Faus	6.55 ± 0.11	6.83 ± 0.13	7.08 ± 0.16	7.33 ± 0.23	7.48 ± 0.27
13	Gölhisar-Çameli, Acıgöl, Tatarlı Kumdanlı Fauls, Dinar Graben	5.83 ± 0.12	6.05 ± 0.13	6.27 ± 0.15	6.55 ± 0.17	6.75 ± 0.18
14	Sultandağı Fault	5.71 ± 0.20	5.96 ± 0.23	6.24 ± 0.26	6.61 ± 0.28	6.86 ± 0.29
15	Kaş ve Beyşehirgölü Faults	6.22 ± 0.14	6.52 ± 0.16	6.78 ± 0.19	7.06 ± 0.26	7.23 ± 0.31

Table 3

The quantiles of the 'true magnitudes' $M_{max}(T)$ estimated for the levels of probability (1) $\alpha = 0.50$, (2) $\alpha = 0.70$, and (3) $\alpha = 0.90$ $T = 5, 10, 20, 50$, and 100 years

Region	Region name	Future years				
		5	10	20	50	100
(1) Quantiles of probability level 0.5						
1	Aliağa Fault	5.24 ± 0.11	5.59 ± 0.13	5.93 ± 0.15	6.34 ± 0.18	6.59 ± 0.22
2	Akhisar Fault	4.92 ± 0.10	5.18 ± 0.13	5.49 ± 0.17	5.89 ± 0.21	6.18 ± 0.23
3	Eskişehir, İnönü Dodurga Fault Zones	5.05 ± 0.11	5.34 ± 0.15	5.69 ± 0.18	6.11 ± 0.21	6.38 ± 0.25
4	Gediz Graben	4.66 ± 0.10	4.88 ± 0.14	5.16 ± 0.17	5.52 ± 0.21	5.78 ± 0.24
5	Simav, Gediz-Dumlupınar Faults	5.48 ± 0.07	5.72 ± 0.08	5.95 ± 0.10	6.21 ± 0.15	6.39 ± 0.21
6	Kütahya Fault Zone	4.73 ± 0.07	4.91 ± 0.10	5.14 ± 0.14	5.40 ± 0.22	5.55 ± 0.31
7	Karova-Milas, Muğla-Yatağan Faults	5.12 ± 0.09	5.43 ± 0.11	5.74 ± 0.13	6.13 ± 0.15	6.39 ± 0.18
8	Büyük Menderes Graben	5.19 ± 0.10	5.52 ± 0.13	5.85 ± 0.15	6.26 ± 0.18	6.55 ± 0.20
9	Dozkırı-Çardak, Sandıklı Faults	4.84 ± 0.13	5.18 ± 0.17	5.54 ± 0.20	5.98 ± 0.23	6.27 ± 0.26
10	Aegean Islands	5.54 ± 0.09	5.88 ± 0.11	6.22 ± 0.12	6.66 ± 0.14	6.97 ± 0.15
11	Aegean Arc	5.69 ± 0.07	6.01 ± 0.08	6.31 ± 0.09	6.70 ± 0.11	6.95 ± 0.14
12	Ege Yayı, Marmaris, Köyceğiz, Fethiye Faults	5.67 ± 0.08	6.00 ± 0.09	6.32 ± 0.10	6.71 ± 0.12	6.97 ± 0.15
13	Gölhisar-Çameli, Acıgöl, Tatarlı Kumdanlı Faults, Dinar Graben	5.19 ± 0.07	5.42 ± 0.09	5.65 ± 0.11	5.94 ± 0.13	6.16 ± 0.15
14	Sultandağı Fault	4.87 ± 0.09	5.11 ± 0.13	5.41 ± 0.17	5.81 ± 0.22	6.11 ± 0.25
15	Kaş ve Beyşehirgölü Faults	5.30 ± 0.10	5.65 ± 0.12	5.98 ± 0.13	6.39 ± 0.16	6.67 ± 0.18
(2) Quantiles of probability level 0.7						
1	Aliağa Fault	5.58 ± 0.13	5.92 ± 0.15	6.23 ± 0.17	6.58 ± 0.21	6.80 ± 0.27
2	Akhisar Fault	5.22 ± 0.13	5.48 ± 0.17	5.79 ± 0.20	6.17 ± 0.23	6.44 ± 0.25
3	Eskişehir, İnönü Dodurga Fault Zones	5.40 ± 0.15	5.68 ± 0.18	6.00 ± 0.20	6.37 ± 0.25	6.60 ± 0.30
4	Gediz Graben	4.93 ± 0.13	5.16 ± 0.17	5.43 ± 0.20	5.77 ± 0.24	6.00 ± 0.28
5	Simav, Gediz-Dumlupınar Faults	5.71 ± 0.08	5.94 ± 0.10	6.14 ± 0.14	6.38 ± 0.21	6.53 ± 0.28
6	Kütahya Fault Zone	4.98 ± 0.10	5.15 ± 0.13	5.34 ± 0.19	5.55 ± 0.31	5.67 ± 0.40
7	Karova-Milas, Muğla-Yatağan Faults	5.42 ± 0.11	5.73 ± 0.12	6.03 ± 0.14	6.38 ± 0.18	6.61 ± 0.22
8	Büyük Menderes Graben	5.52 ± 0.13	5.84 ± 0.15	6.15 ± 0.17	6.54 ± 0.20	6.79 ± 0.22
9	Dozkırı-Çardak, Sandıklı Faults	5.20 ± 0.16	5.53 ± 0.20	5.86 ± 0.22	6.26 ± 0.26	6.51 ± 0.31
10	Aegean Islands	5.87 ± 0.11	6.21 ± 0.12	6.54 ± 0.14	6.96 ± 0.15	7.25 ± 0.15
11	Aegean Arc	5.99 ± 0.08	6.30 ± 0.09	6.60 ± 0.10	6.94 ± 0.13	7.16 ± 0.18
12	Ege Yayı, Marmaris, Köyceğiz, Fethiye Faults	5.99 ± 0.09	6.31 ± 0.10	6.61 ± 0.12	6.96 ± 0.15	7.18 ± 0.19
13	Gölhisar-Çameli, Acıgöl, Tatarlı Kumdanlı Faults, Dinar Graben	5.41 ± 0.09	5.64 ± 0.10	5.86 ± 0.12	6.15 ± 0.15	6.37 ± 0.16
14	Sultandağı Fault	5.16 ± 0.13	5.41 ± 0.17	5.70 ± 0.21	6.10 ± 0.25	6.39 ± 0.27
15	Kaş ve Beyşehirgölü Faults	5.63 ± 0.12	5.97 ± 0.13	6.28 ± 0.15	6.66 ± 0.18	6.90 ± 0.22
(3) Quantiles of probability level 0.9						
1	Aliağa Fault	6.16 ± 0.16	6.44 ± 0.19	6.68 ± 0.24	6.93 ± 0.32	7.06 ± 0.39
2	Akhisar Fault	5.76 ± 0.19	6.02 ± 0.22	6.29 ± 0.24	6.62 ± 0.27	6.84 ± 0.30
3	Eskişehir, İnönü Dodurga Fault Zones	5.99 ± 0.20	6.23 ± 0.23	6.48 ± 0.27	6.73 ± 0.35	6.87 ± 0.42
4	Gediz Graben	5.41 ± 0.19	5.63 ± 0.22	5.87 ± 0.25	6.16 ± 0.31	6.35 ± 0.36
5	Simav, Gediz-Dumlupınar Faults	6.10 ± 0.12	6.29 ± 0.18	6.45 ± 0.24	6.63 ± 0.34	6.73 ± 0.42
6	Kütahya Fault Zone	5.36 ± 0.20	5.48 ± 0.26	5.60 ± 0.35	5.74 ± 0.48	5.81 ± 0.57
7	Karova-Milas, Muğla-Yatağan Faults	5.96 ± 0.14	6.23 ± 0.16	6.49 ± 0.19	6.77 ± 0.26	6.93 ± 0.33
8	Büyük Menderes Graben	6.09 ± 0.17	6.38 ± 0.19	6.65 ± 0.21	6.96 ± 0.25	7.14 ± 0.30
9	Dozkırı-Çardak, Sandıklı Faults	5.82 ± 0.22	6.10 ± 0.24	6.37 ± 0.28	6.66 ± 0.36	6.83 ± 0.42
10	Aegean Islands	6.46 ± 0.13	6.79 ± 0.14	7.09 ± 0.15	7.44 ± 0.15	7.65 ± .016
11	Aegean Arc	6.53 ± 0.10	6.80 ± 0.12	7.05 ± 0.15	7.30 ± 0.22	7.44 ± 0.27
12	Ege Yayı, Marmaris, Köyceğiz, Fethiye Faults	6.54 ± 0.11	6.82 ± 0.13	7.06 ± 0.16	7.32 ± 0.23	7.45 ± 0.29
13	Gölhisar-Çameli, Acıgöl, Tatarlı Kumdanlı Faults, Dinar Graben	5.81 ± 0.12	6.03 ± 0.14	6.25 ± 0.15	6.53 ± 0.17	6.73 ± 0.18
14	Sultandağı Fault	5.70 ± 0.20	5.94 ± 0.23	6.23 ± 0.26	6.59 ± 0.29	6.84 ± 0.29
15	Kaş ve Beyşehirgölü Faults	6.21 ± 0.15	6.51 ± 0.16	6.77 ± 0.20	7.05 ± 0.26	7.21 ± 0.32

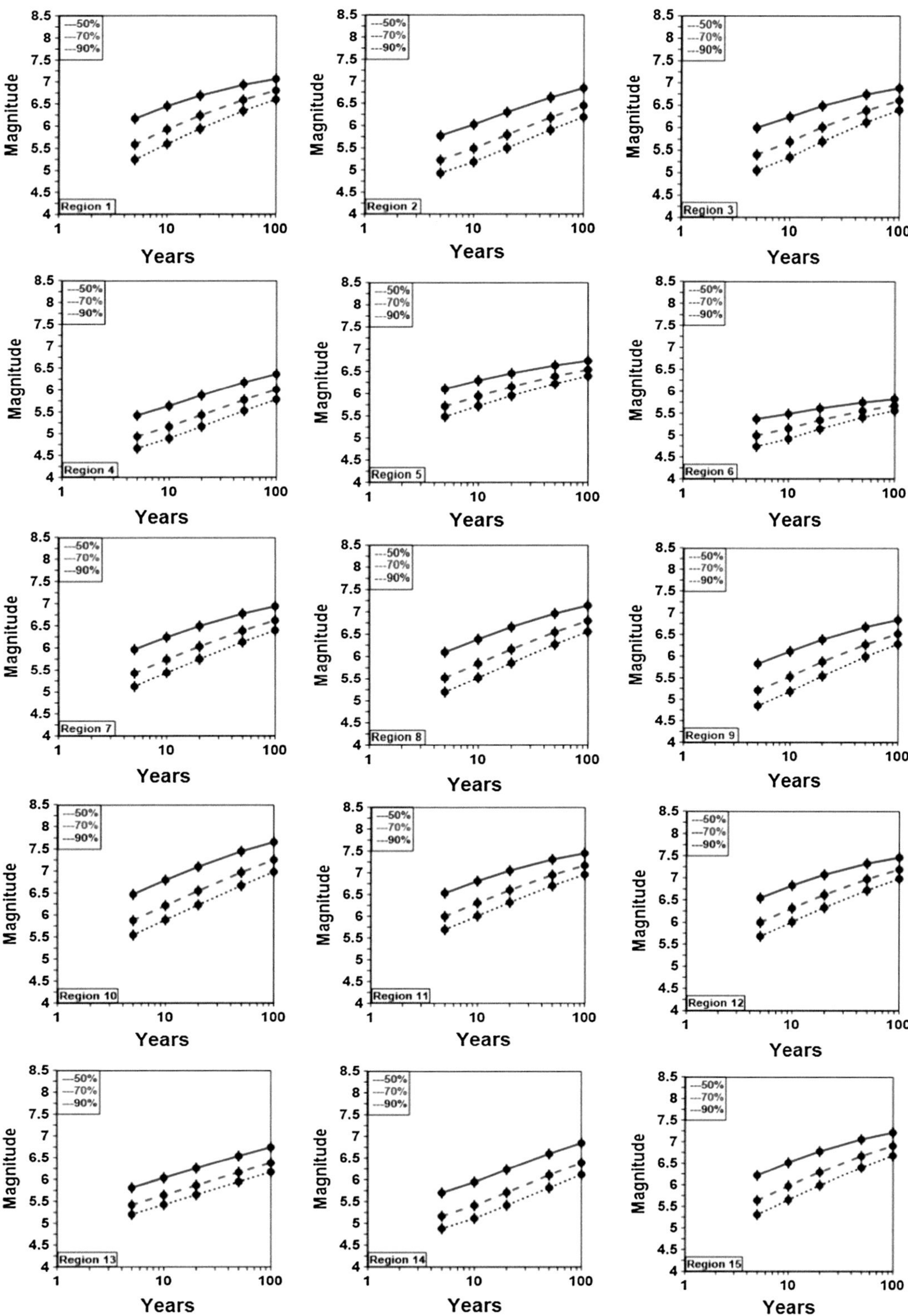

Figure 8
Quantiles for 'true magnitudes' (of 50, 70 and 90 %) of function of distribution of maximum values of M_{max} for a given length T of future time interval for the 15 different source regions of Western Anatolia

100 years are reasonable. The apparent and true magnitudes for 50, 70, and 90 % probability levels within the next 5, 10, 20, 50, and 100 years are calculated for all seismogenic regions. The estimated values are listed in Tables 2 and 3. For the next 100 years, the apparent magnitudes with 90 % probability for the 15 different regions in Western Anatolia are found to be 7.09, 6.85, 6.91, 6.38, 6.78, 5.89, 6.96, 7.16, 6.85, 7.65, 7.47, 7.48, 6.75, 6.86, and 7.23, respectively. The true magnitudes for the same parameters for the regions are computed as 7.06, 6.84, 6.87, 6.35, 6.73, 5.81, 6.93, 7.14, 6.83, 7.65, 7.44, 7.45, 6.73, 6.84, and 7.21, respectively. The highest apparent and true magnitude values in these regions are equal to 7.65 and observed in region 10 comprising the Aegean Islands for the next 100 years. If we compare these two tables, it is easy to observe that the values recorded in Table 3 are less than those in Table 2. This is obvious since Table2 includes the magnitudes (apparent) of Table 3 (true) plus the error ε. The differences between these two values are very low, and we believe that this depends upon the quality of the data, which includes minor errors. Therefore, the efficiency of the data included in Tables 2 and 3 and the results of the analysis are almost the same (Fig. 8).

6. Conclusions

The instrumental earthquake catalog that is homogenous for $M_s \geq 4.0$ was used during the period 1900–2013 to evaluate earthquake hazard parameters for the 15 seismogenic source regions in Western Anatolia using the Bayesian method. For this purpose, maximum regional magnitude (M_{max}), β value, and the seismicity activity rate or intensity (λ) and their uncertainty are computed. The maximum regional magnitude is one of the most important earthquake hazard parameters; therefore, significance is given to the estimation of this parameter as well as to the quantiles of the M_{max} distribution in a future time interval. The computed M_{max} values are between 6.00 and 8.06, while their uncertain values vary between 0.25 and 0.88. While low values are found in the northern part of Western Anatolia, high values are observed in the southern part of Western Anatolia related to the Aegean subduction zone. The largest value is computed in region 10 comprising the Aegean Islands.

The estimated β values for the 15 different regions of Western Anatolia vary between 1.70 and 3.08. In this method, different numbers of earthquakes in different parts of the magnitude-frequency relationship are taken into account for the estimation of β value. Therefore, a significant number of earthquakes are used to estimate lower magnitude and fewer at larger magnitudes. We estimated earthquake probabilities in the next 5, 10, 20, 50, and 100 years. We also computed a posteriori probability densities of $M_{max}(T)$, a posteriori probability functions of $M_{max}(T)$, and 'tail' probabilities $Prob(M_{max}(T) \geq M)$ for the 'apparent' and 'true' magnitude values. In addition, we estimated the quantiles of the 'apparent and true' magnitudes $M_{max}(T)$ for the levels of probability $\alpha = 0.50$, $\alpha = 0.70$, and $\alpha = 0.90$ for the 15 seismic regions of Western Anatolia in the next $T = 5$, 10, 20, 50, and 100 years. Considering the estimated parameters, the results indicate that region 10 comprising the Aegean Islands has a very high probability of experiencing a 7.65 magnitude earthquake within the next century.

Acknowledgments

The authors are thankful to Prof. T. M. Tsapanos for providing computer program, teaching us this program, giving motivation and suggestions to carry out this research. Authors are also thankful to Dr. Pierre Keating, Editor, PAGEOPH and two anonymous reviewers for constructive comments and suggestions which enhanced quality of the manuscript.

References

Ambraseys, N. N., and Finkel, C. F. (1995), *The Seismicity of Turkey and Adjacent Areas, A Historical Review,* Eren Publishing, İstanbul. *240*, 1500–1800.

Aktuğ, B., and Kiliçoğlu, A. (2006), *Recent Crustal Deformation of İzmir, Western Anatolia and Surrounding Regions as Deduced from Repeated GPS Measurements and Strain Field,* Journal of Geodynamics. *41*, 471–484.

Altinok, Y. (1991), *Evaluation of Earthquake Risk in West Anatolia by Semi-Markov Model,* Geophysics. *5*, 135–140.

Anagnos, T., and Kiremidjian, A. S. (1988), *A Review of Earthquake Occurence Models for Seismic Hazard Analysis,* Probabilistic Engineering Mechanics. *1*, 3–11.

Barka, A. A., Reilinger, R. (1997), *Active Tectonics of the Mediterranean Region: Deduced from GPS, Neotectonic and Seismicity Data,* Annali Di Geophis XI. 587–610.

Bağci, G. (1996), *Earthquake Occurrences in Western Anatolia by Markov Model. 10,* Natural Hazards, Geophysics. *10*, 67–75.

Bayrak, Y., Öztürk, S., Çinar, H., Kalafat, D., Tsapanos, T. M., Koravos, G.CH, and Leventakis, G.A. (2009), *Estimating Earthquake Hazard Parameters from Instrumental Data for Different Regions in and Around Turkey,* Engineering Geology. *105*, 200–210.

Bayrak, Y., and Bayrak, E. (2012), *An Evaluation of Earthquake Hazard Potential for Different Regions in Western Anatolia Using the Historical and Instrumental Earthquake Data,* Pure and Applied Geophysics. *169*, 1859–1873.

Bayrak, E., and Bayrak, Y. (2013), *Regional Variation of the w-Upper Bound Magnitude of GIII Distribution in the Different Regions of Western Anatol*ia, 7th Congress of Balkan Geophysical Society, 7-10 October, Tirana, Albania. 18651, doi:10. 3997/2214-4609.20131725.

Bozkurt, E. (2001), *Neotectonics of Turkey-a Synthesis,* Geodynamica Acta. *14*, 3–30.

Bozkurt, E., and Sözbilir, H. (2004), *Tectonic Evolution of the Gediz Graben: Field Evidence for an Episodic, Two-Stage Extension in Western Turkey,* Geological Magazine. *141*, 63–79.

Bozkurt, E., and Sözbilir, H. (2006), *Evolution of the Large-Scale Active Manisa Fault, Southwest Turkey: Implications on Fault Development and Regional Tectonics,* Geodinamica Acta. *19*, 427–453.

Blumenthal, M. M. (1962), *Le Systems Structural Du Taurus Sud Anatolian, Paul Fellot,* 2, Soc. Geol. France. *11*, 611–662.

Brunn, J.H., Dumont, J.F., De Graciansky, P.C., Gutnic, M., Juteau, T., Marcoux, J., and Poisson, A. (1971), *Outline of the Geology of the Western Taurides. In Geology and History of Turkey (ed A.S. Campell),* Petroleum Exploration Society of Libya, Tripoli. 225–257.

Campell, K. W. (1982), *Bayesian Analysis of Extreme Earthquake Occurrences, Part I. Probabilistic Hazard Model,* Seismological Society America. *72*, 1689–1705.

Campell, K. W. (1983), *Bayesian Analysis of Extreme Earthquake Occurrences, Part II. Application to the San Jacinto Fault Zone of Southern California,* Seismol. Soc. Am. *73*, 1099–1115.

Cox, D. R., and Lewis, P. A. W. (1966), *The Statistical Analysis of Series of Events,* Published by Methuen, London.

Dewey, J. F., and Şengör, A. M. C. (1979), *Aegean and Surrounding Regions: Complex Multi-Plate and Continuum Tectonics in a Convergent Zone,* Geol. Soc. America Bull. Part 1, *90*, 84–92.

Dewey, J.F. (1988), *Extensional collapse of orogens,* Tectonics. *7*, 1123–1139.

Dumont, J. F., Uysal, S., Şimşek, S., Karamenderesi, H., and Letouzey, J. (1979), *Formation of the Grabens in Southwestern Anatolia, Bull. Min. Res. Explor.* Ins.Turk. *92*, 7–18.

Frizon De Lamotte, D., Poisson, A., Aubourg, C., and Temiz, H. (1995), *Post-Tortonian Eastward and Southward Thrusting in the Core of the Isparta Recentrant (Taurus, Turkey)*, Geodinamic Implications. Bull. Soc. Geol. France. *166*, 59–67.

Jackson, J.A., and Mckenzie, D. (1984), *Active Tectonics of the Alpine-Himalayan Belt Between Western Turkey and Pakistan.* Geophysical Journal of the Royal Astronomical Society. *77*, 185–264.

Kahraman, S., Baran, T., and Saatçi, İ. A. (2007), *The Effect of Region Border to Determine the Earthquake Hazard, Case study: Western Anatolia,* Sixth National Conference on Earthquake Engineering. 335–346.

Kissel, C., Averbuch, O., Lamotte, D., Monod, O., and Allerton, S. (1993), *First Paleomagnetic Evidence for a Post-Eocene Clockwise Rotation of Western Taurides Thrust Belt East of the Isparta Recentrant (southwestern Turkey).* Earth Planet. Sci. Lett. *117*, 1–14.

Kijko, A., and Sellevoll, M. A. (1992), *Estimation of Earthquake Hazard Parameters from Incomplete Data Files. Part II. Incorporation of Magnitude Heterogeneity.* Bull. Seismol. Soc. Am. *82*, 120–134.

Kijko, A. (2004), *Estimation of the Maximum Earthquake Magnitude, M_{max}.* Pure and Applied Geophysics. *161*, 1655–1681.

Koçyiğit, A., Yusufoğlu, H., Bozkurt, E. (1999), *Evidence from the Gediz Graben Episodic Two-Stage Extension in Western Turkey,* Journal of the Geological Society. *156*, 605–616.

Lamarre, M., Townshed, B., Shah, H. C. (1992), *Application of the Bootstrap Method to Quantify Uncertainty in Seismic Hazard Estimates,* Seismological Society America. *82*, 104–119.

Le Pichon, X., and Angelier, J. (1979), *The Hellenic Arc and Trench System: A Key to the Neotectonic Evolution of the Eastern Mediterranean Area.* Elsevier, *60*, 1–42.

Le Pichon, X., Chamot-Rooke, C., Lallemant, S., Noomen, R., and Veis, G. (1995), *Geodetic Determination of the Kinematics of Central Greece with Respect to Europe: Implications for Eastern Mediterranean Tectonics,* J. Geophys Res. *100*, 12675–12690.

Lyubushin, A. A.,Tsapanos, T. M., Pisarenko, V. F., and Koravos, G. (2002), *Seismic Hazard for Selected Sites in Greece: A Bayesian Estimates of Peak Ground Acceleration,* Natural Hazard. *25*, 83–89.

Lyubushin, A. A., and Parvez, A. I. (2010), *Map of Seismic Hazard of India Using Bayesian Approach,* Nat Hazard. *55*, 543–556.

Marcoux, J. (1987), *Historie et Topologie De La Neo-Tethys. These De Doctor at Det at.* L'Universite Pierre et Marie Curie, Paris. 569.

Mcclusky, S., Balassanian, S., Barka, A., Demir, C., Ergintav, S., Georgiev, I., Gürkan, O., Hamburger, M., Kahle, K.H.H., Kastens, K., Kekelidze, G., King, R., Kotzev, V., Lenk, O., Mahmoud, S., Mishin, A., Nadariya, M., Ouzounis, A., Paradissis, D., Peter, Y., Prilepin, M., Reilinger, R., S¸ Anli, I., Seeger, H., Tealeb, A., Toksöz, M.N., Veis, G. (2000), *Global Positioning System Constraints On Plate Kinematics And Dynamics in the Eastern Mediterranean and Caucasu.* J. Geophys Res. *105*(B3), 5695–5719.

Mueller, S. C. (2010), *The Influence of Maximum Magnitude on Seismic-Hazard Estimates in the Central and Eastern United States.* Bull. of the Seismol. Soc. of *Am. 100*, 699–711.

Morgat, C. P., and Shah, H. C. (1979), *Bayesian Model for Seismic Hazard Mapping,* Seismological Society America. *69*, 1237–1251.

Ocakoğlu, N., Demirdağ, E., and Kuşçu, İ. (2004), *Neotectonic Structures in the Area Offshore of Alacatı, Doğanbey and Kuşadası (Western Turkey): Evidence of Strike-Slip Faulting in Aegean Province,* Tectonophysics. *391*, 67–83.

Ocakoğlu, N., Demirdağ, E., and Kuşçu, İ. (2005), *Neotectonic Structures in İzmir Gulf and Surrounding Regions (Western Turkey): Evidences of Strike-Slip Faulting with Compression in the Aegean Extensional Regime,* Marine Geology. *219*, 155–171.

Oral, M.B., Reilinger, R.E., Toksöz, M.N., Kon, R.W., Barka, A. A., Kinik, I., and Lenk, O. (1995), *Global Positioning System Offers Evidence of Plate Motions in Eastern Mediterranean,* EOS Transac. *76*, 9.

Pisarenko, V. F., Lyubushin, A. A., Lysenko, V. B., and Golubeva, T. V. (1996), *Statistical Estimation of Seismic Hazard Parameters: Maximum Possible Magnitude and Related Parameters,* The Seismological Society of America. *86*, 691–7000.

Pisarenko, V. F., and Lyubushin, A. A. (1997), *Statistical Estimation of Maximal Peak Ground Acceleration at a Given Point of Seismic Region,* J. of Seismology. *1*, 395–405.

Pisarenko, V. F., and Lyubushin, A. A. (1999), *Bayesian Approach to Seismic Hazard Estimation: Maximum Values of Magnitudes and Peak Ground Accelerations,* Earthquake Research in China (English Edition). 1999. *13*, 45–57.

Poisson, A. (1984), *The Extension of the Ionian trough into SW Turkey. In: J. F.Dixon g A. H. Robertson Eds., The Geologic Evolution of the Eastern Mediterranean,* Geol. Soc. LondönSpec. Pub. *17*, 241–249.

Poisson, A. (1990), *Neogene Thrust Belt in Western Taurides. The Imbricate Systems of Thrust Sheets Along a NNW-SSE Transect,* IESCA. 224–235.

Price, S., and Scott, B. (1994), *Fault-Block Rotations at the Edge of a Zone of Continental Extension; Southwest Turkey,* J. Struct. Geol. *16*, 381–392.

Rao, C. R. (1965), *Linear Statistical Inference and Its Application. New York, John Wiley,* Library of Congress Cataloging. 1–618.

Reasenberg, P. (1985), *Second-order moment of central California seismicity*, 1969–82. J Geophys Res. *90*, 5479–5495.

Seyitoğlu, G., and Scott, B. C. (1992), *Late Cenozoic Volcanic Evolution of the Northeastern Aegean region,* Journal of Volcanology and Geothermal Research. *54*, 157–176.

Seyitoğlu, G., Scot, B. C., and Rundle, C. C. (1992), *Timing of Cenezoic Extensional Tectonics in West Turkey,* Journal of the Geological Society London. *149*, 533–538.

Seyitoğlu, G., and Scott, B.C. (1996), *The Age of the Alaşehir Graben (West Turkey) and Its Tectonic Implications,* Geological Journal. *31*, 1–11.

Sözbilir, H. (2002), *Geometry and origin of folding in the Neogene sediments of the Gediz graben,* Geodinamica Acta *15*, 31–40.

Şalk, M., and Sari, C. (2000), *Sediment Thickness of the Western Anatolia Graben Structures Determined by 2D and 3D Analysis Using Gravity Data.* Journal of Asian Earth Sciences. *26*, 39–48.

Şaroğlu, F., Emre, Ö., and Boray, A. (1987), *The Active faults of the Turkey and Earthquakes,* MTA, *8174*, 394, 1987.

Şaroğlu, F., Emre, Ö., and Kuşçu, İ. (1992), *Active Fault Map of Turkey,* Mineral Research and Exploration Institute (MTA) of Turkey Publications, Ankara. *118*, 47–64.

Şengör, A. M. C. (1982), *Kimmerid Orojenik Sisteminin Evrimi, Orta Mesozoyikte Paleo-Tetisin Kapanması Olayı ve Ürünleri:* Türkiye Jeoloji Kurultayı, Şubat, Ankara, Bildiri Özetleri Kitabı, 45–46.

Şengör, A. M. C., Satir, M., and Akkök, R. (1984), *Timing of Tectonic Events in the Menderes Massif, Western Turkey: Implications for Tectonic Evolution and Evidence for Pan-African Basement in Turkey,* Tectonics. *3*, 693–707.

Şengör, A. M. C., Görür, N., and Şaroğlu, F. (1985), *Strike-Slip Faulting and Related Basin Formation in Zones of Tectonic Escape: Turkey as a Case Study, in Strike-Slip Faulting and Basin Formation, Edited by Biddke, K.T. and Christie-Blick, N.,* Society of Econ. Paleont. Min. Sp. Publ. *17*, 227–264.

Şengör, A. M. C. (1987), *Cross-Faults and Differential Stretching of Hanging Walls in Regions of Low-Angle Normal Faulting: Examples from Western Turkey. In: Coward, M. P., Dewey, J. F. and Hancock, P. L. (eds), Continental Extensional Tectonics, Geological Society London,* Special Publications. *28*, 575–589.

Tinti, S., and Mulargia, F. (1985), *Effects of Magnitude Uncertainties in the Gutenberg-Richter Frequency-Magnitude Law. Bull,* Seismol. Soc. Am. *75*, 1681–1697.

Tsapanos, T. M., Lyubushin, A. A., and Pisarenko, V. F. (2001), *Application of a Bayesian Approach for Estimation of Seismic Hazard Parameters in Some Regions of the Circum-Pasific Belt,* Pure And Applied Geophysics. *158*, 859–875.

Tsapanos, T. M., Galanis, O. CH., Koravos, G. CH., and Musson, R. M. W. (2002), *A Method for Bayesian Estimation of the Probability of Local Intensity for Some Cities in Japan,* Annals of Geophysics. *45*, 657–671.

Tsapanos, T. M. (2003), *Appraisal of Seismic Hazard Parameters for the Seismic Regions of the East Circum- Pasific Belt Inferred from a Bayesian Approach,* Natural Hazards. *30*, 59–78.

Tsapanos, T. M., and Christova, C. V. (2003), *Earthquake Hazard Parameters in Crete Island and Its Surrounding Area Inferred from Bayes Statistics: An Integration of Morphology of the Seismically Active Structures and Seismological Data,* Pure and Applied Geophysics. *160*, 1517–1536.

Incorporated Research Institutions for Seismology (IRIS) (2013). http://ds.iris.edu/ieb/index.html. Accessed 15 Feb 2013.

Turknet (2013). http://sismo.deprem.gov.tr/sarbis/Shared/Default.aspx.. Accessed 15 Feb 2013.

Wells, D. L., and Coppersmith, K. J. (1994), *New Empirical Relationships Among Magnitude, Rupture Length, Rupture Width, Rupture Area and Surface Displacement.* Bull. Seismol. Soc. Am. *4*, 975–1002.

Wheeler, J. (2009), *The Preservation of Seismic Anisotropy in the Earth's Mantle During Diffusion Creep. 178*, 1723–1732.

Yadav, R. B. S, Tsapanos, T. M., Bayrak, Y., and Koravos, G. CH. (2012), *Probabilistic Appraisal of Earthquake Hazard Parameters Deduced from a Bayesian Approach in the Northwest Frontier of the Himalayas,* Pure and Applied Geophysics. *170*, 283–297.

Yilmaz, Y., Genç, S. C., Gürer, O. F., Bozcu, A. (2000), *When did the Western Anatolian Grabens Begin to Develop? In: Bozkurt, E., Winchester, J. A. and Piper, J. D. A. (eds), Tectonics and Magmatism in Turkey and The Surrounding Area,* Geological Society London, Special Publications. *173*, 353–384.

(Received February 4, 2015, accepted April 6, 2015, Published online April 22, 2015)

Pure Appl. Geophys. 173 (2016), 221–233

DOI 10.1007/s00024-015-1080-3

Pure and Applied Geophysics

Development of a Combination Approach for Seismic Hazard Evaluation

HUAI-ZHONG YU,[1] FA-REN ZHOU,[1] QING-YONG ZHU,[1,2] XIAO-TAO ZHANG,[1] and YONG-XIAN ZHANG[1]

Abstract—We developed a synthesis approach to augment current techniques for seismic hazard evaluation by combining four previously unrelated subjects: the pattern informatics (PI), load/unload response ratio (LURR), state vector (SV), and accelerating moment release (AMR) methods. Since the PI is proposed in the premise that the change in the seismicity rate is a proxy for the change in the tectonic stress, this method is used to quantify localized changes surrounding the epicenters of large earthquakes to objectively quantify the anomalous areas (hot spots) of the upcoming events. On the short-to-intermediate-term estimation, we apply the LURR, SV, and AMR methods to examine the hazard regions derived from the PI hot spots. A predictive study of the 2014 earthquake tendency in Chinese mainland, using the seismic data from 1970-01-01 to 2014-10-01, shows that, during Jan 01 to Oct 31, 2014, most of the $M > 5.0$ earthquakes, especially the Feb 12 M7.3 Yutian, May 30 M6.1 Yingjiang, Aug. 03 M6.5 Ludian, and Oct 07 M6.6 earthquakes, occurred in the seismic hazard regions predicted. Comparing the predictions produced by the PI and combination approaches, it is clear that, by using the combination approach, we can screen out the false-alarm regions from the PI estimation, without reducing the hit rate, and therefore effectively augment the predictive power of current techniques. This provided evidence that the multi-method combination approach may be a useful tool to detect precursory information of future large earthquakes.

Key words: Pattern informatics, load/unload response ration, state vector, accelerating moment release, seismic hazard.

1. *Introduction*

Large earthquakes usually occur on the fault zones, which should be caused by tectonic activity associated with plate margins and faults, when the crust becomes subjected to strain, and eventually moves (ZHANG *et al.* 2003). With the development of modern digital seismic networks monitoring technologies, the number of earthquake predictability studies has increased significantly using earthquake catalogs and observational constraints on fault slip rates. Many researchers have found that the spatial extent of the stress perturbation caused by an earthquake scales with the moment of the event (YIN *et al.* 2002; ZHANG *et al.* 2008), allowing us to establish a physically reasonable framework to combine different precursory methodologies for seismic hazard estimation. BOWMAN and KING (2001) applied this strategy to examine the $M \geq 6.5$ earthquakes in California from 1950 to 2000. They proposed that the effectiveness for seismic hazard evaluation could be enhanced by combining preseismic acceleration of seismicity with the areas of increased Coulomb stress. In addition, KEILIS-BOROK *et al.* (2004) and YU *et al.* (2006a) also linked this combination model to the evolution of seismicity before large earthquakes.

In recent years, the activities of the Collaboratory for the Study of Earthquake Predictability (CSEP; JORDAN 2006; RHOADES and GERSTENBERGER 2009; ZECHAR and JORDAN 2010), promote the studies of this aspect. Combinations of expected earthquake rates mostly using Bayesian approach were developed by MARZOCCHI *et al.* (2012), RHOADES and GERSTENBERGER (2009) and ZECHAR *et al.* (2010) and others. The physical basis of these pattern recognition methods for combining various precursory phenomena were previously developed by GELFAND *et al.* (1976), KEILIS-BOROK (1982), SOBOLEV *et al.* (1991) and KOSSOBOKOV and CARLSON (1995). Recently SHEBALIN *et al.* (2012, 2014) have suggested a combination method based on differential probability gains; the method gives a tool to combine rate-based and alarm-based forecasting models.

In this paper, an alarm-based earthquake prediction method is introduced that combines four

[1] China Earthquake Networks Center, Beijing 100045, China. E-mail: yuhz750216@sina.com; mcszqy@mail.sysu.edu.cn

[2] School of Engineering, Sun Yat-sen University, Guangzhou 510275, China.

prediction methods: the pattern informatics (PI), load/unload response ratio (LURR), state vector (SV) and accelerating moment release (AMR). This work learns from the study of BOWMAN and KING (2001). However, there still are differences. To determine areas where large earthquake should be expected, the regions where there are noticeable increases of Coulomb stress are replaced by the hotspots derived from the PI method. Moreover, the LURR and SV methods are used to test if the stress change is likely to bring the regional crust closer to failure. YU *et al.* (2013) have applied this approach to the large earthquakes in western China between 2007-01-01 and 2010-01-01. To further show the validity of the approach, the earthquakes in Chinese mainland that occurred during the period 2014-01-01 to 2014-10-31 are chosen as the example in this study.

2. Pattern Informatics

The PI is a long-term earthquake prediction method (RUNDLE *et al.* 2000a, 2002, 2003). TIAMPO *et al.* (2002) pointed out that the PI method has much better predictive power than the "relative intensity" and "random catalogs" models. RUNDLE *et al.* (2002) and TIAMPO *et al.* (2002) found that 25 large earthquakes occurred either on areas of hot-spots or within the margin of error of ± 11 km when they applied the method to examine the 27 $M > 5$ earthquakes in southern California between 2000 and 2010. Similar results were obtained by HOLLIDAY *et al.* (2005) when they studied the $M > 7$ earthquakes between 2000 and 2010. Detailed procedures of the PI can be outlined as follows:

1. To divide the target region into the $\Delta x \times \Delta x$ square bins.
2. To define average rate of occurrence of earthquakes in box i over the period t_b to t

$$I_i(t_b, t) = \frac{1}{t - t_b} \sum_{t'=t_b}^{t} N_i(t'), \tag{1}$$

Here, t_b varies between t_0 and t_1 at a time step of Δt, t_0 is the initial time. The time interval t_b–t_1 is the reference period. N_i (t) is the number of earthquakes with magnitude greater than M_c in the ith box. M_c is the magnitude cutoff.

3. To normalize the activity rate function,

$$\overset{\Lambda}{I}_i(t_b, t) = \frac{I_i(t_b, t) - <I_i(t_b, t)>}{\sigma(t_b, t)}, \tag{2}$$

where $< I_i(t_b,\ t) >$ and $\sigma(t_b,\ t)$ are the average activity rate function and its spatial standard deviation over all boxes at time t.

4. To assess the change of the normalized activity rate function for the time period t_1 to t_2,

$$\Delta I_i(t_b, t_1, t_2) = \overset{\Lambda}{I}_i(t_b, t_2) - \overset{\Lambda}{I}_i(t_b, t_1), \tag{3}$$

5. To calculate the probability of change of activity in the ith box,

$$P_i(t_0, t_1, t_2) = \overline{\Delta I_i(t_0, t_1, t_2)^2}, \tag{4}$$

where $\overline{\Delta I_i(t_0, t_1, t_2)} = \frac{1}{t_1 - t_0} \sum_{t_b=t_0}^{t_1} \Delta I_i(t_b, t_1, t_2)$. In phase dynamical systems, probabilities are related to the square of the associated vector phase function (RUNDLE *et al.* 2000a).

6. To evaluate difference between the $P_i(t_0,\ t_1,\ t_2)$ and its spatial mean $< P_i(t_0,\ t_1,\ t_2) >$, representing the probability of change in activity relative to the background,

$$\Delta P_i(t_0, t_1, t_2) = P_i(t_0, t_1, t_2) - <P_i(t_0, t_1, t_2)>. \tag{5}$$

The "hotspots" are defined to be the boxes (or the regions), where ΔP_i $(t_0,\ t_1,\ t_2)$ is positive. According to the PI method, an earthquake with magnitude $>M_c$ $+2$ will occur in the hotspots during the forecasting time window of 3–10 years. The use of this method implicitly assumes earthquake fault systems are in an unstable equilibrium state and can be treated linearly about their equilibrium points.

3. Load/Unload Response Ratio

The LURR is a short-to-intermediate-term earthquake prediction method developed by Yin and others (e.g., YIN *et al.* 2000; YU and ZHU 2010; ZHANG *et al.* 2004). This method is based on measuring the ratio between Benioff strains released during the time periods of loading and unloading, corresponding to the Coulomb failure stress change induced by earth tides on optimally oriented faults. Before occurrence of a large earthquake ($M > 6.0$), anomalous increase in the time series of LURR within a time frame from months to years, has often been observed. This phenomenon can be used as an important precursor to predict large earthquakes. In retrospective studies, the LURR method has been applied in numerous earthquake prediction practices, in which anomalously high LURR values have often been detected months to 2 years before the large earthquakes (Yin *et al.* 2002).

The LURR values that measure the degree of closeness to instability for a nonlinear system can be defined as

$$Y = \frac{X_+}{X_-}, \tag{6}$$

where '+' and '−' refer to the loading and unloading processes, and X is the response rate. Suppose that P and R are, respectively, the load and response of the nonlinear system, then

$$X = \lim_{\Delta P \to 0} \frac{\Delta R}{\Delta P} \tag{7}$$

can be defined as the response rate, where ΔR denotes the small increment of R, resulted from a small change of ΔP on P.

When the system is in a stable state, $X_+ \approx X_-$ and LURR ≈ 1. When the system evolves beyond the linear state, usually $X_+ > X_-$, and LURR > 1. Thus, the LURR can be used as a criterion to judge the state of stability for a system before its macro-failure.

In earthquake prediction practice using the LURR method, the seismic energy release within certain temporal and spatial windows is usually used as data input. Loading and unloading periods are determined by calculating the earth tide induced Coulomb failure stress change along a tectonically favored rupture direction on a specified fault plane (Hainzl *et al.* 2010; Harris 1998). The Coulomb failure stress is defined as

$$\mathrm{CFS} = \tau_n + f\sigma_n, \tag{8}$$

where f, τ_n and σ_n stand for inner frictional coefficient, shear stress and normal stress (positive in tension), respectively, and n is the normal of the fault plane on which the CFS reaches its maximum. When the change of Coulomb failure stress (ΔCFS) > 0, it is in a loading state; and when ΔCFS < 0, it is in an unloading state. The LURR is thus expressed as a ratio between energy released during loading and that released during unloading periods. Specifically,

$$Y_m = \frac{(\sum_{i=1}^{N+} E_i^m)_+}{(\sum_{i=1}^{N-} E_i^m)_-}, \tag{9}$$

where E_i is seismic energy released by the ith event, and $N+$ or $N-$, represent the numbers of events that occurred during the loading and unloading stages, respectively. When $m = 1/2$, E^m denotes the Benioff strain. Note that the focal mechanisms of the small earthquakes are assumed in agreement with that of the main shock to contribute positively to ΔCFS for the main shock. This assumption is supported by studies of Hauksson (1994), Hauksson *et al.* (2002), and Hardebeck and Hauksson (2001), which demonstrated that the focal mechanisms of regional small earthquakes prior to the Landers and Hector Mine earthquakes were quite consistent with that of the ensuing main shocks. The inner frictional coefficient we adopted to calculate ΔCFS is 0.4 (Yin *et al.* 2000). To avoid volatile fluctuations due to poor statistics, the loading and unloading periods are usually summed over many load–unload cycles within the time window. Generally, the time window is 6 months for prediction of the M6 earthquakes, 1 year for the M7 earthquakes, and 3 years for the M8 earthquakes (Yin *et al.* 2002). Circular region is usually adopted as the spatial window in LURR practice and the optimal critical region scale for LURR evaluation is determined by computing the LURR anomaly within differently sized regions centered at epicenter of the upcoming large event to reach the maximum LURR precursory anomaly (Yin *et al.* 2000). The forecasting time window, from months to years, is determined by magnitude of the detection earthquake: the larger the earthquake, the longer the time (Zhang *et al.* 2006).

4. State Vector

The concept of SV stems from statistical physics, where it is usually used to describe activity patterns of a physical field in the way of coarse-grain (Reichl 1980). We transplanted the idea of state vector from statistical physics into seismology to characterize the spatial and temporal evolution of seismic activities.

The SV is n-dimensional vector which is defined by dividing the target region into n uniform sub-regions (Fig. 1) and the sum of seismic magnitudes in each sub-region within certain temporal window is computed as the component of the n-dimensional vector. Different state vectors at different times can form a trajectory in the phase space. If a series of state vectors at different time steps has been acquired, the temporal and spatial evolution of seismicity may be obtained. Generally, four relevant scalars are defined to quantitatively measure the evolution of SV.

1. The modulus of vector $\mathbf{V}_k$,

$$\mathrm{M} = |\mathbf{V}_\mathrm{k}|, \tag{10}$$

where $\mathbf{V}_k$ is the state vector at time t_k ($k = 1, 2\ldots n$), which slides at a time step of Δt.

2. The angle between vectors $\mathbf{V}_k$ and $\mathbf{V}_{k+1}$,

$$\varphi = \arccos\left(\frac{\mathbf{V}_{k+1}\cdot\mathbf{V}_k}{\mathrm{V}_{k+1}\mathrm{V}_k}\right). \tag{11}$$

3. The modulus of vector increment $\mathbf{V}_{k+1} - \mathbf{V}_k$,

$$\varDelta\mathrm{M} = |\mathbf{V}_{k+1} - \mathbf{V}_k|. \tag{12}$$

4. The angle between vector $\mathbf{V}_k$ and equalized vector $\mathbf{V}_e$,

$$\varphi_c = \arccos\left(\frac{\mathbf{V}_e\cdot\mathbf{V}_k}{\mathrm{V}_e\mathrm{V}_k}\right), \tag{13}$$

where the equalized vector $\mathbf{V}_e$ consists of equal components.

Retrospective tests of this method on the large earthquakes ($M > 6.0$) occurred in Chinese mainland over the last four decades show that the anomalously high values have often been observed in the SV scalar time series months to a year before the earthquakes (Yu *et al.* 2006b).

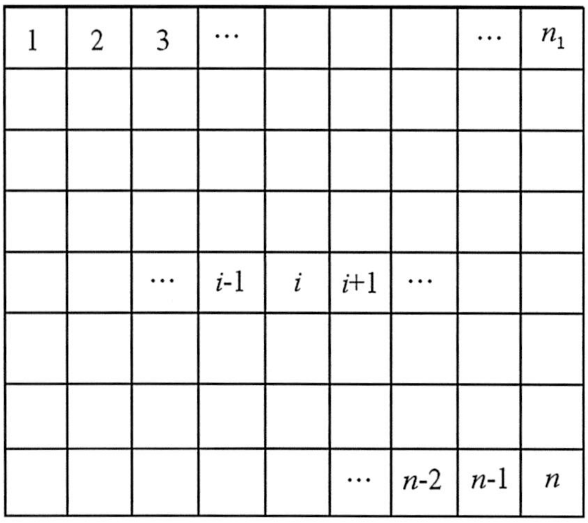

1	2	3	···				···	n_1
		···	i-1	i	i+1	···		
					···	n-2	n-1	n

Figure 1
Division of a continuum into n sub-regions

5. Accelerating Moment Release

Prior to the occurrence of large or great earthquakes the accelerating moment release (AMR) is usually observed (Jaume and Sykes 1999). Bufe and Varnes (1993) suggested that a simple power-law time-to-failure equation derived from damage mechanics could be used to model the observed seismicity. This hypothesis is an outgrowth of efforts to characterize large earthquakes as a critical phenomenon (Rundle 1989). The function has the following form:

$$\varepsilon_p(t) = A + B(t_c - t)^z, \tag{14}$$

where t_c is the time of the large event, B is negative and z is the exponent. A is the value of $\varepsilon(t)$ when $t = t_c$ (i.e., the final Benioff strain up to and including the largest event). The cumulative Benioff strain at time t is defined as

$$\varepsilon(t) = \sum_{i=1}^{N(t)} E_i(t)^{\frac{1}{2}}, \tag{15}$$

where E_i is the energy of the ith event and $N(t)$ is the number of events at time t.

6. Methodology of Combination

Our combination approach is composed of above four prediction methods.

We first apply the PI method to search for the hot spots with a certain magnitude cutoff. Then we use the distribution of PI hotspots to detect areas where large earthquake should be expected by covering the hot spots with circular critical regions from the low to high latitude, and, low–high longitude. The radius of the critical regions is decided by the seismic magnitude of the target earthquake, in which the statistical slope of critical regions and size-magnitude is about 0.36 (Ben-Zion and Lyakhovsky 2002; Yin *et al.* 2002). And, the centers of the critical regions are determined by computing the distances between the PI hotspots. The hotspots that the distance between any 2 of them is less than $2r$ form a potential critical region, and the midpoint coordinate of the hotspots is set to be the center of the circular region. Here, the distance between two critical regions should be:

$$d \geq 2rc, \tag{16}$$

where r is the radius of the critical region, and c is 0.8 (usually $0 < c < 1$), denoting the contact ratio between two critical regions.

Next, we evaluate the short-to-intermediate-term earthquake potential in the critical regions using the LURR and SV methods. In the LURR calculation, the Benioff strain of the earthquakes of $0 < M < 4.0$ within the critical regions was computed as the loading/unloading response, and the internal friction coefficient we adopted to calculate ΔCFS is 0.4. An anomalous signal means the LURR value above the threshold of 1.0. As for the SV calculation, each circular critical region was regarded as an SV target area and subdivided into the 50 × 50 (km^2) sub-regions. It should be pointed out that, before the calculation we have complemented the critical regions to the square areas, i.e., we have built the square areas with side length of $2r$ centered at the epicenters of each critical region, and small earthquakes of $0 < M < 4.0$ in the square areas are used for the SV computation. If LURR and two or more SV scalars changed anomalously during 2012 to 2013, the region is retained, otherwise, removed. Unlike the LURR method, the anomaly in the SV time series is not decided by setting a threshold value, but by subtracting the average value of the whole time series. If the difference is greater than 0, it is an anomalous value.

Finally, the AMR method is used to assess time and magnitude of the impending earthquake in each critical region. We fixed the z value to fit A, B and t_c in Eq. (14). According to the definition, the asymptote time, t_c, is the occurrence time of the target earthquake. A is the cumulative Benioff strain at time t_c, which can be used to assess seismic magnitude of the earthquake:

$$\mathrm{M}_s = \left\{ \lg\left[A - \varepsilon(t_p)\right]^2 - 4.8 \right\} / 1.5, \tag{17}$$

where t_p is time of the prediction made. The relationship between energy and magnitude is derived from Gutenberg and Richter (1956) and Kanamori (1977). The assessed values will be displayed in the final hazard evaluation, if they are consistent with the predictions made by the above three methods. The exponent z is a focus of this stage. Sornette (1992) found that $z = 1/2$ is associated with a critical transition. Rundle *et al.* (2000b) used scaling arguments to show that power-law time-to-failure buildup of cumulative Benioff strain may represent the scaling regime of a spinodal phase transition, with an exponent $z = 1/4$. Ben-Zion and Lyakhovsky (2002) have listed the z value of seismological observations from various authors. They concluded that the exponents fall in the range 0.1–0.55 and $z = 1/3$ for the damage rheology model of Lyakhovsky *et al.* 1997. Similar results were obtained by using the fiber-bundle model of Turcotte *et al.* (2003). Thus, we use $z = 1/3$ to fit all the seismic data.

7. Application to Seismic Data

Using the approach above, we evaluated the seismic hazard in Chinese mainland in 2014. The prediction was made on Oct 1, 2013, and relevant report was submitted to Annual Consultation, a special kind of short-to-intermediate-term earthquake forecast regularly undertaken in China earthquake networks center, on Oct 18th (Yu *et al.* 2013). Seismic data from 1970-01-01 to 2014-10-01 are used

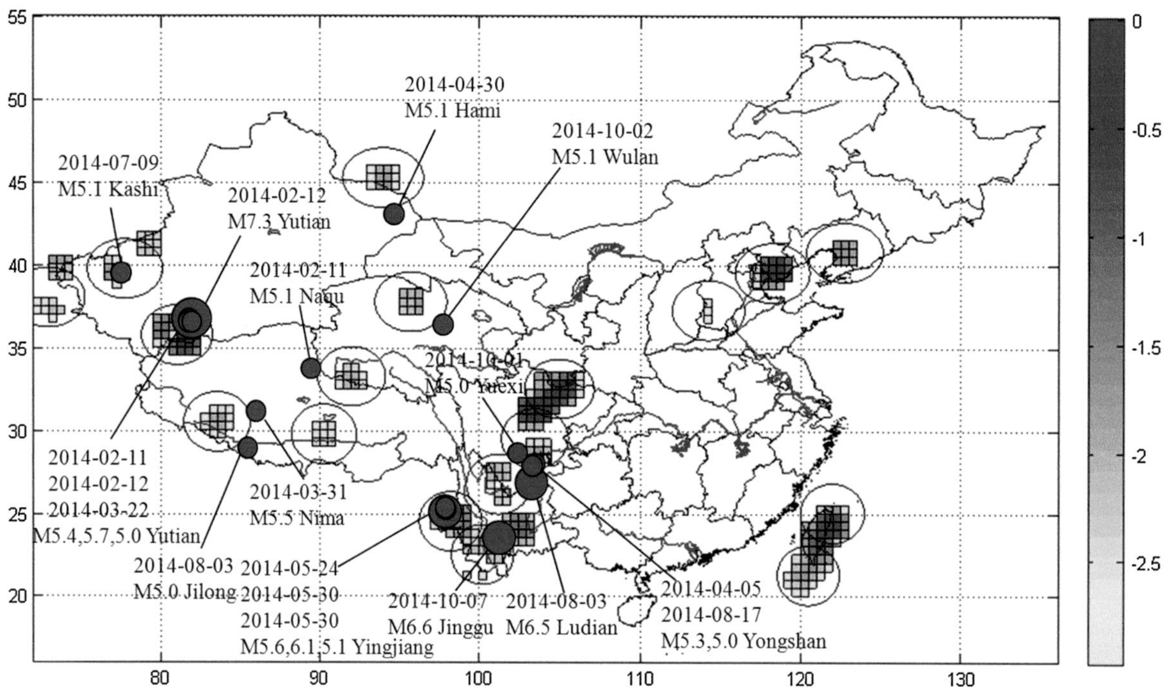

Figure 2
The PI hotspots distribution in China within the forecast period of 2014-01-01 to 2019-01-01. Values of $\log_{10}(\Delta P/\Delta P_{max})$ in the *boxes* are given by *different colors* and the *color code* is explained according to a *color bar right side* of the figure; the deeper the *color*, the higher the value. *Black contours* delineate the seismic hazard regions derived from the PI hotspots. *Red dots* represent the $M > 5$ earthquakes during Jan 1 to Oct 31, 2014

(their catalog retrieved from the China Seismic Network Earthquake Catalogue).

Figure 2 shows the hotspots detected by the PI method, with $\Delta x = 0.5^\circ$ and $\Delta t = 10$ days. According to magnitude of the detection earthquake, the magnitude cutoff is: $M_c = 3.5$. Detailed parameters for PI calculation are: $t_0 =$ 1970-01-01, $t_1 =$ 2009-01-01, $t_2 =$ 2014-01-01, $t_3 =$ 2019-01-01, i.e., the forecast and reference periods are 2014-01-01 to 2019-01-01 and 2009-01-01 to 2014-01-01, respectively. To clearly show the spatial change of the hot spots, value of ΔP (Eq. 5) in each box is calculated using the logarithm of $\log_{10}(\Delta P/\Delta P_{max})$, where ΔP_{max} is the maximum of all the boxes. Also shown in the Fig. 1 are the potential critical regions derived from the PI hotspots. The radius of the seismic hazard areas is 200 km, corresponding to the critical region of a M6 earthquake.

Figure 3 displays the evaluation of LURR anomalies (LURR > 1.0) within the critical regions shown in Fig. 2, with a temporal window of 1 year at a sliding step of 1 month. The scanning radius is 200 km, and the slippage is 25 km. All the insets in Fig. 3 have the same effect in the forecasting. Note that because regions 1 and 2 are not in the Chinese mainland, and regions 12 and 18 are at the China border or outside, they were not discussed in this study. Figure 4 shows the time series of the 4 SV scalars for each critical region, with the same temporal window and sliding step for the LURR evaluation. Corresponding precursory accelerating seismicity and the asymptote of the power-law time-to-failure function are shown in Fig. 5. And the final predictions produced by the combination approach are shown in Fig. 6. All the $M > 5$ earthquakes occurred during Jan 1 to Oct 31, 2014 are also shown in the figure. It is clear that before occurrence of the earthquakes, the LURR and SV time series climbed to the anomalously high values. The asymptote time and magnitude made by the AMR method are consistent with the occurrence time of the events.

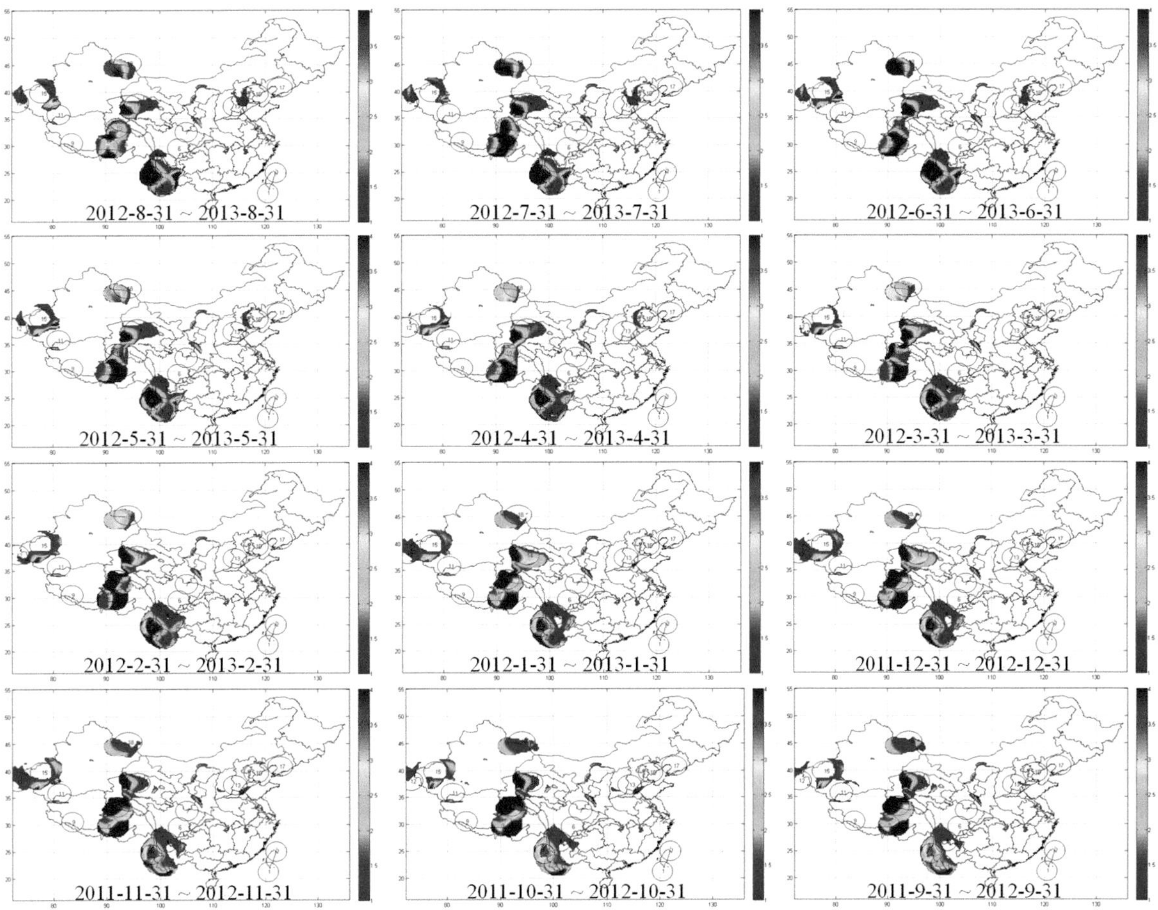

Figure 3
The distribution of LURR anomalies from 2013-8-31 to 2011-9-31 within the critical areas shown in Fig. 1. The *color code bar* at *right side* of the figure shows value of the LURR anomalies

8. Assessment of the Prediction

Our combination approach is more sensitive to detect the earthquakes with larger magnitude. As shown in Fig. 6, all the $M > 6.0$ and part of the $M > 5.0$ earthquakes occurred in the predicted seismic hazard regions during Jan 1 to Oct 31, 2014. Table 1 lists the detailed prediction statistics at different magnitude threshold, in which the H_C, indicating the predictive power of the approach (similar to the R value given by Ma *et al.* 2004), is defined as:

$$H_C = \frac{E_H}{E_T} - \frac{S_T}{S_A}, \tag{18}$$

where E_H and E_T denote, respectively, the number of detected and target earthquakes, and S_T and S_A represent the area of the seismic hazard regions and Chinese mainland, respectively. In fact the value Hc is also equal to $1-\nu-\tau$, one of possible "loss functions" connected to the Molchan error diagram (Molchan 1991), in which ν is the rate of misses and τ is the rate of alarms, representing, respectively, the $1 - E_H/E_S$ and S_T/S_A in Eq. (18). The hit rate seems to highly correlate with the magnitude of detection earthquake (Table 1), suggesting the significant predictive power of the approach: it has moderate predictive power when the magnitude of detection earthquake is greater than 5.0, relatively high predictive power when $M > 5.5$, very significant predictive power when greater than 6.0. In fact, Yu *et al.* (2013) have investigated the predictive power of the approach by testing the large earthquakes in western China between 2007-01-01 and 2010-01-01. Results show that for the $M > 6.5$ earthquakes, value

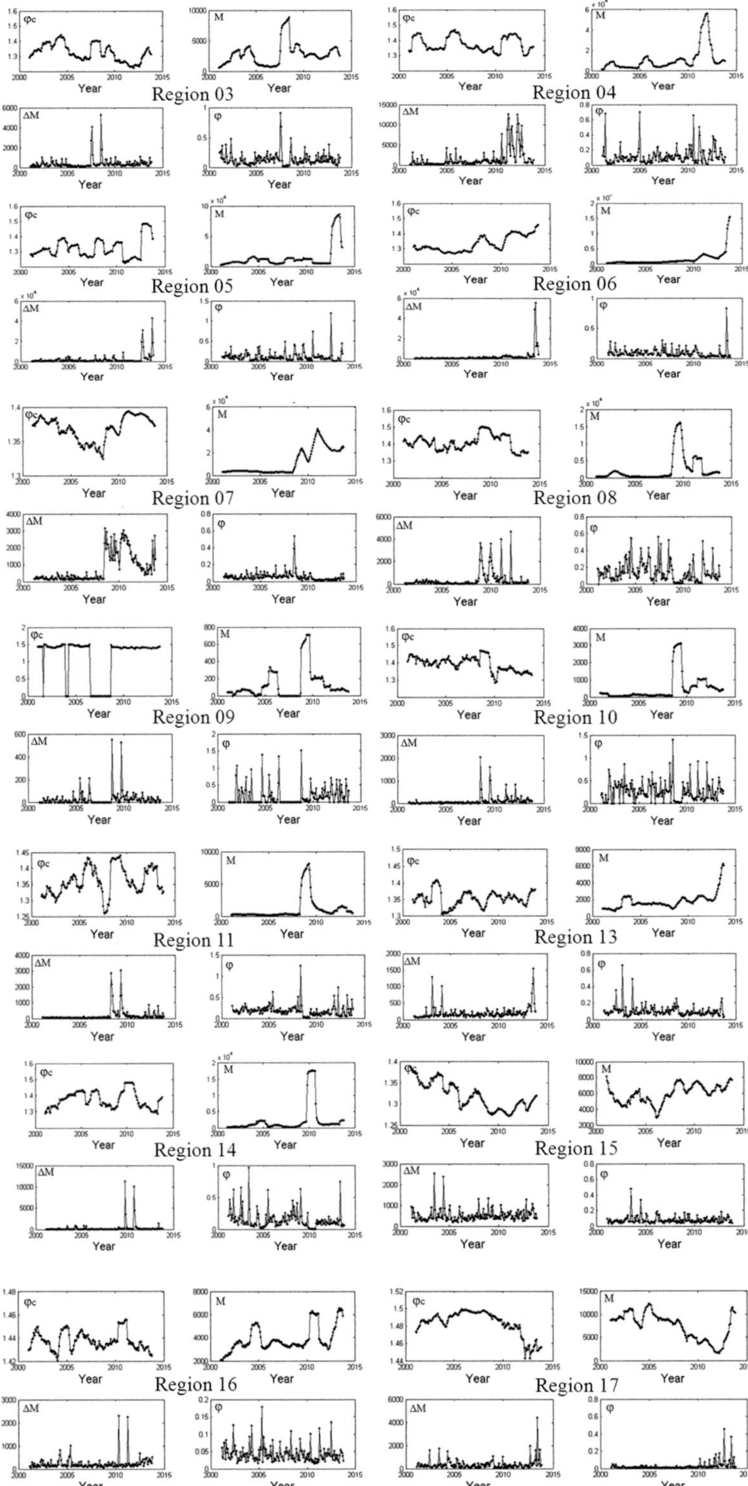

Figure 4

Traces of the 4 SV scalars for the regions 3–11 and 13–17 shown in Fig. 1. The computation time window for each figure is 1 year, and the sliding step is 1 month

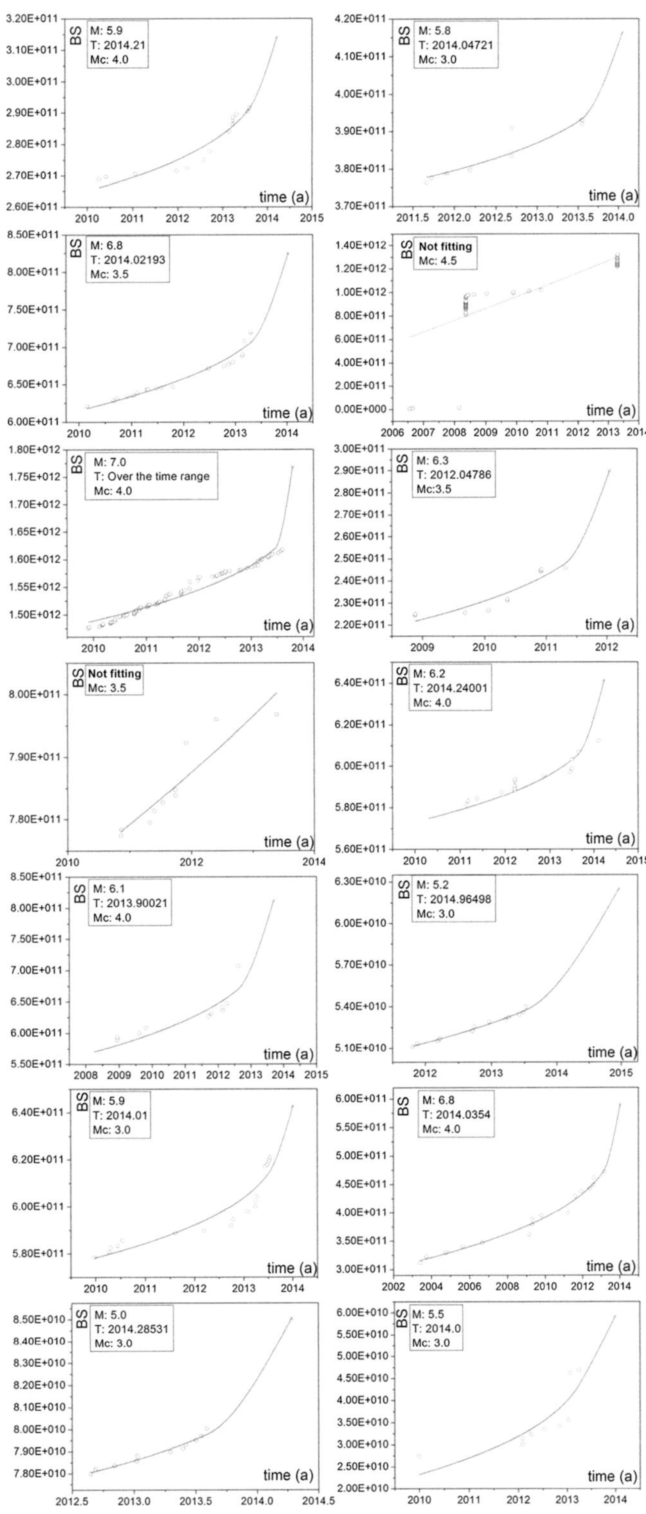

Figure 5

Cumulative Benioff strain of the earthquakes in the regions 3–11 and 13–17 shown in Fig. 1 and the fits of the data to the power-law time-to-failure equation. '×' is the asymptote estimates of the ensuing large earthquakes. Images sequence is the same as Fig. 3. M_c the lowest magnitude in the catalog used in the *curves* fitting, *BS* Benioff strain

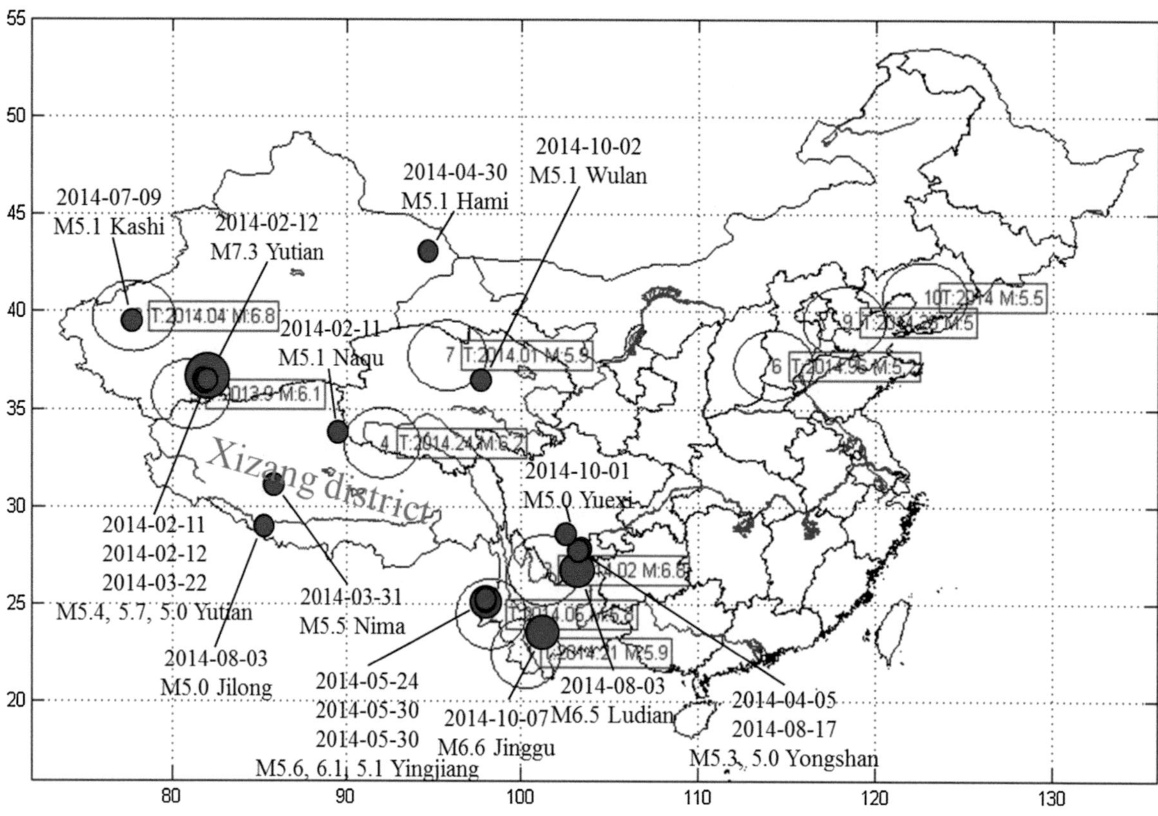

Figure 6

The 2014 seismic hazard regions in Chinese mainland (*black contours*) predicted by the combination approach. *Red boxes* indicate the earthquake time and magnitude. Also shown in the figure are the $M > 5$ earthquakes (*red dots*) during the period of 2014-01-01 to 2014-10-31

Table 1

The predictive power of the combination approach at different magnitude size

Forecast (magnitude)	Observed events			H_c
	Yes	No	Total	
5.0	15	3	18	0.70
5.1	12	2	14	0.73
5.3	8	1	9	0.76
5.4	7	1	8	0.74
5.5	6	1	7	0.73
5.6	6	0	6	0.87
5.7	5	0	5	0.87
6.0	4	0	4	0.87

of H_C is about 0.67, and 0.72 for $M > 7.0$. Although the value is relative lower, the conclusion is basically consistent with our study: the larger the earthquake, the higher the predictive power.

9. Discussion

The PI technique is founded on the premise that the change in the seismicity rate is a proxy for the change in the underlying stress, which is of great sensitivity to detect location of the ensuing large earthquake. In this sense, we first use PI method to search for the seismic hazard regions, and then apply the LURR, SV, and AMR methods to search for the seismic potential evaluation with estimates of the crucial parameters quantitatively, such as earthquake time, magnitude. Here, the magnitudes of the earthquakes assessed by the AMR method should correspond to those predicted by the PI method. Comparing the predictions made by the PI (Fig. 2) and combination approaches (Fig. 6), it is clear that both approaches have detected most of the large events occurred during the forecasting period of 2014-01-01 to 2014-10-31 (such as the Feb 12 M7.3 Yutian, May 30 M6.1 Yingjiang, Aug 03 M6.5

Table 2

The predictive power of the PI approach at different magnitude size

Forecast (magnitude)	Observed events			$H_{\rm pI}$
	Yes	No	Total	
5.0	12	6	18	0.58
5.1	9	5	14	0.56
5.3	7	2	9	0.69
5.4	6	2	8	0.67
5.5	5	2	7	0.63
5.6	5	1	6	0.75
5.7	4	1	5	0.72
6.0	3	1	4	0.67

Ludian and Oct 07 Jinggu earthquakes); just only a few earthquakes are missed: the March 31 M5.5 Nima, Aug 03 M5.0 Jilong, Xizang and Apr 30 M5.1 Hami, Xinjiang earthquakes. In fact, Hami earthquake is near the border, while the Nima and Jilong earthquakes are located in the weak areas of seismic monitoring. The catalogs quality of the earthquake is questionable.

More detailed comparison between the results obtained using the two approaches, however, does show some noticeable differences. The statistics of the predictions made by the PI approach are listed in Table 2, in which the $H_{\rm PI}$ is calculated as $H_{\rm C}$ in Eq. (18). The one produced by the combination approach looks more prominent than the other one produced by the PI method. The reason may be that the numbers of hazard regions are 14 vs. 10 for the PI and combination approaches (Figs. 2, 6). Such contrast makes the prediction stand out more clearly in Fig. 5, and would represent less chance of false alarm for the large events. On the other hand, more target earthquakes were missed using PI method in comparison to our combined method, because we enlarge the selected PI hotspots using the circular regions of 200 km radius. For some earthquakes, such as the Aug 03 M6.5 Ludian earthquake located at northwest Yunnan region, significant anomalies are found if the combination approach is used, and are not found if just the PI method is used for seismic hazard evaluation. The result may suggest that during the establishment of the criticality, seismic risk is not only on the seismogenic fault but also in a region surrounding it (Bowman and King 2001).

10. Conclusion

We found that, by combining the PI method with the LURR, SV and AMR methods, the predictive power of current techniques can be improved. While we illustrate the approach using seismicity in Chinese mainland, the technique is general and can be applied to any tectonically active region. Analysis of the anomalies in corresponding time series may provide us with seismic potential evaluation with estimates of all the crucial parameters provided such as earthquake location, time, and magnitude. Although more data testing is needed to further verify the approach, it is ready to be employed in the real world for quantitative prediction of future large earthquakes.

Acknowledgments

We are grateful to the careful reviews made by the editor and two anonymous reviewers. The research was supported by the Spark Program of Earthquake Science of China (Grant No. XH12058), the National Natural Science Foundation of China (Grant No. 91230114), the International Science & Technology Cooperation Program of China (Grant No. 2010DFB20190), and the State Scholarship Fund of China Scholarship Council.

References

Ben-Zion, Y., Lyakhovsky, V. (2002), *Accelerated seismic release and related aspects of seismicty patterns on earthquake faults.* Pure appl. Geophys, *159* (10), 2385–2412.

Bowman, D. D., and King, G. C. P., (2001), *Accelerating Seismicity and stress Accumulation Before large earthquake*, Geophys. Res. Lett., *28* (21), 4039–4042.

Bufe, C. G., Varnes, D. J. (1993), *Predictive modelling of the seismic cycle of the greater San Francisco bay region.* J. Geophys. Res., *98*, 9871–9883.

Gelfand, I. M., S. A. Guberman, V. I. Keilis-Borok, L. Knopoff, F. Press, E. Y. Ranzman, I. M. Rotwain, and A. M. Sadovsky, (1976), *Pattern recognition applied to earthquake epicenters in California.* Phys. Earth and Planet. Inter., *11*, 227–283.

Gutenberg B and Richter C F. 1956. *Earthquake magnitude, intensity, energy and acceleration.* Bull. Seism. Soc. Am., *46*, 105–145.

Hainzl, S., G. Zoller, and Wang, R. (2010), *Impact of the receiver fault distribution on aftershock activity.* J. Geophys. Res., *115*, B05315, doi:10.1029/2008JB006224.

Hardebeck, J., L., and Hauksson, E. (2001), *Crustal stress field in southern California and its implications for fault mechanics*, J. Geophys. Res., *106*, 21859–21882.

Harris, R. A. (1998), *Introduction to special section: stress triggers, stress shadows, and implication for seismic hazard*, J. Geophys. Res., *103*, 24347–24358.

Hauksson, E. (1994), *State of Stress from Focal Mechanisms Before and After the 1992 Landers Earthquake Sequence*, Bull. Seism. Soc. Am., *84* (3), 917–934.

Hauksson, E. L., Jones, M., and Hutton, K. (2002), *The 1999 Mw7.1 Hector Mine, California, earthquake sequence: Complex conjugate strike-slip faulting*, Bull. Seism. Soc. Am., *92*, 1154–1170.

Holliday, J. R., Rundle, J. B., Tiampo, K. F., Klein, W., and Donnellan, A. (2005), *Systematic procedural and sensitivity analysis of the Pattern Informatics method for forecasting large (M > 5) earthquake events in Southern California*, Pure Appl Geophys., *163* (11–12), 2433–2454.

Jaume, S. C. and Sykes, L. R. (1999), *Evolution toward a critical point: a review of accelerating seismic moment/energy release prior to large great earthquakes*, Pure Appl. Geophys., *155* (2–4), 279–305.

Jordan, T. H. (2006). *Earthquake predictability, brick by brick*, Seismol. Res. Lett. *77*, 3–6.

Kanamori H. 1977. *The energy release in great earthquakes.* J Geophys Res. *82*, 2981–2987.

Keilis-Borok V. I. (1982), *A worldwide test of three long-term premonitory seismicity patterns: A review.* Tectonophysics, *85*, 47–60.

Keilis-Borok V., Shebalin P., Gabrielov A., Turcotte D., (2004), *Reverse tracing of short-term earthquake precursors*, Physics of the Earth and Planetary Interiors,*145* (1–4), 75–85.

Kossobokov V. G. and Carlson, J. M. (1995), *Active zone size vs. activity: A study of different seismicity patterns in the context of the prediction algorithm M8.* J. Geophys. Res., *100*, 0431–0441.

Lyakhovsky V., Ben-Zion Y., Agnon A., (1997), *Distributed damage, faulting, and friction*, J. Geophys. Res. *102* (B12), 27635–27649.

Ma H. S., Liu J., Wu H., and Li J F., (2004), *Scientific evaluation of annual earthquake prediction efficiency based on R-value*, Earthquake, *24*(2), 31–37 (In Chinese).

Marzocchi, W., Zechar, J. D., Jordan, T. H., (2012), *Bayesian Forecast Evaluation and Ensemble Earthquake Forecasting*, Bull. Seismol. Soc. Am., *102* (6), 2574–2584.

Molchan. G. M., (1991), *Structure of optimal strategies in earthquake prediction.* Tectonophysics, *193*, 267–276.

Reichl, L. E., A Modern Course in Statistical Physics (University of Texas, 1980).

Rhoades, D. A., and M. C. Gerstenberger (2009). *Mixture models for improved short-term earthquake forecasting*, Bull. Seismol. Soc. Am. *154*, no. 2A, 636–646.

Rundle, J. B. (1989), *A physical model for earthquakes III*, J. Geophys. Res., *94*, 2839–2855.

Rundle, J. B., Klein, W., Tiampo, K., and Gross, S. (2000a), *Linear pattern dynamics in nonlinear threshold systems*, Phys. Rev., E *61*, 2418–2431.

Rundle, J. B., Klein, W., Tiampo, K., and Gross, S. (2000b), *Precursory seismic Activation and Critical - point Phenomena*, Pure and Appl. Geophs. *157*, 2165–2182.

Rundle, J. B., Tiampo, K. F., Klein, W., and Martins, J. S. S. (2002), *Self-organization in leaky threshold systems: The influence of near-mean field dynamics and its implications for earthquakes, neurobiology, and forecasting*, Proc. Natl. Acad. Sci. USA, *99*, 2514–2521.

Rundle, J. B., Turcotte, D. L., Shcherbakov, R., Klein, W., and Sammis, C. (2003), *Statistical physics approach to understanding the multiscale dynamics of earthquake fault systems*, Rev. Geophys., *41* (4), 1019, doi:10.1029/2003RG000135.

Shebalin P., C. Narteau, and M. Holschneider, (2012), *From Alarm-Based to Rate-Based Earthquake Forecast Models*, Bull. Seismol. Soc. Am., *102* (1), 64–72.

Shebalin P., Narteau C., Zechar J. D. and Holschneider M., 2014, *Combining earthquake forecasts using differential probability gains*, Earth, Planets and Space 2014, *66*: 37.

Sobolev G. A. Chelidze. T. L. and Zavyalov A. D. (1991), *Map of expected earthquakes based on a combination of parameters*, Tectonophysics, *193*, 255–266.

Sornette, D. (1992), *Mean-field Solution of a Block-spring Model of Earthquake*, J. Phys I France, *2*, 2089–2096.

Tiampo, K. F., Rundle, J. B., McGinnis, S. A., and Klein, W. (2002), *Pattern dynamics and forecast methods in seismically active regions*, Pure Appl Geophys., *159* (10), 2429–2467.

Turcotte, D. L., W. I., Newman and R. Shcherbakov, (2003), *Micro- and Macro-scopic models of rock fracture*, Geophys. J. Int. *152*, 718–728.

Yin, X. C., Wang, Y. C., Peng, K. Y., Bai, Y. L., Wang, H. T., and Yin, X. F. (2000), *Development of a New Approach to Earthquake Prediction-Load/unload Response Ratio (LURR) Theory*, Pure Appl Geophys., *157* (11–12), 2365–2383.

Yin, X. C., Mora, P., Peng, K. Y., Wang, Y. C., and Weatherly, D. (2002). *Load-Unload Response Ratio and Accelerating Moment/Energy Release, Critical Region Scaling and Earthquake prediction*, Pure Appl. Geophys., *159*, 2511–2524.

Yu H. Z., Shen Z. K., Wan Y. G. Zhu Q. Y. Yin X. C. (2006a), *Increasing critical sensitivity of the Load/Unload Response Ratio before large earthquakes with identified stress accumulation pattern*, Tectonophysics, *428* (1–4), 87–94.

Yu, H. Z., Yin, X. C., Zhu, Q. Y., and Yan, Y. D. (2006b), *State vector: a new approach to prediction of the failure of brittle heterogeneous media and large earthquakes*, Pure Appl. Geophys., *163* (11–12), 2561–2574.

Yu, H. Z., and Zhu, Q. Y., (2010), *A probabilistic approach for earthquake potential evaluation based on the Load/Unload Response ratio method, Concurrency and Computation: Practice and Experience*, *22*, 1520–1533.

Yu H. Z., Cheng J., Zhang X. T., Zhang L P., Liu J., and Zhang Y X., (2013), *Multi-Methods Combined Analysis of Future Earthquake Potential*, Pure Appl. Geophys. *170* (1–2), 173–183.

Zechar, J., and T. Jordan (2010). *The area skill score statistic for evaluating earthquake predictability experiments*, Pure Appl. Geophys. *167*, 893–906.

Zechar, J.D., M.C. Gerstenberger and D.A. Rhoades (2010). *Likelihood-based tests for evaluating space-rate-magnitude earthquake forecasts*, Bull. Seism. Soc. Am., *100* (3), 1184–1195

Zhang, H. H., Yin, X. C., Liang, N. G., Yu, H. Z., Li, S. Y., Wang, Y. C., Yin, C., Kukshenko, V., Tomiline, and N., Elizarov, S. (2006), *Acoustic emission experiments of rock failure under load simulating the hypocenter condition*, Pure Appl Geophys., *163* (11–12), 2389–2406.

Zhang P. Z., Deng Q. D., Zhang G. M., Ma J., Gan W. J., Min W., Mao F. Y., and Wang Q. (2003), *Active tectonic blocks and strong earthquakes in the continent of China*, Science in China (series D), *46* (supp), 13–24.

ZHANG, Y. X., YIN, X. C., and PENG, K Y. (2004), *Spatial and Temporal Variation of LURR and its Implication for the Tendency of Earthquake Occurrence in Southern California*, Pure Appl Geophys., *161* (11–12), 2359–2367.

ZHANG, Y. X., WU, Y. J., YIN, X. C., *et al.* (2008), *Comparison Between LURR and State Vector Analysis Before Strong Earthquakes in Southern California Since 1980.* Pure Appl Geophys., *165* (3–4), 737–748.

(Received August 27, 2014, accepted April 7, 2015, Published online April 24, 2015)

Pure Appl. Geophys. 173 (2016), 235–244

DOI 10.1007/s00024-015-1079-9

Pure and Applied Geophysics

A Strategy for a Routine Pattern Informatics Operation Applied to Taiwan

LING-YUN CHANG,[1] CHIEN-CHIH CHEN,[1] YI-HSUAN WU,[1] TZU-WEI LIN,[2] CHIEN-HSIN CHANG,[2] and CHIH-WEN KAN[2]

Abstract—We systematically investigated precursory seismic patterns using the pattern informatics (PI) method and suggest an operable procedure for making PI maps for all seasons, in the context of earthquake forecasting. We examined the PI patterns before several inland earthquakes with magnitudes larger than 6, which occurred between 2001 and 2010 in Taiwan. We fixed a cutoff magnitude and a change interval, which is the time span used to calculate the seismicity change. Our results show that locations with high PI anomalies are typically associated with large earthquakes when the cutoff magnitude is 3.2 and the change interval is 4 years. Therefore, the PI method can be utilized as a routine forecasting tool with regular updates, such performing the PI calculation every season. We also conducted random tests, the results of which indicate a significant difference between large events and random, hypothetical events.

Key words: Pattern informatics, earthquake forecasting, phase dynamics, Taiwan seismicity.

1. Introduction

Spatiotemporal seismicity is related to crustal stresses and stress fields that change complexly over various scales of time and space. The occurrence of a large earthquake may be related to the critical stress threshold of a fault plane (CHEN *et al.* 2006b). Thus, detecting whether unusual seismicity occurs before a large earthquake has been proposed as the foundation of earthquake prediction in many studies (TIAMPO and SHCHERBAKOV 2012). This study focuses on pattern informatics, as a branch of earthquake prediction algorithms. Pattern informatics (PI) originated from RUNDLE *et al.* (2000) and has been further developed by TIAMPO *et al.* (2002), NANJO *et al.* (2006), HOLLIDAY *et al.* (2007) and WU *et al.* (2008b) over the past decade. Pattern informatics describes anomalous changes in seismicity (CHEN *et al.* 2005; TIAMPO *et al.* 2002), including activation and quiescence, as a point process in a phase dynamical system. The spatiotemporal evolution of the dynamical system can be characterized by the phase drift. In pattern informatics, the phase drift can be quantified by calculating the seismic activity over a period of time, termed the change interval, and normalizing over the background seismicity.

We applied pattern informatics to the retrospective analyses of several earthquakes with magnitudes greater than 6 and depths shallower than 25 km, which occurred on Taiwan Island from 2001 through 2010. The results of these retrospective pattern informatics analyses allowed further insight into the standard procedure of the routine operation of pattern informatics forecasting. We also randomized the PI parameters, including the cutoff magnitude and the duration of change interval, to examine the significance of PI forecasts.

2. Methods

Pattern informatics can describe changes in seismicity through a phase dynamical system. Changes in seismicity over time are associated with rotations of the state vector in Hilbert space. Stress accumulation and release cause the state vector in the phase dynamical space of the seismic system to change. Excepting the normalized length of the state vector, all information related to a change in the dynamical system can be described by the rotation of the phase angle. Thus, pattern informatics uses the phase drift to express the spatial and temporal

[1] Graduate Institute of Geophysics and Department of Earth Sciences, National Central University, Jhongli 32001, Taiwan, ROC. E-mail: maomaowyh@gmail.com; yhwu@ncu.edu.tw

[2] Seismological Center, Central Weather Bureau, Taipei 10048, Taiwan, ROC.

evolution of the dynamical system. Pattern informatics calculates the angle of drift over a period of time, the change interval, to predict whether the area is at seismic risk in the coming period of time, the predict interval.

The procedures for the PI method are described as follows:

1. The study area is binned into many grids of size $0.1° \times 0.1°$.
2. Four time parameters are defined. First, t_0 denotes the beginning time of the catalog. Second, t_1 and t_2 mark the start and end of the change interval, respectively. Finally, a sampling reference time, t_b, shifts between t_0 and t_1. The t_b is shifted by 3 days to remove some of the clustering aftershock and background fluctuations.
3. The seismic intensity, $I(x_i, t_b, t)$ is defined as the average number of earthquakes between t_b and t with magnitudes larger than the cutoff magnitude, M_C, that occur in the grid box, x_i, and its eight neighboring boxes. Thus, the change in intensity during the change interval, i.e., $\Delta I(x_i, t_b, t_1, t_2) = I(x_i, t_b, t_1) - I(x_i, t_b, t_2)$, can be computed, and is denoted by $\Delta I(x_i, t_b)$.
4. For each x_i, shifting by t_b produces a time series of $\Delta I_{x_i}(t_b)$. We then perform a temporal normalization and obtain a temporally normalized intensity change, $\Delta \tilde{I}_{x_i}(t_b)$, for each location, x_i. After temporal normalization, we spatially normalize the spatial series for each t_b to obtain a spatio-temporally normalized intensity change $\Delta \hat{I}_{x_i}(x_i, t_b)$.
5. The normalized intensity change, $\Delta \hat{I}_{x_i}(x_i, t_b)$, is computed at each location. To consider both activation and quiescence, we take the absolute value of the change. The temporal average of $|\Delta \hat{I}(x_i, t_b)|$, denoted by $\overline{|\Delta \hat{I}(x_i)|}$, is then computed for each location. Finally, the mean squared change, $P(x_i) = \overline{|\Delta \hat{I}(x_1)|^2}$, which indicates the relative possibility of large-threshold events, is computed.
6. In a PI map, the mean-subtracted index of the probability of occurrence, i.e., $\Delta P(x_i) = P(x_i) - \mu_P$, is color-coded and plotted. Here, μ_P is the mean of $P(x_i)$ over all boxes and can be considered to be the average background probability for the entire area.

3. Data

In this study, we analyzed the data catalog from the Central Weather Bureau Seismic Network (CWBSN) in Taiwan. The CWBSN has been responsible for monitoring regional seismic activity since 1991. It currently consists of a central recording system with 71 telemeter stations, which are equipped with 3-component Teledyne/Geotech S13 seismometers. The CWBSN has had enhanced earthquake monitoring capability in Taiwan since the end of 1993 (Wu and Chiao 2006, Wu *et al.* 2008b). Figure 1 shows the distribution of earthquake depth versus time. The plot shows a considerable increase in the number of earthquakes after 1994. Therefore, considering the consistency of the earthquake catalog, we set 01/01/1994 as the starting point of the data, t_0. In Fig. 1, we denote the large earthquakes, those with a magnitude larger than 6, which occurred from 2001 to 2010 with red circles, and those before 2001 with white circles.

The ratio of the cumulative number of earthquakes shallower than a given depth to the total number of earthquakes is shown in Fig. 2. We see that approximately 80 % of total earthquakes occur above a depth of 25 km. Therefore, for analyzing the seismicity change using the PI method, we chose earthquakes with depths less than 25 km within Taiwan and within 20 km of its coastline to account for the location error (the yellow area in Fig. 3). There are 539 study cells shown in Fig. 3. To investigate the precursors of large earthquakes, we used earthquakes with magnitudes larger than 6 and depths less than 25 km from 2001 to 2010. There is a total of 7 of these, which represent the target earthquakes in this study (Table 1; Fig. 3, red circles).

The end of the change interval, t_2, is defined as the end of the last season before each target earthquake. For instance, if the target earthquake occurred on 12/10/2003, t_2 should be 09/30/2003. Based on prior experience, we set the change interval to 4 years, so t_1 is 4 years before t_2.

A cutoff magnitude (M_C) is given to confirm the quality of data. Earthquakes with magnitudes smaller than the cutoff magnitude were not considered in the calculation. According to previous studies (Chen

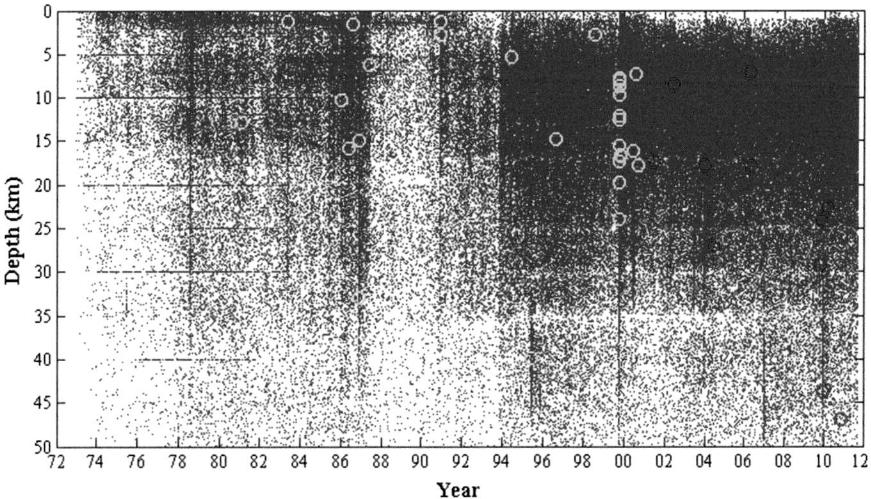

Figure 1
The distribution of earthquake depth versus time. The distribution shows an increase in the number of earthquakes after 1994. The *white circles* indicate earthquakes with magnitudes larger than 6 before 2001. The *red circles* indicate earthquakes with magnitudes larger than 6 from 2001 to 2010, and 7 *red circles* are at depths shallower than 25 km

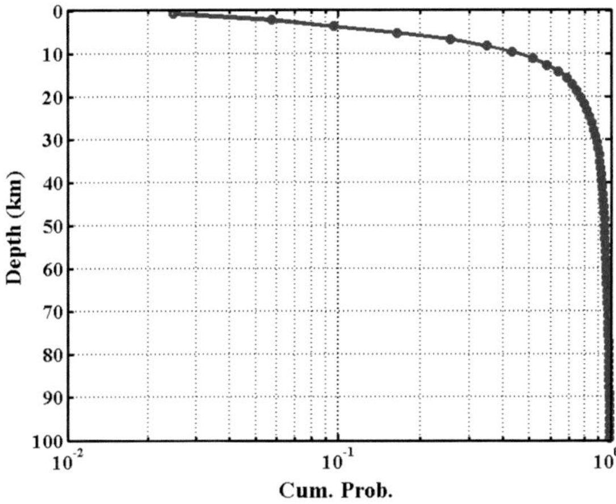

Figure 2
The ratio of the cumulative number of earthquakes that occurred at depths shallower than each given depth, to the total number of earthquakes. Eighty percent of earthquakes occur above 25 km

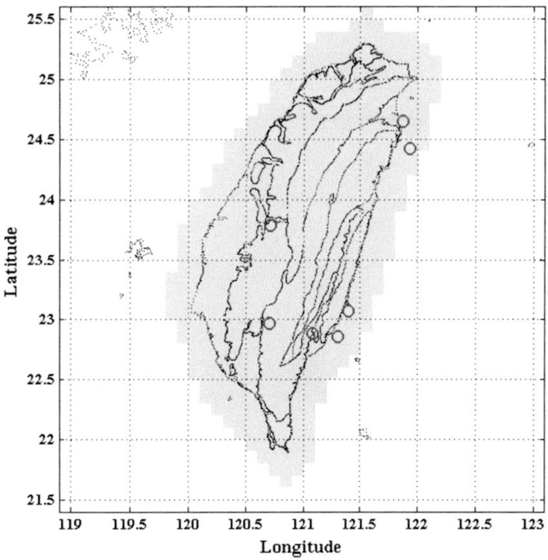

Figure 3
The study area. The *yellow grid* indicates the area analyzed in PI calculations. The *red circles* indicate the epicenters of the 7 target events

et al. 2005, 2006a; WU *et al.* 2008a, b; MIGNAN *et al.* 2011), we use $M_C = 3.2$ for this study.

4. Results

In our study, we chose 7 target earthquakes (Table 1) to investigate the precursor phenomena. Figure 4 is the hotspot map for the December 10, 2013 Chengkung earthquake (the third one in Table 1), which had a magnitude of 6.4 and hypocenter depth of 17.7 km. We colored the 100 boxes with the highest PI values, with darker colors representing higher values. We plotted blue circles to indicate the epicenter of the Chengkung earthquake, and inverted gray triangles for the locations of earthquakes with magnitudes larger than 5.5 that

Table 1

Parameters of the 7 target events in this study

#	Longitude	Latitude	Year	Month	Day	Depth	M_W	t_1	t_2	M_C	%
1	121.9280	24.4188	2001	6	14	17.29	6.30	1997/3/31	2001/3/31	3.2	61
2	121.8718	24.6510	2002	5	15	8.52	6.20	1998/3/31	2002/3/31	3.2	91
3	121.3982	23.0667	2003	12	10	17.73	6.42	1999/9/30	2003/9/30	3.2	88
4	121.0807	22.8835	2006	4	1	7.20	6.23	2001/12/31	2005/12/31	3.2	85
5	121.3035	22.8555	2006	4	15	17.90	6.04	2002/3/31	2006/3/31	3.2	87
6	120.7187	23.7890	2009	11	5	24.08	6.15	2005/9/30	2009/9/30	3.2	42
7	120.7066	22.9691	2010	3	4	22.64	6.42	2005/12/31	2009/12/31	3.2	94

Time t_2 is the final day of the last season before the earthquake occurred, and time t_1 is 4 years before t_2, when we set the fixed change interval M_C is the cutoff magnitude, from which we calculate the PI hotspot map using fixed parameters, and the percentile (%) is the percentage of the PI value of the epicenter for the entire map

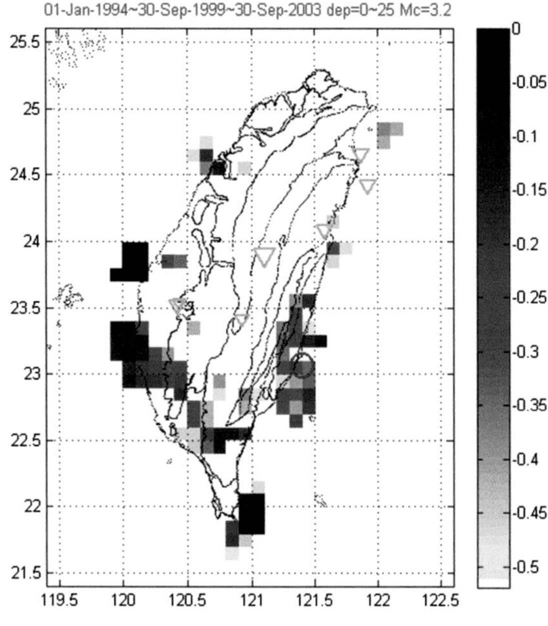

Figure 4
The hotspot map from the Chengkung earthquake (121.40°E, 23.07°N), which occurred on 12/10/2003. The 100 boxes with the highest PI value are colored, with *darker colors* indicating higher values. The *blue circles* indicate the locations of earthquakes with magnitudes larger than 5.5 in the prediction interval. The *inverted triangle* indicates the location of earthquakes with magnitudes larger than 5.5 in the change interval

occurred during the change interval from September 30, 1999 to September 30, 2003. The epicenter of the Chengkung earthquake is located in an anomalously high-PI area. However, a weakness of the current hotspot map is that it is noisy. The noise is most likely caused by a single event throughout the discretized temporal windows and lacks statistical significance.

Using the same process, we can produce hotspot maps of other target earthquakes. Figure 5a is the hotspot map for the first target earthquake, which occurred on 6/14/2001 and had a magnitude of 6.3 and hypocenter depth of 17.3 km. Figure 5a shows that there are many anomalies in southern Taiwan, although the target epicenter (121.93°E, 24.42°N) is located near an anomalous area. Figure 5b is the hotspot map for the second target earthquake (121.87°E, 24.65°N), which occurred on 5/15/2002, and had a magnitude of 6.2 and hypocenter depth of 8.5 km. The epicenter was formed with a high PI anomaly before the target event. Additionally, when compared to Fig. 4, we notice that the anomalously high-PI area that appeared during the Chengkung earthquake, 4 years earlier, migrated from the surroundings to the center. This migration is similar to that of the 1999 Chi-Chi earthquake and the 2006 Pingtung earthquake (Wu *et al.* 2008a, b, 2011).

Figure 5c is the hotspot map for the fourth target earthquake (121.08°E, 22.88°N), which occurred on 4/1/2006 in Taitung and had a magnitude of 6.2 and hypocenter depth of 7.2 km. Fifteen days later, this area experienced another earthquake (121.30°E, 22.86°N) with a magnitude larger than 6, which occurred on 4/15/2006 and had a hypocenter depth of 17.9 km. Per the operational definition, both events share the same PI hotspot map. However, we are curious about what caused the PI map to have a

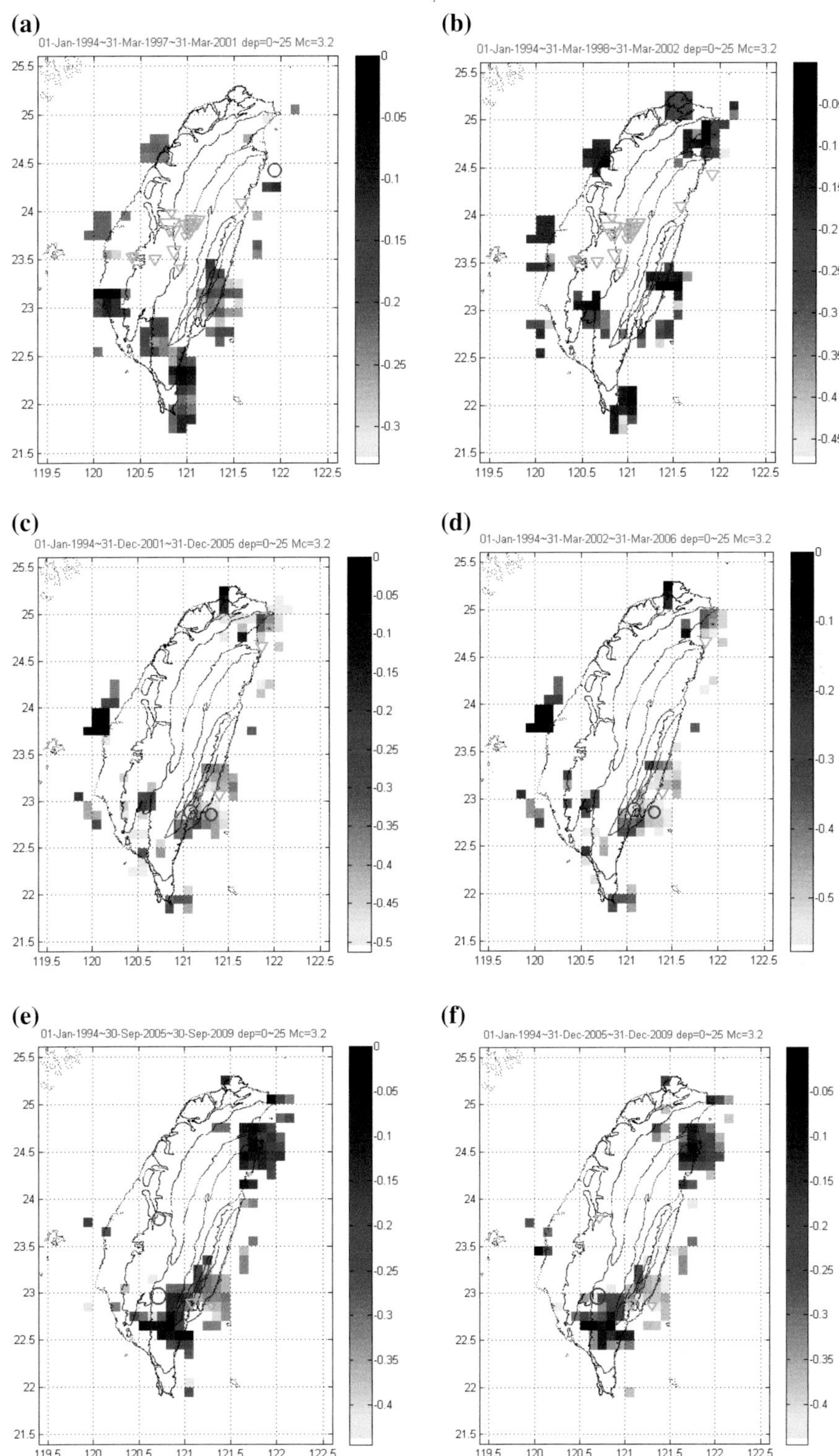

Figure 5
The hotspot maps from the other six target events. The *symbols* are the same as in Fig. 4. The hotspot map from **a** the first, **b** the second, **c** the fourth, **d** the fifth, **e** the sixth and **f** the seventh target event

slightly downward-shifted change interval, as shown in Fig. 5d. The patterns of Fig. 5c, d are very similar, and both epicenters are located in the anomalously high-PI area.

Figure 5e is the hotspot map for the sixth target earthquake (120.72°E, 23.79°N), which occurred on 11/5/2009, and had a magnitude of 6.1 and hypocenter depth of 24.08 km. The anomalies are distributed in the northeast and southern parts of Taiwan, and the epicenter is located outside the anomalous area. The anomalies may be strongly associated with the southern area where the Jiasian earthquake, the seventh target earthquake, occurred. It had a magnitude of 6.4 on March 4, 2010. Figure 5f is the PI hotspot map for the Jiasian earthquake (120.71°E, 22.97°N), with a hypocenter depth of 22.6 km.

After analyzing all 7 target earthquakes listed in Table 1, we have evidence that the anomalous locations that resulted from the PI calculations are closely associated with earthquakes. It is best to monitor anomalous areas for approximately 4 years, constituting the change interval, before an earthquake. Therefore, the results suggest a standard PI-forecasting process to regularly examine the anomalous areas, which may be potentially associated with high seismic risks in the future.

5. *Discussion and Random Test*

To confirm the statistical significance of the above PI calculation, we performed three random tests under different conditions. Because the range of PI values in each PI hotspot map may vary, we express the PI value by percentile in the following random tests.

In the first test, we examined the sensitivity of the PI index to the times and locations of target earthquakes by picking 7 PI indices at different times and locations. We also examined the reality of the PI index, that is, whether the PI index reflects the anomalous seismicity, by randomizing the catalog.

To examine whether the PI index occurred around the target earthquakes, we randomly picked 10,000 times between 01/01/2000 and 09/30/2011, at 10,000 locations in our study region. To avoid the effect of large, target events in Table 1, we blanked out the locations with distances to the epicenter of target earthquakes shorter than 0.5°, as well as the times preceding target events within 1 year when sampling the 10,000 random points. The blue dots in Fig. 6 show the 10,000 randomly chosen times and their spatial grid numbers. We consider the randomly chosen times and locations to be imaged events, for comparing with target events, and call them "random events" after comparison. The red stars indicate the 7 target large events, which, by definition, are associated with a blank area both near and before them. Among the 10,000 random events, we randomly chose 7, for which we calculated the PI hotspot maps using fixed PI parameters ($M_C = 3.2$ and 4 years for the change interval). The cumulative probability of the PI index for the 7 random events is shown as the black line in Fig. 9.

Figure 7 is the normalized hotspot map of the Chengkung earthquake. We only colored those cells with percentiles larger than the Chengkung epicenter. Because the percentile at the epicenter of the Chengkung earthquake is approximately 0.9, there are approximately 60 colored cells. Figure 8 shows the hotspot maps of other target earthquakes, recolored by percentile. The percentile of each target earthquake is also shown in the last column of Table 1. The percentiles of large target events are

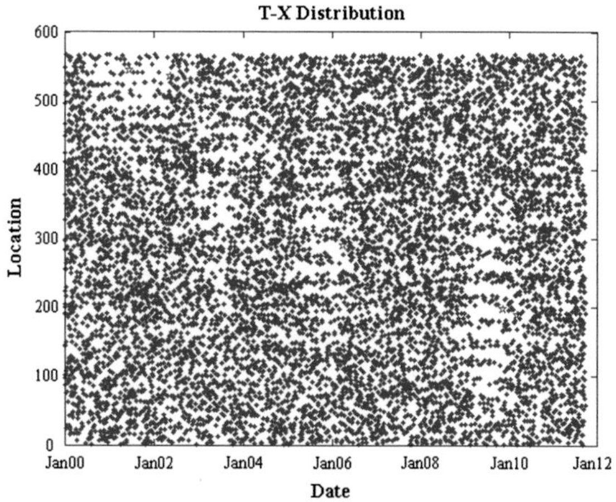

Figure 6
The 10,000 imaged events, shown over time, versus cell numbers. We call them "random events." The *blue dots* represent the random events, whose occurrence in time and space are chosen randomly. The *red stars* indicate the target events. Random events that are near and before the target events are removed to eliminate the effects of the target events

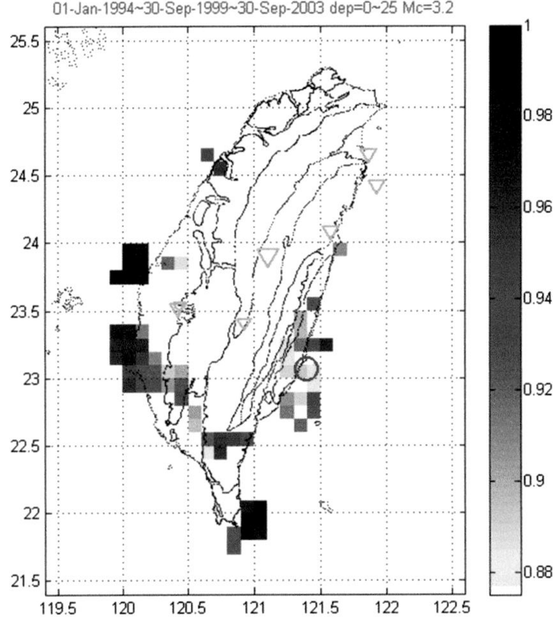

Figure 7
The hotspot map of the Chungkung earthquake that occurred on 12/10/2003, recolored by percentile. Cells with percentiles larger than that of the epicenter are *colored*

typically high and, therefore, the cumulative probability curve of these 7 target events, represented by the blue line in Fig. 9, is initially flat but quickly increases after percentile 0.8.

In order to show that the PI indices reflect the seismic anomalies, we randomized the location of the data using the Poisson model, with the 7 target earthquakes unchanged. First, we counted the number of events for each grid from the real data set and calculated the mean value. Then, we gave event numbers to each grid, which followed the Poisson distribution. Finally, we kept the times of the original data set, but randomly picked a location from the last step. The red line in Fig. 9 shows the cumulative probability for the 7 target earthquakes with randomized data. The cumulative probability for the 7 random events (black line in Fig. 9), as well as that for the 7 target earthquakes with randomized data (red line in Fig. 9), trend diagonally. This denotes that high and low PI values are chosen approximately equally often in these two cases. Accordingly, the Kolmogorov–Smirnov test concludes that the high-quality performance of the PI forecasts for those 7 large target events was not obtained by chance.

In the second test, we wanted to confirm the effect of the chosen parameters, cutoff magnitude and length of the change interval. We made PI hotspot maps for 7 target events, and 20 random events, using random PI parameters. For this analysis, M_C was chosen between 3 and 4, and the length of the change interval was chosen between 1 and 5 years. For each event, 10,000 PI hotspot maps with different parameters were generated, and the PI percentiles of each event were calculated. These parameters may affect the PI hotspot map and hence the probability distribution of PI percentiles. As an example, Fig. 10 shows the probabilities of the percentiles for the Chengkung earthquake. The percentiles of the Chengkung earthquake in the 10,000 PI hotspot maps, which are generated with different parameters, cluster at approximately 0.9 and 0.72. The blue arrow indicates the result from the fixed parameter analysis, 3.2 for M_C and 4 years for the change interval. We sum the probability to obtain the cumulative curve shown in Fig. 11. The cumulative curves of 7 target events are shown by the gray and black lines in Fig. 11, while the solid circles indicate the results from the fixed parameters for every target. All 7 cumulative curves occur in the area between the diagonal line and the lower right corner, and the probabilities of having percentiles larger than 60 % at the epicenter are above 0.6 for all 7 target earthquakes. Conversely, the average of the results, which is calculated 10,000 times for each of the 20 random events, is shown by blue line, with the light blue dashed lines representing one standard deviation of the percentile. Although the blue curve is initially flat, it increases linearly from 0.2 to 0.8 and becomes flat again at the end. The curve is almost diagonal, indicating a uniform percentile. All the curves for the target events exceed one standard deviation. This indicates that regardless of the randomly chosen parameters, the epicenter of large target earthquakes has a reliably high PI percentile, which is caused by anomalous seismicity.

In the third test, we calculated the PI percentile for 10,000 random events using fixed (the red solid line in Fig. 11) and random (the red dot-dashed line in Fig. 11) PI parameters. Both cumulative distributions are nearly diagonal. These diagonal curves show that random samples are ergodic, in a statistical

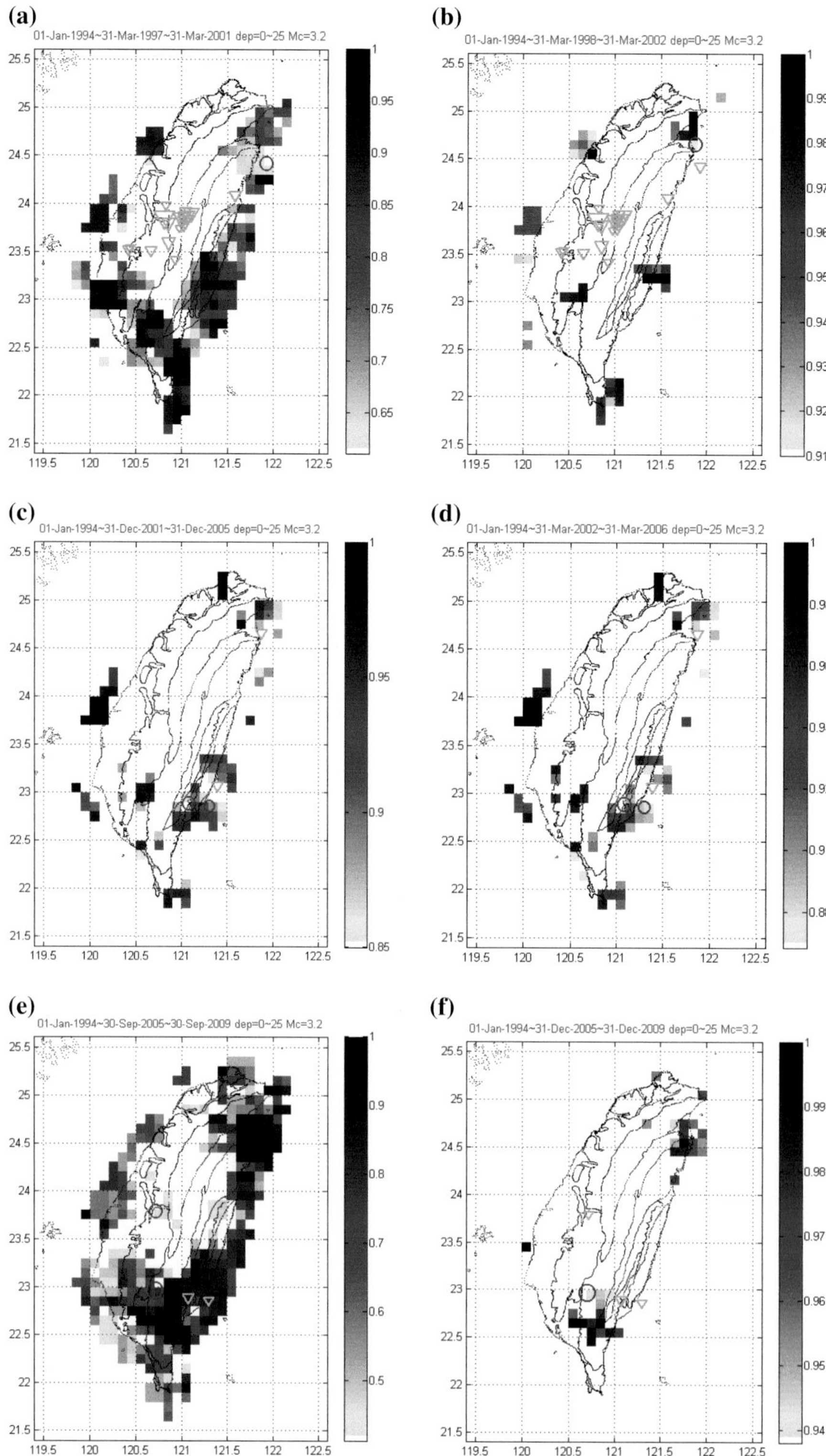

Figure 8

The hotspot map of the other target earthquakes by percentile. The hot spot map from **a** the first, **b** the second, **c** the fourth, **d** the fifth, **e** the sixth and **f** the seventh target event

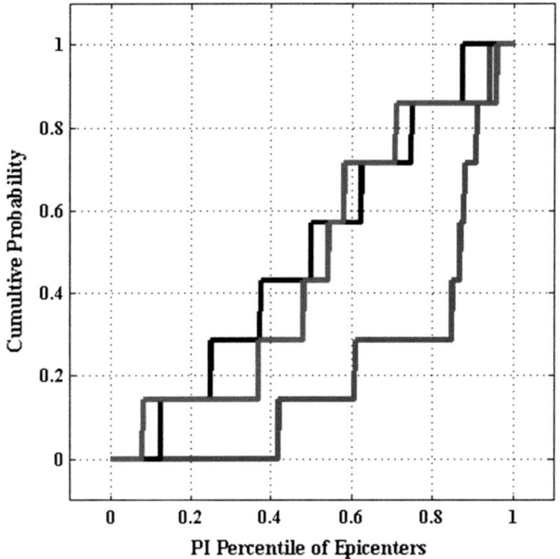

Figure 9
The cumulative probability of percentiles at given events. The *blue line* indicates the cumulative probability of percentile at the target events. The *black line* is the cumulative probability for 7 random events. The *red line* shows the cumulative probability of percentile for 7 target events from the re-calculated PI map with randomized seismicity given by the Poisson model

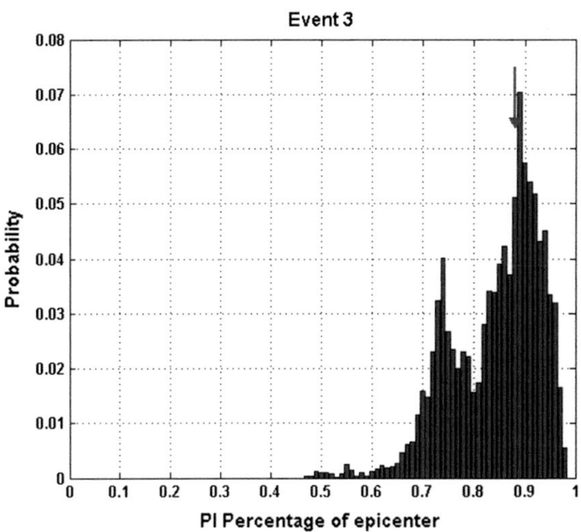

Figure 10
The probability distribution of the percentile for the Chengkung earthquake with random test parameters. We calculated the hotspot maps using 10,000 pairs of random parameters and counted the percentile at the epicenter in each map. The *blue arrow* indicates the result from the fixed parameters

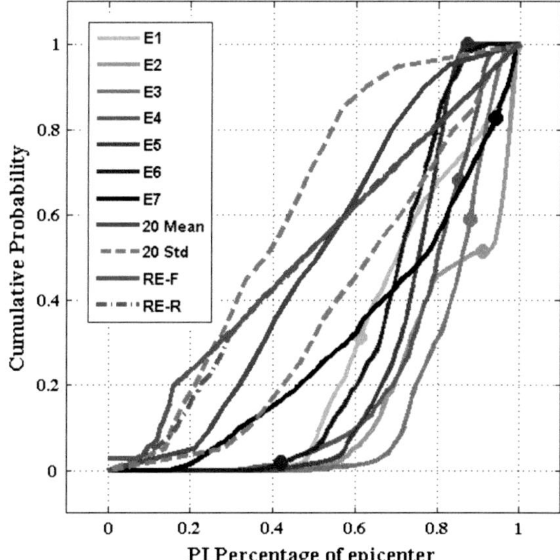

Figure 11
The cumulative probability of PI percentiles for 7 target events and 20 random events using 10,000 random PI parameter sets, and for 10,000 random events using fixed and random parameters. The lines (*E1–E7*) shown in *gray* and *black* indicate the cumulative probabilities of the PI percentiles for the 7 target events, and the *solid circle* on the line denotes the PI percentile of each event obtained from the PI hotspot map using the fixed parameters in Table 1. The *blue line* shows the mean of cumulative probabilities of the PI percentiles for the 20 random events, and the *dashed blue lines* show a range of one standard deviation. The *solid red line* (RE-F) and *dashed-dotted red line* (RE-R) are the cumulative probability distributions of 10,000 random events with fixed and random parameters, respectively

sense (Tiampo *et al.* 2010). However, the curves of the 7 target events, which are significantly different from the diagonal curves of the random samples, suggest that the epicenter area could experience a significant and reliable PI anomaly before large earthquakes.

6. *Summary*

In this study, we demonstrate how PI can be implemented to detect the likely epicenters of large earthquakes using a seasonal routine operation. Free parameters often and inevitably affect the performance in scientific simulations/predictions. In some cases, they can represent important physical meanings, and in other cases, simply mathematical tricks. In PI, the cutoff magnitude can be associated with the relationship between larger and smaller earthquakes, and the change interval with the time scale of nucleation processes of larger events. Nevertheless, they need to be assigned a priori for the routine operation.

For the PI parameters, we suggest a cutoff magnitude of 3.2 and a duration of change interval of 4 years. These values reveal anomalous seismic areas, which are typically associated with the forthcoming occurrence of $M > 6$ events in the Taiwan region. As demonstrated by the 7 targets listed in Table 1, the PI calculations with fixed parameters exhibit excellent potential for the epicenter detection of future large events. Although a visual inspection of the PI maps from the second random test does not produce clear precursory anomalies, statistical tests show that there is a precursory correlation. Furthermore, the statistical tests provide verification for the results of the routine operation. One drawback, which appeared in many PI hotspot maps, is the presence of many single and scattered patches. We regard these patches as noise, but they should be addressed in future studies.

Acknowledgments

The authors thank the Central Weather Bureau (CWB) for providing the earthquake data. CCC is grateful for research support from the Ministry of Science and Technology (ROC) and the Department of Earth Sciences at National Central University (ROC).

References

CHEN, C.-C., RUNDLE, J.B., HOLLIDAY, J.R., NANJO, K.Z., TURCOTTE, D.L., LI, S.C., and TIAMPO, K.F. (2005), *The 1999 Chi-Chi, Taiwan, earthquake as a typical example of seismic activation and quiescence*, Geophys. Res. Lett., *32*(22), L22315, doi:10.1029/2005GL023991.

CHEN, C.-C., and WU, Y.X. (2006), *An improved region-time-length algorithm applied to the 1999 Chi-Chi, Taiwan earthquake*, Geophys. J. Int., *166*, 1144–1147.

CHEN, C.-C., RUNDLE, J.B., LI, H.C., HOLLIDAY, J.R., NANJO, K.Z., TURCOTTE, D.L., and TIAMPO, K.F. (2006a), *From tornadoes to earthquakes: Forecast verification for binary events applied to the 1999 Chi-Chi, Taiwan, earthquake*, Terr. Atmos. Ocean. Sci., *17*, 503–516.

CHEN, C.-C., RUNDLE, J.B., LI, H.C., HOLLIDAY, J.R., TURCOTTE, D.L., and TIAMPO, K.F. (2006b), *Critical point theory of earthquakes: Observation of correlated and cooperative behavior on earthquake fault systems*, Geophys. Res. Lett., *33*(18), L18302, doi:10.1029/2006GL027323.

HOLLIDAY, J.R., CHEN, C.-C., TIAMPO, K.F., RUNDLE, J.B., TURCOTTE, D.L., and DONNELAN, A. (2007), *A RELM earthquake forecast based on pattern informatics*, Seismol. Res. Lett., *78*(1), 87–93.

LEE, Y.T., TURCOTTE, D.L., HOLLIDAY, J.R., SACHS, M.K., RUNDLE, J.B., CHEN, C.-C., and TIAMPO, K.F. (2011), *Results of the Regional Earthquake Likelihood Models (RELM) test of earthquake forecasts in California*, Proc. Natl. Acad. Sci. USA, *108*, 16533–16538, doi:10.1073/pnas.1113481108.

LI, H.C., and CHEN, C.-C. (2011), *Characteristics of long-term regional seismicity before the 2008 Wen-chuan, China, earthquake using pattern informatics and genetic algorithms*, Nat. Hazards Earth Syst. Sci., *11*, 1003–1009.

MIGNAN, A., WERNER, M.J., WIEMER, S., CHEN, C.-C., and WU, Y.M. (2011), *Bayesian estimation of the spatially varying completeness magnitude of earthquake catalogs*, Bull. Seismol. Soc. Am., *101*, 1371–1385.

NANJO, K.Z., HOLLIDAY, J.R., CHEN, C.-C., RUNDLE, J.B., and TURCOTTE, D.L. (2006), *Application of a modified pattern informatics method to forecasting the locations of future large earthquakes in the central Japan*, Tectonophysics, *424*, 351–366.

RUNDLE, J.B., KLEIN, W., TIAMPO, K. and GROSS, S. (2000), *Linear Pattern Dynamics in Nonlinear Threshold Systems*, Phys. Rev. E, *61*, 2418–2432.

SCHORLEMMER, D., ZECHAR, J.D., WERNER, M.J., FIELD, E.H., JACKSON, D.D., JORDAN, T.H., and the RELM Working Group (Chen, C.-C. included) (2010), *First results of the Regional Earthquake Likelihood Models experiment*, Pure Appl. Geophys., *167*, 859–876.

TIAMPO, K.F., RUNDLE, J.B., MCGINNIS, S., GROSS, S.J. and KLEIN, W. (2002), *Mean-field threshold systems and phase dynamics: An application to earthquake fault systems*, Europhys. Lett., *60*(3), 481–487.

TIAMPO, K.F. and SHCHERBAKOV, R. (2012), *Seismicity-based earthquake forecasting techniques: Ten years of progress*, Tectonophysics, Volumes *522–523*, 89–121.

TIAMPO, K.F., KLEIN, W., LI, H.C., MIGNAN, A., TOYA, Y., KOHEN-KADOSH, S.Z.L., RUNDLE, J.B., and CHEN, C.-C. (2010), *Ergodicity and earthquake catalogs: Forecast testing and resulting implications*, Pure Appl. Geophys., *167*, 763–782.

TOYA, Y., TIAMPO, K.F., RUNDLE, J.B., CHEN, C.-C., LI, H.C., and KLEIN, W. (2010), *Pattern informatics approach to earthquake forecasting in 3D*, Concurrency and Computation Practice and Experience, *22*, 1569–1592.

WU, Y.H., CHEN, C.-C., and RUNDLE, J.B. (2008a), *Detecting precursory earthquake migration patterns using the pattern informatics method*, Geophys. Res. Lett., *35*, L19304, doi:10.1029/2008GL035215.

WU, Y.H., CHEN, C.-C., and RUNDLE, J.B. (2008b), *Precursory seismic activation of the Pingtung (Taiwan) offshore doublet earthquakes on 26 December 2006: A pattern informatics analysis*, Terr. Atmos. Ocean. Sci., *19*, 743–749.

WU, Y.H., CHEN, C.-C., and RUNDLE, J.B. (2011), *Precursory small earthquake migration patterns*, Terra Nova, *23*, 369–374.

WU, Y.M., and CHIAO, L.Y. (2006), *Seismic quiescence before the 1999 Chi-Chi, Taiwan, M_w 7.6 earthquake*, Bull. Seism. Soc. Am., *96*(1), 321–327.

(Received August 19, 2014, accepted April 7, 2015, Published online April 18, 2015)

Pure Appl. Geophys. 173 (2016), 245–254
© 2015 Springer Basel
DOI 10.1007/s00024-015-1123-9

Pure and Applied Geophysics

The Central China North–South Seismic Belt: Seismicity, Ergodicity, and Five-year PI Forecast in Testing

SHENGFENG ZHANG,[1] ZHONGLIANG WU,[1] and CHANGSHENG JIANG[1]

Abstract—Instrumentally recorded seismicity from 1970/01/01 to 2014/01/01 of the central China north–south seismic belt (21.0°–41.5°N, 97.5°–107.5°E) was analyzed, emphasizing the applicability of the predictive algorithms based on the assumptions of meta-stable equilibrium. The seismicity in this region was shown to exhibit ergodicity from 1980 to the present, with sub-region dependence, and interrupted by the 2008 Wenchuan earthquake. pattern informatics algorithm, a statistical physics-based predictive model for five-year time scale, is put to forward forecast test for the period 2014/01/01 to 2019/01/01.

Key words: The central China north–south seismic belt, Ergodicity, Pattern informatics (PI) algorithm.

1. Introduction

In the reduction of earthquake disaster risk in China, one of the focuses of special attention is the so-called central China north–south seismic belt. Historical documentations left numerous recordings of disastrous earthquakes in this belt, including the 1920 Haiyuan great earthquake with fatality up to about 200,000 (Department of Earthquake Disaster Prevention of the State Seismological Bureau 1995, 1999). The recent memory of a great earthquake is the 2008 Wenchuan earthquake (CHEN and BOOTH 2011). Tectonically, this belt lies in between the Tibetan Plateau, the Ordos block, the Sichuan basin, and the south China block (ZHANG *et al.* 2003). The region is characterized by variation of crust thickness from 30–46 km in the east to 46–74 km in the west, reflecting the rising and spreading of the Tibetan plateau with the collision between the India plate and the Eurasia plate (LI *et al.* 2006).

Directly related to statistical seismology, this region is also featured by complicated fault systems and heterogeneous seismicity (XU and DENG 1996; YI *et al.* 2002; XU *et al.* 2005), preventing from the simple application of the existing recurrence models such as the time-predictable and/or magnitude-predictable models. It is somehow challenging, therefore, that this region is suggested (MIGNAN *et al.* 2013) as one of the testing regions of the Collaboratory for the Study of Earthquake Predictability (CSEP).[1] As a 'baseline' analysis of the predictability of seismicity, a series of retrospective forecasts were made (MIGNAN *et al.* 2013) using the 'simple smoothed seismicity model', or the 'TripleS model', of ZECHAR and JORDAN (2010), smoothening the locations of past earthquakes to construct a predictive density of future seismicity. To better characterize the seismicity and its predictability taking into account the phenomenology of activation and quiescence in a unified framework as the 'fluctuation', in this paper, we analyze the ergodicity of seismicity and the 'hotspots' of the pattern informatics (PI) algorithm (TIAMPO *et al.* 2002; RUNDLE *et al.* 2003; HOLLIDAY *et al.* 2006). Meanwhile, to put the PI algorithm in forward forecast test in China, we present the forecast for the period 2014/01/01 to 2019/01/01, and discuss how the PI forecast be coupled with the regional practises of the assessment of time-dependent seismic hazard, or, earthquake forecast in a general sense.

[1] Institute of Geophysics, China Earthquake Administration, Beijing 100081, China. E-mail: wuzhl@ucas.ac.cn

[1] http://www.cseptesting.org/.

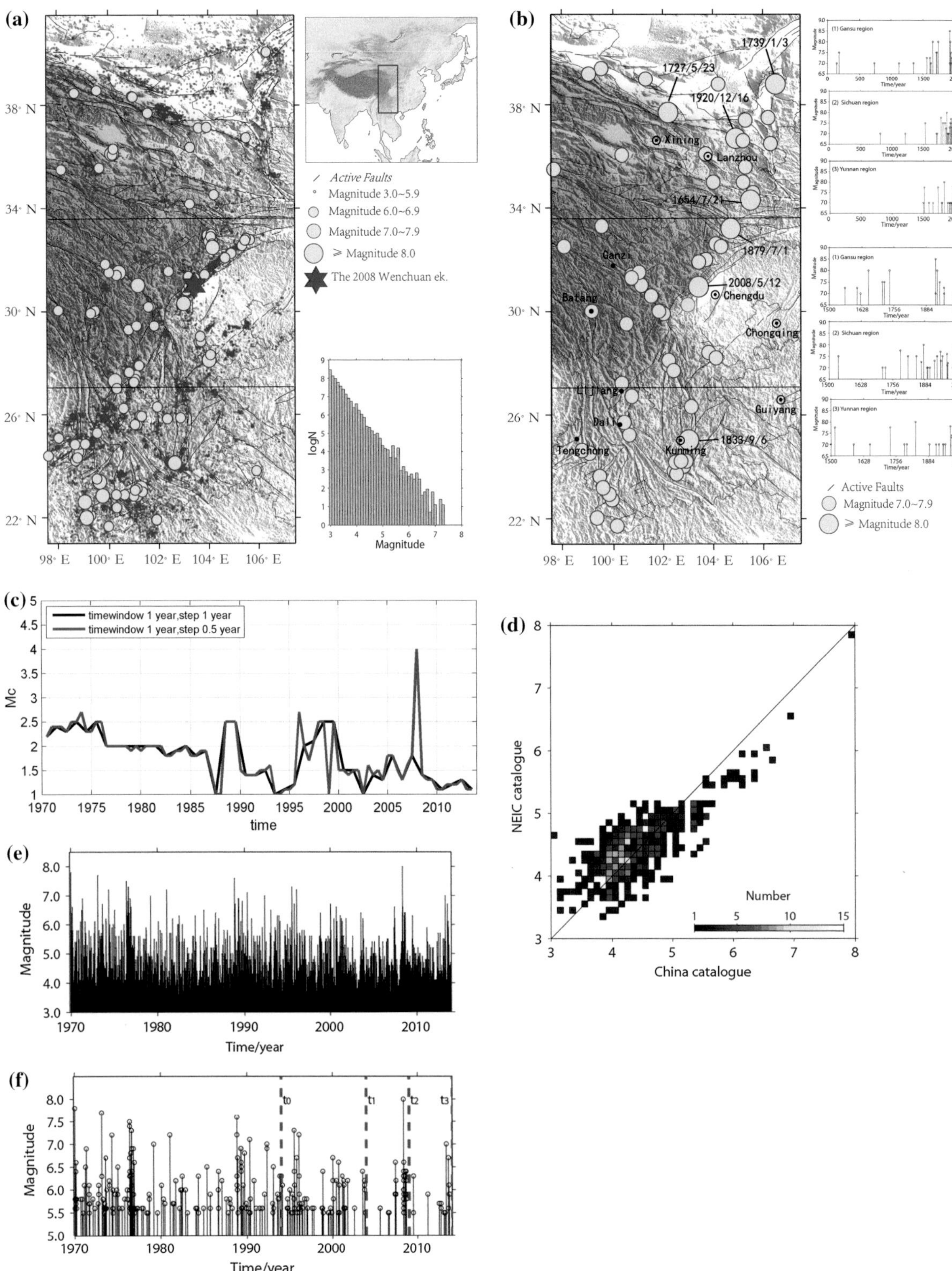
(a)
(b)
(c)
(d)
(e)
(f)
Active Faults
Magnitude 3.0~5.9
Magnitude 6.0~6.9
Magnitude 7.0~7.9
≥ Magnitude 8.0
The 2008 Wenchuan ek.
logN
Magnitude
1739/1/3
1727/5/23
1920/12/16
Xining
Lanzhou
1654/7/21
1879/7/1
Ganzi
2008/5/12
Chengdu
Batang
Chongqing
Lijiang
Guiyang
Dali
Tengchong
Kunming
1833/9/6
(1) Gansu region
(2) Sichuan region
(3) Yunnan region
Time/year
timewindow 1 year,step 1 year
timewindow 1 year,step 0.5 year
Mc
time
NEIC catalogue
China catalogue
Number
Magnitude
Time/year
t0
t1
t2
t3

◀ Figure 1
a Spatial distribution of earthquakes for the period from 1970/01/01 to 2014/01/01 in the central China north–south seismic belt. *Black lines* denote active faults. The red hexagon represents the epicentre of the M_W7.9/M_S8.0 Wenchuan earthquake on May 12, 2008. The indexing figure to the *top right* shows the region under discussion, in which the Tibetan plateau can be seen from the topography. To the bottom right is the non-cumulative frequency–magnitude distribution (that is, the number N of events with certain magnitude) of the earthquakes above 3.0 in the due period. **b** Spatial distribution of earthquakes on record from 0 BC to AD 2013 with $M \geq 7.0$. Dates are marked to the earthquakes with magnitude larger than 8. The whole region is divided into 3 sub-regions, namely the Gansu region (*top*), the Sichuan region (*middle*), and the Yunnan region (*bottom*), separated by the solid thin lines with latitude 27°N and 33.5°N, respectively. To the right the subplots show the magnitude–time diagram for different sub-regions within different time periods. **c** Time-dependent completeness magnitude for the region under study, using the method of Mignan and Woessner (2012), with different sliding steps. **d** Comparison of the magnitude in the CENC catalogue (horizontal axis, noted as China catalogue) and the NEIC catalogue (vertical axis) for the events in the central China north–south seismic belt. The time span is from 1980/01/01 to 2014/01/01. See text for details. **e** Temporal distribution of the earthquakes with magnitude above 3.0 as shown in (**a**). **f** Temporal distribution of the earthquakes with magnitude above 5.5 as shown in Fig. 1(e). The *red vertical dashed lines* stand for the times used in the PI algorithm. See text for details

2. The Earthquake Catalogue

Somehow arbitrarily, the study region is delimitated by the ranges of latitude and longitude 21.0°–41.5°N and 97.5°–107.5°E, with the reference of the CSEP testing region proposed by Mignan *et al.* (2013). Figure 1a shows the background seismicity of this region, highlighting the well-known May 12, 2008, Wenchuan M_W7.9/M_S8.0 earthquake (Chen and Booth 2011). Figure 1b further shows the historical earthquakes in the region, where ancient Chinese literatures recorded several magnitude 8 + earthquakes (Earthquake data from: Department of Earthquake Disaster Prevention of the State Seismological Bureau 1995, 1999).[2]

The data used in the present study is the Monthly Earthquake Catalogue from 1970/01/01 to 2014/01/01 provided by the China Earthquake Networks Center (CENC). As shown in Fig. 1c, the completeness magnitude changed with time, depending on the development of observational facilities. The 2008 Wenchuan earthquake sequence exhibits a significant effect on the overall completeness of the catalogue, due to the limitation of routine seismological observation and interpretation, which is to be dealt with in future. As an overall estimate, a homogeneous monitoring capability of completeness magnitude 3.0 (Su *et al.* 2003; Mignan *et al.* 2013) has been seen since 1970. Figure 1a, in its subplot to the bottom right, gives the non-cumulative frequency–magnitude distribution of the events under discussion. The linearity of such a log–log distribution shows a good overall completeness for the events above 3.0, with the Gutenberg–Richter b value 0.93.

The catalogue uses a 'mixed' magnitude system. For earthquakes below 4.0 the Chinese local magnitude M_L is used. For earthquakes above 5.5 the Chinese surface wave magnitude M_S is used. For the events in between, the transition from the Chinese M_L to the Chinese M_S is not clearly defined. Figure 1d shows the magnitude in the present catalogue and that in the NEIC catalogue. From the figure the two magnitudes can be compared to each other. In the comparison, the events in both catalogues, with the differences of latitude and longitude less than 0.1° and origin time less than 5 s, are regarded as the identical events with different parameters determined by different networks. Similar to Bormann *et al.* (2007), colour bar is used to indicate the number of events, and diagonal line indicates the equality of the two magnitudes.

Considering the ambiguity in the magnitude, in this study, we use only 'magnitude' for the size of the events in the catalogue. For the analysis based on number counting, such as the calculation of the Pattern Informatics (PI) 'hotspots', this ambiguity does not affect the result. This is the case especially when the cutoff magnitude is taken as below 4.0 (that is, the magnitude is simply the local M_L). According to convention, the cutoff magnitude of the earthquake catalogue under study should be no less than the completeness magnitude and is generally taken as about 2 magnitude units less than that of the 'target' earthquakes. In our analysis, we take the cutoff magnitude as 3.0 and the 'target' magnitude as above 6.0. However, for other algorithms which consider not only the number counting but also the size of

[2] http://www.csndmc.ac.cn/newweb/data.htm.

earthquakes, the issues of magnitude transition, as well as of magnitude uncertainty, has to be taken into account.

3. The Ergodic Behavior of Seismicity

Figure 1e, f shows the temporal complexity of the small earthquakes in this region. To describe the degree of its heterogeneity and in turn the statistical predictability of strong events, we consider the ergodic behavior taking the approach of Tiampo *et al.* (2004, 2007) who used the Thirumalai-Mountain diagram (the TM diagram) as a measure of the ergodicity (Thirumalai *et al.* 1989; Thirumalai and Mountain 1993). In the analysis, the function $\mathrm{TM_n}(t)$ is defined by

$$\mathrm{TM_n}(t) = \frac{1}{L}\sum_{i=1}^{L}[n_i(t) - \overline{n}(t)]^2 \tag{1}$$

in which L is the number of the grids in the spatial range under consideration, n_i the number of events in the ith grid within time 0–t, and $\overline{n}(t)$ the mean number of events located in all of the grids within time 0–t,

$$n_i(t) = \frac{1}{t}\int_0^t N_i(t')dt' \tag{2}$$

$$\overline{n}(t) = \frac{1}{L}\sum_i^L n_i(t) \tag{3}$$

According to Tiampo *et al.* (2004, 2007), if the system exhibits the 'effective ergodic behavior' in a long period, then the function 1/TM will be proportional to time t, or, $\mathrm{TM_n}(t) \propto 1/t$.

Figures 2 and 3 show the TM diagrams of the seismicity in different sub-regions (defined in Fig. 1a, b), with grid size 0.2° × 0.2° and cutoff magnitude 3.0. From the figure it may be seen that different regions and their different combinations show different ergodic behavior. The Gansu region has not a good ergodicity, at least before 1990. Previous results (e.g., Tiampo *et al.* 2004, 2007) concluded that, once variations in the catalogue data resulting from technical and network issues are considered, natural seismicity displays ergodicity. In this regard, and

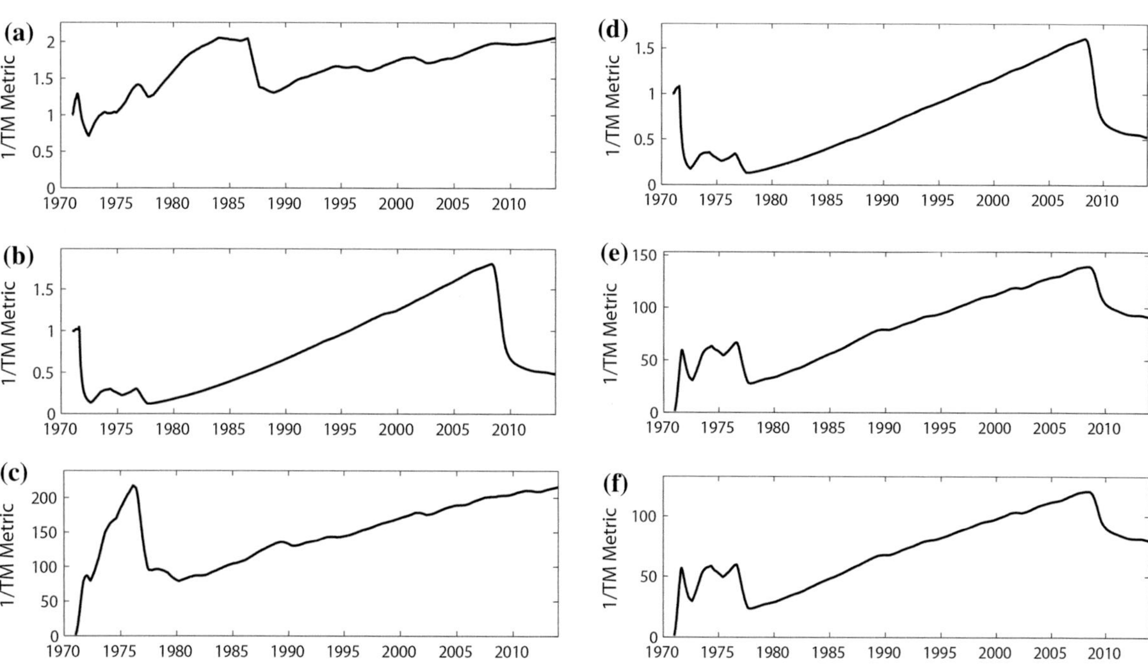

Figure 2

The TM diagram to show the ergodicity of the seismicity in different sub-regions and different combinations of the sub-regions. **a** Gansu region; **b** Sichuan region; **c** Yunnan region; **d** Gansu and Sichuan region; **e** Sichuan and Yunnan region; **f** the whole region of the central China north–south seismic belt

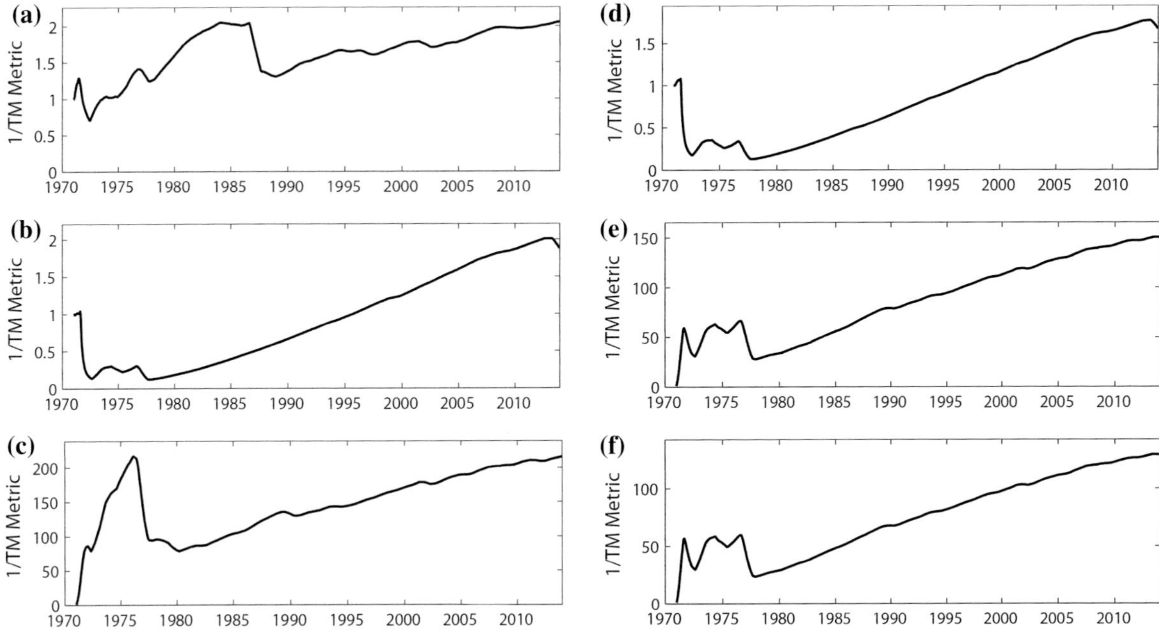

Figure 3
The TM diagram, with the Wenchuan aftershocks removed, to show the ergodicity of the seismicity in different sub-regions and different combinations of the sub-regions. **a** Gansu region; **b** Sichuan region; **c** Yunnan region; **d** Gansu and Sichuan region; **e** Sichuan and Yunnan region; **f** the whole region of the central China north–south seismic belt

considering the history of seismic networks in China, the sub-region dependence in the central China north–south seismic belt may be attributed to the regionalized characteristics of seismological observation and interpretation.

In general, since about 1980, the seismicity in the whole region, the Sichuan region and the Yunnan region shows good ergodicity. The 2008 Wenchuan earthquake seems to have significant influence on the ergodicity, as comparing Figs. 2 and 3, in which the disruption of the TM diagram around 2008 is predominately contributed by the Sichuan region (Fig. 2), and disappears when the Wenchuan sequence is removed (Fig. 3). This is in consistence with the previous results (e.g., Tiampo *et al.* 2007) that large event disrupts the ergodicity. But here it may be argued that such an influence may simply be the contribution of the aftershock sequence which obeys a different statistics (that is, the Omori's law).

In the statistical physics of earthquakes, the importance of ergodicity lies in that effective ergodicity generally implies stationary periods of metastable equilibrium, which guaranteed the applicability of the analysis models developed based on the theoretical assumptions such as mean-field and spatio-temporal correlations, being applied to a large class of physical systems, driven threshold systems in general and earthquake fault systems in particular (Tiampo *et al.* 2004, 2007). Our result shows that for the central China north–south seismic belt, availability of seismic data to test such models (for example, the PI algorithm) is from 1980.

4. The PI Algorithm and its Parameter Settings

In recent years, the PI algorithm has been applied to different places, such as California (Tiampo *et al.* 2002; Rundle *et al.* 2003; Holliday *et al.* 2006, 2007), central Japan (Nanjo *et al.* 2006a, b), Taiwan (Chen *et al.* 2005; Wu and Chen, 2007; Wu *et al.* 2008), and south-west China (Jiang and Wu 2008, 2010, 2011). In many aspects, the algorithm has also been discussed (e.g., Zechar and Jordan, 2008), evaluated (e.g., Cho and Tiampo 2012), improved (e.g., Wu *et al.* 2008), and extended (e.g., Wu *et al.*

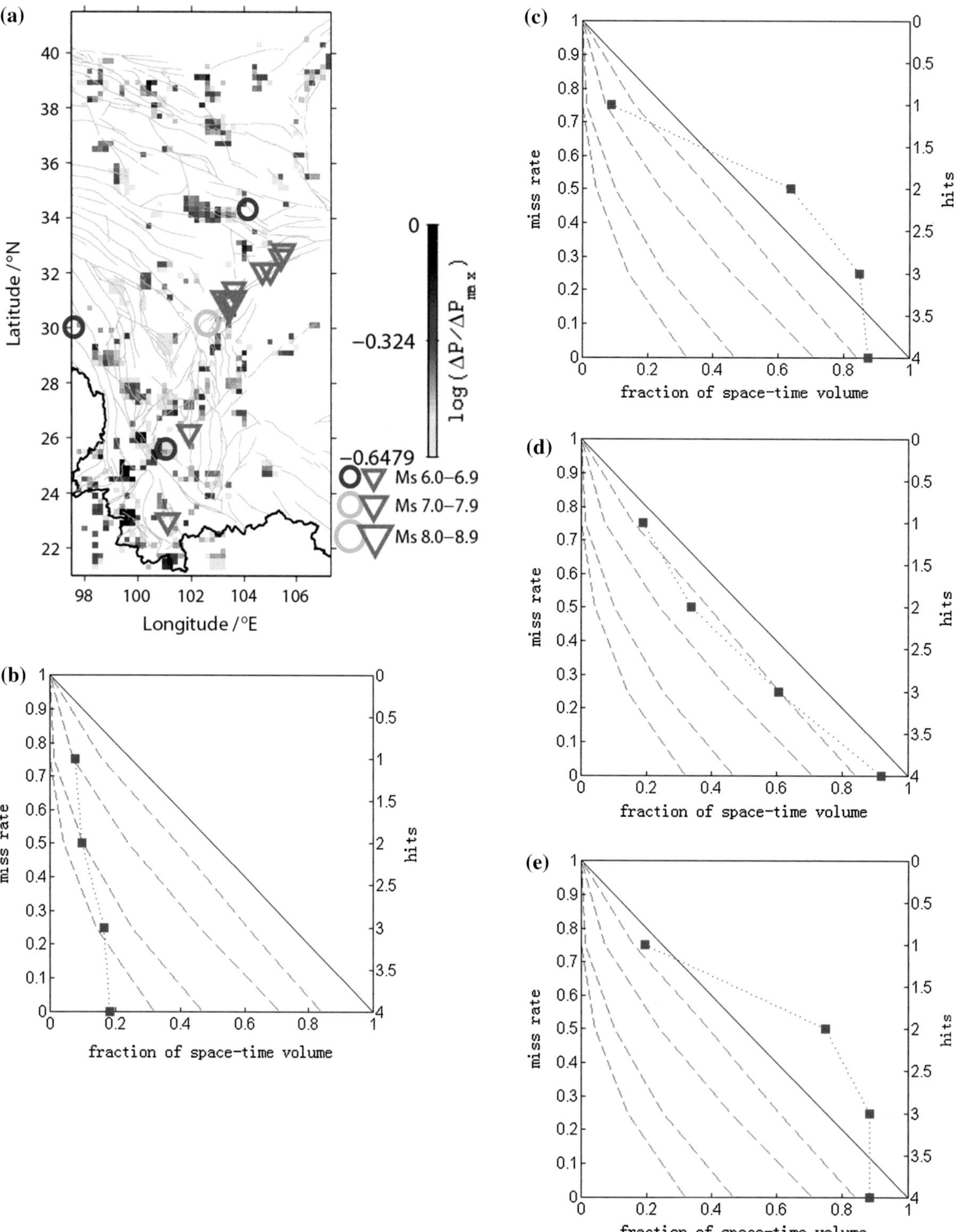

2011). Operationally, the PI approach uses instrumentally recorded earthquake catalogues, aiming at the detection of the intermediate-term variation of local seismicity by comparing with the long-term background level. The detected spatio-temporal fluctuations of seismicity, such as anomalous

◀ Figure 4
Output of the PI algorithm for the central China north–south seismic belt, with forecast window 2009/01/01 to 2014/01/01, referring to Fig. 1f. **a** PI hotspot distribution and 'target earthquakes' (shown by *blue* and *cyan* circles, respectively). The reverse triangles in grey stand for the events above magnitude 6.0 occurred within the 'anomaly training window'. Cutoff magnitude is 3.0. **b** Molchan error diagram of the forecast in (**a**). The *diagonal* shows the random guess with gain = 1. *Dashed lines* show the theoretical curves with different standard significant values, that from the *left* to the *right*: 1, 5, 25, and 50 %, respectively. **c** Molchan error diagram for the relative intensity (RI) forecast with identical parameter settings. **d** Molchan error diagram of the PI forecast using cutoff magnitude 3.5. **e** Molchan error diagram of the PI forecast using cutoff magnitude 4.0

activation or quiescence, are connected with the variation of the underlying stress and used as an indication of the increase of the probability of strong earthquakes (Tiampo and Shcherbakov 2012).

Considering background seismicity and the local needs for the time-dependent seismic hazard assessment (Wu *et al.* 2013), the 'target earthquakes' in the central China north–south seismic belt are selected as the events no less than magnitude 6.0. Parameters in use are as follow: a 10-year 'background window' from t_0 to t_1, a 5-year 'anomaly identification window' from t_1 to t_2, a 5-year 'forecast window' from t_2 to t_3, as shown in Fig. 1f as a showcase example; the grids dividing the whole region into different cells being 0.2° in both latitude and longitude; and cutoff magnitude 3.0. As in previous works (Keilis-Borok and Rotwain 1990; Holliday *et al.* 2007), only shallow earthquakes are used.

The result of the PI forecast, for the recent time period from 2009/01/01 to 2014/01/01, is shown in Fig. 4a, with the selections of the time windows shown in Fig. 1f. Different from the previous works in which the diagram of the Receiver Operating Characteristic (ROC) test (Swets 1973) was used, we use the Molchan error diagram (Molchan 1997) to evaluate the performance of the PI forecast, as shown in Fig. 4b. To compare with the RI forecast, Fig. 4c provides the Molchan error diagram of the RI forecast with the same parameter settings. From these figures it may be seen that PI forecast outperforms not only random guess but also RI forecast. To investigate the effect of cutoff magnitude on the forecast, Fig. 4d, e show the Molchan error diagrams for the forecasts using different cutoff magnitudes, 3.5 and 4.0, respectively. It may be seen that the cutoff magnitude does have influence on the forecast. With the cutoff magnitude increases, the forecast capability decreases. For the case that the cutoff magnitude is 4.0, the PI forecast has almost no difference with the random forecast. This is in consistence with the previous result (e.g., Jiang and Wu 2010) and is easy to understand, since the using of lower cutoff magnitude means to use more events in the calculation, and thus more information is included in the forecast.

5. *The PI Forecast put to Test: 2014/01/01 to 2019/01/01*

Putting the PI forecast in testing, Fig. 5a shows the distribution of the PI hotspots, for the forecast window 2014/01/01 to 2019/01/01. The parameter settings are the same as those of Fig. 4. One of the issues in need of further discussion is, in practise, how to use the seemingly wide-spreading 'hotspots' to determine the places of future strong earthquakes.

One of the practical measures might be the combination of the PI 'hotspots' with other forecasts (e.g., Zhao *et al.* 2010). Figure 5b shows the same hotspot distribution as in Fig. 5a, with the earthquakes above magnitude 6.0 in 2014 shown by green circles. As a reference, in the figure, the closed circles in red and blue show the annually hazardous regions in 2014 as an output of the Annual Consultation Meeting.[3] These regions were identified by the expert panel in the Annual Consultation Meeting by combing seismological, geodetic, geophysical, and geochemical data. The Annual Consultation Meeting has been a persistent forward forecast test organized by the China Earthquake Administration (CEA) since the 1970s (Zhao *et al.* 2010; Wu *et al.* 2013). Analysis showed that the performance of such annual forecast outperforms random guess (Shi *et al.* 2001), while to much extent contributed by the statistics of seismicity (Zhuang and Jiang 2012). From the comparison shown in Fig. 5b it can be seen that the regions with successful forecast (or, 'hit') in the annual consultation (the red circles) seem to be associated with a wide coverage of PI hotspots

[3] Data from the Department of Earthquake Monitoring and Prediction, China Earthquake Administration, 2015.

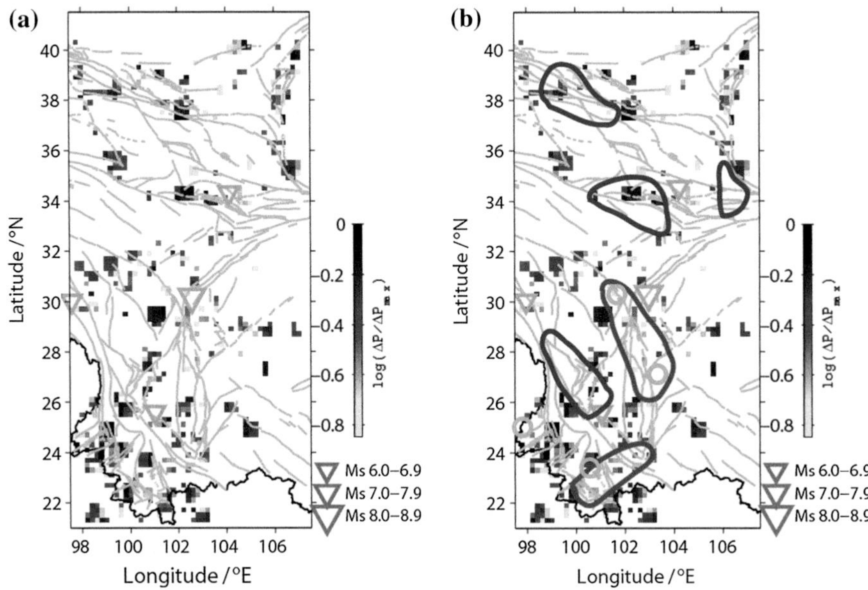

Figure 5
a Distribution of the PI hotspots, with forecast window 2014/01/01 to 2019/01/01, put to forward forecast test. See text for details. **b** The same as (**a**), with the earthquakes above magnitude 6.0 in 2014 shown by *green circles*. The *red* and *blue* circles show the annually hazardous region in 2014 as an output of the Annual Consultation Meeting, in which *blue* indicates false alarms, while *red* indicates successful forecasts

within the region. Other regions (the blue circles), with poor coverage by the PI hotspots, are false alarms. The annual consultation missed a strong earthquake (to the southwest), which located near a cluster of PI hotspots.

6. Conclusions and Discussion

Regional forecast tests with time-independent and/or time-dependent models in the framework of CSEP (Jordan 2006) are now underway in California, Italy, Japan, New Zealand, northwest and southwest Pacific, and other places (e.g., Schorlemmer *et al.* 2010; Marzocchi *et al.* 2010; Gerstenberger and Rhoades 2010; Nanjo *et al.* 2010). The central China north–south seismic belt is a late comer (Mignan *et al.* 2013) in the stage of preparation. Beyond the analysis of the completeness of earthquake catalogues (e.g., Nanjo *et al.* 2010; Mignan *et al.* 2013), as a preparatory stage of the testing, some input models for demonstrating the predictability of seismicity, and acting as the baseline forecast for comparison, are needed.

As a development of the work of Mignan *et al.* (2013), we analyzed the earthquake catalogue in this region, emphasizing the applicability of the predictive algorithms based on the assumptions of meta-stable equilibrium. In this region, ergodicity was shown to exist for most of the time periods from 1980 to the present, with sub-region dependence that the Sichuan and Yunnan region, to the south of the seismic belt, have better ergodicity. Such ergodicity is apparently disrupted by the 2008 Wenchuan earthquake. A different idea proposed is that it may be unnecessary to relate the Wenchuan earthquake to an equilibrium-breaking event, since the disruption to the ergodicity seems simply from the clustering property of the aftershock sequence. In considering the sub-region dependence, it is apparently not persuasive to suggest to shrink the area of the testing region to the Sichuan-Yunnan region, that is, the south part. But in any case, the difference in the characteristics of seismicity in different sub-regions is in need of careful discussion.

The pattern informatics (PI) algorithm, which outperforms random forecast and the relative intensity (RI) statistics, can act as a model for comparison and testing. Previous works on the PI forecast in this region were mainly concentrated on the retrospective analysis, while this paper tries to put the algorithm in forward forecast test, although not yet in the CSEP standard. The present test uses the parameters as follow: cutoff magnitude 3.0, target magnitude 6.0+, grid size 0.2° × 0.2°, 'background window' 10 years, 'anomaly identification window' 5 years, and 'forecast window' 5 years. As a matter of fact, such a test is not only a test of the algorithm but also a test of the parameter settings.

Acknowledgments

Thanks to Prof. J. B. Rundle and Prof. C.-C. Chen for providing the code of the PI algorithm. Comments of the anonymous referees as well as the guest editor helped much to improve the results and the text of the manuscript. The earthquake catalogues are from the China Earthquake Networks Center (CENC).

REFERENCES

BORMANN, P., LIU, R. F., REN, X., GUTDEUTSCH, R., KAISER, D. and CASTELLARO, S. (2007) *Chinese national network magnitudes, their relation to NEIC magnitudes, and recommendations for new IASPEI magnitude standards*. Bull. Seismol. Soc. Amer. *97*, 114–127, doi:10.1785/0120060078.

CHEN, C. C., RUNDLE, J. B., HOLLIDAY, J. R., NANJO, K. Z., TURCOTTE, D. L., LI, S. C. and TIAMPO, K. F. (2005) *The 1999 Chi-Chi, Taiwan, earthquake as a typical example of seismic activation and quiescence*, Geophys. Res. Lett. *32*, L22315, doi:10.1029/2005GL023991.

CHEN, Y and BOOTH, D. C. (2011) The Wenchuan Earthquake of 2008: Anatomy of a Disaster. Science Press cooperating with Springer, Beijing.

CHO, N. F. and TIAMPO, K. F. (2012) *Effects of location errors in pattern informatics*. Pure Appl. Geophys. *170*, 185–196, doi:10.1007/s00024-011-0448-2.

Department of Earthquake Disaster Prevention of the State Seismological Bureau (1995) The Catalogue of Chinese Historical Strong Earthquakes. Seismological Press, Beijing **(in Chinese with English abstract)**.

Department of Earthquake Disaster Prevention of the China Earthquake Administration (1999) The Catalogue of Chinese Modern Earthquakes. China Science and Technology Press, Beijing **(in Chinese with English abstract)**.

GERSTENBERGER, M. C. and RHOADES, D. A. (2010). *New Zealand Earthquake Forecast Testing Centre*. Pure Appl. Geophys. *167*, 877–892, doi:10.1007/s00024-010-0082-4.

HOLLIDAY, J. R., RUNDLE, J. B., TIAMPO, K. F., KLEIN, W. and DONNELLAN, A. (2006) *Modification of the pattern informatics method for forecasting large earthquake events using complex eigenfactors*. Tectonophysics *413*, 87–91, doi:10.1016/j.tecto.2005.10.008.

HOLLIDAY, J. R., CHEN, C. C. TIAMPO, K. F., RUNDLE, J. B., TURCOTTE, D. L. and DONNELLAN, A. (2007) *A RELM earthquake forecast based on pattern informatics*. Seismol. Res. Lett. *78*, 87–93, doi:10.1785/gssrl.78.1.87.

JIANG, C. S. and WU, Z. L. (2008) *Retrospective forecasting test of a statistical physics model for earthquakes in Sichuan-Yunnan region*. Sci. China Ser. D Earth Sci. *51*, 1401–1410, doi:10.1007/s11430-008-0112-6.

JIANG, C. S. and WU, Z. L. (2010) *PI forecast for the Sichuan-Yunnan region: retrospective test after the May 12, 2008, Wenchuan earthquake*. Pure Appl. Geophys. *167*, 751–761, doi:10.1007/s00024-010-0070-8.

JIANG, C. S. and WU, Z. L. (2011) *PI forecast with or without declustering: an experiment for the Sichuan-Yunnan region*. Nat. Hazard. Earth Sys. *11*, 697–706, doi:10.5194/nhess-11-697-2011.

JORDAN, T. H. (2006) *Earthquake predictability, brick by brick*. Seismol. Res. Lett. *77*, 3–6.

KEILIS-BOROK, V. and ROTWAIN, I. (1990) *Diagnosis of time of increased probability of strong earthquakes in different regions of the world: algorithm CN*. Phys. Earth Planet. Interi. *61*, 57–72, doi:10.1016/0031-9201(90)90095-F.

LI, S. L., MOONEY, W. D. and FAN, J. C. (2006) *Crustal structure of mainland China from deep seismic sounding data*. Tectonophysics *420*, 239–252. doi:10.1016/j.tecto.2006.01.026.

MARZOCCHI, W., SCHORLEMMER, D. and WIEMER, S. (2010) *Collaboratory of the study of earthquake predictability. Preface*, Ann. Geophys. *53*, iii–viii, doi:10.4401/ag-4851.

MIGNAN, A., JIANG, C. S., ZECHAR, D. J., WIEMER, S., WU, Z. L. and HUANG, Z. B. (2013) *Completeness of the mainland China earthquake catalog and implications for the setup of the China earthquake forecast testing center*. Bull. Seismol. Soc. Amer. *103*, 845–859, doi:10.1785/0120120052.

MIGNAN A. and WOESSNER J (2012) *Estimating the magnitude of completeness for earthquake catalogs*, Community Online Resource for Statistical Seismicity Analysis. doi:10.5078/corssa-00180805. Available at http://www.corssa.org.

MOLCHAN, G. M. (1997) *Earthquake prediction as a decision making problem*. Pure Appl. Geophys. *149*, 233–247, doi:10.1007/BF00945169.

NANJO, K. Z., RUNDLE, J. B., HOLLIDAY, J. R. and TURCOTTE, D. L. (2006a) *Pattern informatics and its application for optimal forecasting of large earthquakes in Japan*. Pure Appl. Geophys. *163*, 2417–2432, doi:10.1007/978-3-7643-8131-8_12.

NANJO, K. Z., HOLLIDAY, J. R., CHEN, C. C., RUNDLE, J. B. and TURCOTTE, D. L. (2006b) *Application of a modified pattern informatics method to forecasting the locations of future large earthquakes in the central Japan*, Tectonophysics *424*, 351–366, doi:10.1016/j.tecto.2006.03.043.

NANJO, K. Z., ISHIBE, T., TSURUOKA, H., SCHORLEMMER, D., ISHIGAKI, Y. and HIRATA, N. (2010) *Analysis of the completeness magnitude and seismic network coverage of Japan*. Bull. Seismol. Soc. Am. *100*, 3261–3268, doi:10.1785/0120100077.

RUNDLE, J. B., TURCOTTE, D. L., SHCHERBAKOV, R., KLEIN, W. and SAMMIS, C. (2003) *Statistical physics approach to understanding the multiscale dynamics of earthquake fault systems*. Rev. Geophys. *41*, 1019, doi:10.1029/2003RG000135.

SCHORLEMMER, D., ZECHAR, J. D., WERNER, M. J., FIELD, E. H., JACKSON, D. D. and JORDON, T. H. (2010) *First results of the regional earthquake likelihood models experiment*, Pure Appl. Geophys. *167*, 859–876, doi:10.1007/s00024-010-0081-5.

SHI, Y. L., LIU, J., and ZHANG, G. M. (2001) *An evaluation of Chinese Annual Earthquake Predictions, 1990∼1998*, J. Appl. Probab., 38A: 222–231.

SU, Y. J., LI, Y. L., LI, Z. H., YI, G. X. and LIU, L. F. (2003) *Analysis of minimum complete magnitude of earthquake catalog in Sichuan–Yunnan region*, J. Seismol. Res. *26*(Suppl), 10–16 **(in Chinese with English Abstract)**, doi:10.3969/j.issn.1000-0666.2003.z1.003.

Swets, J. A. (1973) *The relative operating characteristic in psychology*. Science *182*, 990-1000.

Tiampo, K. F., Rundle, J. B., Klein, W. and Martins, J. Sá (2004) *Ergodicity in nature fault systerms*. Pure Appl. Geophys. *161*, 1957–1968. doi:10.1007/s00024-004-2542-1.

Tiampo, K. F., Rundle, J. B., Klein, W., Holliday, J., Martins, J. S. Sá and Fergusan, C. D. (2007) *Ergodicity in nature earthuquake fault networks*. Phys. Rev. E *75*, 066107. doi:10.1103/PhysRevE.75.066107.

Tiampo, K. F., Rundle, J. B., McGinnis, S., Gross, S. and Klein, W. (2002), *Mean-field threshold systems and phase dynamics: An application to earthquake fault systems*. Europhys. Lett. *60*, 481–487, doi:10.1209/epl/i2002-00289-y.

Tiampo, K. F. and Shcherbakov, R. (2012) *Seismicity-based earthquake forecasting techniques: Ten years of progress*. Tectonophysics *522–523*, 89–121. doi:10.1016/j.tecto.2011.08.019.

Thirumalai, D., Mountain, R. D. and Kirkpatrick, T. R. (1989) *Ergodic behavior in supercooled liquids and in glasses*. Phys. Rev. A *39*, 3563–3574.

Thirumalai, D. and Mountain, R. D. (1993) *Activated dynamics, loss of ergodicity, and transport in supercooled liquids*. Phys. Rev. E *47*, 479–489.

Wu, A. X., Zhang, Y. X., Zhou, Y. Z., Zhang, X. T. and Li, G. J. (2011) *On the spatial-temporal characteristcs of ionospheric parameters before Wenchuan earthquake with the MPI method*. Chinese J. Geophys. *54*, 2445–2457, doi:10.3969/j.issn.0001-5733.2011.10.002.

Wu, Y. M. and Chen, C. C. (2007) *Seismic reversal pattern for the 1999 Chi-Chi, Taiwan, M_W 7.6 earthquake*. Tectonophysics *429*, 125–132, doi:10.1029/2008GL035215.

Wu, Y. H., Chen, C. C. and Rundle, J. B. (2008) *Detecting precursory earthquake migration patterns using the pattern informatics method*. Geophys. Res. Lett. *35*, L19304, doi:10.1029/2008GL035215.

Wu, Z. L., Ma, T. F., Jiang, H. and Jiang, C. S. (2013) *Multi-scale seismic hazard and risk in the China mainland with implication for the preparedness, mitigation, and management of earthquake disasters: An overview*. Int. J. Disaster Risk Reduction *4*, 21–33.

Xu, X. W. and Deng, Q. D. (1996) *Nonlinear characteristics of paleoseismicity in China*. J. Geophys. Res. *101*, 6209–6231, doi:10.1029/95JB01238.

Xu, X. W., Zhang, P. Z., Wen, X. Z., Qin, Z. L., Chen, G. H. and Zhu, A. L. (2005) *Features of active tectonics and recurrence behaviors of strong earthquakes in the western Sichuan Province and its adjacent regions*. Seismol. Geol. *27*, 446–461 **(in Chinese with English abstract)**.

Yi, G. X., Wen, X. Z. and Xu, X. W. (2002) *Study on recurrence behaviors of strong earthquakes for several entireties of active fault zones in Sichuan-Yunnan region*. Earthq. Res. China *18*, 267–276 **(in Chinese with English abstract)**.

Zechar, J. D. and Jordan, T. H. (2008) *Testing alarm-based earthquake predictions*. Geophys. J. Int. *172*, 715–724, doi:10.1111/j.1365-246X.2007.0367.x.

Zechar, J. D. and Jordan, T. H. (2010) *Simple smoothed seismicity earthquake forecasts for Italy*. Ann. Geophys. *53*, 99–105, doi:10.4401/ag-4845.

Zhang, P. Z., Deng, Q. D., Zhang, G. M., Ma, J., Gan, W. J. Min W., Mao, F.,Y., and Wang, Q. (2003) *Active tectonic blocks and strong earthquakes in the continent of China*. Sci China Ser D *46*(Suppl.), 13–24, doi:10.136./03dz0002.

Zhao, Y. Z., Wu, Z. L., Jiang, C. S. and Zhu, C. Z. (2010) *Reverse tracing of precursors applied to the annual earthquake forecast: Retrospective test of the annual consultation in the Sichuan-Yunnan region of southwest China*. Pure Appl. Geophys. *167*, 783–800, doi:10.1007/s00024-010-0077-1.

Zhuang, J. C. and Jiang, C. S. (2012) *Evaluation of the prediction performance of the Annual Consultation Meeting on earthquake tendency by using the gambling score*. Chinese J. Geophys. *55*, 1695–1709, doi:10.6038/j.issn.0001-5733.2012.05.026 **(in Chinese with English abstract)**.

(Received August 15, 2014, revised June 2, 2015, accepted June 9, 2015, Published online June 27, 2015)

Pure Appl. Geophys. 173 (2016), 255–268

DOI 10.1007/s00024-015-1043-8

Pure and Applied Geophysics

Characteristics of Seismoelectric Wave Fields Associated with Natural Microcracks

Yukio Fujinawa[1] and Yoichi Noda[2]

Abstract—Properties of seismoelectric waves in relation to natural earthquakes have been investigated. The electromagnetic disturbances were analyzed to test the hypothesis that pulse-like electric variations are directly related to microcracks as source. Because variation is very difficult to detect, there have been few quantitative field investigations. We used selected events with clear S and P phases from the data catalog obtained before the Tohoku earthquake in 2011. The electric strength of the fast P wave (P_f), S wave (S), and electromagnetic wave (EM) associated with formation of cracks of tensile mode were estimated. The co-seismic electric signal accompanied by the S wave has the largest strength, well above the noise level, and the EM wave has the lowest strength. Analytical estimation of the ratio of the strengths of the P_f and EM phases to that of the S phase by use of Pride's equations gave results partially in agreement with observation (the order was $Ap_f > A_s > A_{em}$). The strength of the observed electromagnetic mode is approximately two orders of magnitude larger than that estimated from the theory. We suggest this greater strength can be attributed to the converted modes at layer contracts or to the effect of the boundary between free atmosphere and crust. Overall agreement between observations and theoretical estimates suggests that electromagnetic anomalies, crustal deformation, and groundwater changes can be investigated on the basis of the unified equations for the coupled electromagnetics, acoustics, and hydrodynamics of porous media.

Key words: Seismoelectric waves, natural earthquakes, microcracks, electrokinetic effect, electromagnetic precursors, converted modes, strength.

1. Introduction

Numerous reports indicate that electromagnetic signals are associated with earthquakes. Many of these reports reveal co-seismic field variation is associated with the seismic wave, but there are few field observations of waves propagating faster than the P-wave. The first field observation obtained by use of magnetic measurement was reported by Belov *et al.* (1974). An electromagnetic wave arriving before the P wave, but not starting at the time of origin, has frequently been reported (ERG 1997; Huang 2002; Iyemori 1996; Honkura *et al.* 2002). For example, the Network-MT observatories recorded seismoelectric signals from the 1995 Kobe earthquake approximately 8–18 s after the origin time of the main shock (ERG 1997). A model based on the piezoelectric effect and the dislocation theory of faulting was developed by Huang (2002) for theoretical investigation of the generation of co-seismic electric signals. By analogy with the seismic rupture model, a double-couple source model was first tried by Huang (2002), who assumed the interaction force was a piezoelectric effect. This was one of the earliest discussions of controversial earthquake electromagnetic precursors on the basis of the fault-rupture model.

Field detection of co-rupture electromagnetic anomalies is very rare, because of the very small field strength compared with environmental noise. The arrival time of such signals at the observation site hardly coincides with the origin time of the earthquake, which case is the most useful for application to the earthquake early warning (Fujinawa *et al.* 2009). As far as we are aware, there are only two examples: one is at the time of the great Hokkaido Earthquake in 1994 (Fujinawa and Takahashi 1998), the other is electric field detection starting from the time of origin of two moderate earthquakes (Fujinawa *et al.* 2011). Higher-speed recording and use of a highly sensitive sensor revealed electric signals starting from the instant of the time of origin (± 1 s) of a crustal M4.9 earthquake with an epicenter distance of approximately 100 km. Moreover the primary and second phases of the electrogram were found to coincide with the seismic P and S waves, respectively.

[1] Mieruka Bousai Inc., Shibuya-ku, Tokyo, Japan. E-mail: hjfujinawa@cl.cilas.net; yukio.fujinawa@mieruka.co.jp
[2] Tierra Tecnica Ltd, Musashi-murayama, Tokyo, Japan.

The mechanism of generation of the seismoelectromagnetic variation was suggested as being the electrokinetic effect (Gershenzon *et al.* 1994; Gershenzon and Bambakidis 2001) by order estimation of several possible effects. The much faster propagation than the seismic P wave suggests it is an electromagnetic mode but not of co-seismic mode. The co-rupture electric variation was suggested as originating from a converted electric mode (EM) of a surface wave trapped to propagate along the boundary between materials of different conductivity (Gershenzon and Bambakidis *et al.* 2001; Fujinawa *et al.* 2011).

Study of seismoelectric variation began a new era after the work of Pride (1994). The physics of wave generation and propagation in porous media saturated with fluid was formulated from a microscopic perspective treating motion of the main rock substance and of the fluid phase. He derived equations for the coupled poroelastic and EM fields on the basis of Biot's poroelastodynamic equations and Maxwell's equations of electrodynamics. The wave mode solution for fixed frequency was shown to have four modes, fast and slow compressional P-waves (P_f and Ps) and two transverse modes, the S–wave and the electromagnetic mode (S and EM).The first three modes are co-seismic electric field changes accompanying seismic waves. The EM mode is nothing but an electromagnetic wave in fully saturated porous media with dynamic electric permittivity depending on frequency and dynamic conductivity. The analytical form was given by Pride (1994) and Pride and Haartsen (1996). Seismoelectric waves were detected in numerical and laboratory experiments (Gao and Hu 2010; Garambois and Dietrich 2001; Araji *et al.* 2012; Ren *et al.* 2012; Schoemaker *et al.* 2012; Haas *et al.* 2013) and indicated three phases, P_f, S, and EM modes, were present, although the EM mode was too small to be observed in the field (Gao and Hu 2010; Garambois and Dietrich 2001; Araji *et al.* 2012).

The converted EM fields generated at the interfaces were investigated in detail from the perspective of geophysical prospecting (Araji *et al.* 2012 and references cited therein). They investigated mainly the co-seismic electric and magnetic fields. However, the properties of seismoelectric variations propagating at EM velocity radiating from the natural source were not investigated in detail (Ren *et al.* 2012).

The electrokinetic coupling model was analyzed in detail by Gao and Hu (2010) for double-couple sources and provided a useful means of multi-disciplinary investigation of earthquake precursors by use of electromagnetic and seismological approaches. They discussed the relative strength of different phases and revealed the P_f-phase is larger than the S phase with exceptional cases of the opposite behavior under specific physical conditions.

We have previously reported that special antenna could be used to observe pulse-like variations related to microcracks in the nucleation stage of the Tohoku earthquake (Fujinawa *et al.* 2013). Of those variations, one type (the B-type) was suggested as being induced by microcracks which define the stage of nucleation of the rupture, as confirmed by rock experiments. The finding confirmed the practical utility of this method of observation and enabled detection of the characteristics of electric variations occurring in the nucleation stage only. It is, however, difficult to discuss their intrinsic mechanism of generation because of the very rare field detection of electric variations related to earthquakes.

The electric field variations associated with microcracks sometimes have clear P-phase with the dominant S-phase (Fig. 11 in Fujinawa *et al.* 2013). It is also found that some events have a weak phase which is assumed to be EM mode. Here we further analyzed the properties of this electric variation by use of electric events with clear traces of S, P_f, and possible EM mode. Mahardika *et al.* (2012) developed a useful method of waveform inversion by using localized seismic sources to analyze source characteristics. This method, however, could not be applied because of the limited data from this observation. We compared the observed and estimated amplitudes of each mode both quantitatively and qualitatively by using the electric field induced by a double-couple point source in the isotropic medium (Pride and Haartsen 1996; Gao and Hu 2010).

2. Method and Data

We have investigated electromagnetic phenomena from the perspective of short-term earthquake prediction. A borehole antenna (Fujinawa and Takahashi

1990) was used at all sites to measure the vertical electric field by means of a casing pipe of length from 150 to 1200 m in deep borehole (Fig. 1); this was based on an original idea TAKAHASHI and TAKAHASHI (1989). The steel casing pipe functions as a monopole of the antenna in the conductive subsurface, and naked copper wire buried underground, surrounding the casing pipe with radius of 10–50 m, is used as the antenna reflector. The wire is set in the midst of coke bottom layers 10 cm wide and 10 cm high in a ditch 1 m deep (TAKAHASHI *et al.* 2000; FUJINAWA *et al.* 2001). The copper wire and belt of coke act as the antenna reflector.

The antenna is similar to submarine antennas used for communication in the conductive crustal medium (FUJINAWA *et al.* 2001). Thirteen observation sites were constructed in central Japan. Observation has proved the sensor is highly robust to environmental noise, enabling discrimination, without difficulty, anomalous signals associated with earthquakes and volcanic eruptions in the noise-rich environment (FUJINAWA *et al.* 2001). We used data at the site Hasaki situated around the south–west corner of the rupture area of the Tohoku earthquake. The borehole at Hasaki is 804 m deep (FUJINAWA *et al.* 2013).

A 16-bit second-phase detector with an increased frequency range (0–18 kHz) and dynamic range was installed on March 3, 2011, 8 days before the great Tohoku earthquake on March 11, 2011. Analysis of the first-phase observation data from approximately a month, on April 7, 2011, showed there are three types of pulse-like electric variation. These were denoted A, B, and C-types depending on the dominant frequencies (Fig. 7 in FUJINAWA *et al.* 2013). The frequencies of the A, B, and C-types were approximately 0.01 Hz, 500 Hz, and 5 kHz, respectively, and we suggest they characterize successive pre and post stages of the main shock, i.e the nucleation stage and the strain

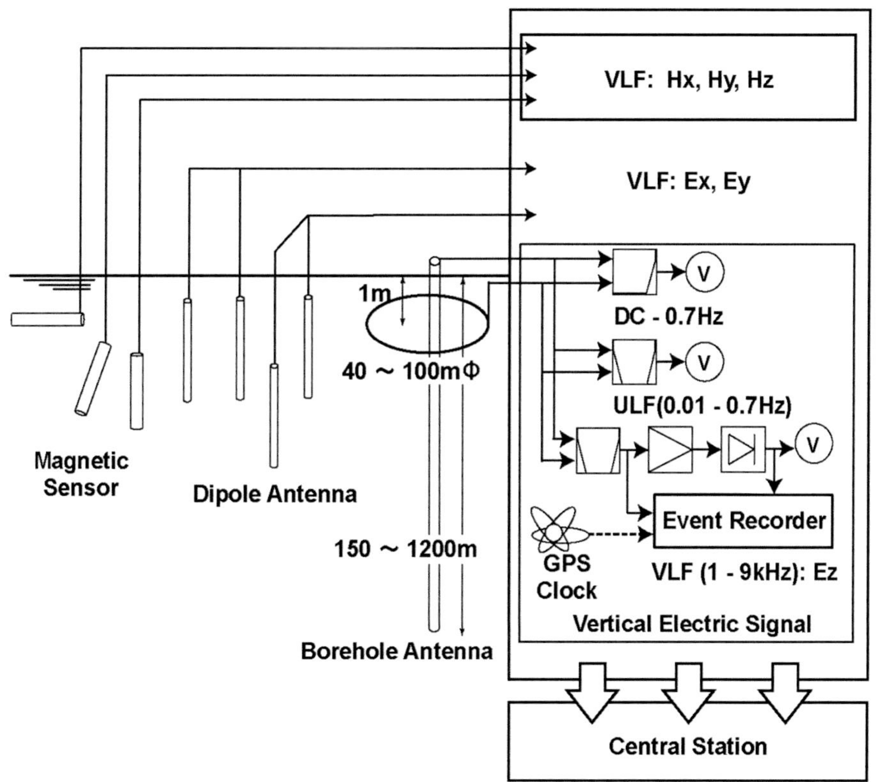

Figure 1
Block diagram of electromagnetic field observations (FUJINAWA and TAKAHASHI 1998; FUJINAWA *et al.* 2001, 2013). The electric field is measured by use of a pair of horizontal dipoles and a vertical mono-pole of 600–1200 m. The observation at Hasaki was conducted by using the simplest model using only a borehole sensor and continuous data acquisition at a sampling rate of 4.5 kHz before March 11, 2011 and of 18 kHz after this date

relaxation stage after the main shock (Fujinawa *et al.* 2013). We suspect the B-type variation is induced by microcracks, identifying the nucleation period of the preparatory process before the main shock. The basis of this inference was that they appeared only in the period immediately before the earthquake (Fig. 8 in Fujinawa *et al.* 2013), and their activity evolved in accordance with the modified Omori's law (Fig. 9 in Fujinawa *et al.* 2013). B-type events are further grouped into three types on the basis of waveforms corresponding to the rupture modes tensile mode, shear mode, and mixed mode.

Some of the B-type electric variations have clear S phase and P phases, as suggested elsewhere (Fig. 11 in Fujinawa *et al.* 2013). For the other two types, A and C, it is difficult to detect the P, S phases. B-type events are the topic of this paper. We suggest there are some examples in which the appears signal near the time of crack occurrence (Fig. 8 in Fujinawa *et al.* 2013). The phase was suggested to be the EM mode. Here we make more in detailed analysis of the EM mode to investigate properties of those phases in the electrogram of B-type events. The twelve events with more or less apparent traces of EM mode at approximately the time of origin are shown in Table 1.

We determined the magnitude of the B-type events by using duration time or rupture time to investigate the statistics of occurrence and to estimate moment magnitude. By using the arrival times of P and S waves we estimated the focal distance by assuming seismic P and S velocities. EM phases were identified by use of a semi-quantitative method. The observed strengths of the P_f, S, and EM modes were then compared with those estimated on the basis the expression of Gao and Hu (2010). A comparison was also performed for the different phases to check the hypotheses that the pulse-like events are phenomena occurring in the nucleation period, and that they can be understood by use of the formula of Pride and successive researchers.

3. Results

3.1. Magnitude

We have already provided much evidence that pulse-like electric variations are induced by three kinds of rupture before and after the earthquake (Fujinawa *et al.* 2013). Almost all type-A events occurred after the main shock. The type-B variation only occurred immediately before the main shock and the cumulative number of events followed the modified Omori law. The C-type variations occurred only after the main shock during the after-effect stage of the seismic cycle. On the basis of these observations and relevant knowledge of seismology and seismoelectromagnetics we suggested that those signals were activated in the preparatory stage and the after-effect relaxation stage of earthquake occurrence. Here further evidence will be presented for evaluation of the hypothesis on the basis of magnitude and frequency of occurrence of the events: the Gutenberg–Richter relationship and the *b* value.

Several methods are used for estimation of magnitude, for example the empirical method relying on the distance dissipation law of seismic waves and using the focal position and observed amplitude. Here we used the relationship between magnitude and seismic wave duration, Td, because one observation only was used in this work. Tsumura (1967) derived a magnitude determination formula by using Td and the epicenter distance, Δ:

$$M = -2.4 + 2.9 \log \mathrm{Td} + 0.0014\Delta \quad (1)$$

In this work the third term on the right is negligibly small compared with other terms, because the epicenter distances of the events studied are less than 200 m. The characteristics are listed in Table 1.

From the magnitudes of all 243 B-type events (Fujinawa *et al.* 2013) we know they are distributed in the range:

$$-9.0 < M < -5.8 \quad (2)$$

with mean value Mav:

$$\mathrm{Mav} = -7.0 \quad (3)$$

The rupture length scales are estimated on the basis of Otsuka (1965), with the resulting mean value:

$$\mathrm{Lm} = 0.5\ \mathrm{cm} \quad (4)$$

The length is in the range of acoustic emission (AE) in the framework of the grouping of earthquakes by Ogasawara (2009).

Table 1

Characteristics of the 12 B-type events (Fujinawa et al. 2013)

Day	s	h	PS (ms)	R (m)	R* (m)	Δ S_p	S_p*	V_p*	V_s*
2011/3/3	76994.65	21.39	6.7	48	38	0.055	0.29	3.48	2.19
2011/3/4	81148.25	22.54	13.0	94	75				
2011/3/6	64,179.25	17.83	6.6	48	38				
2011/3/6	83,206.25	23.11	10.1	73	58				
2011/3/6	84,237.5	23.4	3.5	26	20	0.078	0.31	3.22	2.03
2011/3/8	17,343	4.82	18.7	135	108	0.061	0.29	3.41	2.14
2011/3/9	71,162.05	19.77	13.3	96	76				
2011/3/9	73,619.15	20.45	20.7	150	119	0.052	0.28	3.52	2.21
2011/3/9	76,353	21.21	14.1	102	81	0.067	0.30	3.34	2.10
2011/3/9	78,908.15	21.92	22.0	160	127	0.055	0.29	3.48	2.19
2011/3/9	82,632.3	22.95	26.9	195	155				
2011/3/11	15,506.45	4.31	4.7	34	27	0.069	0.30	3.31	2.08
					Mean	0.062	0.29	3.39	2.13
					STD	0.009	0.01	0.11	0.07

P–S time (ms), focal distance *R* (m) based on the initial assumption of the seismic velocities of the P-wave and S-wave, corrected focal distance *R** (m), corrected slowness of P-wave *Sp** (s/km), and corrected P-wave velocity V_p* (km/s) and corrected S-wave velocity *Vs** (km/s) are shown

The correction could not be done for some events because of too vague trace of EM phase

3.2. *b* Value

The accumulated number N (A) below the amplitude A (mV) is shown in Fig. 2. The distribution follows the Gutenberg–Richter relationship with the *b* value.

$$b \fallingdotseq 0.7 \tag{5}$$

Here the *b* value was estimated by using the formula of Ishimoto and Iida (1937) and Asada *et al.* (1950). The *b* value is known to be $b \fallingdotseq 1.0$, 0.35 for the foreshock, and 0.75 for the aftershock (Nanjo *et al.* 2012). This value is in reasonable agreement with previous results. This result is more evidence that the events are induced in the process of earthquake occurrence, and are not urban noise.

Another method of magnitude estimation uses the rupture time τ of the fracture (Uchide *et al.* 2009). The rupture time is assumed to be the width of the envelope of the absolute value of the electric time series. The moment, M_0 (τ), of the double-couple rupture is estimated from:

$$M_0(\tau) = 2.0 \times 10^{17}\tau^3 \tag{6}$$

Observed moments are listed in Table 2.

The moment magnitude M_w is obtained by use of the relationship:

$$\mathrm{Log}M_0 = 1.5M_w + 9.1 \tag{7}$$

It is difficult to estimate the rupture time for the tensile mode, because the mode has the waveform of relaxation decaying exponentially (Fig. 7 in Fujinawa *et al.* 2013). We limit our interest to the shear mode in comparison of the amplitudes of different modes. The "M_w" values are listed in Table 2. In this case:

$$-0.98 < M_w < -0.53 \tag{8}$$

The original formula by Uchide *et al.* (2009) is for the magnitude range $M_w > 1.5$. Here we assumed it could be extrapolated to the smaller values of the magnitude, on the basis of the scaling law of seismic activity.

3.3. *Identification of EM phases*

Detection and identification of the EM mode is very difficult for natural earthquakes. We have previously reported that some B-type events have slight trace of this phase at approximately time of origin of the micro rupture (Fujinawa *et al.* 2013). Here we analyze that in more detail.

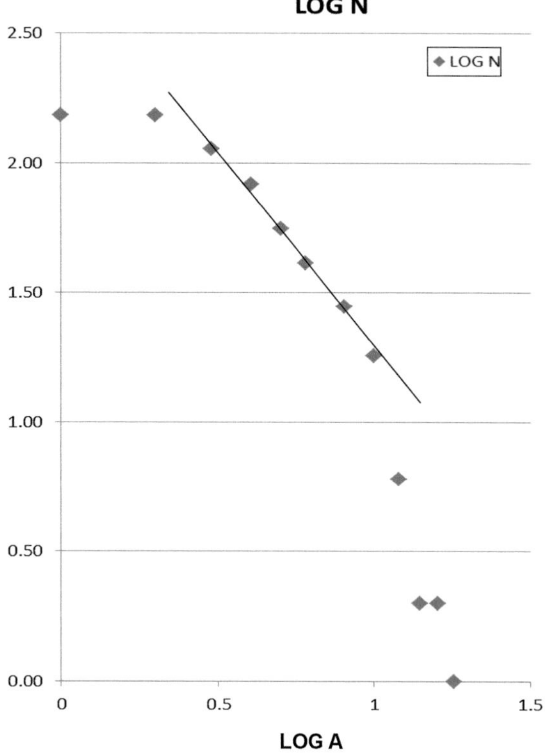

Figure 2
Plot of the correlation between electric strength A and the cumulative number of B-type microcracks detected in the period March 3 to March 11, 2011, just before the Tohoku earthquake. The dominant part is seen to follow the Gutenberg–Richter law with a b value of 0.7

The EM mode is supposed to propagate at a speed (approx. 100 km/s for 500 Hz) much larger than the seismic wave speed. It is expected that the EM mode arrives just after the origin time in the record of electrogram. Numerical simulations of wave propagation (Jardani *et al.* 2010; Gao and Hu 2010) and laboratory experiments (Araji *et al.* 2012) show that the EM phase arrives much earlier than the coseismic P and S phases, immediately after the origin time. Those phases are assumed to be co-seismic P and S electric variations, on the basis of very good agreement between the seismogram and electrogram (Fujinawa *et al.* 2011) and because of electromagnetic P_f and S modes have nearly the same phase velocities as the P and S waves of the bulk elastic modulus.

Figure 3 shows three typical cases having clearest EM phases in the electrogram. The candidate for the EM phase is not so convincing for the other examples. More detailed analysis was then conducted. The P–S time furnishes the origin time, T_0, by assuming seismic P and S wave V_p and V_s as:

$$V_p = 4.3\ \text{km/s} \tag{9a}$$

$$V_s = 2.7\ \text{km/s} \tag{9b}$$

Those values are based on the previous work by Suzuki (2002) and Hayashi *et al.* (2006) indicating that

Table 2

Physical characteristics of events: focal mechanism, magnitude, and moment

Day	s	h	R* (m)	Mechanism	τ (s)	M_0 (τ) (N/m)	M_w	M	M_0 (N/m)
2011/3/6	84,237.5	23.4	20	Shear	0.00073	7.9E+07	−0.80	−6.9	6.5E−02
2011/3/11	15,506.45	4.31	27	Tensile				−6.2	8.0E−01
2011/3/6	64,179.25	17.83	38	Tensile				−6.3	4.8E−01
2011/3/3	76,994.65	21.39	39	Shear	0.00067	5.9E+07	−0.88	−6.7	1.4E−01
2011/3/6	83,206.25	23.11	58	Tensile				−6.3	4.8E−01
2011/3/4	81,148.25	22.54	75	Shear	0.00073	7.9E+07	−0.80	−8.1	1.3E−03
2011/3/9	71,162.05	19.77	76	Shear	0.00100	2.0E+08	−0.53	−6.8	1.1E−01
2011/3/9	76,353	21.21	81	Shear	0.00060	4.3E+07	−0.98	−8.1	1.3E−03
2011/3/8	17,343	4.82	108	Shear	0.00067	5.9E+07	−0.88	−6.5	2.7E−01
2011/3/9	73,619.15	20.45	119	Tensile				−6.3	4.8E−01
2011/3/9	78,908.15	21.92	127	Tensile				−8.1	1.3E−03
2011/3/9	82,632.3	22.95	155	Tensile				−6.7	1.4E−01

τ is the rupture time of events of shear mode, *Mo* (τ) is the seismic moment, and M_w is the moment magnitude

The M and M_0 in the last two columns are the magnitude estimated from the duration of the signal and corresponding moment, respectively

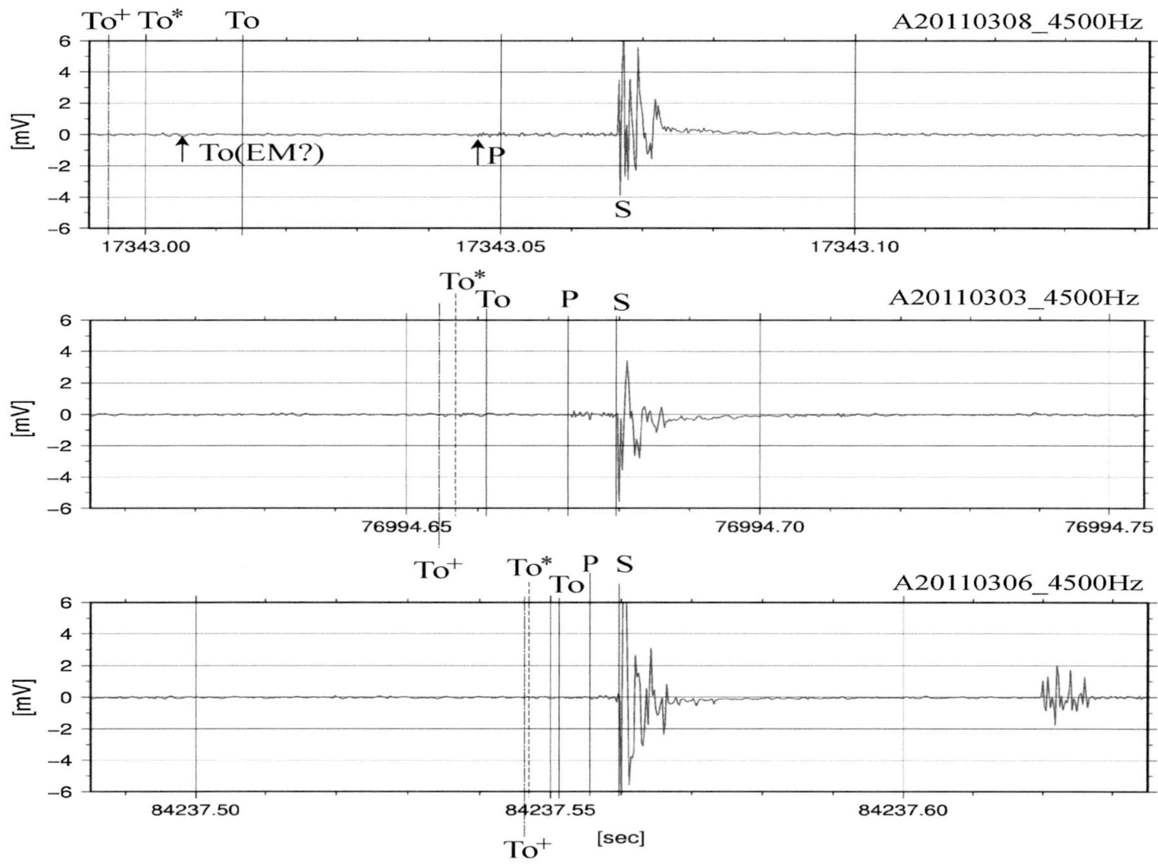

Figure 3
Three sampled waveforms of events having the primary phase ("P" in the figure) before the main phase ("S" in the figure), and a weak EM phase near the origin time T_0 (indicated "?"). The EM mode is suspected at approximately the origin time T_0 (EM) taking into consideration the very large speed of more than one hundred times that of the fast P phase. The T_0 is corrected to T^* by considering the difference between the origin time and the apparent time of EM as indicated by the eye. $T+$ denotes the origin time updated by using the corrected seismic velocity

the upper boundary of the pre-Neogine basement, with speeds of 4.5–5.4 km/s situated approximately 500 m beneath the Hasaki observation site in the northeastern corner of the Kanto Plain. The values are also very near to those assumed by GAO and HU (2010).

The assumed origin time is indicated by T_0 in Fig. 3. Initial inspection of the electrogram suggests there is more than one candidate for the EM phases at approximately T_0. More suitable origins than T_0 were sought by eye checking, and were denoted by T^* for each event. The corrected seismic P wave velocity V_p* is obtained by using difference between T^* and T_0. The corrected slowness is estimated as an average for seven examples (Table 1) to provide the second approximation of seismic velocities, assuming the ratio V_p/V_s is constant (=1.6). The corrected origin time is denoted $T+$, as shown in Fig. 3. I it is apparent that the results are in more suitable positions. The focal distances based on the updated propagation speeds are shown as R^* in Table 1. Correct velocities are:

$$V_p = 3.4 \pm 0.1 \text{ km/s} \quad (10a)$$

$$V_s = 2.7 \pm 0.1 \text{ km/s} \quad (10b)$$

The values are larger for the shallow part of the area than the value calculated for the observation site (IKEDA 1987). The corrected values approach the velocities of the shallower part than the pre-Neogine basement around the site. There seems to be some effect of effective higher speed hidden in the process of the above procedures used to search for the focal distance.

3.4. Electric Strength of the P_f, S, and EM Phases

3.4.1 Observed Values

The observed absolute amplitude A of two co-seismic phases S and P_f, and direct the EM are denoted A_s (O), A_p (O), and A_{em} (O), respectively; the values are given in Table 3.

From this result we know that the order of the strength is:

$$A_s(O) > A_p(\mathrm{O}) > A_{em}(\mathrm{O}) \qquad (11)$$

without exception. Our previous observations of co-rupture electric variations for fifty moderate earthquakes in central Japan have also the property:

$$A_s(\mathrm{O}) \gg A_p(\mathrm{O}) \qquad (12)$$

without exception (Fujinawa *et al.* 2011). This is the first feature to be noted, in contrast with results from simulation (Gao and Hu 2010). Estimated strengths were also compared by using related values almost the same as those of Gao and Hu (2010). This will be discussed below.

First we investigate the ratio of pairs among three amplitudes. The value $A_s(O)$ ranges from 2.2 to 9.7 mV. To compare the strength of different phases in average we use not only the real strength but to use ratio of the strength of different phases. In the Table 4 the observed value of the ratio are shown as $A_p(O)/A_s(O)$ and $A_{em}(O)/A_s(O)$ with the mean values being 0.078 (±0.078), and 0.033 (±0.016). It is shown that the S phase has far large strength than P and EM phases. The whole events of 243 have same ordering ($A_p/A_s(O) \ll 1.0$) without exception (Fujinawa *et al.* 2013) as is illustrated in Fig. 2. The feature is the same for the cases of natural earthquakes (Fujinawa *et al.* 2011). This is contrary with the results of numerical simulations as seen afterward and those of previous works (e.g., Gao and Hu 2010; Araji *et al.* 2012; Garambois and Dietrich 2001; Ren *et al.* 2012).

3.5. Calculated Values of the Electric Strength

B-type events were previously interpreted as being induced by microcracks associated with the 3.11 Tohoku Earthquake (Fujinawa *et al.* 2013). We use the basic formulation of Pride (1994) and the subsequent formulation for the double-couple source of Gao and Hu (2010) to compare observed amplitudes with calculated values.

Gao and Hu (2010) showed that the electric field **E** induced by the double couple of the dipole moment M_0 is described by the far field, intermediate, and near field parts, as for the seismic waves in the elastic medium. In the far field approximation, the ratio of amplitudes leads to:

$$|\mathbf{E}\mathrm{pf}|/|\mathbf{E}_s| = \left(L^F_{E,pf}/T^F_{E,S}\right)\left(S_{pf}/S_s\right)e^{i\omega r(S_{pf}-S_s)}|\mathbf{e}^{\mathbf{far}}_{\mathbf{L}}|/|\mathbf{e}^{\mathbf{far}}_{\mathbf{T}}| \qquad (13)$$

(Pride and Haartsen 1996; Gao and Hu 2010), where r is the distance from the source location to the observation point, ω is the frequency in radians, $\mathbf{e}^{\mathbf{far}}_{\mathbf{L}}$ and $\mathbf{e}^{\mathbf{far}}_{\mathbf{T}}$ are, respectively, the far field polarization unit vectors of the longitudinal and transverse modes for the electric field, and $\mathbf{e}^{\mathbf{far}}_{\mathbf{L}}$ is along the direction of wave propagation. S_{pf}, S_{ps}, S_s, and S_{em} are the complex slowness of the fast P wave (P_f), slow P wave (P_s), shear wave (S), and EM wave, respectively. The complex amplitudes of the transverse and longitudinal components are $T^F_{E,em}$ and $L^F_{E,pf,}$ respectively (Eqs. 154–157 of Pride and Haartsen 1996).

The ratio $|\mathbf{E}\mathrm{m}|/|\mathbf{E}\mathrm{s}|$ is independent of the direction of the observation points, but the amplitude ratio $|\mathbf{E}\mathrm{pf}|/|\mathbf{E}\mathrm{s}|$ depends on the direction of the observation point in reference to the fault plane. For this example we have no data for direction. Here we calculate the basic value assuming the averaged direction factor $|\mathbf{e}^{\mathbf{far}}_{\mathbf{L}}/\mathbf{e}^{\mathbf{far}}_{\mathbf{T}}| = 1$. Cases having a clear P phase are supposed to have a focal distance between the source position and observation point of, at most, 150 m (Table 1).

We estimated the slowness for four waves by using the methods of Pride and Haartsen (1996; Eqs. 92 and 93) and Gao and Hu (2010; Appendix A). The physical properties listed in Table 5 are those of Gao and Hu (2010). We estimated near, intermediate, and far field cases, and picked the largest at the focal distance of the events listed in Table 1. The far field term is dominant for the S and P modes, and the near field is dominant for the EM mode.

Table 3

Amplitudes of P_f, S, and EM from observation, analytical results obtained by use of the formula of PRIDE *(1994) and* PRIDE *and* HAARTSEN *(1996), and results obtained by estimation for the double-couple model (*GAO *and* HU *2010)*

Day	s	h	R*(m)	Mechanism	A_p(mV)	A_p*(TP) V/m	A_p*(τ) V/m	$A_p(C)$	A_p*(τ)/$A_p(C)$
2011/3/6	84,238	23.4	20	Shear	0.13	1.56E−07	1.98E−15	4E−13	5.30E−03
2011/3/11	15,506	4.31	27	Tensile	0.29	3.68E−07		3E−13	
2011/3/6	64,179	17.83	38	Tensile	0.32	3.95E−07		2E−13	
2011/3/3	76,995	21.39	39	Shear	0.17	2.13E−07	3.59E−15	2E−13	1.78E−02
2011/3/6	83,206	23.11	58	Tensile	0.55	6.84E−07		1E−13	
2011/3/4	81,148	22.54	75	Shear	0.21	2.63E−07	3.34E−15	1E−13	3.30E−02
2011/3/9	71,162	19.77	76	Shear	0.12	1.49E−07	7.44E−16	1E−13	7.52E−03
2011/3/9	76,353	21.21	81	Shear	0.40	5.00E−07	1.16E−14	9E−14	1.24E−01
2011/3/8	17,343	4.82	108	Shear	0.17	2.11E−07	3.55E−15	7E−14	5.07E−02
2011/3/9	73,619	20.45	119	Tensile	0.42	5.26E−07		6E−14	
2011/3/9	78,908	21.92	127	Tensile				6E−14	
2011/3/9	82,632	22.95	155	Tensile				5E−14	
		Mean			0.26	3.38E−07	4.13E−15	2E−13	3.97E−02
		STD			0.14	1.93E−07	3.81 E-15	1E−13	4.47E−02

Day	A_s(mV)	A_s*(TP) V/m	A_s*(τ) V/m	$A_s(C)$	A_s*(τ)/$A_s(C)$	A_{em}(mV)	A_{em}*(TP) V/m	A_{em}*(τ) V/m	$A_{em}(C)$	A_{em}* (τ)/$A_{em}(C)$
2011/3/6	8.30	1.04E−05	1.32E−13	1E−14	12.4	0.125	1.56E−07	1.981 E−15	1.18E−13	1.68E−02
2011/3/11	4.32	5.39E−06		8E−15		0.063	7.89E−08		3.94E−14	
2011/3/6	3.26	4.08 E-06		6E−15		0.063	7.89E−08		9.3E−15	
2011/3/3	5.58	6.97E−06	1.18E−13	6E−15	20.9	0.149	1.86E−07	3.142E−15	9.3E−15	3.38E−01
2011/3/6	2.21	2.76E−06		4E−15		0.105	1.32E−07		1.76E−15	
2011/3/4	3.01	3.76E−06	4.77E−14	3E−15	16.6	0.147	1.84E−07	2.336E−15	6.37E−16	3.67E+00
2011/3/9	2.74	3.42E−06	1.71E 14	3E−15	6.1	0.133	1.67E−07	8.333E−16	5.81 E-16	1.43E + 00
2011/3/9	9.70	1.21 E-05	2.81E−13	3E−15	106.3	0.292	3.65E−07	8.451E−15	4.59E−16	1.84E + 01
2011/3/8	8.00	1.00E−05	1.69E−13	2E−15	85.2	0.168	2.11E−07	3.553E−15	1.47E−16	2.42E + 01
2011/3/9	2.84	3.55E−06		2E−15		0.168	2.11 E-07		9.77E−17	
2011/3/9				2E−15					7.64E−17	
2011/3/9				1E−15					3.45E−17	
	5.24	5.82E−06	1.27E−13	4E−15	41.3	0.139	1.52E−07	3.383E−15	1.79E−14	8.00E + 00
	2.79	3.00E−06	9.36E−14	3E−15	43.0	0.068	5.26E−08	2.658E−15	3.71 E−14	1.05E+01

A_p, A_s, and A_{em} are the input electric potentials of the P-phase, S-phase, and EM phase, respectively

A_p*(TP), A_s*(TP), and A_{em}*(TP) are the electric fields normalized by the moment M_0, and A_p*(τ), A_p*(τ), and A_p*(τ) are the electric fields normalized by the moment Mo(τ). $A_p(C)$, $A_s(C)$, and $A_{em}(C)$ are calculated electric field for unit moment

A_p*(τ)/$A_p(C)$ is the ratio of the observed electric field to the estimated field of the P-phase

3.6. Comparison of Electric Strength

The observed and calculated column amplitude ratios A_p/A_s and A_{em}/A_s for each event are listed in Table 4. The order of the observed strengths is $A_s(O) > A_{Pf}(O) > A_{em}(O)$. However, the order of the estimated strengths is $A_{Pf}(C) \gg A_{em}(C)$, $A_s(C)$. EM mode has a larger amplitude for small epicenter distances whereas S mode has a larger amplitude for larger distances. The critical distance depends on the frequency, approximately 63 and 48 m for 300 and 500 Hz, respectively. The results in the two lowest rows of the table are the means and standard deviations. The standard deviations of $A_{pf}(C)$ and $A_s(C)$ are approximately half of the mean values; that for $A_{em}(C)$ is nearly the same, which is indicative of greater dependence on distance. The A_p/A_s ratio is approximately 30 times larger; the strength of P_f wave is much larger than that of the S phase.

The observed ratio shows that the ratio of the strength of EM mode to that of the S mode is 0.030 ± 0.02. The calculated value does, however, depend strongly on focal distance, reflecting the larger dissipation with distance of the EM mode—approximately one and half orders of magnitude, from 11 to 0.02, or the frequency of 500 Hz for the focal range 20–160 m.

The value depends on the wave frequency and on the properties of the medium. For the 300 Hz frequency, the

Table 4

Ratios of amplitudes of pairs among P_f, S and EM modes

Day	s	h	R*(m)	Mechanism	$A_p/A_s(O)$	$A_p/A_s(C)$	$A_{em}/A_s(O)$	$A_{em}/A_s(C)$
2011/3/6	84,237.5	23.4	20	Shear	1.51 E−02	3.51 E+01	1.51 E−02	1.11E+01
2011/3/11	15,506.45	4.31	27	Tensile	6.83E−02	3.51 E+01	1.46E−02	4.87E+00
2011/3/6	64,179.25	17.83	38	Tensile	9.68E−02	3.57E+01	1.94E−02	1.65E+00
2011/3/3	76,994.65	21.39	39	Shear	3.05E−02	3.57E+01	2.67E−02	1.65E+00
2011/3/6	83,206.25	23.11	58	Tensile	2.48E−01	3.52E+01	4.76E−02	4.76E−01
2011/3/4	81,148.25	22.54	75	Shear	6.99E−02	3.53E+01	4.90E−02	2.22E−01
2011/3/9	71,162.05	19.77	76	Shear	4.35E−02	3.53E+01	4.87E−02	2.07E−01
2011/3/9	76,353	21.21	81	Shear	4.12E−02	3.53E+01	3.01 E−02	1.74E−01
2011/3/8	17343	4.82	108	Shear	2.11 E−02	3.54E+01	2.11 E−02	7.43E−02
2011/3/9	73,619.15	20.45	119	Tensile	1.48E−01	3.55E+01	5.93E−02	5.47E−02
2011/3/9	78,908.15	21.92	127	Tensile		3.55E+01		4.55E−02
2011/3/9	82,632.3	22.95	155	Tensile		3.56E+01		2.51 E−02
				Mean	3.69E−02	3.54E+01	3.18E−02	2.23E+00
				STD	1.96E−02	1.99E−01	1.42E−02	4.36E+00

Observed and calculated values are shown

A_p/A_s(O) is the ratio of the observed amplitude of the P-phase to that of the S-phase

A_p/A_s(C) is the ratio of the estimated amplitude of the P-phase to that of the S-phase

Table 5

Properties of the porous medium used for calculation of the electric strengths of the P_f, S, and EM phases, in accordance with Gao and Hu (2010)

Porosity	$\phi = 0.15$	Units
Permeability	$\kappa_0 = 0.1$	mD
Tortuosity	$\alpha_\infty = 3$	
Solid grain bulk modulus	$K_s = 35.7$	GPa
Frame bulk modulus	$K_b = 17.9$	GPa
Shear modulus	$G = 17.8$	GPa
Fluid bulk modulus	$K_f = 2.25$	GPa
Solid grain density	$\rho_s = 2650$	kg m^{-3}
Fluid density	$\rho_f = 1\,000$	kg m^{-3}
Fluid viscosity	$\eta = 0.001$	Pas
Salinity	CO = 0.001	mol L^{-1}
Solid relative permittivity	$\varepsilon_s = 4$	–
Fluid relative permittivity	$\varepsilon_f = 80$	–
Velocity of P wave	$V_P = 4320$	ms^{-1}
Velocity of S wave	$V_S = 2721$	ms^{-1}

A_{pf}/A_s ratio remains almost the same whereas the A_{em}/A_s ratio decreases to a third of its original value (Table 4). The permeability difference of 2 orders of magnitude ($k_0 = 10^{-14}$ m^2, $k_0 = 10^{-12}$ m^2) resulted in almost the same values for A_{pf}/A_s, and a decrease of approximately 10 % for E_{em}/E_s.

We compared the absolute values of the electric field associated with the elastic mode of P_f and S, and the EM mode, instead of comparing the ratio. For this purpose we used the shear mode events only (Table 4). Observed electric strengths were divided by the moment magnitude of the event by using two methods of magnitude estimation, one based on the duration of the seismic waves and the other on the rupture time. The electric strength obtained by use of the rupture time is shown in the columns A_p*, A_s*, and A_{em}* in units of V/m. The observed electric strength was divided by the assumed antenna length, 800 m, to obtain the electric field strength. The results shown in the bottom two rows are the mean values and the standard deviations.

We see there is considerable agreement between the observed and estimated electric strength. At the same time it is noted that observed values for the S and EM modes are an order of magnitude larger than the estimated values; for the P_f mode the observed value was approximately 1/50 of the estimated value. The exceptionally large estimated values are discussed below.

The result obtained by use of the moment based on the duration time is approximately 8 orders of magnitude larger than the estimated value. It is suggested that the magnitude can be more reasonably estimated by use of the rupture time. We can correct the magnitude estimation formula (Eq. 1) for the electrogram by adjusting the constant −2.4 as:

$$M = 4.0 + 2.9 \log \mathrm{Td} \quad (14)$$

4. Discussion

Estimation by numerical simulation results in a larger amplitude for the P_f mode than for the S mode, in agreement with previous work (GARAMBOIS and DIETRICH 2001; GAO and HU 2010; ARAJI *et al.* 2012). GARAMBOIS and DIETRICH (2001) and HU and GAO (2009) noted the possibility of making the term $(1 - \rho C/\rho_f\ H)$ very small or even zero with the result of larger strength of A_s than A_p for particular types of sandstone.

GARAMBOIS and DIETRICH (2001) performed a very elaborate laboratory experiment by observing the seismic and electromagnetic fields. They focused, especially, on the relationship between seismic cancellation and electric field, as noted by GAO and HU (2010), and showed that the theoretical ratio can be matched to the observed values by correct choice of fluid conductivity σ_f.

Another very likely reason is presented and discussed by REN *et al.* (2012). Their numerical simulation using the half-space model limited by the free air surface shows that the S wave mode is stronger than the P_f mode, in agreement with observed results. The whole space model with the same physical conditions furnishes the same ordering of strength of S and P_f as our numerical results and those of GAO and HU (2010). They explain the phenomena on the basis of P to SV and SV to P conversions at the ground surface when the receiver is very near the surface. Our sensor extends downward from the ground surface to satisfy the situation as needed for their explanation. For any events, including those shown in Fig. 1, the P_f and S phases do not seem to be mixed. The P_f phases, especially, are very small, without any trace of mixing with the other phase. It is usual that the first arrival at the receiver is the direct P_f phase, followed by the reflected P and converted SV phases, with the resulting appearance of a large-amplitude P_f mode. Numerical simulations by HUANG *et al.* (2015) showed that complex structures generate more complicated waveforms, because of the effects of multiple reflections. Further investigations are needed understand this.

They also analyzed effects of related conditions on the relative strengths of the phases. We compared the values listed in Table 5 with those deduced from the logging data at the Hasaki site (IKEDA 1987). Conditions used other than those listed in Table 5 were, $V_p = 3.4$ km, $V_s = 2.0$ k/m, $K_s = 36$ GPa, and $\rho_s = 2000$ kg/m^3. We obtained $A_e/A_{em} = 0.37$ instead of 0.48, and $Ap_f/A_s = 5.3 \times 10^5$ instead of 2.3×10^5. We found that the choice of conditions does not provide better result in the limit of whole-space simulation.

In this work we compared ratios for different phases and obtained highly consistent values for A_{em}/A_s but a value for A_p/A_s which was very different from the theoretical estimate. This may be because of the frequency difference compared with previous authors. Further work is needed to explain this contradiction. Observed values for the EM mode were extremely large compared with analytical estimates. One possibility is the converted wave effect, as discussed elsewhere (JARDANI *et al.* 2010; HU *et al.* 2002; HU and GAO 2009).

An explanation is suggested by numerical simulations of the layered structure of the crust. SCHOEMAKER *et al.* (2012), ZHANG *et al.* (2013), and HUANG *et al.* (2015) showed that effects of the medium tend to amplify the co-seismic EM signals for shallow focal positions. Detailed comparison based on the layered model (HAARTSEN and PRIDE 1997; GARAMBOIS and DIETRICH 2001; WHITE and ZHOU 2006; ZHANG *et al.* 2013; GROBBE and SLOB 2013; HUANG *et al.* 2015) will be conducted in future work.

The comparison was conducted by use of Pride's formula assuming saturated water conditions and streaming potential coefficient. Differences between observed and estimated amplitudes are possibly caused by unsaturation and neglect of the induced polarization (IP) effect (REVIL *et al.* 2014). The is also a target for future investigation.

In the comparison of the amplitudes, wave attenuation is an important factor (MUELLER *et al.* 2010). Description of the electric wave field by the double-couple model takes into account the intrinsic attenuation during wave propagation through uniform porous media. The effects of geometrical spreading and fluid flow in the pore channels were taken into account in the calculation. The Q^{-1} factor for the three modes with frequency 500 Hz of interest in the comparison are 5.6×10^{-4}, 2.0, and 1.32×10^{-4}, for P_f, EM, and S wave, respectively. The very small values of Q^{-1} for the P_f and S modes indicate the effects of fluid viscosity are very small. The large Q^{-1} for the EM mode is for a large change of focal distance, as seen in Table 3. But differences between estimated and observed values cannot be attributed to this intrinsic effect. The P_f and S mode electric fields are accompanied by the seismic wave of the P_f and S modes, suggesting the Q values are almost the same as for seismic waves.

A recent investigation of seismic attenuation in the porous media suggested the squirt flow model is large enough to explain P wave attenuation of the order $1/Q = 0.1$–0.02 (Pride *et al.* 2004). The model does not seem to explain S wave attenuation. Dissipation of the seismic wave is, however, known to be largely as a result of scattering because of the inhomogeneity of seismic wave velocities at the interface between mesoscale and macroscale inhomogeneity (Mueller *et al.* 2010). Further investigations will be conducted on scattering by the mesoscale heterogeneity of the porous rocks.

5. Concluding Remarks

Characteristics of the three dominant phases (P_f, S, and EM) of seismoelectric waves have been investigated by use of electrogram waveforms believed to be induced by microcracks in the nucleation stage of the Tohoku Earthquake in 2011. Twelve B-type events with clear P and S phases and an apparent EM phase with a dominant frequency of 500 Hz were selected for analysis from the catalog of B-type events (Fujinawa *et al.* 2013). Slight traces of EM phases were sought at approximately the origin time. We estimated the focal distances of those events by using the S–P time and assuming the P and S wave velocity, which are corrected on the basis of time of arrival of the identified EM phase. The corrected velocities are in good agreement with measured logging data at the borehole. We can conclude that the EM mode of the seismo-electromagnetic signal (SES) related to the natural earthquake can be observed by use of highly sensitive sensor. The accumulated number of events was found to follow the Gutenberg–Richter relationship with the b value equal to 0.7, indicating that the B-type events are natural micro-earthquakes.

The electric strength of the fast-P, S, and EM modes of shear-type events were compared. Averages of observed amplitudes were in the order $A_s > A_{pf} > A_{em}$, which is different from results obtained from current and previous numerical simulations. Analytical estimation was conducted on the basis of the seismoelectromagnetic theory of Pride (1994), assuming an isotropic homogeneous medium filled with confined solvent. The force field is the double-couple model of Gao and Hu (2010), which is consistent with the hypothesis that the source is microcrack. Typical values for the physical properties of elastic material and solvent, and the streaming coefficient for seismic frequency result in the order $A_{pf} > A_s > A_{em}$, slightly different from results from observation. This contradiction between observed and numerical simulation results may be attributed to the assumed whole-space model rather than the half-space model. Overall agreement between the observed and estimated results implies that:

(1) pulse-like B-type events from the electrogram before the Tohoku Earthquake in 2011 are induced by rupture of microcracks in the period of nucleation of the main shock;

(2) observed phases of the P, S, and EM waveform are confirmed to be co-seismic P_f, S, and free electromagnetic waves, as assumed by Pride (1994); and

(3) SES phenomena in the field can be analyzed by the method of Pride (1994) and successive progress by successors (e.g., André Revil).

Precursory phenomena were investigated by means of seismic activity, crustal deformation, groundwater anomalies, and electromagnetic anomalies. Our finding of overall agreement between observation and estimation for seismoelectromagnetic phenomena suggests that seismological, crustal, groundwater, and electromagnetic phenomena affect the preparatory process of earthquake occurrence and can be analyzed and discussed on the basis of the theory of behavior of seismic activity, and hydrodynamic and electromagnetic phenomena induced by the variety of the cracks, taking electrokinetic effects into account. We already have impressive observation of magnetic ULF band anomalies (Han 2012) occurring simultaneously with the slow slip at the deep plate boundary by means of a highly sensitive seismometer (Ozawa *et al.* 2003).

Many of electric anomalies observed (Hayakawa and Fujinawa 1994) can be interpreted more satisfactorily on the basis of a unified scheme. A multidisciplinary approach on the basis of a unified theory on the effects induced by varieties of crack is expected to lead to advances in practical earthquake prediction. For instance, we can investigate the fluid motion associated with fracture in a porous medium as conducted in the laboratory by Haas *et al.* (2012).

Acknowledgments

The authors thanks Mr Takamatsu, for cooperation with observations, and Mr Takehiko Aruga of Mieruka Bousai Inc., for continuous encouragement. Many thanks are expressed to Professors André Revil, Qinghua Huang, and Dr Niels Grobbe for their valuable comments.

References

Araji, A. H., Revil, A., Jardani, A., Minsley, B. J., and Karaoulis, M. (2012) *Imaging with cross-hole seismoelectric tomography*, Geophys. J. Int. *188*, 1285–1302.

Asada, T., Suzuki, Z., Tomoda, Y. (1950) *On energy and frequency of earthquakes*, J. Seismol.Soc. Japan (Zisin), *3*, 11–15. (In Japanese with English abstract).

Belov, S.V., Migunov, N.I., and Sobolev, G.A. (1974) *Magnetic effects accompanying strong Kamchatkan earthquakes*, Geomagnetism and Aeronomy, *14*, No. 2, 380–382, (in Russian).

ERG (Electormagnetic Research Group for the 1995 Hyogo-ken Nanbu Earthquake) (1997) *Tectonoelectric signal related with the occurrence of the 1995 Hyogo-ken Nanbu earthquake (M7.2) and preliminary results of electromagnetic observation around the focal area*, J. Phys. Earth., *45*, 91–104.

Fujinawa, Y. and Takahashi, K. (1990) *Emission of electromagnetic radiation preceding the Ito seismic swarm of 1989*, Nature, *347*, 376–378.

Fujinawa, Y. and Takahashi, K. (1998) *Electromagnetic radiations associated with major earthquakes*, Phys. Earth and Planet. Inter., *105*, 249–259.

Fujinawa, Y. and Takahashi, K., Matsumoto, T., Iitaka, H., Nakayama, T., Sawada, T., Sakai, H. (2001) *Electric Field Variations Related with Seismic Swarms*, Bull Earthq. Res. Instit., *76*, 391–415.

Fujinawa, Y., Rokugo, Y., Noda, Y. Mizui, Y., Kobayashi, M., and Mizutani, E. (2009) *Development of application systems for earthquake early warning*, J. Disaster Res. *4*(4).

Fujinawa, Y. and Takahashi, K., Noda, Y., Iitaka, H., and Yazaki, S. (2011) *Remote detection of the electric field change induced at the seismic wave front the start of fault rupturing*, Intern. J. Geophys., Article ID 752193, 11 pages, doi:10.1155/2011/752193.

Fujinawa, Y., Noda, Y., Takahashi, K., Kobayashi, M., Takamatsu, K., and Natsumeda, J. (2013), *Field detection of microcracks to define the nucleation*, International Journal of Geophysics, 2013, Article ID 651823, 18 pages 10.1155/2013/651823.

Garambois, S., and Dietrich, M. (2001) *Full waveform numerical simulations of seismoelectromagnetic wave conversions in fluid-saturated stratified porous media*, J. geophys. Res., *107*(B7), 2148, doi:10.1029/2001JB00316.

Gao Y. X., and Hu, H. S. (2010) *Seismoelectromagnetic waves radiated by a double couple source in a saturated porous medium*, Geophys. J. Int., *181*, 873–896.

Gershenzon,N. I., and Bambakidis, G. (2001) *Modeling of seismo-electromagnetic phenomena. Russian*, J. Earth Sci. *3*, no. 4, 247–275.

Gershenzon, N. I., Gokhberg, M. B., Gul'yel'mi, A.V. (1994) *Electromagnetic field of seismic pulses*, Phys. Solid Earth (English Translation), *29*, 789–794.

Grobbe, N. and Slob, E. (2013) *Validation of an electroseismic and seismoelectric modeling code, for layered earth models, by the explicit homogeneous space solutions*, SEG Technical Program Expanded Abstracts 2013: pp. 1847–1851. doi:10.1190/segam2013-1208.1.

Haartsen M. W., and Pride, S. R. (1997) *Electroseismic waves from point sources in layered media*, J. Geophys. Res., *102*, 24745–24769.

Haas, A. K., Revil, A., Karaoulis, M., Frash, L., Hampton, J., Gutierrez, M., and Mooney M.(2012) *Electric potential source localization reveals a borehole leak during hydraulic fracturing*, Geophysics, *78*, No. 2; P. D93–D113. doi:10.1190/GEO2012-0388.1,

Han, P. (2012) *Investigation of ULF seismo-magnetic phenomena in Kanto*, Japan during 2000–2010 [Ph.D. Thesis], Chiba University.

Hayakawa M., and Fujinawa Y. (Eds.) (1994) *Electromagnetic Phenomena Related to Earthquake Prediction*, Terra Scientific Publishing Company, Tokyo, Japan.

Hayashi, H., Kasahara, K., and Kimura, H. (2006) *Pre-Neogine basement rocks beneath the Kanto Plain, central Japan*, J. Geol. Soc. Japan, *112*, 2–13. (In Japanese with English abstract).

Honkura, Y., Matsushima, M., Oshiman, N., Tuncer, M.K., Baris, S., Ito, A., Iio, Y., and Isikara, A. M.(2002) *Small electric and magnetic signals observed before the arrival of seismic wave*, Earth Planets Space, *54*, 9–12.

Hu, H., and Gao, Y.(2009) *The electric field induced by the fast P-wave and its nonexistence in a dynamically compatible porous medium*, SEG Technical Program Expanded Abstracts, RP P1, 2170–2174.

Hu, H., Liu, J., and Wang, K. (2002) *Attenuation and seismo-electric characteristics of dynamically compatible porous media*, SEG Technical Program Expanded Abstracts, 1817–1820.

Huang, Q. (2002). *One possible generation mechanism of co-seismic electric signals*, Proc. Japan. Acad., *78*(B7), 173–178.

Huang, Q., and Liu, T. (2006) *Earthquakes and tide response of geoelectric potential field at the Niijima station*, Chin. J. Geophys., *49*(6), 1745–1754.

Huang, Q., Ren, H., Zhang, D., and Chen, Y. J. (2015) *Medium effect on the characteristics of the coupled seismic and electromagnetic signals*, Proc. Jpn. Acad., Ser. B *91*, 17–24.

Ikeda, R. (1987) *Groundwater pressure variation at Chikura,Chiba Prefecture Associated with the eruption of Izu-Oshima, Coordinating Comittee for Earthquake Prediction*, *38*, 216–218.

Ishimoto, M., and Iida K. (1937), *Determination of elastic constants of soils by means vibration methods, Part 2,Modulus of rigidity and Poisson's ratio*, Bull. Earthquake Res., Univ. Tokyo,*15*,67–85.

Iyemori, T. et al. (1996) *Co-seismic geomagnetic variations observed at the 1995 Hyogoken-Nanbu earthquake*, J. Geomag. Geoelectr. *48*, 1059–1070.

Jardani, A., Revil, A., Slob, E., and Sollner, W. (2010) *Stochastic joint inversion of 2D seismic and seismoelectric signals in linear poroelastic materials*, Geophysics, *75*(1), N19–N31, doi:10. 1190/1.3279833.

Mahardika1, H., Revil, A., and Jardani, A. (2012) *Waveform joint inversion of seismograms and electrograms for moment tensor characterization of fracking events*, Geophysics, *77*(5), P. ID23–ID39, doi:10.1190/GEO2012-0019.1.

Mueller, T. M., Gurevich, B., and Lebedev, M. (2010) *Seismic wave attenuation and dispersion resulting from wave-induced flow in porous rocks - A review*, Geophysics. *75*(5): pp. 75A147-75A164.

Nagao, T., Orihara, Y., Yamaguchi, T., Takahashi, I., Hattori, K., Noda, Y. Sayanagi, K., and Uyeda, S. (2000) *Co-seismic geoelectric potential changes observed in Japan*, Geophys. Res. Lett., *27*, 1535-1538.

Nanjo, K. Z., Hirata, N.,Obara, K., and Kasahara K. (2012) *Decade scale decrease in b- value prior to the M9-class 2011 Tohoku and 2004 Sumatra quakes*, Geophys. Res. Lett., doi:10. 1029/2012GL052997.

Ogasawara, H., *et al.*, (2009) *Semi-controlled earthquake-generation experiments in deep gold mines, South Africa—Monitoring at closest proximity to elucidate seismogenic process—*, J. Seismol. Soc. Japan (Zisin), *61*, S563–S573. (In Japanese with English abstract).

Otsuka M. (1965) *Earthquake magnitude and surface fault formation*, J. Seismol. Soc. Japan (Zisin), *18*, 1–8.

Ozawa S, Miyazaki, S., Hatanaka, Y., Imakiire, T., Kaidzu, M., and Murakami, M. (2003) Characteristic *silent earthquakes in the eastern part of the Boso Peninsula, Central Japan*, Geophys. Res. Lett. 30:1283.

Pride, S. (1994) *Governing equations for the coupled electromagnetics and acoustics of porous media*, Phys. Rev. B, *50*, no. 21, 15678–15696.

Pride, S. R., and Haartsen, M.W. (1996) *Electroseismic wave properties*, J. Acoust. Soc. Am., *100*, 1301–1315.

Pride, S. R., Berryman, J. G., and Harris J. M. (2004) *Seismic attenuation due to wave-induced flow*, J. Geophys. Res., *109*, B01201, doi:10.1029/2003JB002639.

Ren, H. X., Chen, X.F., and Huang, Q. H. (2012) *Numerical simulation of coseismic electromagnetic fields associated with seismic waves due to finite faulting in porous media*, Geophys. J. Int.,*188*, 925–944 doi:10.1111/j.1365-246X.2011.05309.x.

Revil A., Barnier, G., Karaoulis M., and Sava, P.(2014) *Seismoelectric coupling in unsaturated porous media: Theory, petrophysics, and saturation front localization using an electroacoustic approach*, Geophysical Journal International, *196*(2): 867–884, doi:10.1093/gji/ggt440 .

Schoemaker, F. C., Grobbe, N., Schakel, M. D., de Ridder, S. A. L. Slob, E. C., and Smeulders, D. M. J., *Experimental Validation of the Electrokinetic Theory and Development of Seismoelectric Interferometry by Cross-Correlation*, International Journal of Geophysics, 2012, Article ID 514242, 23 pages, doi:10.1155/2012/514242.

Suzuki, H. (2002) Underground geological structure beneath the Kanto Plain, Japan, Report of the National Research Institute for Earth Science and Disaster Prevention, No.63, 1–19 (In Japanese with English abstract).

Takahashi, H., Takahashi, K. (1989) *Tomography of seismo-radio wave source region predicting imminent earthquakes*, Phys. Earth and Planet. Inter., *51*, 40–44.

Takahashi, K., Fujinawa, Y., Matsumoto, T., Nakayama, T., Sawada, T., Sakai, H., and. Iitaka, H. (2000) *An anomalous electric field variation with the seismic swarm(1)-Underground electric field observation at Hodaka station (1995-1999)*, Tec. Note Nat. Res. Inst. Earth Science and Disaster Prevention, No. 204, 1-224.

Tsumura, K. (1967) *Determination of earthquake magnitude from duration of oscillation*, J. Seismol.Soc. Japan (Zisin) 2, 20, 1, 30–40.

Uchide, T, Ide,S., and Beroza, G. C. (2009) *Dynamic high-speed rupture from the onset of the 2004 Parkfield, California, earthquake*, Geophys. Res. Lett., *36*, L04307, doi:10.1029/ 2008GL036824.

White, B.S. & Zhou, M.(2006) *Electroseismic prospecting in layered media*, Soc. Indust. Appl. Math., *67*(1), 69–98.

Zhang, D., Ren, H.X. and Huang, Q.H. (2013) *Numerical simulation study of co-seismic electromagnetic signals in porous media*, Chin. J. Geophys. *56*, 2739–2747.

(Received September 12, 2014, revised January 12, 2015, accepted January 16, 2015, Published online March 5, 2015)

Pure Appl. Geophys. 173 (2016), 269–284

DOI 10.1007/s00024-015-1055-4

Pure and Applied Geophysics

Shifting Correlation Between Earthquakes and Electromagnetic Signals: A Case Study of the 2013 Minxian–Zhangxian M_L 6.5 (M_W 6.1) Earthquake in Gansu, China

Feng Jiang,[1] Xiaobin Chen,[1] Yan Zhan,[1] Guoze Zhao,[1] Hao Yang,[1] Lingqiang Zhao,[1] Liang Qiao,[1] and Lifeng Wang[1]

Abstract—The shifting correlation method (SCM) is proposed for statistical analysis of the correlation between earthquake sequences and electromagnetic signal sequences. In this method, the two different sequences were treated in units of 1 day. With the earthquake sequences fixed, the electromagnetic sequences were continuously shifted on the time axis, and the linear correlation coefficients between the two were calculated. In this way, the frequency and temporal distribution characteristics of potential seismic electromagnetic signals in the pre, co, and post-seismic stages were analyzed. In the work discussed in this paper, we first verified the effectiveness of the SCM and found it could accurately identify indistinct related signals by use of sufficient samples of synthetic data. Then, as a case study, the method was used for analysis of electromagnetic monitoring data from the Minxian–Zhangxian M_L 6.5 (M_W 6.1) earthquake. The results showed: (1) there seems to be a strong correlation between earthquakes and electromagnetic signals at different frequency in the pre, co, and post-seismic stages, with correlation coefficients in the range 0.4–0.7. The correlation was positive and negative before and after the earthquakes, respectively. (2) The electromagnetic signals related to the earthquakes might appear 23 days before and last for 10 days after the shocks. (3) To some extent, the occurrence time and frequency band of seismic electromagnetic signals are different at different stations. We inferred that the differences were related to resistivity, active tectonics, and seismogenic structure.

Key words: Shifting correlation method (SCM), seismic electromagnetic signals (SEMS), Minxian–Zhangxian earthquake, magnetotelluric.

1. Introduction

Seismic electromagnetic signals (SEMS) are electric, magnetic, and electromagnetic signals related to the genesis and occurrence of earthquakes and the structural recovery of seismogenic region. SEMS have been reported in a large amount of literature (Eftaxias *et al.* 2001, 2009; Fujinawa *et al.* 1998; Han *et al.* 2014; Hattori *et al.* 2012; Huang *et al.* 2011b; King 1983; Park *et al.* 1993; Zhang *et al.* 2011; Zhao *et al.* 2009). Abundant indoor and/or outdoor experiments and numerical simulations have been conducted to verify the existence of SEMS phenomena (Huang *et al.* 1998; Kuo *et al.* 2014; Zhao *et al.* 2009; Enomoto *et al.* 2012; Ren *et al.* 2012; Potirakis *et al.* 2012; Huang 2011a). For example, radiation of ultra-low-frequency (ULF) electromagnetic signals was observed in the early, middle, and late stages of a rock-fracture experiment (Hao *et al.* 2003). At the end of last century, the Greek physicists proposed the VAN method for earthquake monitoring on the basis of seismic electrical signals (SES). It was claimed this method could be used to predict earthquakes above magnitude 5 (Uyeda *et al.* 2009; Varotsos *et al.* 1991), which generated much interest (Huang 2005). Since the 1960s, China has established a nationwide network of earthquake-monitoring stations based on geoelectric resistivity, geoelectric fields, and electromagnetic waves. In recent years, in China, much research has been conducted on the development of a method for monitoring extremely low-frequency electromagnetic radiation and an earthquake electromagnetic satellite (Zhao *et al.* 2007, 2012). The development of these monitoring techniques reflects the attention given to SEMS-based earthquake prediction methods by the scientific community.

To identify and extract SEMS effectively, such approaches as maximum entropy estimation (Liu

Electronic supplementary material The online version of this article (doi:10.1007/s00024-015-1055-4) contains supplementary material, which is available to authorized users.

[1] State Key Laboratory of Earthquake Dynamics, Institute of Geology, China Earthquake Administration, Beijing, China. E-mail: cxb@ies.ac.cn

et al. 2012; Fan et al. 2010), time–frequency analysis (Eftaxias et al. 2001), wavelet transform (Zhang et al. 2013; Xie et al. 2013; Han et al. 2011), and principle-component analysis (Uyeda et al. 2002; Han et al. 2009), have been used. The occurrence time, frequency band, and propagation path of SEMS have been studied (Fujinawa et al. 1998). However, most of these methods compare electromagnetic signals (or treated data) within a period of time of a single earthquake to seek earthquake-related anomalies and analyze the potential SEMS characteristics. In fact, the SEMS may not be obviously abnormal signals because of low-intensity and/or mixing with the background field and interfering noise. This may be why no SEMS anomalies have been observed before and after some earthquakes, and why some abnormal electromagnetic signals cannot be related to the corresponding earthquakes (Orihara et al. 2012; Fan 2010; Han et al. 2014; Hattori et al. 2012). Although many methods have been used to investigate and extract SEMS on the basis of ground and/or ionosphere measurements, the correlation between earthquakes and electromagnetic signals is still difficult to quantify solely on the basis of electromagnetic anomalies before and after a single event, not to mention the general features of complex and inconstant SEMS. Therefore, although many studies in this field have been reported, the nature of SEMS remains elusive, not to mention the temporal–spatial distribution of SEMS and their variations.

For a specific earthquake, SEMS may arise during the seismogenic process, in the course of rock rupture during the co-seismic moment, and during post-seismic recovery of the seismogenic structure. By taking the time of occurrence of an earthquake as a reference point, the time axis can be divided into three stages, pre, co, and post-seismic, in which SEMS may or may not exist. When multiple earthquakes occur successively, the SEMS in the three stages may overlap and be submerged by noise. Therefore, the specific characteristics of SEMS in the three stages cannot be determined with certainty by analyzing a single seismic event before the physical mechanisms of seismic electromagnetic radiation are clear. Recently, statistical study by superpose epoch analysis (SEA) has been used to investigate the relationship between earthquakes and geomagnetic variations. The results of statistical analysis preformed in Japan suggested there was no correlation between earthquakes and geomagnetic anomalies, and that ULF geomagnetic anomalies were probably more sensitive to earthquakes which were larger and closer to geomagnetic monitoring stations (Han et al. 2014; Hattori et al. 2012). As an alternative, in this paper, we introduce a new statistical method for study and investigation of the direct correlation between earthquakes and electromagnetic signals.

Earthquakes are caused by tectonic movement with similar dynamic processes. For all earthquakes, or all earthquakes of a specific type, the corresponding SEMS may share similar temporal distribution characteristics. This means we could choose several earthquakes for statistical analysis and extract the common features of SEMS. Correlation analysis is a basic statistical tool. For simultaneous recognition of pre, co, and post-seismic SEMS, we propose use of the shifting correlation method (SCM) for SEMS and earthquake series. Continuous shifting of two sequences and calculation of correlation coefficients can result in a plot of correlation coefficients over the entire time axis. By shifting correlation analysis of electromagnetic signals of different frequency, a diagram based on the correlation coefficients can be obtained, and these diagrams clearly show the time–frequency distribution characteristics of SEMS.

In this study, we first introduce the basic principle of the shifting correlation method (SCM) and then verify the performance and functional characteristics of the SCM by use of synthetic data. Later, we discuss the application of the SCM to SEMS recognition by applying it to sequences of the main-after shock of the Minxian–Zhangxian earthquake and electromagnetic monitoring data.

2. Basic Principle of SCM

In mathematics, the correlation coefficient between two discrete sequences can be used to measure the correlation between two variables, as shown by Eq. (1).

$$R_{ME} = \frac{\sum_{k=1}^{m} (M_k - \overline{M})(E_k - \overline{E})}{\sqrt{\sum_{k=1}^{m} (M_k - \overline{M})^2}\sqrt{\sum_{k=1}^{m} (E_k - \overline{E})^2}} \quad (1)$$

where R_{ME} is the correlation coefficient between the two discrete sequences M and E. $\overline{M}$ and $\overline{E}$ are the means of the two sequences, and m is the number of samples.

Under actual conditions, the variations of the two variables may not be synchronized, i.e. the effect or action of one variable on another may happen in advance or lag behind. These phenomena with asynchronous correlation may be extracted by relative shifting then calculation of the correlation between the two sequences, i.e. before calculation of correlation coefficients by use of eq (1), one sequence is fixed and the other sequence is continually shifted. The correlation coefficient between the two sequences after each shift is then calculated. On the basis of the amount of shifting corresponding to the maximum correlation coefficient, we can determine not only whether the two variable sequences are significantly correlated but also the time difference between two asynchronous variables. This is highly important for analysis of the temporal distribution characteristics of SEMS. Figure 1 shows a schematic diagram of interpretation of this calculation process by use of the so-called shifting correlation method (SCM).

M_i is a sequence derived from a long sequence; as an example, $i = 6$ is shown in Fig. 1. E is the short sequence and determines the sample size involved in the shifting correlation calculation. m denotes the sample size involved in calculation; $m = 14$ is used as an example in Fig. 1. M_i used to calculate the correlation coefficients with the sequence E changed by continual variation of a positive integer, i, which causes the invariable sequence E to be shifted to the right or left relative to the long sequence on the horizontal axis during the process of calculation of the correlation coefficients. The range of variation of i is $[1, n - m + 1]$ when the sample size of the long sequence defined as n. The above process can be expressed as Eq. (2):

$$R_{M_iE} = \frac{\sum_{k=1}^{m} (M_{i,k} - \overline{M_i})(E_k - \overline{E})}{\sqrt{\sum_{k=1}^{m} (M_{i,k} - \overline{M_i})^2}\sqrt{\sum_{k=1}^{m} (E_k - \overline{E})^2}},$$
$$i \in [1, n - m + 1] \tag{2}$$

where M_i and E are the two equal-length sequences for correlation calculation (M_i is obtained from the long sequence, and E is the short sequence), $M_{i,k}$ and E_k are the kth sample values of the two sequences, $\overline{M_i}$ and $\overline{E}$ are the respective means of the two sequences, i is the serial number of the long sequence, m is the sample size involved in calculation, and R_{M_iE} is the correlation coefficient. In comparison with Eq. (1), M is replaced by M_i in Eq. (2).

The correlation coefficients between the two sequences, calculated by shifting correlation method, are not single value but a correlation coefficient sequence, i.e. the curve of variation of the correlation coefficient when the short sequence shifts to the left or right relative to the long sequence.

3. Effectiveness of the SCM Verified with Synthetic Data

First, recognition of the single correlated signals is confirmed. A series of discrete random numbers is generated to form a long sequence, as shown in

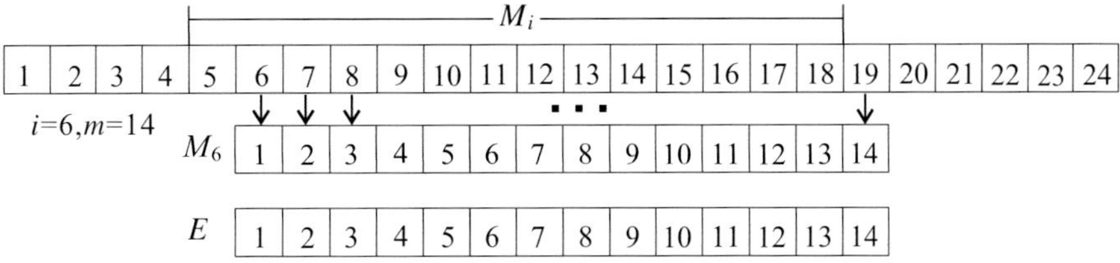

Figure 1

Schematic diagram of the SCM calculation process. Three sequences are given in the figure. The upper sequence M_i is derived from a long sequence. $i = 6$ is shown in the middle. E is a short sequence which determines the sample size involved in the shifting correlation calculation. m is the sample size involved in calculation; $m = 14$ is used in this example. The range of variation of i is $[1, n - m + 1]$ when the sample size of the long sequence is defined as n. The linear correlation coefficient between M_6 and E will be calculated when the value of i is 6. This means that, with continual variation of the positive integer i, different M_i will be obtained from the long sequence for calculation of linear correlation coefficients with the sequence E during the calculation process. Different i correspond to different M_i and to different correlation coefficients

Fig. 2a. Samples are derived from different sections of the long sequence and added to a specific amount of random noise to form the short sequences. As shown in Fig. 2a, three sequences with the sample size of 30 are selected from the blocks M1, M2, and M3 (representing different sections). Then approximately 30 % random noise was added to form the three sequences E1, E2, and E3. E1 and M1 are synchronously correlated sequences, E2 is a post-correlated sequence obtained by left shifting M2 by 20 samples, and E3 is a pre-correlated sequence obtained by right shifting M3 by 10 samples.

As is apparent from Fig. 2b, the shifting correlation coefficients are calculated when the three short sequences are shifted relative to the long sequence M. Relative to the starting point of the shifting, the numbers of the shifted samples are denoted by positive and negative values on the horizontal axis when the short sequences are shifted right or left. It is apparent that strong correlation occurs at positions of 0,

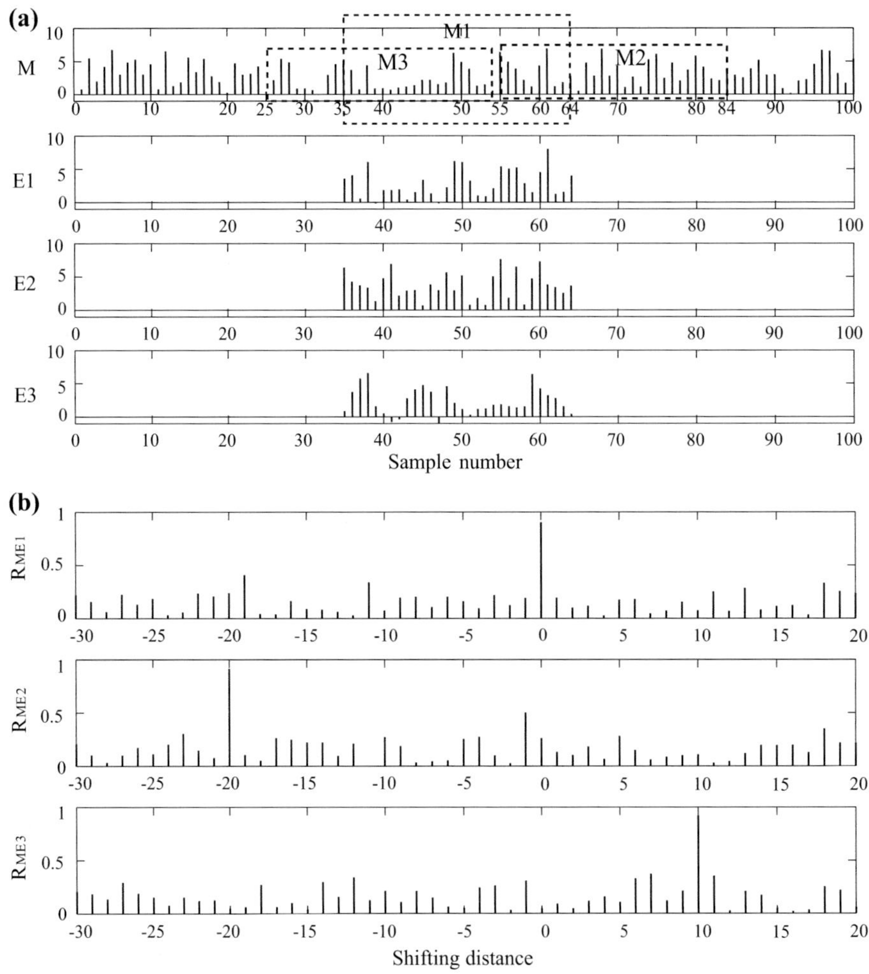

Figure 2
Shifting correlation coefficients obtained from synthetic data. **a** Long sequence and three short sequences. The *horizontal axis* represents the serial number of the sample; the *vertical axis* represents the preset amplitude, varying from 0 to 7. E1, E2, and E3 are sequences M1, M2 and M3 after addition of 30 % noise. The sample size is 30. E1 and M1 are synchronously correlated. E2 is derived by left shifting M2 by 20 samples; it is the post-correlated sequence. E3 is derived from right shifting M3 by 10 samples; it is the pre-correlated sequence. **b** R_{ME1}, R_{ME2}, and R_{ME3} are plots of the variation of the correlation coefficients between the three short sequences (E1, E2, and E3) and the long sequence. The absolute positive and negative values on the horizontal axis represent the number of samples in which the short sequences shift to the left and right

−20, and 10, respectively, on the three curves which coincides completely with the preset signals. This indicates that the SCM can accurately distinguish asynchronous correlation between two sequences. This in indicative preliminary validation of the effectiveness of the method.

The effectiveness of the SCM was validated by simulating the single correlation through the synthetic data as above. However, if we superimposed sequences E1, E2 and E3 in equal proportion in Fig. 2a, so a short sequence of equal length was synthesized, and then performed shifting correlation calculation with the corresponding long sequence, could the SCM distinguish the correlation at the three different positions simultaneously? Following this idea, further verification of the effectiveness of the SCM was conducted.

As shown in Fig. 3, a long sequence containing 200 samples was randomly generated and the method for short sequence generation was the same as for Fig. 2, but the three short sequences (E1, E2, and E3) were superimposed into one short sequence in equal proportion. In the calculation of shifting correlation coefficients on the basis of the different sample size, three cases of m ($m = 30$, 50, and 100) as short sequences were analyzed. Figure 3b shows an example of a short sequence for which the sample size was 50; it was located at a position synchronous with the long sequence. Comparison of the curve shapes in Figs. 2 and 3 reveals that although E1, E2, and E3 contain random interference noise relative to M1, M2, and M3, some similarities could be still observed by pairwise comparison, although the curves in Fig. 3a, b seems to be completely uncorrelated.

In the same way, the shifting correlation calculation is performed for the two sequences in Fig. 3a, b. In Fig. 3c, it is apparent that three significant correlations have been distinguished simultaneously at 0, −20, and 10, respectively. To improve the universality of this phenomenon, one hundred tests were performed for the whole of the process above. The results are shown in Fig. 4a–c for when m (the sample size of the short sequence) is 30, 50, and 100, respectively. There is a "column-shaped" significant correlation at −20, 0, and 10 on the horizontal axis. This means that the SCM could simultaneously distinguish several hidden correlations which may occur at different times. Figure 4a–c also indicate that the

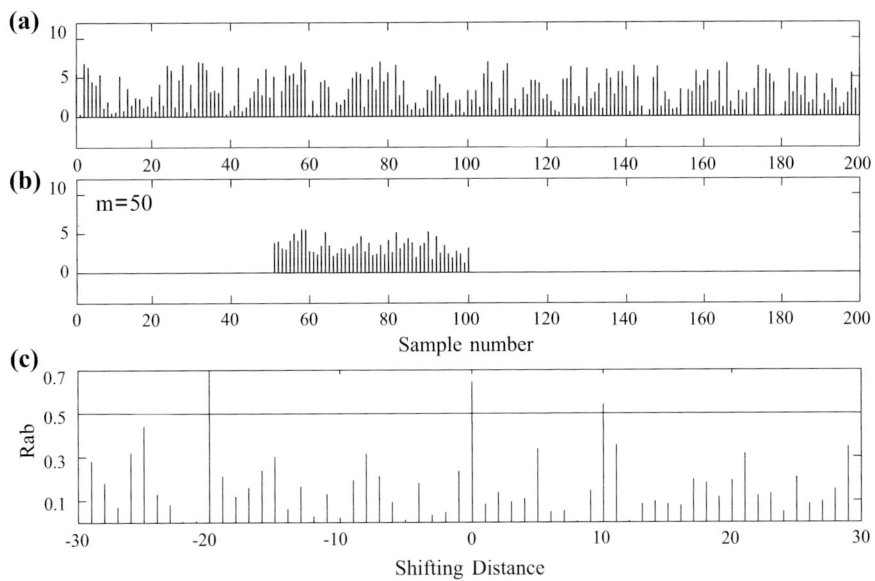

Figure 3
Result of recognition of several superimposed correlated signals. **a** Long sequence with a sample size of 200, randomly fluctuating in the range 0–7. **b** Short sequence formed by superimposition of three sequences. Take the sample size of 50 as an example. For the range 51–100, the short sequence is synchronous with the long sequence, i.e. the starting point. **c** Plot of variation of the correlation coefficient; there are three positions with significant correlation. Relative to the long sequence, the short sequence shifted to the left and right by 30 steps

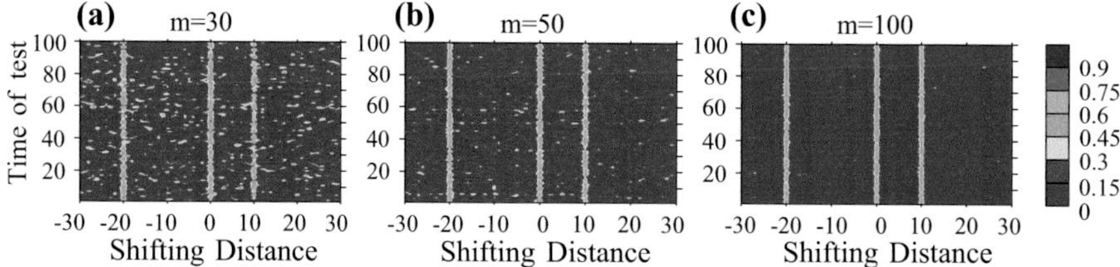

Figure 4
One hundred experiments of shifting correlation calculation. The *vertical axis* represents the times of the experiments; the *horizontal axis* represents that the largest sample size by which the short sequence shifts to the left and right, which is 30. The negative and positive specification is the same as above; the color code is the magnitude of the correlation coefficient. **a–c** Show the results when the sample size of the short sequence (*m*) is 30, 50, and 100, respectively. The "*column-shaped*" significant correlation occurs at −20, 0 and 10 on the abscissa

resolution of the calculation increases with increasing sample size of the short sequence.

These calculations using synthetic data indicate that the SCM can clearly distinguish simple, hidden, and complex correlations between two sequences. The larger the sample size of the short sequence, the higher the resolution. This method can be used to investigate the correlation between two physical quantities and the correlation characteristics. As an example, the method was used to analyze the correlation between an earthquake and electromagnetic signals.

4. Case Study

Electromagnetic signals are regarded as among the most sensitive physical responses to earthquake (Zhao *et al.* 2007) but are vulnerable to interference from several sources. If SEMS indeed exist, they may be mixed with a variety of electromagnetic signals and noise, which makes it difficult to distinguish then. If the earthquake is correlated with the electromagnetic signals, the correlation can be recognized by the shifting correlation method, as indicated by the verification above.

If the existence of SEMS is manifest as a co-seismic effect, the earthquake sequence may have a strong correlation with the electromagnetic sequences monitored synchronously; but if it is a pre or post-seismic effect, the correlation between the two physical quantities cannot be observed unless the electromagnetic sequences are shifted relative to earthquake sequences, by a corresponding distance backward (left) or forward (right). Therefore, it is possible for us to extract the pre, co, and post-seismic SEMS simultaneously when performing the SCM on electromagnetic signals the and earthquake sequence. On the basis of the results calculated we could further analyze the time–frequency feature of SEMS and its spatial distribution characteristics. It should be noted that the SCM is a statistical method based on several seismic events, not simply analysis of the correspondence between one or several seismic events with the anomalies in SEMS analysis.

4.1. General Information about the Minxian–Zhangxian Earthquake

The M_L 6.5 earthquake (after correction by Fang *et al.*), with the focal depth of 20 km, occurred at the boundary of Minxian and Zhangxian in Dingxi City, Gansu Province (34.5°N, 104.2°E) at 07:45 on July 22, 2013 Beijing Time. According to the Institute of Geophysics, CEA. (see Data and Resources Section) the earthquake was a thrusting sinistral strike-slip earthquake. The earthquake was closely associated with the Lintan–Tanchang fault belt (F2) in the northeast. This fault is clamped between the East Kunlun fault system in the south and the northern margin of the western Qinling fault in the north and the Diebu–Bailongjiang fault (F3) and the Guanggaishan–Dieshan fault belt (F1). The three faults (F1–F3) were involved in tectonic transformation and regional stress redistribution (Zheng *et al.* 2013). After the earthquake, four monitoring stations, Zhujiawan (ZJW), Majiagou (MJG), Shimen (SHM), and Shuguang (SHG) (Fig. 5), were arranged within

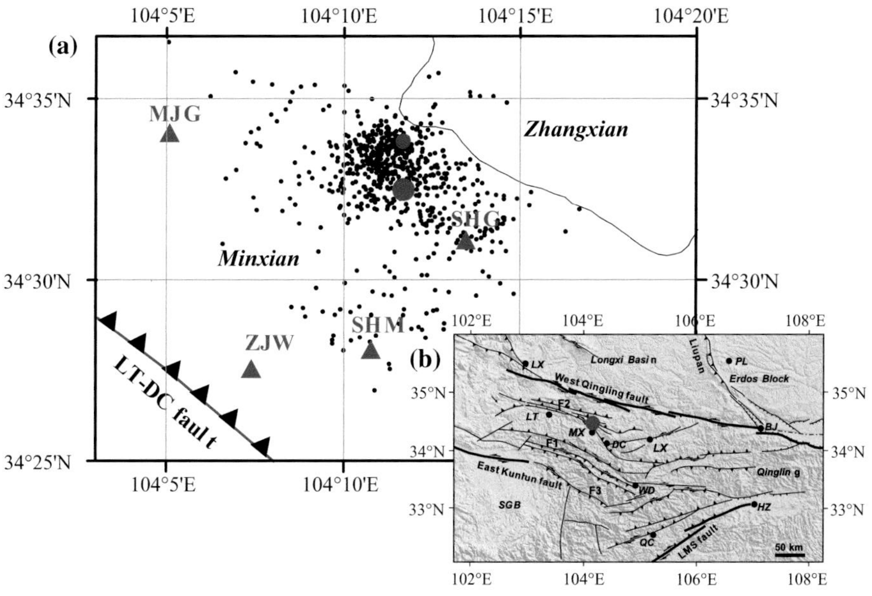

Figure 5

Schematic maps of emergency electromagnetic monitoring sites and positions of the main shock and aftershocks (**a**) and active tectonics of the seismic region (**b**) (revised from ZHENG *et al.* 2013). In **a** the *big solid circle* indicates the main shock, and the *smaller solid circle* indicates the largest aftershock. In **b** F1 is the Guanggaishan–Dieshan fault, F2 is the Lintan–Tanchang fault, and F3 is the Diebu–Bailongjiang fault

Table 1

Basic time and data information for electromagnetic monitoring of the Minxian–Zhangxian earthquake

Monitoring station	Epicenter distance (km)	Measured time (day)	Data lost	Data used for analysis (day)
Majiagou (MJG)	10.4	23/7/2013–19/8/2013	2th–3th August	26
Zhujiawan (ZJW)	11.3	25/7/2013–19/8/2013	30th July, 6th–7th August	23

"Data lost" indicates no data were available or the data were not used for analysis because of cutoff of power to the devices or failure of sensors

15 km of the epicenter to perform continuous electromagnetic monitoring for nearly a month. The V5-2000 magnetotelluric device by Canada Phoenix was used.

4.2. Shifting Correlation Analysis

The shifting correlation method was used to analyze electromagnetic monitoring data from the Minxian–Zhangxian earthquake. The main procedures were:

- first, the main shock and aftershock sequence of the earthquake was processed into a long sequence used for correlation analysis;
- second, the electromagnetic monitoring data was processed into the short sequence; and
- third, final treatment, calculation, and analysis of the shifting correlation coefficient between the two sequences.

Note that road construction and mining were in progress near SHM and SHG stations, causing strong electromagnetic interference; as a result, data from these two stations were not of sufficient quality and data from ZJW and MJG stations, only, were used in the SCM analysis. Basic information about the ZJW and MJG stations are listed in Table 1.

4.2.1 Earthquake Sequence Treatment

Here, first, we define the earthquake magnitude in different ways. The original conventional earthquake sequences were first divided by equal time intervals [we use local time (LT) rather than universal time]. Then, within each time interval, the sum of seismic energy released was estimated as conventional magnitude M and converted to "equivalent magnitude" (Meq) to represent the seismic energy sequence. Then, using Eq. (3), the original conventional magnitude sequence (main-after shocks sequence from 07:45 on July 22 to 21: 50 on September 26) was converted to energy sequence and the summation was performed within the time interval. The distribution of all the earthquake events was less than 26 km from both stations, and most were within 20 km of the stations.

$$E = \sum_{i=1}^{s} 10^{(A \times M_i + B)} \quad (3)$$

where E is the seismic energy (Joules) released, M_i is the magnitude of the ith earthquake, and s is the number of seismic events within the time interval. A and B are constants, with values of 1.96 and 2.05, respectively (Gutenberg *et al.* 1956). To ensure the effectiveness of the method, when determining the time interval, factors such as data amount, potential duration of SEMS for one event, and the continuity of seismic energy release should be taken into account. After repeated calculation and analysis, 1 day was selected as the time interval for the sequence of the Minxian–Zhangxian earthquake, that is, one energy value per day. The energy sequence was then converted to Meq sequence by use of Eq. (4):

$$M_{eq} = {[\lg(E) - B]}/{A} \quad (4)$$

M_{eq} is Meq. For days in which no earthquakes were detected, cubic spline interpolation was performed. (This was also performed on synthetic data to validate its suitability. The result of the calculation did not differ significantly from Fig. 4, so no results are given in this paper.) Figure 6a and b are the original earthquake sequence and the Meq sequence, after treatment, respectively. It is apparent the two sequences do not differ significantly before and after treatment. The Meq sequence in Fig. 6b was used for subsequent calculation.

4.2.2 Treatment of Electromagnetic Monitoring Data

The robust time series processing software SSMT2000 provided by Phoenix was improved so the time series observed continuously could be divided into equal time intervals and the batch

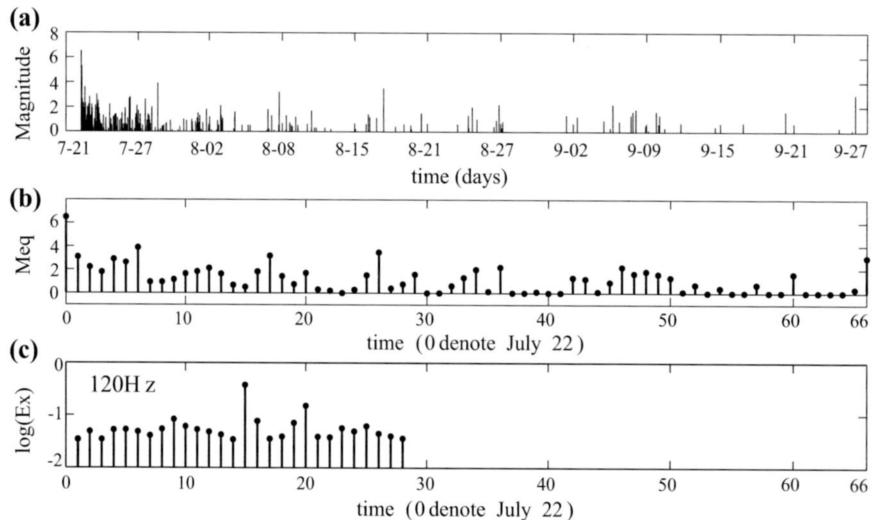

Figure 6
Earthquake sequences (**a**, **b**) before and after treatment and electromagnetic signal sequence (**c**). **a** Original earthquake sequence. The main shock occurred on July 22. The last major aftershock occurred on September 26; **b** Meq sequence after treatment to convert it into 1-day time intervals. The *horizontal axis* is continuously marked with positive integers; "0" represents the time of occurrence of the main shock, which was on July 22, and "66" represents September 26. During this period, the days are marked consecutively; **c** daily variation of the electric field Ex at the frequency of 120 Hz at station MJG. The relative status of the two sequences in **b** and **c** is non-shifting

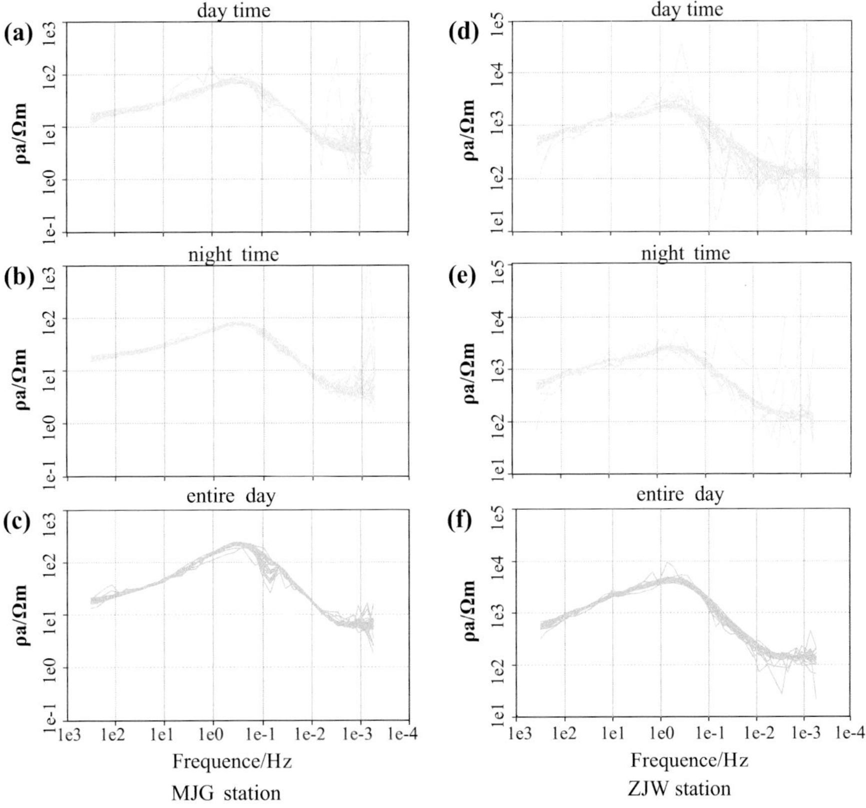

Figure 7
Apparent resistivity (*Rxy*) curves for two stations. **a–c** are daytime, nighttime, and entire day apparent resistivity curves, respectively, from MJG station; **d–f** are the same as **a–c** but for ZJW station

processing was performed automatically. In accordance with Meq sequence, the time series (Table 1) of the electromagnetic signal was also divided into time intervals of a day (LT 00:00 A.M–24:00 P.M.). SSMT2000 was then used to obtain the full-band (0.00055–320 Hz) electromagnetic spectrum. The treatment result for each day was output in the form of an EDI file. The EDI files of the power spectrum, after auto-edit by MTEDITOR (software provided by Phoenix), were managed by MT-Pioneer software (Chen *et al.* 2004), and output as a diurnal curve of the power spectrum values with different frequencies (a total of forty frequency points). As a result, forty curves of electromagnetic response were obtained. For data missing from Table 1, cubic spline interpolation was performed to form a complete electromagnetic signal sequence at the different frequency points. Figure 6c shows the sequence of Ex at a frequency of 120 Hz at MJG station.

The noise lever of electromagnetic data can be judged by the quality of the apparent resistivity curves. In Fig. 7, panels (c) and panels (f) show the daily apparent resistivity curves for stations MJG and ZJW, respectively. Except for a few small and individual disturbances, it is apparent that the apparent resistivity curves of the two stations are very smooth, which indicated that the artificial noise from electromagnetic signals is very small. To further investigate whether the electromagnetic disturbance around the stations was stronger during the daytime than at nighttime, we divided the 1 day time series into days and nights (LT, daytime is 06:00 a.m–06:00 p.m., the other time is nighttime) and the corresponding apparent resistivity was then calculated. The results are given in Fig. 7. Panels (a) and (b) show the daytime and nighttime curves, respectively, for MJG station. Panels (d) and (f) show the daytime and nighttime curves, respectively, for ZJW station. By comparing panels (a) and (b), and panels (d) and (f), we found that the quality of apparent resistivity at nighttime was slightly better than in the daytime. However, in the relatively low frequency bands, neither day nor nighttime was good enough at the two stations, and the results for entire day were much better than the single daytime or nighttime results. Furthermore, it is a fact that whenever an earthquake occurs, whether in the daytime or nighttime, its pre, co, and/or post-seismic electromagnetic responses may have arisen in the daytime and/or nighttime. Much SEMS information would probably be lost if only one time period was used in the statistical analysis. Using the mean of the entire day power spectrum is, therefore, the best choice for SCM calculation, and was used in the work discussed in this paper.

4.2.3 Calculation of Shifting Correlation Coefficients

By using the Meq sequence and electromagnetic signal sequence divided into time intervals of 1 day, the curve of the shifting correlation coefficient between the two physical quantities was calculated. During the shifting process, the number of days shifted was determined by the length of the electromagnetic signal sequence, m, after treatment, i.e. controlling the magnitude of n (in Fig. 1). To ensure satisfactory statistical analysis, the largest sample size was used when calculating the pre-seismic correlation, i.e. when the electromagnetic signal sequence shifts to the right, for MJG station. For example, $m = 28$, and the number of days shifted was marked by negative value, representing pre-seismic correlation. In this study, because the earthquake sequences before July 22 were not included into the calculation, the number of samples used for calculation of the correlation coefficient are reduced when the electromagnetic signal was shifted to the left relative to the Meq sequence. The number of left shifted days was marked by a positive value, representing post-seismic correlation. Finally, we calculated the shifting correlation coefficients between full-band electromagnetic signals and the Meq sequence and obtained nephrograms of the time–frequency distribution of the correlation coefficients.

5. Calculation Results and Analysis

It was found that the correlation coefficients between Ex and Hy and between Ey and Hx at the two stations were higher than 0.99, and the nephrograms of the correlation with Meq were also similar. Therefore, only the shifting correlation nephrograms between the Ex, Ey, and Meq sequences at the two stations are given in this paper (Figs. 8 and 9).

5.1. MJG Station

After a series of treatments, the length of electromagnetic signal sequence and Meq sequence at MJG station was 28 and 67, respectively. In most previous studies, the SEMS probably arose within 2 or 3 weeks before the earthquakes (Zhang *et al.* 2011; Uyeda *et al.* 2002; Orihara *et al.* 2012; Hattori *et al.* 2012; Han *et al.* 2009, 2014), so to ensure sufficient

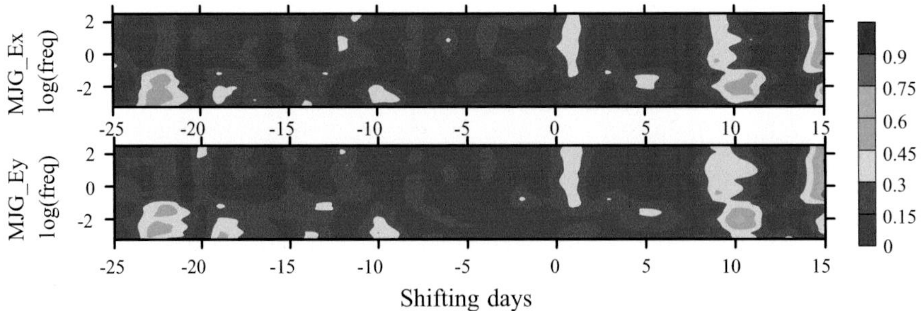

Figure 8

Calculation results for MJG station. The electromagnetic signal sequences in Ex (*upper*) and Ey (*lower*) shifted by 15 and 25 sample sizes (days) relative to the Meq sequence resulted in the correlation nephrograms. The *marks on the horizontal axis* are the same as those in Fig. 2a. The *vertical axis* represents the logarithm of frequency. The *color* represents the magnitude of correlation coefficient, as in the *colored bar on the right*

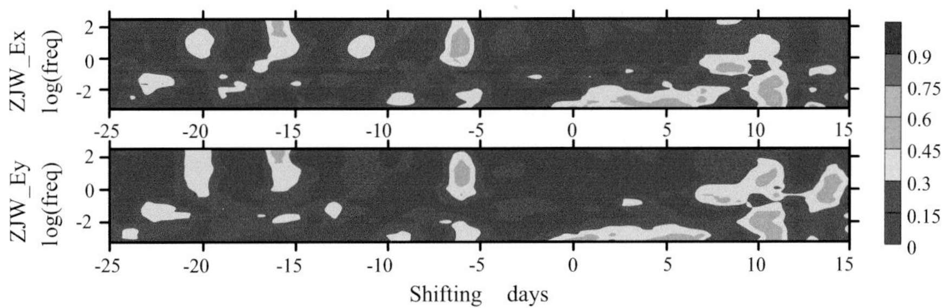

Figure 9
Calculation results for ZJW station. The meanings of the *symbols* are the same as in Fig. 8

sample size, the electromagnetic signal sequence was shifted to the left and right relative to the Meq sequence by a maximum of 15 and 25 days, respectively. In the former situation, when left shifted 15 days the number of samples for SCM calculation was 14. Thus, in Fig. 8, the strong correlation at 15 days after may be attributed to reduction of the amount of data available for calculation during the process of shifting to the left.

For the two anomaly regions with very prominent correlation at low-frequency, 23 days before the earthquake and 11 days after the earthquake, the number of samples involved in the calculation was 28 and 19, respectively. The correlation coefficient at 0.01 Hz was 0.5 and 0.7, respectively. The lead time of the pre-seismic anomaly was approximately similar to that in studies of the Wenchuan, Lushan, and Izu Island earthquake by use of different methods (Uyeda *et al.* 2002; Ma *et al.* 2013; Fan *et al.* 2010). Moreover, the correlation of the high-frequency component was not significant, especially preseismically. This is in agreement with the results of Fan *et al.* (2010), Fujinawa *et al.* (1998) and Park *et al.* (1993) who performed studies on the frequency of electromagnetic anomalies.

It is apparent from the nephrogram that during the period from 22 days before to 10 days after the earthquake, the correlation coefficient was very small. There was almost no co-seismic correlation. A weak correlation coefficient of approximately 0.38 appeared at high frequency approximately 1 day after the earthquake. Eftaxias *et al.* (2001) reported a failure to record the co-seismic anomaly; in contrast, (Orihara *et al.* 2012; Contoyiannis *et al.* 2010; Tang *et al.* 2010) reported that they had observed co-seismic electromagnetic signals. The missing co-seismic signals are discussed in detail below.

5.2. ZJW Station

It is apparent that 20 days before the earthquake, a moderately strong correlation was observed. A significant correlation occurred at a relatively high frequency (approx. 15 Hz) six days before earthquake, with the correlation coefficient 0.67. Sixteen and six days before the earthquake, correlation appeared intermittently at high frequency, with the highest on 6 days before the earthquake. The anomaly in the low-frequency was quite continuous from the co-seismic stage to six days after the earthquake. Starting from seven days after, the correlation extended to the medium frequency until 11 days after earthquake. The correlation coefficient was 0.56 in the co-seismic stage and increased to 0.65 five days after the earthquake. The continuous variation of correlation coefficient from the co-seismic stage to 11 days after the earthquake may be indicative of post-seismic stress adjustment. This result corroborates the hypothesis concerning the disappearance of the anomaly for the impending earthquake and that the post-seismic anomaly does not disappear immediately (Tang *et al.* 1998; Eftaxias *et al.* 2001). It can be seen from Fig. 7f that the quality of low-frequency data is still satisfactory at this station, which indicates the high-reliability of this phenomenon. It should be also noted that strong correlation appeared at medium frequency at Ey

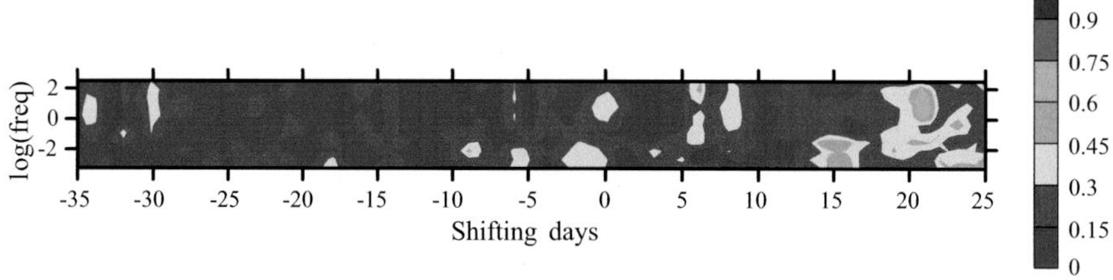

Figure 10
Result of shifting correlation coefficients between the disordered Meq sequence and the electromagnetic signal sequence in the Ex from MJG station

approximately 14 days after. This might be for the same reason as the strong correlation 15 days after earthquake at MJG station—a reduction in the amount of data.

6. Discussion

6.1. Correlation Between EM Sequence and Random Sequences

Considering the limited sample size of the electromagnetic signal sequences, the reliability of the results using shifting correlation method may be in doubt. Is it possible that similar correlation can be found between the electromagnetic signal sequence and a random sequence? To obtain an answer, the shifting correlation calculation was performed between the electromagnetic signal sequence and the "random sequence" of the Meq sequence. The electromagnetic signal sequence was the same, but the Meq sequences in Fig. 6b were rearranged randomly. Thus a new sequence with the same amplitude but different order was generated as the "random sequence". A nephrogram with the best correlation was finally chosen from among 10 experiments, as shown in Fig. 10. The shifting was to the left and right by 25 and 35 days, respectively. It is apparent from the figure that the correlation coefficients indeed increase systematically after shifting to the right by 14 days. This demonstrates that the reduction in sample size has a significant effect on the correlation coefficients, which is consistent with our preconceived idea mentioned above. When the sample size was 28 and no shifting was performed, or the shifting was to the right, the correlation coefficients were below 0.4, as shown in Fig. 10. No correlation coefficient exceeded 0.5 but some were equal to 0.3, as for the measured data. Therefore, in this paper, we mainly focus on correlation coefficients larger than 0.5 in real measured data.

6.2. Comparative Analysis of Two Stations

In this paper, absolute correlation coefficients were used in the nephrograms (Figs. 4, 8, 9). Therefore, we could not discriminate whether the two physical quantities are positive or negative correlations. Figure 11 shows an example for MJG station. These are the variation curves for the Ex sequence and Meq sequence over time after shifting to the right by 23 days and shifting to the left by 11 days at a frequency about 0.05 Hz (curve smoothing by cubic spline interpolation). As seen from the curves, the correlation is primarily positive and negative before and after the earthquake, respectively, and correlation coefficients are 0.53 and −0.67, respectively. The same phenomenon is observed in the high-value area of the correlation coefficient for ZJW station. We also found the correlation was negative in the co-seismic stage, with correlation coefficient −0.56. These phenomena are very interesting, but the reason for this must be investigated further.

Analysis of the data at the two stations shows that the correlation nephrograms reveal the occurrence of similarities and differences. First, from the nephrograms there is no co-seismic correlation in the high-frequency component at the two stations. We

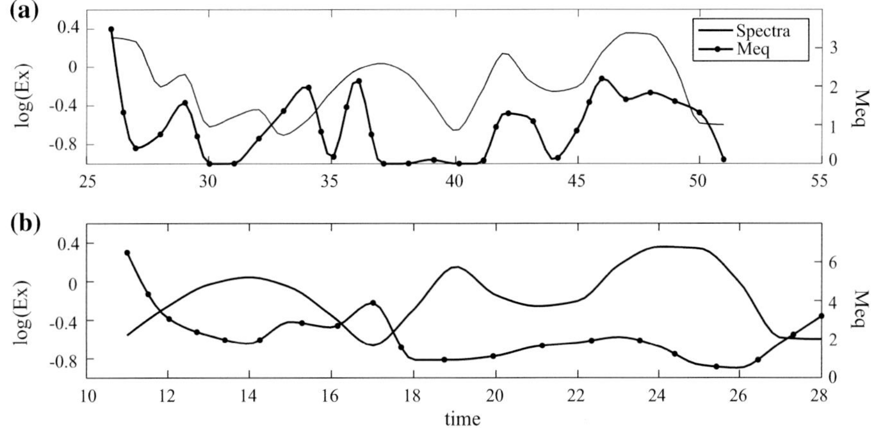

Figure 11
Comparison of curves of electromagnetic signal sequence in the Ex direction and the Meq sequence over time at a frequency of 0.05 Hz. The two *vertical axes* on the *left* and *right* are the logarithms of the electromagnetic spectrum value and Meq. The meaning of the *horizontal axis* is the same as in Fig. 6; **a** comparison of the two sequences after shifting to the *right* by 23 days; **b** comparison of the two sequences after shifting to the *left* by 11 days

speculate that there are two different interpretations for the absence of co-seismic high-frequency signals:

1 Because of the excessively high frequency of co-seismic electromagnetic radiation, the device fails to record the signals. In other reports, the frequency range of the co-seismic electromagnetic anomaly is of kHz or MHz magnitude (Eftaxias *et al.* 2009). However, the highest frequency of the electromagnetic monitoring instrument we used in this article is 320 Hz (V5-2000, Phoenix), thus the high-frequency signals has not been recorded.
2 The high-frequency signals in the observed frequency range are absorbed by crustal media. The seismic electromagnetic signals may come from two sources: some electromagnetic signals are released through the hypocenter and transmitted by the earth medium, whereas others are transmitted by the waveguide between the ionosphere and earth (Fujinawa *et al.* 1998). The monitoring stations, which were within 12 km of the swarm of earthquakes, were located near the epicenter in this study. No surface ruptures were caused by this shock (Zheng *et al.* 2013). Therefore, we infer that most of observed SEMS may directly come from the hypocenter through the earth medium, the high-frequency SEMS may be absorbed by the crust, as suggested by Eftaxias *et al.* (2001). At ZJW station, highly correlated SEMS appeared below a frequency of 0.001 Hz, and the ribbon-like signals lasted 10 days after the shock. To some extent, this phenomenon seems to indicate regularity in seismogenic and post-seismic activity.

Second, although the distance of the two stations from the epicenter almost the same, statistical significance arose on different days and for different frequencies. The reasons for the different results at MJG and ZJW are:

- because the number of samples in the electromagnetic sequence is limited, the result is vulnerable to local noise, and cannot reflect the time–frequency characteristics of SEMS steadily, which may lead to differences at the two stations; or
- the distance between the epicenter and the two stations is only 12 km, but the apparent resistivity curves (Fig. 7) and the results from one-dimensional magnetotelluric inversions showed their deep resistivity structures are quite different (the results as seen in the attachment). The resistivity at MJG is significantly lower than that at ZJW. Moreover, previous studies have suggested that different deep structure may result in different recording of SEMS (Varotsos *et al.* 1991; Hattori *et al.* 2012; Huang *et al.* 2010); different distributions of tectonic deformation fields, stress and strain field, and active tectonics may also affect the results.

Because this study focuses mainly on the theory and realization of the SCM, further studies of the exact reason for the above phenomenon will be conducted in the future.

7. Conclusions

The shifting correlation method (SCM) is proposed for analysis of the correlation between earthquakes and electromagnetic signals. We assumed that seismic electromagnetic signals (SEMS) could exist in the pre, co, and post-seismic stages. After continuous shifting of one sequence, the correlation coefficients between the two physical quantities were calculated. Thus, we could seek information about SEMS along the entire time axis (the position with high correlation) with the time of occurrence of the earthquakes as the origin. Synthetic data were first used to estimate the efficacy of the SCM for recognition of signals with asynchronous correlation. SCM was then used for analysis of electromagnetic monitoring data from the Minxian–Zhangxian earthquake, to obtain the preliminary temporal–frequency distribution characteristics of SEMS.

From the synthetic study we found that SCM could suppress noise to some extent. The larger the sample size involved in SCM, the more effectively the noise was minimized and the higher the resolution of the correlating signals. Thus, SCM is very suitable for treatment and analysis of long-term monitoring data obtained by use of seismic station networks. Moreover, as was apparent from the analytical procedure, SCM is not confined to correlation analysis between earthquake and electromagnetic signals. It is also suitable for correlation analysis of other precursory physical quantities.

The results of a case study of the Minxian–Zhangxian earthquake corroborate the belief that SEMS precede earthquakes. In the frequency range involved in this study, SEMS may appear within 23 days before the shock, and disappear 5 days before the shock. Strongly correlated SEMS appear at low frequency in the co-seismic and post-seismic stages, and may disappear 10 days after the earthquake. We also found that the time of occurrence of SEMS varied for the different stations and the frequency band of SEMS was also different at different stages. However, the case study had some limitations, for example the limited number of samples of observed electromagnetic data and we only considered linear correlation between earthquakes and electromagnetic signals. Non-linear correlation with sufficient samples is worthy of study.

In general, a new method has been proposed for investigation of the relationship between earthquakes and electromagnetic signals, and some results are in agreement with those from previous studies. The relationship between SEMS characteristics, position of monitoring stations, active tectonics, and seismic rupture are worthwhile being further profound studied.

Acknowledgments

The authors thank two anonymous reviewers for their constructive comments on the original manuscript. The authors also thank Liu Ming, who contributed to the field work, and Fang Lihua, associated professor from the Institute of Geophysics, CEA, who provided the original main-after shocks catalogue for this paper. This research was supported by the Special Fund for Basic Scientific Research of the Chinese National Nonprofit Institutes (grant no. IGCEA1306) and the Natural Science Foundation of China (grant no. 41174058).

References

Chen, X., G. Zhao and Y. Zhan (2004). *Visual system of MT data processing and interpretation*. Oil Geophysical Prospecting. *39*, 11–16.

Contoyiannis, Y. F., C. Nomicos, J. Kopanas, G. Antonopoulos, L. Contoyianni and K. Eftaxias (2010). *Critical features in electromagnetic anomalies detected prior to the L'Aquila earthquake*. Physica A: Statistical Mechanics and its Applications. *389*, 499–508.

EFTAXIAS, K., L. ATHANASOPOULOU, G. BALASIS, M. KALIMERI, S. NIKOLOPOULOS, Y. CONTOYIANNIS, J. KOPANAS, G. ANTONOPOULOS and C. NOMICOS (2009). *Unfolding the procedure of characterizing recorded ultra low frequency, kHZ and MHz electromagetic anomalies prior to the L'Aquila earthquake as pre-seismic ones-part 1*. Natural Hazards & Earth System Sciences. *9*.

EFTAXIAS, K., P. KAPIRIS, J. POLYGIANNAKIS, N. BOGRIS, J. KOPANAS, G. ANTONOPOULOS, A. PERATZAKIS and V. HADJICONTIS (2001). *Signature of pending earthquake from electromagnetic anomalies*. Geophysical Research Letters. *28*, 3321–3324.

ENOMOTO, Y. (2012). *Coupled interaction of earthquake nucleation with deep Earth gases: a possible mechanism for seismo-electromagnetic phenomena*. Geophysical Journal International. *191*, 1210–1214.

FAN, Y., X. DU, J. ZLOTNICKI, D. TAN, J. LIU, Z. AN, J. CHEN, G. ZHENG and T. XIE (2010). *The electromagnetic phenomena before the Ms8.0 Wenchuan Earthquake*. Chinese Journal of Geophysics. *53*, 2887–2898.

FUJINAWA, Y. and K. TAKAHASHI (1998). *Electromagnetic radiations associated with major earthquakes*. Physics of the Earth and Planetary Interiors. *105*, 249–259.

GUTENBERG, B. and C. F. RICHTER (1956). *Magnitude and energy of earthquakes*. Annals of Geophysics. *9*, 1–15.

HAN, P., K. HATTORI, M. HIROKAWA, J. ZHUANG, C.-H. CHEN, F. FEBRIANI, H. YAMAGUCHI, C. YOSHINO, J.-Y. LIU and S. YOSHIDA (2014). *Statistical analysis of ULF seismomagnetic phenomena at Kakioka, Japan, during 2001–2010*. Journal of Geophysical Research: Space Physics. *119*, 2014JA019789.

HAN, P., K. HATTORI, Q. HUANG, T. HIRANO, Y. ISHIGURO, C. YOSHINO and F. FEBRIANI (2011). *Evaluation of ULF electromagnetic phenomena associated with the 2000 Izu Islands earthquake swarm by wavelet transform analysis*. Natural Hazards and Earth System Science. *11*, 965–970.

HAN, P., Q. HUANG and J. XIU (2009). *Principal component analisis of geomagnetic duurnal variation associated with earthquakes: case study of the M6.1 Iwate-ken Nairiku Hokubu earthquake*. Chinese Journal of Geophysics. *52*, 1556–1563.

HAO, J., S. QIAN, J. GAO, J. ZHOU and T. ZHU (2003). *ULF electric and magnetic anomalies accopanying the cracking of rock sample*. Acta Seismologica Sinica. *25*, 102–111.

HATTORI, K., P. HAN, C. YOSHINO, F. FEBRIANI, H. YAMAGUCHI and C.-H. CHEN (2012). *Investigation of ULF Seismo-Magnetic Phenomena in Kanto, Japan During 2000–2010: Case Studies and Statistical Studies*. Surveys in Geophysics. *34*, 293–316.

HUANG, Q. (2005). *The State-of-the-art in Seismic Electromagnetic Observation*. Recent Developments in Word Seismology. *11*, 2–5.

HUANG, Q. (2011a). *Rethinking earthquake-related DC-ULF electromagnetic phenomena: towards a physics-based approach*. Natural Hazards and Earth System Science. *11*, 2941–2949.

HUANG, Q. (2011b). *Retrospective investigation of geophysical data possibly associated with the Ms8.0 Wenchuan earthquake in Sichuan, China*. Journal of Asian Earth Sciences. *41*, 421–427.

HUANG, Q. and M. IKEYA (1998). *Seismic electromagnetic signals (SEMS) explained by a simulation experiment using electromagnetic waves*. Physics of the earth and planetary interiors. *109*, 107–114.

HUANG, Q. and Y. LIN (2010). *Selectivity of seismic electric signal (SES) of the 2000 Izu earthquake swarm: a 3D FEM numerical simulation model*. Proceedings of the Japan Academy, Series B. *86*, 257–264.

KING, C.-Y. (1983). *Earthquake prediction: electromagnetic emissions before earthquakes*. Nature. *301*, 377.

KUO, C. L., L. C. LEE and J. D. HUBA (2014). *An improved coupling model for the lithosphere-atmosphere-ionosphere system*. Journal of Geophysical Research: Space Physics. *119*, 3189–3205.

LIU, J., X. DU, Y. FAN and Z. AN (2012). *The changes of the ground and ionosphere electric/magnetic fields before several great earthquakes*. Chinese Journal of Geophysics. *54*, 2885–2897.

MA, Q., G. FANG, W. LI and J. ZHOU (2013). *Electromagnetic anomalies before the 2013 Lushan Ms7.0 earthquake*. Acta Seismologica Sinica. *35*, 717–730.

ORIHARA, Y., M. KAMOGAWA, T. NAGAO and S. UYEDA (2012). *Preseismic anomalous telluric current signals observed in Kozu-shima Island, Japan*. Proceedings of the National Academy of Sciences. *109*, 19125–19128.

PARK, S. K., M. J. JOHNSTON, T. R. MADDEN, F. D. MORGAN and H. F. MORRISON (1993). *Electromagnetic precursors to earthquakes in the ULF band: A review of observations and mechanisms*. Reviews of Geophysics. *31*, 117–132.

POTIRAKIS, S. M., G. MINADAKIS and K. EFTAXIAS (2012). *Relation between seismicity and pre-earthquake electromagnetic emissions in terms of energy, information and entropy content*. Nat. Hazards Earth Syst. Sci. *12*, 1179–1183.

REN, H., X. CHEN and Q. HK (2012). *Numerical simulation of coseismic electromagnetic fields associated with seismic waves due to finite faulting in porous media*. Geophysical Journal International. *188*, 925–944.

TANG, J., Y. ZHAN, L. WANG, Z. DONG, G. ZHAO and J. XU (2010). *Electromagnetic coseismic effect associated with aftershock of Wenchuan Ms8.0 earthquake*. Chinese Journal of Geophysics. *53*, 526–534.

TANG, J. and G. ZHAO (1998). *Variation and analysis of resistivity before and after the Zhangbei-Shangyi earthquake*. Seismology and Geoloy. *20*, 164–171.

UYEDA, S., M. HAYAKAWA, T. NAGAO, O. MOLCHANOV, K. HATTORI, Y. ORIHARA, K. GOTOH, Y. AKINAGA and H. TANAKA (2002). *Electric and magnetic phenomena observed before the volcano-seismic activity in 2000 in the Izu Island Region, Japan*. Proc Natl Acad Sci U S A. *99*, 7352–7355.

UYEDA, S., T. NAGAO and M. KAMOGAWA (2009). *Short-term earthquake prediction: Current status of seismo-electromagnetics*. Tectonophysics. *470*, 205–213.

VAROTSOS, P. and M. LAZARIDOU (1991). *Latest aspects of earthquake prediction in Greece based on seismic electric signals*. Tectonophysics. *188*, 321–347.

XIE, T., X. DU, J. LIU, Y. FAN, Z. AN, J. CHEN and D. TAN (2013). *Wavelet power spectrum analysis of the electromagnetic signals of Wenchuan Ms8.0 and Haiti Mw7.0 earthquake*. Acta Seismologica Sinica. *35*, 61–71.

ZHANG, J., L. JIAO, X. LIU and X. MA (2013). *The study on the characteristics of ULF electromagnetic spectrum before and after the Wenchuan Ms8.0 earthquake*. Chinese Journal of Geophysics. *56*, 1253–1261.

ZHANG, X., H. CHEN, X. MA, X. LIU, L. YAO and Y. YUAN (2011). *Preliminary analysis of characteristics of anomalous variation of geoelectric field before Yushu Ms7.0 earthquake*. Journal of Geodesy and Geodynamics. *31*, 38–42.

ZHAO, G., X. CHEN and J. CAI (2007). *Electromagnetic observationby satellite and earthquake prediction*. Chinese Journal of Geophysics. *22*, 667–673.

ZHAO, G., L. WANG, Y. ZHAN, J. TANG, Q. XIAO, J. WANG, J. CAI, X. WANG and J. YANG (2012). *A new electromagnetic technique for earthquake monitoring: CSELF and the first observation network.* Seismology and Geoloy. *34*, 0235–4967.

ZHAO, G., Y. ZHAN, L. WANG, J. WANG, J. TANG, Q. XIAO and X. CHEN (2009). *Electromagnetic anomaly before earthquakes measured by electromagnetic experiments.* Earthquake Science. *22*, 395–402.

ZHENG, W., D. YUAN, W. MIN, Z. REN, X. LIU, A. WANG, C. XU, W. GE and F. LI (2013). *Geometric pattern and active tectonics in Southeastern Gansu province: Discussion on seismogenic mechanism of the Minxian–Zhangxian Ms6.6 earthquake on July 22, 2013*. Chinese Journal of Geophysics. *56*, 4058–4071.

(Received September 4, 2014, revised February 2, 2015, accepted February 9, 2015, Published online May 16, 2015)

Pure Appl. Geophys. 173 (2016), 285–303

DOI 10.1007/s00024-015-1116-8

Pure and Applied Geophysics

Long-Term RST Analysis of Anomalous TIR Sequences in Relation with Earthquakes Occurred in Greece in the Period 2004–2013

ALEXANDER ELEFTHERIOU,[1] CAROLINA FILIZZOLA,[2] NICOLA GENZANO,[3] TEODOSIO LACAVA,[2] MARIANO LISI,[3] ROSSANA PACIELLO,[2] NICOLA PERGOLA,[2,3] FILIPPOS VALLIANATOS,[1] and VALERIO TRAMUTOLI[2,3,4]

Abstract—Real-time integration of multi-parametric observations is expected to accelerate the process toward improved, and operationally more effective, systems for time-Dependent Assessment of Seismic Hazard (t-DASH) and earthquake short-term (from days to weeks) forecast. However, a very preliminary step in this direction is the identification of those parameters (chemical, physical, biological, etc.) whose anomalous variations can be, to some extent, associated with the complex process of preparation for major earthquakes. In this paper one of these parameters (the Earth's emitted radiation in the Thermal InfraRed spectral region) is considered for its possible correlation with $M \geq 4$ earthquakes occurred in Greece in between 2004 and 2013. The Robust Satellite Technique (RST) data analysis approach and Robust Estimator of TIR Anomalies (RETIRA) index were used to preliminarily define, and then to identify, significant sequences of TIR anomalies (SSTAs) in 10 years (2004–2013) of daily TIR images acquired by the Spinning Enhanced Visible and Infrared Imager on board the Meteosat Second Generation satellite. Taking into account the physical models proposed for justifying the existence of a correlation among TIR anomalies and earthquake occurrences, specific validation rules (in line with the ones used by the Collaboratory for the Study of Earthquake Predictability—CSEP—Project) have been defined to drive a retrospective correlation analysis process. The analysis shows that more than 93 % of all identified SSTAs occur in the prefixed space–time window around ($M \geq 4$) earthquake's time and location of occurrence with a false positive rate smaller than 7 %. Molchan error diagram analysis shows that such a correlation is far to be achievable by chance notwithstanding the huge amount of missed events due to frequent space/time data gaps produced by the presence of clouds over the scene. Achieved results, and particularly the very low rate of false positives registered on a so long testing period, seems already sufficient (at least) to qualify TIR anomalies (identified by RST approach and RETIRA index) among the parameters to be considered in the framework of a multi-parametric approach to t-DASH.

Key words: Robust Satellite Techniques (RST), TIR anomalies, short-term seismic hazard assessment, earthquake precursors and forecast, Greece seismicity, t-DASH.

1. Introduction

A renewed interest on the study of preparatory phases of earthquakes has been solicited in recent years by the, everyday more evident, weakness of traditional approaches to seismic hazard assessment as well as from the significant consequences of their failures in terms of human and economic losses (e.g., WYSS *et al.* 2012). For instance GELLER (2011) reports that "...since 1979, earthquakes that caused 10 or more fatalities in Japan actually occurred in places assigned a relatively low probability". KOSSOBOKOV and NEKRASOVA (2012) measured quantitatively, the errors of the Global Seismic Hazard Project (GSHAP) maps by the difference between observed and expected earthquake intensities. They found that GSHAP accelerations significantly underestimated the observed ones. No better results were achieved in other regions where seismic hazard assessment is based mostly (if not exclusively) on the study of earthquakes catalogs. This is for instance the case of Italy where, on the basis of probabilistic seismic hazard analysis (PSHA) methods (e.g., CORNELL 1968), all the events that caused 10 or more fatalities in the past 20 years were largely unexpected and/or underestimated in terms of magnitude or peak ground acceleration (PGA). For instance the San Giuliano earthquake (31 October 2002, $M_L = 5.4$) that killed 27 kids in a school, occurred in an area that was previously considered of minor concern. Particularly

[1] Department of Environmental and Natural Resources Engineering, School of Applied Sciences, Technological Educational Institute of Crete, Chania, Crete.
[2] Institute of Methodologies for Environmental Analysis of the National Research Council, Tito Scalo, PZ, Italy. E-mail: valerio.tramutoli@unibas.it
[3] School of Engineering, University of Basilicata, Via dell'Ateneo Lucano 10, 85100 Potenza, Italy.
[4] International Space Science Institute, Bern, Switzerland.

enlightening is the explanation given by Chiarabba *et al.* (2005): "…Seismic hazard for the region had not been previously retained high and the earthquake was mostly unexpected by seismologists. The reason was that neither historical or instrumental events had been previously reported in seismic catalogues for that area".

In the case of recent Emilia earthquake (20 May 2012, $M_W = 5.8$), the observed PGA (>0.25 g) in the epicentral zone (Panza *et al.* 2014) was significantly higher than the one (<0.175 g) predicted by the PSHA map assumed as reference by the Italian Ministry of Infrastructures (see Zuccolo *et al.* 2011 and reference herein) in defining the new rules for building in seismic areas.

Also for these reasons, an everyday increasing interest of scientific community, has been addressed to alternative observational techniques and data analysis methods suitable for improving our present capability to assess seismic hazard in the short–medium term. In this context a renewed role could be played by the research on earthquake precursors if it is addressed to develop/improve systems for time-Dependent Assessment of Seismic Hazard (t-DASH, Tramutoli *et al.* 2014) instead to the deterministic earthquake predictions. Several geophysical parameters (see for instance Tronin 2006 and Cicerone *et al.* 2009, and reference herein) have been proposed, since decades, as possible earthquake precursors. Although a large scientific documentation exists about the occurrence (in apparent relationship with earthquake preparation phases) of anomalous space–time transients of their measurements (see for instance Tronin 2006 and reference herein; Cicerone *et al.* 2009), no one single measurable parameter and no one observational methodology have demonstrated, until now, to be sufficiently reliable and effective for the implementation of an operational earthquake prediction system (see also Geller 1997).

A multi-parametric approach seems, instead, to be the most promising approach in order to increase reliability and precision (e.g., Huang 2011a, b; Ouzounov *et al.* 2012; Tramutoli *et al.* 2012a) of short-term seismic hazard forecast.

To this aim several research programs have been initiated—like the EU-FP7 project PRE-EARTHQUAKES (www.pre-earthquakes.org; Tramutoli *et al.* 2012b) and the Italian INGV-S3 project, https://sites.google.com/site/ingvdpc2012progettos3/)—or reiterated (e.g., the iSTEP—integrated Search for Taiwan Earthquake Precursors; Tsai *et al.* 2006) which apply a real-time integration of multi-parametric observations to develop improved systems for t-DASH and earthquake short-term (from days to weeks) forecast.

However, a very preliminary step of whatever multi-parametric approach is to identify those parameters (chemical, physical, biological, etc.) whose anomalous variations can be, to some extent, associated with the complex process of preparation for a big earthquake.

To this aim physical models (e.g., Scholz *et al.* 1973; Tronin 1996; Freund 2007; Pulinets and Ouzounov 2011; Huang 2011b; Tramutoli 2013a) that are able to justify the reason why such parameter should/could exhibit significant variations in relation with the preparation phases of an earthquake surely can help but, in some case, just statistical correlation analyses have been proposed in order to establish if or not, and in which measure, such parameter anomalous variations (identified with some specific data analysis technique) can be or not related to an impending earthquake.

In both cases, a convincing demonstration that a not casual relationship exists, among anomalous variations of a candidate precursor and earthquake occurrence, should be provided before deciding to include it in a multi-parametric t-DASH scenario.

Such long-term correlation analyses have been already successfully performed for several parameters like plasma frequency at the ionospheric F2 peak *foF2* (e.g., Liu *et al.* 2006), ionospheric ion density recorded by the detection of electro-magnetic emissions transmitted from earthquake region (DEMETER) satellite (e.g., Li and Parrot 2013), and ultra low frequency (ULF) geomagnetic signal (e.g., Han *et al.* 2014; Xu *et al.* 2013). Studies like these represent good exempla of what we can establish as the minimum starting point of whatever multi-parametric observational approach: *to identify suitable candidate parameters by measuring their level of correlation with earthquake occurrence on a sufficiently long time series of measurements.* This preliminary step will also serve to determine the relative weight to be attributed to each parameter (in comparison with the other contributing parameters) in a multi-parametric system for dynamically updating seismic hazard estimates.

However, such long-term correlation analyses, even if necessary, are not always possible due to the lack of long-enough time series of consistent observations. In this context satellite-based measurements could represent a unique reservoir of long-term (more than 30 years in some case), global, continuous dataset.

The fluctuations of Earth's thermally emitted radiation, as measured by satellite sensors operating in the thermal infrared (TIR, e.g., Wang and Zhu 1984; Gorny *et al.* 1988; Qiang *et al.* 1991; Tronin 1996; Tramutoli *et al.* 2001, 2015a; Ouzounov and Freund 2004 and reference herein) as well as in the longwaves (OLR, e.g., Ouzounov *et al.* 2007 and reference herein) spectral range, have been proposed since long time as potential earthquake precursors. However, very refined data analysis techniques are required in order to isolate residual TIR variations, potentially associated with earthquake occurrence, from the normal variability of TIR signal due to other causes (see for instance Tramutoli *et al.* 2005).

More than 10 years (since 2001) of applications of the general Robust Satellite Techniques (RST; Tramutoli 1998, 2005, 2007) methodology to this issue, have shown the ability of this approach to discriminate anomalous TIR signals possibly associated with seismic activity (hereafter referred simply as *TIR anomalies* or TAs) from normal fluctuations of Earth's thermal emission related to other causes (e.g., meteorological) independent of the earthquake occurrences.

Being based on a statistical definition of *TIR anomalies* and on a suitable method for their identification even in very different local (e.g., related to atmosphere and/or surface) and observational (e.g., related to the time/season or satellite view angles) conditions, RST approach has been widely applied to tens of earthquakes, covering a wide range of magnitudes (from 4.0 to 7.9) and those occurred in very different geo-tectonic contexts (compressive, extensional, and transcurrent) in four different continents.

RST intrinsic exportability permitted its implementation on TIR images acquired by sensors on board of different polar (see Di Bello *et al.* 2004; Filizzola *et al.* 2004; Lisi *et al.* 2010; Pergola *et al.* 2010; Tramutoli *et al.* 2001) and geostationary satellites (see Aliano *et al.* 2007a, b, 2008a, b, c, 2009; Corrado *et al.* 2005; Genzano *et al.* 2007, 2009a, b, 2010, 2015; Tramutoli *et al.* 2005).

In this paper the level of correlation among *TIR anomalies* identified by the RST methodology (Tramutoli *et al.* 2005) and earthquake occurrence, is evaluated for the first time in a quite long time period (10 years). To this aim 10 years (from May 2004 to December 2013) of TIR satellite records, collected over Greece by the geostationary satellite sensor Meteosat Second Generation–Spinning Enhanced Visible and Infrared Imager (MSG–SEVIRI), have been retrospectively analyzed using the RST approach, and time and location of *TIR anomalies* occurrences were compared with the ones of earthquakes with $M \geq 4$. Achieved results will be discussed in the perspective of a multi-parametric approach for a t-DASH.

2. RST Methodology

The RST methodology is based on the general approach Robust AVHRR Technique (RAT; Tramutoli 1998), which, being exclusively based on satellite data at hand (do not require whatever ancillary data), is intrinsically exportable on different satellite packages, the reason why the original name RAT was changed in the more general Robust Satellite Techniques (RST, Tramutoli 2005, 2007).

The RST approach is based on a multi-temporal analysis of historical data set of satellite observations acquired in similar observational conditions (e.g., same month of the year, same hour of the day, same sensor, etc.).

Such preliminary analysis is devoted to characterize the measured signal (in terms of its expected value and variation range) for each pixel of the satellite image to be processed. In this way space–time anomalies are identified always by comparison with a preliminarily computed signal behavior. On this basis, *TIR anomalies* possibly that are related to earthquake occurrences could be defined and isolated from those signal variations which are related to known (but also unknown) natural and/or observational factors (see TRAMUTOLI *et al.* 2005) that could be responsible for *false alarm* proliferation.

Since the first RST application to the thermal monitoring of earthquake prone areas, TIR fluctuations were identified using the *RETIRA* index (Robust Estimator of TIR Anomalies, FILIZZOLA *et al.* 2004; TRAMUTOLI *et al.* 2005), which can be computed as follows:

$$\otimes_{\Delta T}(x,y,t) = \frac{\Delta T(x,y,t) - \mu_{\Delta T}(x,y)}{\sigma_{\Delta T}(x,y)}, \quad (1)$$

where:

- x,y represent the coordinates of the center of the ground resolution cell corresponding to the pixel under consideration on a satellite image;
- t is the time of the measurement acquisition with $t \in \tau$, where τ defines the homogeneous domain of multi-annual satellite imagery collected in the same time slot of the day and period (month) of the year;
- $\Delta T(x,y,t) = T(x,y,t) - T(t)$ is the value of the difference between the punctual value of TIR brightness temperature $T(x,y,t)$ measured at the location x, y, acquisition time t, and its spatial average $T(t)$ computed on the investigated area considering only cloud-free locations, all belonging to the same, land or sea, class (i.e., considering only sea pixels if x, y is located on the sea and only land pixels if x, y is located on the land). Note that the choice of such a differential variable $\Delta T(x,y,t)$ instead of $T(x,y,t)$ is expected to reduce possible contributions (e.g., occasional warming) due to day-to-day and/or year-to-year climatological changes and/or season time drifts;
- $\mu_{\Delta T}(x,y)$ time average value of $\Delta T(x,y,t)$ at the location x, y computed on cloud-free records belonging to the selected data set ($t \in \tau$);
- $\sigma_{\Delta T}(x,y)$ standard deviation value of $\Delta T(x,y,t)$ at the location x, y computed on cloud-free records belonging to the selected data set ($t \in \tau$).

In this way $\otimes_{\Delta T}(x,y,t)$ gives the local excess of the current $\Delta T(x,y,t)$ signal compared with its historical mean value and weighed by its historical variability at the considered location. Both, $\mu_{\Delta T}(x,y)$ and $\sigma_{\Delta T}(x,y)$ are computed, once and for all, for each location x, y processing several years of historical satellite records acquired in similar observational conditions. They are two reference images describing the normal behavior of the signal and of its variability at each location x, y in observational conditions as similar as possible to the images at hand.

Excess $\Delta T(x,y,t) - \mu\Delta T(x,y)$ then represents the Signal (S) to be investigated for its possible relation with seismic activity. It is always evaluated by comparison with the corresponding natural/observational Noise (N), represented by $\sigma_{\Delta T}(y,x)$ which describes the overall (local) variability of S including all (natural and observational, known and unknown) sources of its variability as historically observed at the same site in similar observational conditions (sensor, time of day, month, etc.). This way, the relative importance of the measured TIR signal (or the intensity of anomalous TIR transients) can naturally be evaluated in terms of S/N ratio by the RETIRA index. It should be noted that prescriptions on the temporal domain τ, which select TIR images acquired in the same period of the year (e.g., month) and in the same hour of the day (e.g., midnight), guarantee the reduction of the signal variability due to the daily (diurnal variation of the temperature) and annual (seasonal variation of the temperature but also of emissivity, which is mainly related to the different vegetation coverage) solar cycles.

3. Data Analysis

3.1. A Refined Implementation of RST Approach

In this work all (3151) TIR images acquired by MSG/SEVIRI satellite sensor in the IR10.8 channel (at 9.80–11.80 μm) in the first time slot of the day (00:00 ÷ 00:15 GMT; i.e., 02:00 ÷ 02:15 LT) since May 2004 up to December 2013 have been analyzed. Night-time TIR images were preferred, as usual, because less influenced by effects related to soil–air temperature differences (which are normally higher during other hours of the day) and less sensitive to local variations (due for instance to cloud cover or shadows) of solar illumination which could represent a further element of variability of TIR signal, independent from seismicity.

For each month a multi-annual homogeneous data set of TIR satellite images was built to compute 24 reference fields (two, $\mu_{\Delta T}$ and $\sigma_{\Delta T}$, for each of 12 months) for a testing area (top-left 42.1°N—19.2°E; top-right 42.7°N—30.4°E; bottom-right 33.9°N—26.1°E; bottom-left 33.5°N—16.7°E) which includes the whole Greek peninsula. Following RST prescriptions, the computation of reference fields was made considering only the cloud-free pixels. In fact thick meteorological clouds are not transparent to the passage of the Earth's emitted TIR radiation so that measured signal in those pixels refers to the cloud top temperature (usually very low) and not to the near surface conditions. Errors in the identification (and the consequent non-exclusion) of cloudy pixels, could heavily condition quality of reference fields which, in general, will result biased toward lower values of averages $\mu_{\Delta T}(x,y)$ and higher values of standard deviations $\sigma_{\Delta T}(x,y)$. Even if the latter effect—increasing the denominator of expression (1)—could compensate the first one (increasing the numerator of the same expression)—avoiding, thanks to the robustness of RETIRA index, a proliferation of false positives—a significant reduction of the overall sensitivity can be however observed as a consequence of cloudy pixel identification errors.

In order to identify (and discard from reference field computation) cloudy affected pixels, the OCA (One-channel Cloudy-radiance-detection Approach; Pietrapertosa *et al.* 2001; Cuomo *et al.* 2004) method (still RST based)—devoted to identify *cloudy radiances* (i.e., radiances which deviate significantly from the expected values for a specific place and time of observation)—was preferred to traditional *cloud detection* methods (devoted to identify pixels containing clouds) which are much more exposed to commission (i.e., classifying pixels as cloudy independently if clouds affect or don't affect the measured radiance in the considered spectral band) and omission (mostly because of the use of a fixed threshold approach) errors (see for instance Pietrapertosa *et al.* 2001; Tramutoli *et al.* 2000).

However, in this paper, due to their importance, particular attention has been paid to cloudy pixel handling, introducing the following refinement of the standard RST pre-processing phases.

(a) In order to be sure that only cloud-free radiances contribute to the computation of reference fields, after cloudy pixels have been identified by OCA, not only those pixels but (like in Eneva *et al.* 2008) also the 24 ones in a 5 × 5 box around it (very often belonging to cloud edges) have been excluded by the following computations of reference fields.

(b) As shown by Aliano *et al.* (2008a) and Genzano *et al.* (2009a), spatial distribution of clouds, over a thermally heterogeneous scene, can significantly change the value of the measured signal $\Delta T(x,y,t) = T(x,y,t) - T(t)$ in the remaining (cloud-free) pixels of the scene belonging to the same land/sea class. In fact the same $T(x,y,t)$ value of measured TIR signal can be associated with an higher or lower $\Delta T(x,y,t)$ values depending on the spatial average $T(t)$ computed on the remaining cloud-free pixels (belonging to the same land/sea class of the pixel centered at the x, y coordinates). It has been shown—firstly by Aliano *et al.* (2008a) and then by Genzano *et al.* (2009a) who named it *cold spatial average effect*—that, if clouds mostly cover the warmer part of the land (or sea) portion of the scene, the spatial average $T(t)$ will result lower than expected in clear sky conditions. As a consequence anomalously higher values of the signal $\Delta T(x,y,t) = T(x,y,t) - T(t)$ can be measured over the remaining, cloud-free, land (or sea), portion of the scene which are due only to such an anisotropic distribution of clouds along the North–South direction. If not properly taken into account, such a pure meteorological phenomenon not only could (occasionally) introduce false positives in the interpretation phases but will also strongly affect $\mu_{\Delta T}(x,y)$ and $\sigma_{\Delta T}(x,y)$ reference fields which will be both biased toward higher values with a strong reduction of the overall sensitivity of RETIRA index. To face the problem TIR images suffering from such a *cold spatial average effect* have been automatically identified and excluded from the computation of reference fields. In order to identify them the values of the temporal average μ_T of $T(t)$ and its standard deviation σ_T have been computed (using all the dataset of SEVIRI images collected over

Greece in between May 2004 and December 2013) and for those scenes having $T(t) < \mu_T - 2\sigma_T$ [being $T(t)$, μ_T and σ_T all computed separately for land or sea pixels] the corresponding land (or sea) portions of image have been excluded from reference field computation.

(c) Even if not producing a *cold spatial average effect*, an extended cloud coverage can determine values of $T(t)$ and then of the considered signal $\Delta T(x,y,t)$ scarcely representative of the actual conditions of cloud-free pixels. So, when the cloudy fraction of land (or sea) portion of the scene was >80 %, then that portion (i.e., all pixels belonging to that land or sea class) have been excluded from the computation of the reference fields $\mu_{\Delta T}(x,y)$ and $\sigma_{\Delta T}(x,y)$.

On this basis $\mu_{\Delta T}(x,y)$ and $\sigma_{\Delta T}(x,y)$ reference fields, more reliable than the ones achievable using the standard RST pre-processing phases, have been computed (Fig. 1).

In Fig. 2 it is shown, separately for land and sea classes, the day-by-day results of the analysis performed on the whole data set of TIR images in order to apply tests (b) and (c).

RETIRA indexes $\otimes_{\Delta T}(x,y,t)$ have been then computed for all MSG-SEVIRI TIR images belonging to the dataset producing one *Thermal Anomaly Map* (TAM) $\otimes_{\Delta T}(x,y,t)$ for each day t in between May 1st 2004 and December 31st 2013. Locations with $\otimes_{\Delta T}(x,y,t) \geq 4$—i.e., with signal excess $\Delta T(x,y,t) - \mu_{\Delta T}(x,y) \geq 4\sigma_{\Delta T}(x,y)$—will be particularly addressed in this paper and hereafter we will refer to them simply as *Thermal Anomalies (TAs)*.

3.2. Identification of Significant Sequences of TIR Anomalies (SSTAs)

As already widely discussed in previous papers (see for example Filizzola *et al.* 2004; Tramutoli *et al.* 2005), the RETIRA index, being based on time-averaged quantities, is intrinsically not protected from the abrupt occurrence of signal outliers related to particular natural (see Aliano *et al.* 2008a; Genzano *et al.* 2009a) or observational (see Filizzola *et al.* 2004; Aliano *et al.* 2008b) conditions. The particular spatial distribution of this kind of TA and their transitory character in the temporal domain, normally allows to identify them and, in any case, to distinguish them from the spatially and temporally persistent ones possibly related to an impending earthquake, even in the case where they have similar intensity.

This is the reason why (together with relative intensity) spatial extension and persistence in time are requirements to be satisfied in order to preliminarily identify what we call significant thermal anomalies (STAs). Like in all the previous applications to thermal monitoring of earthquake prone area (Aliano *et al.* 2007a, b, 2008a, b, c, 2009; Bonfanti *et al.* 2012; Corrado *et al.* 2005; Di Bello *et al.* 2004; Filizzola *et al.* 2004; Genzano *et al.* 2007, 2009a, b, 2010, 2015; Lisi *et al.* 2010; Pergola *et al.* 2010; Pulinets *et al.* 2007; Tramutoli *et al.* 2001, 2005, 2009, 2012a, b, 2013a, b, 2015b) TA highlighted by RST methodology have been subjected to such a preliminary space–time persistence analysis before it was qualified among STAs.

However, other well-known (see for instance Filizzola *et al.* 2004; Aliano *et al.* 2008a; Genzano *et al.* 2009a) spurious effects exist that prevent to include among STAs some, even space–time persistent, sequence of TAs. The ones already identified are:

- *TAs due to meteorological effects* (*Fig.* 3) These are all anomalous pixels appearing in the TIR scenes affected by a wide cloudy cover or in the TIR scenes affected by an asymmetrical distribution of clouds mainly over the warmest portions of a scene, which expose the remaining clear portions of the scene to the appearance of spurious anomalies (*cold spatial average effect*, Aliano *et al.* 2008a; Genzano *et al.* 2009a). Such a circumstance could appear in the portions of TIR scene having the daily spatial average $\langle T(t)\rangle \leq \langle\mu_T\rangle - \langle\sigma_T\rangle$ (being $\langle\mu_T\rangle$ and $\langle\sigma_T\rangle$ the monthly average and corresponding standard deviation of $T(t)$ computed for the same month of the image at hand using the whole historical dataset of TIR images) or having a cloudy coverage ≥80 % of total pixels of the same

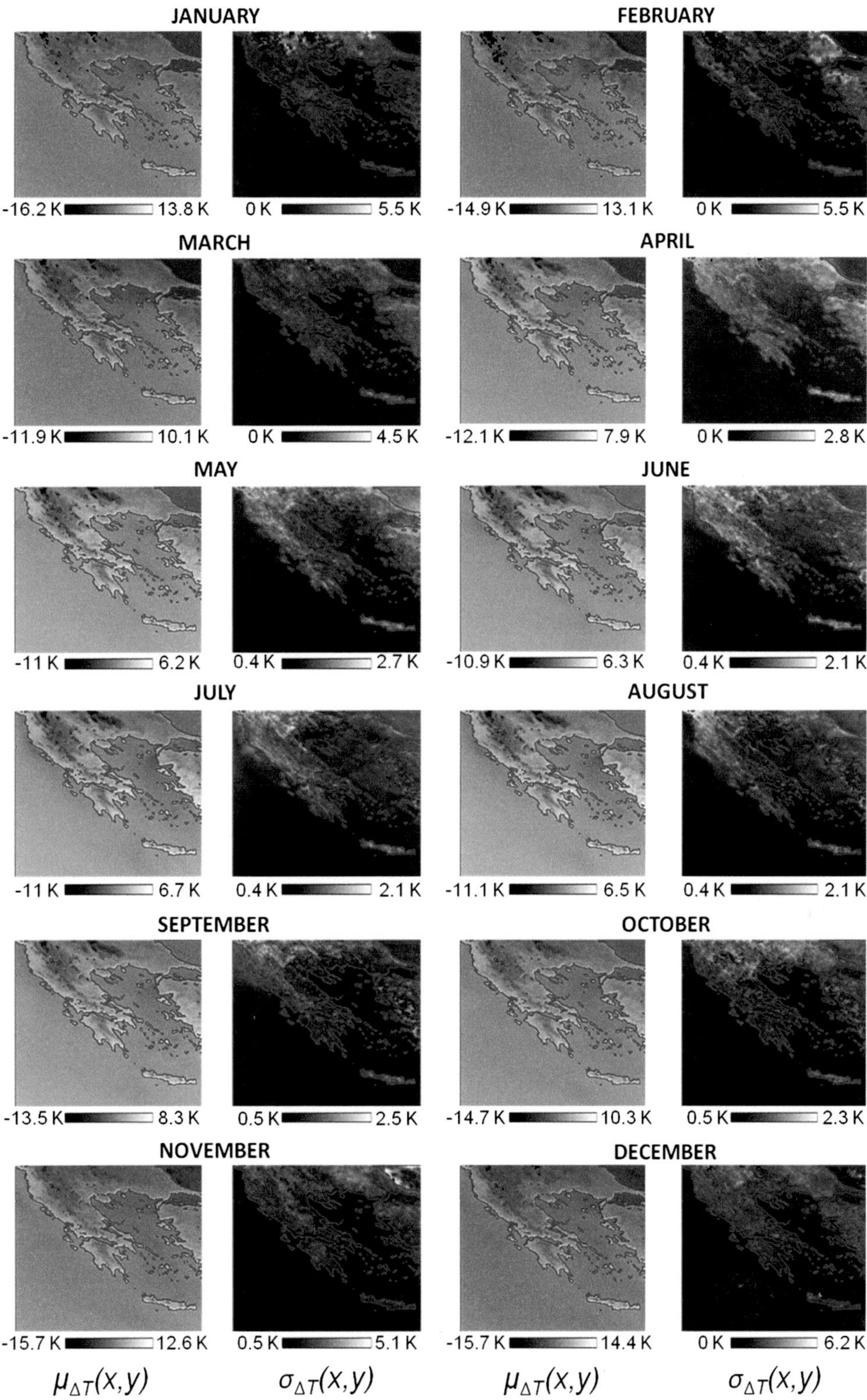

Figure 1
Monthly reference fields $\mu_{\Delta T}(x,y)$ and $\sigma_{\Delta T}(x,y)$ computed for all the months of the year on the basis of SEVIRI TIR observations acquired over Greece from May 2004 to December 2013 (see text)

classes (land/sea). Moreover, also TAs generated by local warming due to night-time cloud passages have been recognized as artifacts of the meteorological effects (see ALIANO *et al.* 2008a).

- *TAs due to errors in image navigation/co-location process* (*Fig.* 3) Although, this artifact is not rare for polar platforms, also in the cases of geostationary platforms a wrong navigation may cause intense TAs where sea pixels turn out to be erroneously co-located over land portions (see FILIZZOLA *et al.* 2004; ALIANO *et al.* 2008b).
- *Space–time persistent TAs due to extreme events* Usually, these TAs can be observed in relation with particularly rare (over decades) events increasing (for more than 1 day) measured TIR signal because of an increase of surface temperature (e.g., in the case of extremely extended forest fires) or emissivity (e.g., extremely extended floods).

By this way an operational definition of STA can be given by considering a location *x, y* affected by a STA at the time *t* if the following requirements are satisfied:

(a) *Relative intensity* $\otimes_{\Delta T}(x,y,t) > K$ (with $K = 4$ in our case)
(b) *Control on spurious effects* Absence of known sources of spurious TAs (see above)
(c) *Spatial persistence* It is not isolated being part of a group of TAs covering at least 150 km^2 within an area of $1° \times 1°$
(d) *Temporal persistence* Previous conditions (i.e., the existence of a group of TAs covering at least 150 km^2 within an area of $1° \times 1°$ around *x, y*) are satisfied at least one more time in the 7 days preceding/following *t*.

After applying the above-mentioned rules to the whole data set of 3151 SEVIRI scenes over Greece 62 Significant Sequences of Thermal Anomalies (SSTAs) were identified where each one is composed by several STAs spanned on 2 or more TAMs.

3.3. Long-Term Correlation Analysis

In order to evaluate the possible correlations existing among the appearance of SSTAs and time, location and magnitude of earthquakes, empirical rules were applied which were mostly based on the long-term (more than 14 years) experience on TAM analyses (ALIANO *et al.* 2007a, b, 2008a, b, c, 2009; BONFANTI *et al.* 2012; CORRADO *et al.* 2005; DI BELLO *et al.* 2004; FILIZZOLA *et al.* 2004; GENZANO *et al.* 2007, 2009a, b, 2010, 2015; LISI *et al.* 2010; PERGOLA *et al.* 2010; PULINETS *et al.* 2007; TRAMUTOLI *et al.* 2001, 2005, 2009, 2012a, b, 2013a, 2015b) performed by authors in four different continents, different tectonic settings, for tens of earthquakes with magnitudes ranging from 4.0 to 7.9.

By this way each single STA observed at the time *t* in the location (*x,y*) will be considered possibly related to seismic activity if:

- It belongs to a previously identified SSTA;
- An earthquake of $M \geq 4$ occurs 30 days after its appearance or within 15 days before[1] (*temporal window*)
- An earthquake with $M \geq 4$ occurs within a distance *D*, from the considered STA, so that $150 \text{ km} \leq D \leq R_D$ being $R_D = 10^{0.43M}$ the DOBROVOLSKY *et al.* (1979) distance (*spatial window*).

By this way, starting from each STA belonging to an SSTA, different possibly affected areas can be built for different possible magnitudes of future/past

Figure 2 ▶
Results of the analysis performed to reduce effects related to cloud coverage extension and/or distribution across a scene. Each *rhomb* represents the spatial average *T*(*t*) computed on a TIR image collected over Greece on the day *t* at 00:00 UTC considering only cloud-free pixels. *Blue rhombs* correspond to scenes used for the computation of reference fields, the *yellow* ones to scenes removed from the used data sets. *Red lines* and *light red bands* represent, respectively the temporal averages ($\langle\mu_T\rangle$) and ±2 sigma bounds ($2\langle\sigma_T\rangle$) computed considering the images collected during the same month in the past between May 2004 and December 2013. *Vertical gray bars* represent the percentage of cloudy pixels identified over each scene. The *dashed horizontal blue line* indicates the cloudiness limit of 80 % adopted to exclude cloudy scenes from reference field computation (see text). *Top* only over land; *bottom* only over sea

[1] On the models which foresee the occurrence of similar anomalies also immediately after the quake; see for instance SCHOLZ *et al.* (1973) and, with reference to TIR anomalies, TRAMUTOLI *et al.* (2005, 2013a).

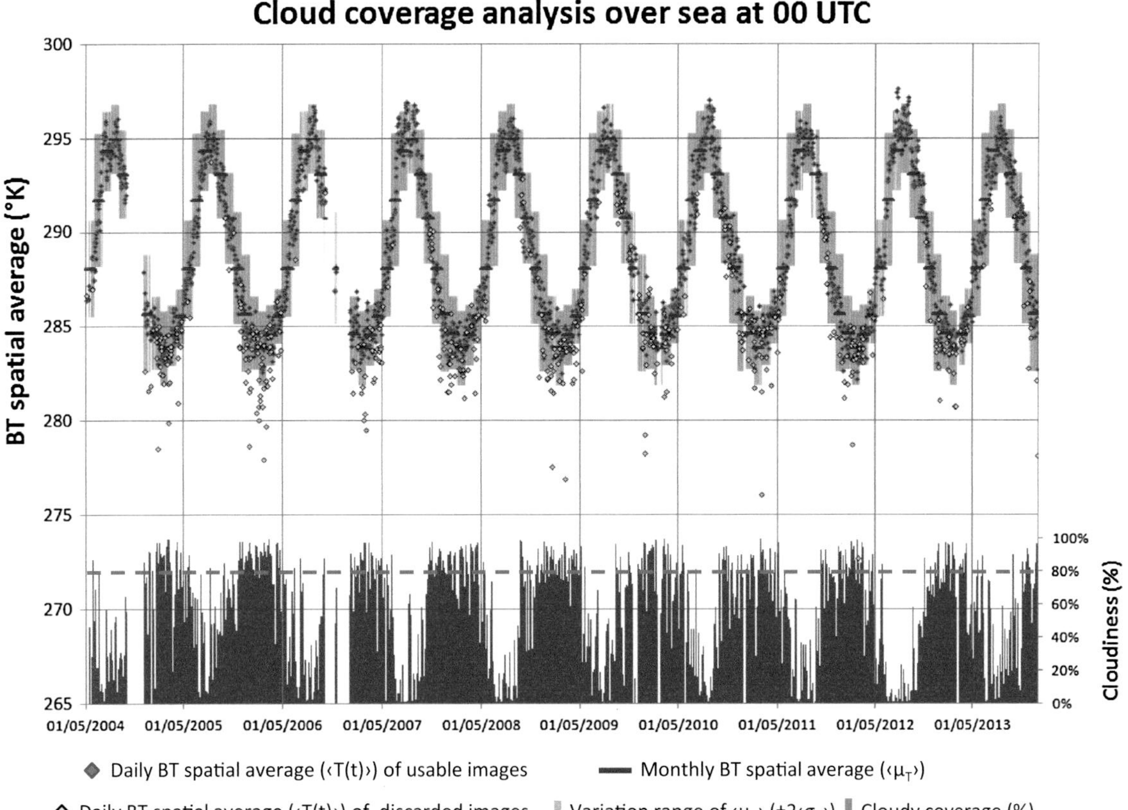
Cloud coverage analysis over sea at 00 UTC
BT spatial average (°K)
300
295
290
285
280
275
270
265
100%
80%
60%
40%
20%
0%
Cloudiness (%)
01/05/2004
01/05/2005
01/05/2006
01/05/2007
01/05/2008
01/05/2009
01/05/2010
01/05/2011
01/05/2012
01/05/2013
Daily BT spatial average (‹T(t)›) of usable images
Monthly BT spatial average (‹μT›)
Daily BT spatial average (‹T(t)›) of discarded images
Variation range of ‹μT› (±2‹σT›)
Cloudy coverage (%)

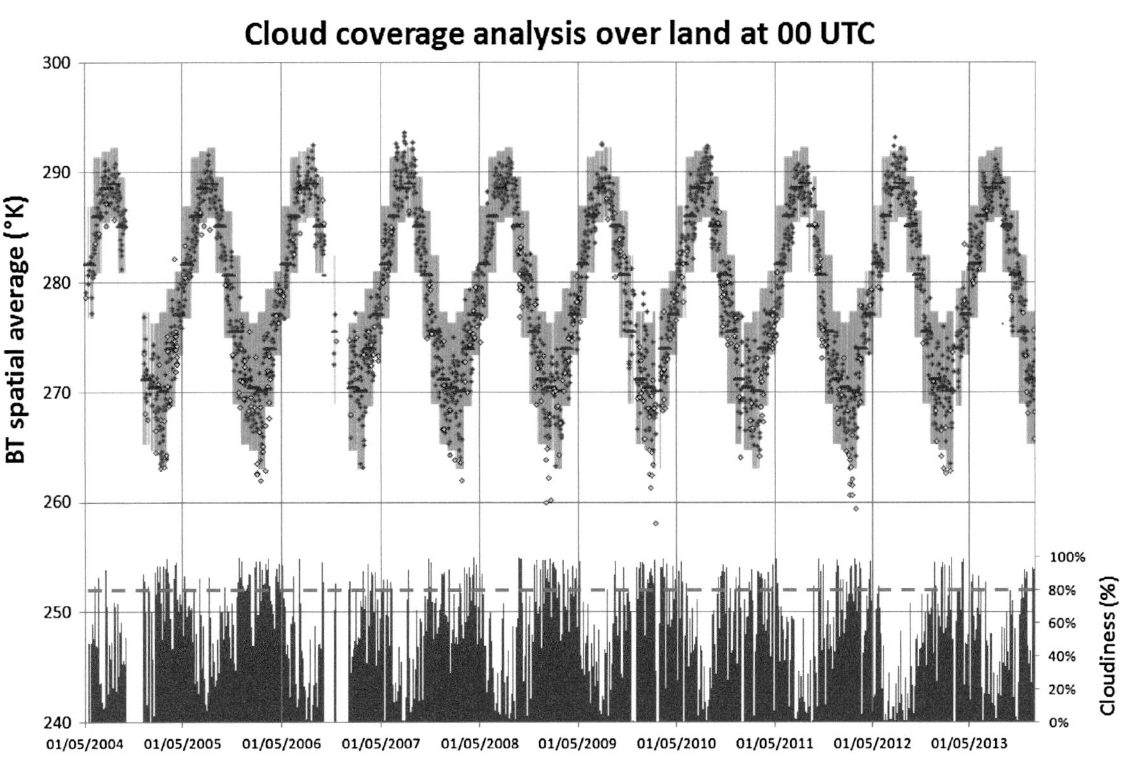
Cloud coverage analysis over land at 00 UTC
BT spatial average (°K)
300
290
280
270
260
250
240
100%
80%
60%
40%
20%
0%
Cloudiness (%)
01/05/2004
01/05/2005
01/05/2006
01/05/2007
01/05/2008
01/05/2009
01/05/2010
01/05/2011
01/05/2012
01/05/2013

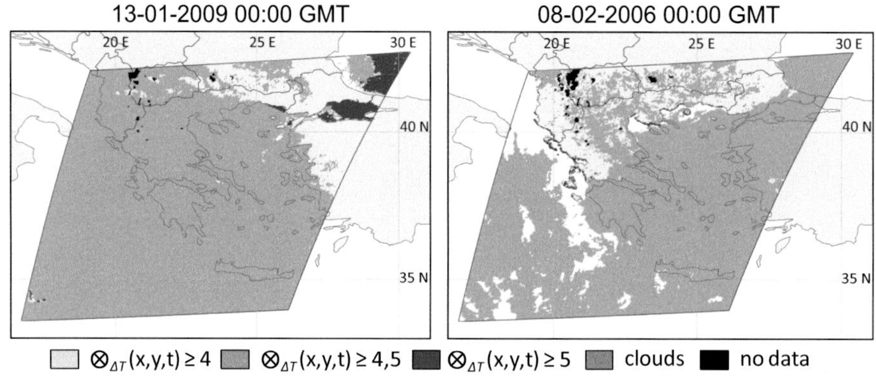

Figure 3
Left side an example of artifacts due to the *cold spatial average effect* (see text) in the SEVIRI TIR image of the 13 January 2009 at 00:00 GMT. *Right side* an example of artifacts due to navigation/co-location errors in the processing of SEVIRI TIR image of the 8 February 2006 at 00:00 GMT

earthquakes. The convolution of the contours drawn for all the STAs belonging to the same SSTA, allow to draw the contours of the areas (different for different magnitudes) possibly affected by future/past earthquakes.

The possible correlation among previously identified SSTAs and earthquake occurrence was investigated considering all earthquakes[2] with magnitude $M \geq 4$ occurred from April 1st 2004 to January 31st 2014 in:

- The area (contoured in red in Fig. 3) of TAMs (top-left 42.1°N—19.2°E; top-right 42.7°N—30.4°E; bottom-right 33.9°N—26.1°E; bottom-left 33.5°N—16.7°E) using the seismic catalog of National Observatory of Athens (NOA 2014) for the Greek territory
- The area extending up to 1° from its borders (top-left 43.1°N—18.2°E; top-right 43.7°N—31.04°E; bottom-right 32.9°N—27.1°E; bottom-left 32.5°N—15.7°E) using the seismic catalog of National Earthquake Information Center of U.S. Geological Survey (USGS 2014).

At first glance it is noticeable that most of STAs respecting the above-described correlation rules appear few days around the time of the earthquake

Figure 4 ▶
Examples of SSTAs identified in the period 26–29 June 2007. Significant TAs (STAs) with $\otimes_{\Delta T}(r,t) \geq 4$ (depicted in different colors according to the corresponding RETIRA values) appear in the Peloponnesus area on 26 and 27 June 2007 (*upper part*) and in the Crete island on 27 and 29 June 2007 (*bottom part*) before and after $M \geq 4$ seismic events. In addition to the magnitude of earthquakes, also the temporal gap from the first appearance of STAs is indicated by a number (*N*) in parentheses ($\pm N$ means that the earthquake occurred *N* days after/before the first appearance of STAs). Contours in different *colors* correspond to different space/magnitude windows (see text). The *red contoured box* indicates the limits of analyzed SEVIRI TIR scenes (Thermal Anomaly Map area)

occurrence being generally localized near main tectonic lineaments of the epicentral area.

Looking for instance at the example shown in Fig. 4 it is possible to note, in the period 26–29 June 2007, the presence of two different SSTAs:

1. Peloponnesus region (upper part of Fig. 4): several STAs appear near tectonic lineaments from June 26th, up to June 29th, 2007. A seismic event with $M = 5.2$ occurred on 29 June 2007 (3 days after the first STA appearance) and various other with $M \geq 4$ (before and after the first STA appearance), all well within the corresponding spatial correlation windows.
2. Crete island (bottom part of Fig. 4): STAs appear in the western part of the island on 27 June 2007 and, with a greater spatial extension, on 29 June 2007 in the eastern part just few

[2] 1083 events with $M \geq 4$, 80 events with $M \geq 5$ and 8 events with $M \geq 6$.

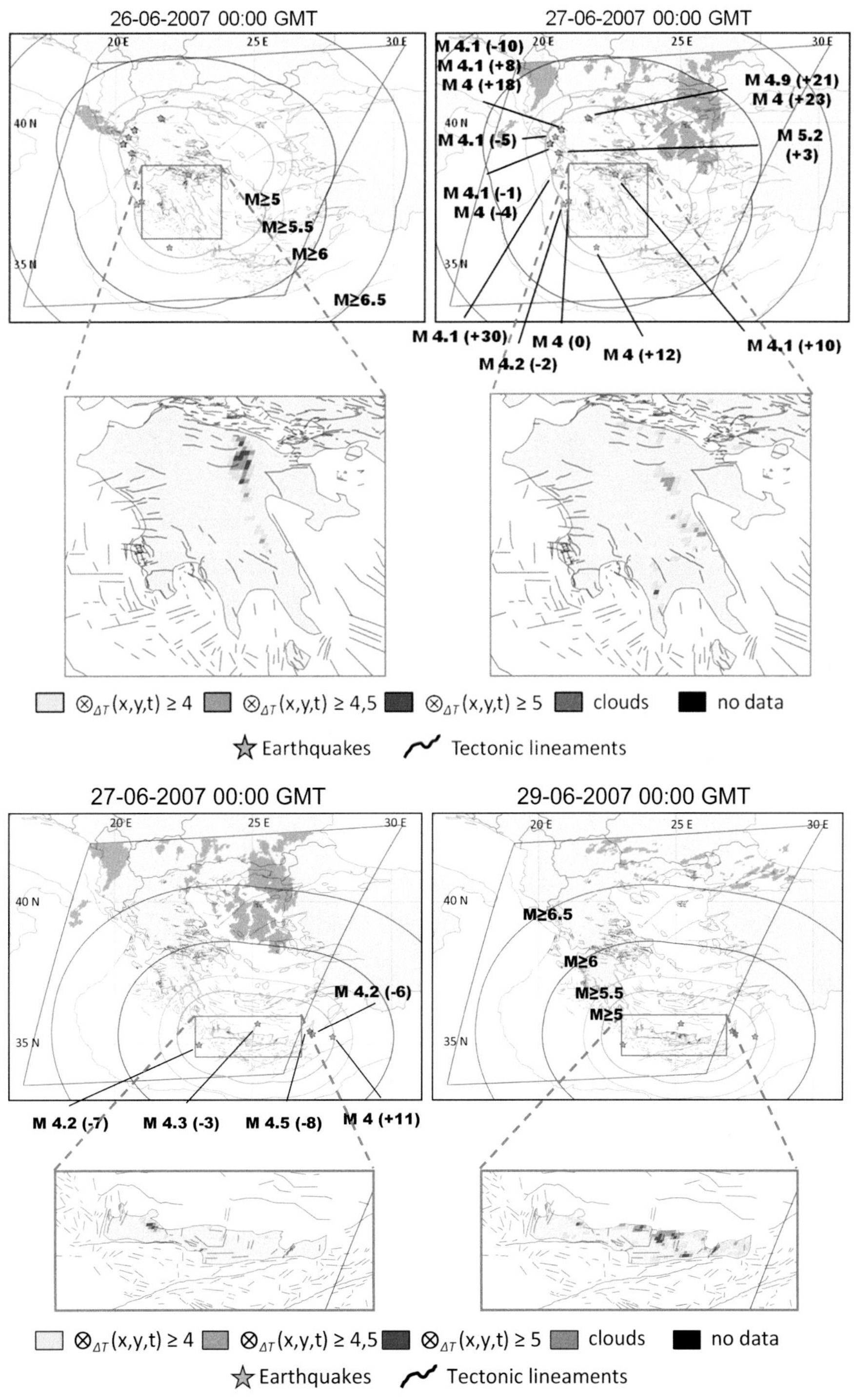

days after the occurrence of several low magnitude earthquakes (i.e., from 4.2 to 4.5) in the Sea of Crete.

A more comprehensive view of the correlation analysis performed for all the 62 SSTAs identified on the whole SEVIRI TIR data set is

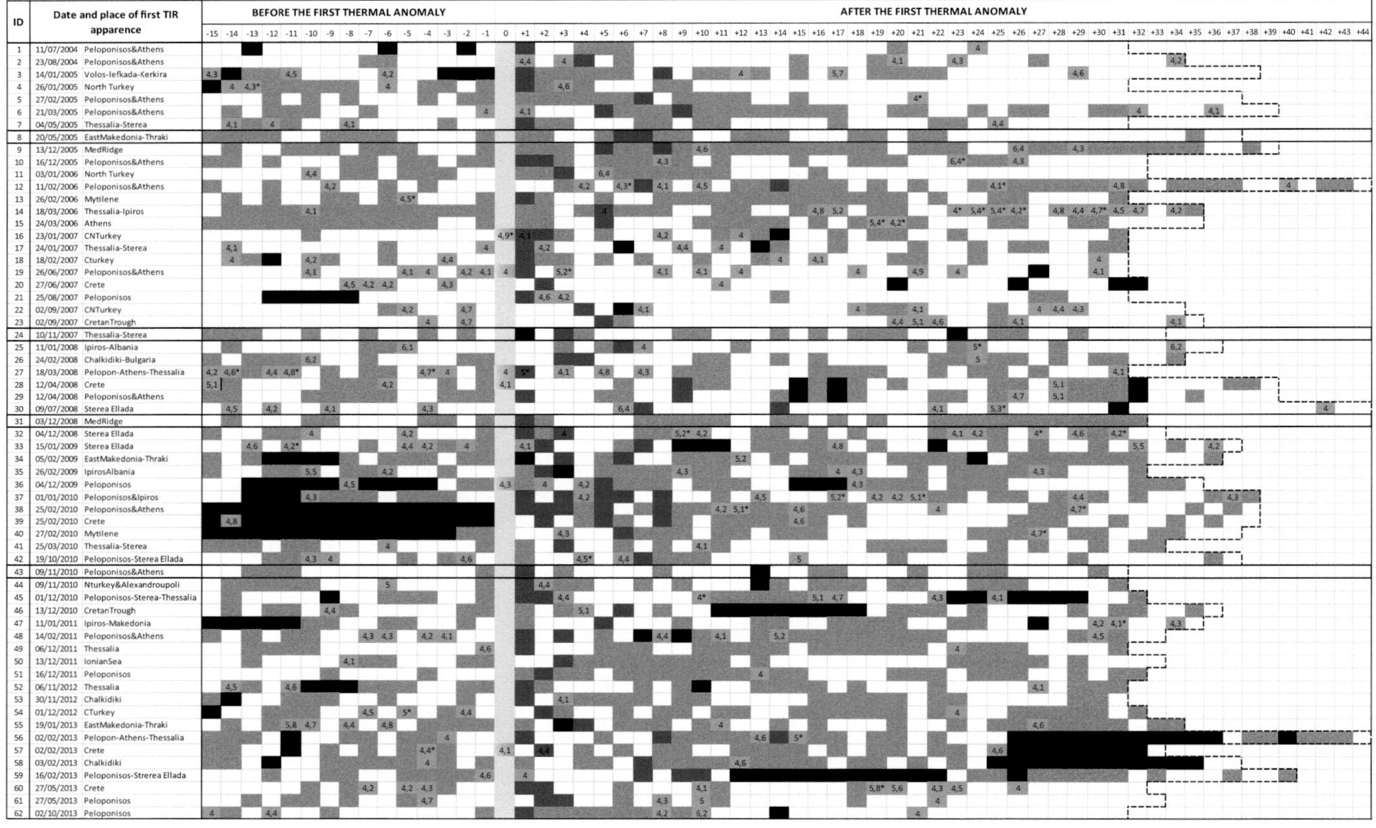

Figure 5

Correlation analysis among Thermal Anomalies and earthquakes with $M \geq 4$ occurred over Greece from May 2004 until December 2013 (see text). All 62 identified SSTAs have been reported one per each row (affected area and actual start date are reported on the *left*). The cells corresponding to the day of the first STA is reported in *yellow* (in correspondence to the day zero) each following persistence is depicted in *red*. *Black* and *gray* cells indicate, respectively, the absence of available satellite data and days with a wide cloud coverage (not usable data) in the investigated area. Cells with numbers indicate days of occurrence, and magnitude, of seismic events (*asterisk* indicates that the maximum magnitude is reported in the case of more than one event per day). For each SSTA the considered period (i.e., 30 days after last STA and 15 days before the first STA) is bounded by *dashed black line*. The 4 SSTAs apparently not associated with earthquakes (rows 8, 24, 31, 43) are bounded by a continuous *black line* reported in Fig. 5, where each row corresponds to one SSTA.

Considering the examples reported in Fig. 4, which are summarized in rows 19 (i.e., Peloponnesus region) and 20 (i.e., Crete island) of Fig. 5, the cells corresponding to the day of appearance of the first STAs (which are part of the same SSTA) are colored in yellow, in that case 26 and 27 June 2007 respectively, and they are located at the day zero on the temporal (horizontal) axis. Days of further appearances of STAs, i.e., 28 June for Peloponnesus region and 29 June for Crete island, in the same area (i.e., persistence) are depicted in red in the same row.

Observational gaps (no data) due to data missing or to overcast conditions are indicated respectively with black and gray cells. Only earthquakes with $M \geq 4$ occurred within the prefixed correlation space/magnitude windows (bordered by colored lines in Fig. 4) are reported with numbers indicating their magnitude inside the corresponding cells (days of occurrence).

The analysis performed by applying previously established correlation rules to all the 62 SSTAs identified on the whole time series of SEVIRI TIR observations in the period May 2004–December 2013 highlighted 58 SSTAs ($\sim$93 % of the 62 previously identified) in apparent space–time relations with earthquake occurrence and only 4 SSTAs ($\sim$7 %) apparently not related to documented seismic activity. Looking at Fig. 6 it is possible moreover to note that:

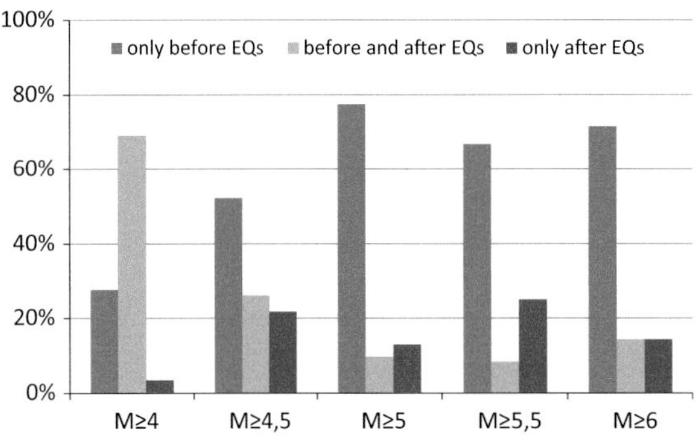

Figure 6
Distribution of the SSTAs with respect to the earthquake occurrence for different class of magnitude. It is possible to note that, for $M \geq 5$, SSTAs mainly appear only before the occurrences of the seismic events

- As foreseen by general (e.g., Scholz *et al.* 1973) and specific (Tramutoli *et al.* 2013a) physical models, SSTAs appear mostly before but also after the occurrence of seismic events, showing however an increasing tendency to appear mostly before (more than 66 % of the total) in the cases of medium–high magnitude earthquakes ($M \geq 5$);
- The presence of meteorological clouds (gray cells in Fig. 6) prevents to guarantee continuity to the observations and to fully appreciate possible space–time persistence of TAs. This is for instance is the case of STAs observed on 2nd and 3rd October 2013 in the Peloponnesus area (sequence 62 in Fig. 5) 10 and 9 days before a large event ($M = 6.2$) occurred in the Peloponnesus–Cretan Ridge on 12 October 2013 (Fig. 7) after 8 overcast days.

3.4. Comparison with a Random Alarm Function: Molchan Diagram

In order to better qualify the possible contribution of the use of SSTAs in the framework of a multi-parametric system for a t-DASH, it is important to verify its actual added value in comparison with a random alarm function (see for instance Zechar and Jordan 2008 and reference herein). It is, in fact, particularly important to understand to which extent the very low rate of false positives observed in the considered case of Greece (2004–2013) is due to the high prognostic capability of SSTAs or to the very high seismicity of the considered region coupled with an eventually too large used correlation space/time window.

To this aim the Molchan approach (Molchan 1990, 1991, 1997; Molchan and Kagan 1992) has been preferred (even if it usually applies in the absence of observational gaps that, because of clouds, we cannot avoid) to likelihood tests (e.g., Zechar and Jordan 2008) which cannot be applied to models with a non-probabilistic alarm function (Shebalin *et al.* 2014). In our case the *Molchan error diagram* was implemented (like in Shebalin *et al.* 2006) by plotting the fraction ν of *missed* earthquakes (i.e., apparently non preceded/followed by SSTAs) against the fraction of *alerted* space–time volume τ:

$$\nu(M) = \frac{\text{Number of EQs with magnitude } \geq M \text{ outside the correlation window (missed)}}{\text{Total number of EQs with magnitude } \geq M \text{ occurred within the whole space } \times \text{ time volume}}, \quad (2)$$

$$\tau(M) = \frac{\text{Alerted space } \times \text{ volume for EQs with magnitude } \geq M}{\text{Whole investigated space} \times \text{time volume}}. \quad (3)$$

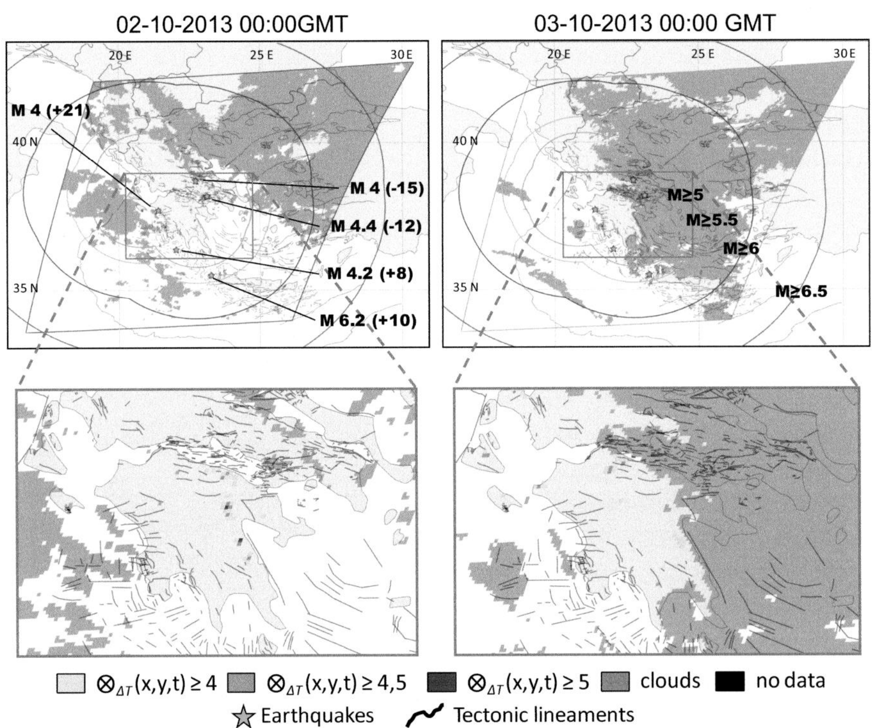

Figure 7
Example of SSTA identified on 2 and 3 October 2013. As in Fig. 3, TIR anomalies [$\otimes_{\Delta T}(r,t) \geq 4$] are depicted in different colors according to RETIRA index values and *the colored contours* correspond to different space/magnitude correlation windows. Significant TIR anomalies appeared in the Peloponnesus area 10 days before a large seismic event ($M = 6.2$) happened on 12 October 2013 and before/after some earthquakes with lesser magnitude. As in Fig. 4 the temporal gap of earthquakes from the first appearance of corresponding SSTAs is indicated by a *number* in *parentheses*

For the computation of $\nu(M)$ all earthquakes (with magnitude $\geq M$) occurred within the investigated area augmented for an area extending up to 1° from its borders, were counted into the denominator of (2). The numerator of (2) reports instead the number of earthquakes (with magnitude $\geq M$) not preceded (followed) by an STA in the previous 30 (following 15) days (*missed*) independently on the actual observation conditions (clouds, data gaps) possibly affecting continuity of observations (see par. 3.3). For this reason computed values of $\nu(M)$ have to be considered as an upper limit due to the used observational technology which is not able to monitor with continuity the occurrence of all possible STAs.

The numerator of (3) has been computed, for each class of earthquake magnitude, as the sum of all

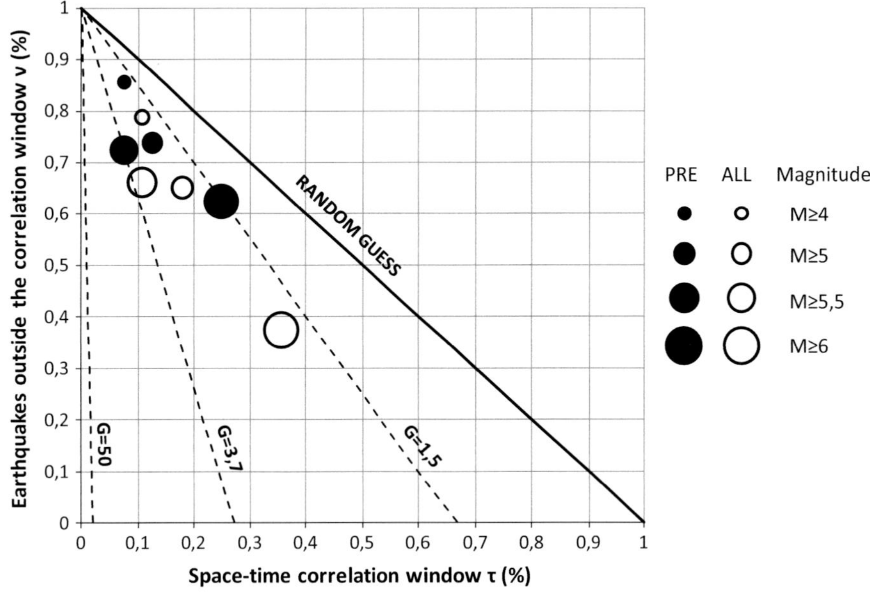

Figure 8
Molchan error diagram analysis computed for different class of magnitude and SSTAs on the whole study period (2004–2013). *Full circles* refer to earthquakes occurred only after the appearances of SSTAs (pre-seismic anomalies). *Empty circles* refer to earthquakes occurred before or after the appearances of SSTAs (earthquake-related anomalies). *Dashed lines* indicate points with equal probability gains *G* (reported as labels) in comparison with random guess results which are represented by the *black continuous line*

space × time volumes alerted by the 62 identified SSTAs:

$$\text{Alerted space} \times \text{time volume} = \sum_{SSTA=1}^{SSTA=62} \text{Alerted Area}_{SSTA} \times (t_f - t_i)_{SSTA},$$

where:

- t_f is the day of last STA appearance plus 30 days,
- t_i is the day of the first STA appearance minus 15 days,
- Alerted Area (for each SSTA) is the convolution of all the potentially affected areas (by past/future earthquakes with magnitude $\geq M$) corresponding to all STAs belonging to the considered SSTA.

For instance, in the case of the SSTA over Crete island shown in paragraph 3.3 (line 20 in Fig. 5) the potentially affected (alerted) space × time volume has been computed by multiplying the total possible affected areas by future/past earthquakes obtained by the convolution of the contours drawn in Fig. 4 for all STAs belonging to that SSTA—for a time period corresponding to the temporal range since 15 days before the first STA appearance (i.e., 27 June 2007) up to 30 days after the last STA appearance (i.e., 29 June 2007) so, in the considered case, the temporal period was 48 days.

To evaluate the alerted space × time volume only before the possible occurrence of an earthquake (prevision mode) the numerator of (3) has been computed as follows:

$$\text{Alerted space} \times \text{time volume} = \sum_{SSTA=1}^{SSTA=62} \text{Alerted Area}_{SSTA} \times (t_f - t_0)_{SSTA},$$

where t_f is the day of last STA appearance plus 30 days and t_0 is the day of first STA appearance.

In both cases, the denominator of (3) is the total space × time volume computed by multiplying the whole area covered by TAMs (contoured in red in Fig. 3) plus 1° in neighboring areas for all the analyzed days (i.e., 3531).

In Fig. 8 are reported the results achieved for different earthquake magnitudes (and spatial correlation window), separately considering the cases of earthquake preceded by SSTAs (full circles) from earthquakes preceded or followed by SSTAs (empty circles). In addition, the probability gain (computed as $G = (1 - \nu)/\tau$, e.g., Aki 1989; Molchan 1991; McGuire *et al.* 2005) has been also computed and plotted.

Notwithstanding previous considerations on the systematic underestimation of ν values (which reflects in a corresponding underestimation of measured gains) from Fig. 8 it is easy to recognize that:

- A non-casual correlation actually exists among observed SSTAs and earthquake ($M \geq 4$) occurrence with a probability gain (compared with a random guess) in between 1, 8 and 3, 2;
- The prognostic value of SSTAs is also non-casual with a probability gain which is in between 1, 5 (for $M \geq 6$ earthquakes) and 3, 7 (for $M \geq 5$ earthquakes) if only pre-seismic SSTAs are considered;
- In both previous cases the maximum gain is achieved when earthquakes with $M \geq 5$ are considered.

4. Conclusions

In this paper the Earth's thermally emitted radiation measured from TIR satellite sensors has been evaluated, on a 10-year long testing period, for its possible relations with earthquake occurrence. 10 years (2004–2013) of daily TIR images acquired by the Spinning Enhanced Visible and Infrared Imager (SEVIRI) on board the Meteosat Second Generation (MSG) satellite, were used to identify TIR fluctuations in a possible space–time relation with $M \geq 4$ earthquakes occurred in Greece in the same period. A refined RST data analysis approach and RETIRA index were used to preliminarily define and then to identify SSTAs. On the base of specific validation rules (in line with the ones used by the Collaboratory for the Study of Earthquake Predictability—CSEP—Project), based on physical models and previous results obtained by applying RST approach to several earthquakes all around the world, the correlation analysis showed that more than 93 % of all identified SSTAs occur in the prefixed space–time window around time and location of occurrence of earthquakes (with $M \geq 4$), showing a prevalence for SSTAs (i.e., ≥66 %) to appear only before the occurrences of earthquakes with $M \geq 5$. The overall false positive rate is <7 %. Even if the presence of meteorological clouds in the TIR scenes do not allow to give continuity to the observations (producing, in this way, a possible overestimation of missed events), Molchan error diagram analysis gave a clear indication of non-casualty of such a correlation. A probability gain (compared with a random guess) from 1.5 up to 3.7 was achieved as far as only SSTAs preceding earthquakes with $M \geq 5.5$ are considered. Such a non-casual correlation together with an extraordinary low rate of false positives, both registered on a so long testing period, seems already sufficient to qualify SSTAs among the parameters to be seriously considered (for now at least for Greece) within a multi-parametric system for time-Dependent Assessment of Seismic Hazard (t-DASH, Tramutoli *et al.* 2014) able to improve short-term (from days to weeks) seismic hazard forecasting.

Acknowledgments

The research leading to these results has received funding from the European Union Seventh Framework Programme (FP7/2007–2013) under Grant agreement no 263502—PRE-EARTHQUAKES project: Processing Russian and European EARTH observations for earthQUAKE precursors Studies. The document reflects only the author's views and the European Union is not liable for any use that may be made of the information contained herein.The research leading to these results has been also supported by the International Space Science Institute in the framework of the Project "Multi-instrument Space-Borne Observations and Validation of the Physical Model of the Lithosphere–Atmosphere–Ionosphere–Magnetosphere Coupling" realized by the International Team 298/2013.The authors wish to thank the Aeronautica Militare Italiana for its support to have access to MSG-SEVIRI data used in this work.The research leading to these results has been supported by INGV-DPC S3 project "Short Term Earthquake prediction and preparation".The research leading to these results has been also supported by FSE Basilicata 2007/2013 in the framework of the Project SESAMO (Sviluppo E Sperimentazione di tecnologie integrate Avanzate per il MOnitoraggio della pericolosità sismica).

References

Aki K (1989) *Ideal probabilistic earthquake prediction.* Tectonophysics *169*:197–198. doi:10.1016/0040-1951(89)90193-5.

Aliano C, Corrado R, Filizzola C, Genzano N, Pergola N, Tramutoli V (2007a) *From GMOSS to GMES: Robust TIR Satellite Techniques for Earthquake active regions monitoring.* Glob Monit Secur Stab - Integr Sci Technol Res Support Secur Asp Eur Union, EU-report EUR 23033 EN 320–330. doi:10.2788/53480.

Aliano C, Corrado R, Filizzola C, Pergola N, Tramutoli V (2007b) *Robust Satellite Techniques (RST) for Seismically Active Areas Monitoring: the Case of 21st May, 2003 Boumerdes/Thenia (Algeria) Earthquake.* 2007 Int. Work. Anal. Multi-temporal Remote Sens. Images. IEEE, pp 1–6.

Aliano C, Corrado R, Filizzola C, Genzano N, Pergola N, Tramutoli V (2008a) *Robust TIR satellite techniques for monitoring earthquake active regions: limits, main achievements and perspectives.* Ann Geophys *51*:303–317.

Aliano C, Corrado R, Filizzola C, Pergola N, Tramutoli V (2008b) *Robust satellite techniques (RST) for the thermal monitoring of earthquake prone areas: the case of Umbria-Marche October, 1997 seismic events.* Ann Geophys *51*:451–459.

Aliano C, Martinelli G, Filizzola C, Pergola N, Genzano N, Tramutoli V (2008c) *Robust Satellite Techniques for monitoring TIR anomalies in seismogenic areas. 2008 Second Work.* Use Remote Sens. Tech. Monit. Volcanoes Seism. Areas. IEEE, pp 1–7.

Aliano C, Corrado R, Filizzola C, Genzano N, Lanorte V, Mazzeo G, Pergola N, Tramutoli V (2009) *Robust Satellite Techniques (RST) for monitoring thermal anomalies in seismically active areas.* 2009 IEEE Int. Geosci. Remote Sens. Symp. IEEE, pp III–65–III–68.

Bonfanti P, Genzano N, Heinicke J, Italiano F, Martinelli G, Pergola N, Telesca L, Tramutoli V (2012) *Evidence of CO_2-gas emission variations in the central Apennines (Italy) during the L'Aquila seismic sequence (March-April 2009).* Boll Di Geofis Teor ed Appl *53*:147–168. doi:10.4430/bgta0043.

Chiarabba C, De Gori P, Chiaraluce L, Bordoni P, Cattaneo M, De Martin M, Frepoli A, Michelini A, Monachesi A, Moretti M, Augliera GP, D'Alema E, Frapiccini M, Gassi A, Marzorati S, Bartolomeo P Di, Gentile S, Govoni A, Lovisa L, Romanelli M, Ferretti G, Pasta M, Spallarossa D, Zunino E (2005) *Mainshocks and aftershocks of the 2002 molise seismic sequence, southern Italy.* J Seismol 9:487–494. doi:10.1007/s10950-005-0633-9.

Cicerone RD, Ebel JE, Britton J (2009) *A systematic compilation of earthquake precursors.* Tectonophysics *476*:371–396. doi:10.1016/j.tecto.2009.06.008.

Cornell CA (1968) *Engineering seismic risk analysis.* Bull Seismol Soc Am *58*:1583–1606.

Corrado R, Caputo R, Filizzola C, Pergola N, Pietrapertosa C, Tramutoli V (2005) *Seismically active area monitoring by robust TIR satellite techniques: a sensitivity analysis on low magnitude earthquakes in Greece and Turkey.* Nat Hazards Earth Syst Sci *5*:101–108.

Cuomo V, Filizzola C, Pergola N, Pietrapertosa C, Tramutoli V (2004) *A self-sufficient approach for GERB cloudy radiance detection.* Atmos Res *72*:39–56.

Di Bello G, Filizzola C, Lacava T, Marchese F, Pergola N, Pietrapertosa C, Piscitelli S, Scaffidi I, Tramutoli V (2004) *Robust Satellite Techniques for Volcanic and Seismic Hazards Monitoring.* Ann Geophys *47*:49–64.

Dobrovolsky IP, Zubkov SI, Miachkin VI (1979) *Estimation of the size of earthquake preparation zones.* Pure Appl Geophys PAGEOPH *117*:1025–1044. doi:10.1007/BF00876083.

Eneva M, Adams D, Wechsler N, Ben-Zion Y, Dor O (2008) *THERMAL PROPERTIES OF FAULTS IN SOUTHERN CALIFORNIA FROM REMOTE SENSING DATA. 71.*

Filizzola C, Pergola N, Pietrapertosa C, Tramutoli V (2004) *Robust satellite techniques for seismically active areas monitoring: a sensitivity analysis on September 7, 1999 Athens's earthquake.* Phys Chem Earth *29*:517–527. doi:10.1016/j.pce.2003.11.019.

Freund FT (2007) *Pre-earthquake signals – Part I: Deviatoric stresses turn rocks into a source of electric currents.* Nat. Hazards Earth Syst. Sci., *7*, 535–541.

Geller RJ (2011) *Shake-up time for Japanese seismology.* Nature *472*:407–409. doi:10.1038/nature10105.

Geller RJ (1997) *Earthquake prediction: a critical review.* Geophys J Int *131*:425–450. doi:10.1111/j.1365-246X.1997.tb06588.x.

Genzano N, Aliano C, Filizzola C, Pergola N, Tramutoli V (2007) *Robust satellite technique for monitoring seismically active areas: The case of Bhuj-Gujarat earthquake.* Tectonophysics *431*:197–210.

Genzano N, Aliano C, Corrado R, Filizzola C, Lisi M, Mazzeo G, Paciello R, Pergola N, Tramutoli V (2009a) *RST analysis of MSG-SEVIRI TIR radiances at the time of the Abruzzo 6 April 2009 earthquake.* Nat Hazards Earth Syst Sci *9*:2073–2084.

Genzano N, Aliano C, Corrado R, Filizzola C, Lisi M, Paciello R, Pergola N, Tsamalashvili T, Tramutoli V (2009b) *Assessing of the robust satellite techniques (RST) in areas with moderate seismicity.* Proc. Multitemp 2009. Mistic, Connecticut, USA, 28-30 July 2009, pp 307–314.

Genzano N, Corrado R, Coviello I, Grimaldi CSL, Filizzola C, Lacava T, Lisi M, Marchese F, Mazzeo G, Paciello R, Pergola N, Tramutoli V (2010) *A multi-sensors analysis of RST-based thermal anomalies in the case of the Abruzzo earthquake.* 2010 IEEE Int. Geosci. Remote Sens. Symp. IEEE, pp 761–764.

Genzano N, Filizzola C, Paciello R, Pergola N, Tramutoli V (2015) *Robust Satellite Techniques (RST) for monitoring Earthquake prone areas by satellite TIR observations: the case of 1999*

Chi-Chi earthquake (Taiwan). J Asian Earth Sci. doi:10.1016/j.jseaes.2015.02.010.

Gorny VI, Salman AG, Tronin AA, Shilin BB (1988) *The Earth outgoing IR radiation as an indicator of seismic activity.* Proceeding Acad. Sci. USSR 301. pp 67–69.

Han P, Hattori K, Hirokawa M, Zhuang J, Chen C-H, Febriani F, Yamaguchi H, Yoshino C, Liu J-Y, Yoshida S (2014) *Statistical analysis of ULF seismomagnetic phenomena at Kakioka, Japan, during 2001-2010.* J Geophys Res Sp Phys *119*:4998–5011. doi:10.1002/2014JA019789.

Huang QH (2011a) *Retrospective investigation of geophysical data possibly associated with the Ms8.0 Wenchuan earthquake in Sichuan, China.* J Asian Earth Sci, *41*:421–427. doi:10.1016/j.jseaes.2010.05.014.

Huang QH (2011b) *Rethinking earthquake-related DC-ULF electromagnetic phenomena: towards a physics-based approach.* Nat Hazards Earth Syst Sci, *11*(11): 2941–2949, doi:10.5194/nhess-11-2941-2011.

Kossobokov V, Nekrasova A (2012) *Global seismic hazard assessment program (GSHAP) maps are Erroneous.* Seismic Instruments *48*(2):162-170.

Li M, Parrot M (2013) *Statistical analysis of an ionospheric parameter as a base for earthquake prediction.* J Geophys Res Sp Phys *118*:3731–3739. doi:10.1002/jgra.50313.

Lisi M, Filizzola C, Genzano N, Grimaldi CSL, Lacava T, Marchese F, Mazzeo G, Pergola N, Tramutoli V (2010) *A study on the Abruzzo 6 April 2009 earthquake by applying the RST approach to 15 years of AVHRR TIR observations.* Nat Hazards Earth Syst Sci *10*:395–406. doi:10.5194/nhess-10-395-2010.

Liu JY, Chen YI, Chuo YJ, Chen CS (2006) *A statistical investigation of preearthquake ionospheric anomaly.* J Geophys Res *111*:A05304. doi:10.1029/2005JA011333.

McGuire JJ, Boettcher MS, Jordan TH (2005) *Erratum: Foreshock sequences and short-term earthquake predictability on East Pacific Rise transform faults.* Nature *435*:528–528. doi:10.1038/nature03621.

Molchan GM (1990) *Strategies in strong earthquake prediction.* Phys Earth Planet Inter *61*:84–98.

Molchan GM (1991) *Structure of optimal strategies in earthquake prediction.* Tectonophysics *193*:267–276. doi:10.1016/0040-1951(91)90336-Q.

Molchan GM (1997) *Earthquake prediction as a decision-making problem.* Pure Appl Geophys PAGEOPH *149*:233–247. doi:10.1007/BF00945169.

Molchan GM, Kagan YY (1992) *Earthquake prediction and its optimization.* J Geophys Res *97*:4823. doi:10.1029/91JB03095.

NOA (2014) *Earthquake Catalogues.* http://www.gein.noa.gr/en/seismicity/earthquake-catalogs.

Ouzounov D, Freund F (2004) *Mid-infrared emission prior to strong earthquakes analyzed by remote sensing data.* Adv Sp Res *33*:268–273. doi:10.1016/S0273-1177(03)00486-1.

Ouzounov D, Liu D, Chunli K, Cervone G, Kafatos M, Taylor P (2007) *Outgoing long wave radiation variability from IR satellite data prior to major earthquakes.* Tectonophysics *431*:211–220. doi:10.1016/j.tecto.2006.05.042.

Ouzounov D, Pulinets S, Davidenko D, Kafatos M, Taylor P (2012) *Space-borne observations of atmospheric pre-earthquake signals in seismically active areas.* Case study for Greece 2008-2009. Proceedings of Aristotle University of Thessaloniki, Greece, 259-265.

Panza G, Kossobokov VG, Peresan A, Nekrasova A (2014) *Earthquake Hazard, Risk and Disasters.* Max Wyss ed., Academic Press. 309–357. doi:10.1016/B978-0-12-394848-9.00012-2.

Pergola N, Aliano C, Coviello I, Filizzola C, Genzano N, Lacava T, Lisi M, Mazzeo G, Tramutoli V (2010) *Using RST approach and EOS-MODIS radiances for monitoring seismically active regions: a study on the 6 April 2009 Abruzzo earthquake.* Nat Hazards Earth Syst Sci *10*:239–249. doi:10.5194/nhess-10-239-2010.

Pietrapertosa C, Pergola N, Lanorte V, Tramutoli V (2001) Self adaptive algorithms for change detection: OCA (the one-channel cloud-detection approach) an adjustable method for cloudy and clear radiances detection. In: Le Marshall, J.F., Jasper JD (Eds.. (ed) Tech. Proc. Elev. Int. (A)TOVS Study Conf. Budapest, Hungary, 20-26 Sept. 2000, Bur. Meteorol. Res. Cent. Melbourne, Australia., pp 281–291.

Pulinets SA, Biagi P, Tramutoli V, Legen'ka AD, Depuev VK (2007) *Irpinia Earthquake 23 November 1980 - Lesson from Nature reviled by joint data an analysis.* Ann Geophys *50*:61–78.

Pulinets SA, Ouzounov D (2011) *Lithosphere–Atmosphere–Ionosphere Coupling (LAIC) model - An unified concept for earthquake precursors validation.* Journal of Asian Earth Sciences *41*, 371-382.

Qiang ZJ, Xu XD, Dian CG (1991) *Thermal infrared anomaly precursor of impending earthquakes.* Chinese Sci Bull *36*:319–323.

Scholz CH, Sykes LR, Aggarwal YP (1973) *Earthquake prediction: a physical basis.* Science *181*(4102):803–810. doi:10.1126/science.181.4102.803.

Shebalin P, Keilis-Borok V, Gabrielov a., Zaliapin I, Turcotte D (2006) *Short-term earthquake prediction by reverse analysis of lithosphere dynamics.* Tectonophysics *413*:63–75. doi:10.1016/j.tecto.2005.10.033.

Shebalin PN, Narteau C, Zechar J, Holschneider M (2014) *Combining earthquake forecasts using differential probability gains.* Earth, Planets Sp *66*:37. doi:10.1186/1880-5981-66-37.

Tramutoli V (1998) *Robust AVHRR Techniques (RAT) for Environmental Monitoring: theory and applications.* In: Zilioli E (ed) Proc. SPIE. pp 101–113.

Tramutoli V (2005) *Robust Satellite Techniques (RST) for natural and environmental hazards monitoring and mitigation: ten year of successful applications.* In: Liang S, Liu J, Li X, Liu R, Schaepman M (eds) 9th Int. Symp. Phys. Meas. Signatures Remote Sensing, IGSNRR, Beijing, China,XXXVI. pp 792–795.

Tramutoli V (2007) *Robust Satellite Techniques (RST) for Natural and Environmental Hazards Monitoring and Mitigation: Theory and Applications.* 2007 Int. Work. Anal. Multi-temporal Remote Sens. Images. IEEE, pp 1–6.

Tramutoli V, Lanorte V, Pergola N, Pietrapertosa C, Ricciardelli E, Romano F (2000). *Self-adaptive algorithms for environmental monitoring by SEVIRI and GERB: a preliminary study.* Proc. EUMETSAT Meteorol. Satell. data User's Conf. Bol. Italy, 29 May - 2 June, 2000. Bologna, Italy, pp 79–87.

Tramutoli V, Di Bello G, Pergola N, Piscitelli S (2001) *Robust satellite techniques for remote sensing of seismically active areas.* Ann di Geofis *44*:295–312.

Tramutoli V, Cuomo V, Filizzola C, Pergola N, Pietrapertosa C (2005) *Assessing the potential of thermal infrared satellite surveys for monitoring seismically active areas: The case of Kocaeli (İzmit) earthquake, August 17, 1999.* Remote Sens Environ *96*:409–426. doi:10.1016/j.rse.2005.04.006.

Tramutoli V, Aliano C, Corrado R, Filizzola C, Genzano N, Lisi M, Lanorte V, Tsamalashvili T (2009) *Abrupt change in greenhouse gases emission rate as a possible genetic model of*

TIR anomalies observed from satellite in Earthquake active regions. Proc. ISRSE 2009. pp 567–570.

Tramutoli V, Inan S, Jakowski N, Pulinets S, Romanov A, Filizzola C, Shagimuratov I, Pergola N, Ouzounov D, Papadopoulos G, Genzano N, Lisi M, Corrado R, Alparslan E, Wilken V, Tsybulia K, Romanov A, Paciello R, Coviello I, Zakharenkova I, Romano G, Cherniak Y (2012a) *The PRE-EARTHQUAKES EU-Fp7 Project: Preliminary Results of the PRIME Experiment for a Dynamic Assessment of Seismic Risk (DASR) by Multiparametric Observations.* 31° Convegno Naz. Grup. Naz. di Geofis. della Terra Solida (GNGTS). 20-22 novembre, Potenza. Potenza, pp 384–388.

Tramutoli V, Inan S, Jakowski N, Pulinets SA, Romanov A, Filizzola C, Shagimuratov I, Pergola N, Genzano N, Serio C, Lisi M, Corrado R, Grimaldi CSL, Faruolo M, Petracca R, Ergintav E, Çakir Z, Alparslan E, Gurol S, Mainul Hoque M, Missling KD, Wilken V, Borries C, Kalilnin Y, Tsybulia K, Ginzburg E, Pokhunkov A, Pustivalova L, Romanov A, Cherny I, Trusov S, Adjalova A, Ermolaev D, Bobrovsky S, Paciello R, Coviello I, Falconieri A, Zakharenkova I, Cherniak Y, Radievsky A, Lapenna V, Balasco M, Piscitelli S, Lacava T, Mazzeo G (2012b) *PRE-EARTHQUAKES, an Fp7 Project for Integrating Observations and Knowledge on Earthquake Precursors: Preliminary Results and Strategy.* 2012 IEEE Int. Geosci. Remote Sens. Symp. IEEE, Munich, pp 3536–3539.

Tramutoli V, Aliano C, Corrado R, Filizzola C, Genzano N, Lisi M, Martinelli G, Pergola N (2013a) *On the possible origin of thermal infrared radiation (TIR) anomalies in earthquake-prone areas observed using robust satellite techniques (RST).* Chem Geol *339*:157–168. doi:10.1016/j.chemgeo.2012.10.042.

Tramutoli V, Corrado R, Filizzola C, Genzano N, Lisi M, Paciello R, Pergola N, Sileo G (2013b) *A decade of RST applications to seismically active areas monitoring by TIR satellite observations.* 2013 EUMETSAT Meteorol. Satell. Conf. &19th Am. Meteorol. Soc. Satell. Meteorol. Oceanogr. Climatol. Conf. p 8 pp.

Tramutoli V, Jakowski N, Pulinets S, Romanov A, Filizzola C, Shagimuratov I, Pergola N, Ouzounov D, Papadopulos G, Genzano N, Lisi M, Alparslan E, Wilken V, Romanov A, Zakharenkova I, Paciello R, Coviello I, Romano G, Tsybulia K, Inan S, Parrot M (2014) *From PRE-EARTQUAKES to EQUOS: how to exploit multi-parametric observations within a novel system for time-dependent assessment of seismic hazard (T-DASH) in a pre-operational Civil Protection context.* Prooceding of Second European Conference on Earthquake Engineering and Seismology (2ECEES) Turkey 24-29 August, 2014.

Tramutoli V, Corrado R, Filizzola C, Genzano N, Lisi M, Pergola N (2015a) *From visual comparison to Robust Satellite Techniques: 30 years of thermal infrared satellite data analyses for the study of earthquakes preparation phases.* Boll Di Geofis Teor ed Appl. doi:10.4430/bgta0149.

Tramutoli V, Corrado R, Filizzola C, Genzano N, Lisi M, Paciello R, Pergola N (2015b) *One year of RST based satellite thermal monitoring over two Italian seismic areas.* Boll Di Geofis Teor ed Appl. doi:10.4430/bgta0150.

Tronin AA (1996) *Satellite thermal survey—a new tool for the study of seismoactive regions.* Int J Remote Sens *17*:1439–1455. doi:10.1080/01431169608948716.

Tronin AA (2006) *Remote sensing and earthquakes: A review.* Phys Chem Earth *31*:138–142. doi:10.1016/j.pce.2006.02.024.

Tsai Y-B, Liu J-Y, Ma K-F, Yen H-Y, Chen K-S, Chen Y-I, Lee C-P (2006) *Precursory phenomena associated with the 1999 Chi-Chi earthquake in Taiwan as identified under the iSTEP program.* Phys Chem Earth, Parts A/B/C *31*:365–377. doi:10.1016/j.pce.2006.02.035.

USGS (2014) *Earthquake Archive Search & URL Builder.* http://earthquake.usgs.gov/earthquakes/search/.

Wang L, Zhu C (1984) *Anomalous variations of ground temperature before the Tangshan and Haicheng earthquakes.* J Seismol Res *7*:649–656.

Wyss M, Nekrasova A, Kossobokov V (2012) *Errors in expected human losses due to incorrect seismic hazard estimates.* Nat Hazards *62*:927–935. doi:10.1007/s11069-012-0125-5.

Xu GJ, Han P, Huang QH, Hattori K, Febriani F, Yamaguchi H (2013) *Anomalous behaviors of geomagnetic diurnal variations prior to the 2011 off the Pacific coast of Tohoku earthquake (Mw9.0).* J Asian Earth Sci, *77*: 59-65. doi:10.1016/j.jseaes.2013.08.011.

Zechar JD, Jordan TH (2008) *Testing alarm-based earthquake predictions.* Geophys J Int *172*:715–724. doi:10.1111/j.1365-246X.2007.03676.x.

Zuccolo E, Vaccari F, Peresan A, Panza GF (2011) *Neo-Deterministic and Probabilistic Seismic Hazard Assessments: a Comparison over the Italian Territory.* Pure Appl Geophys *168*:69–83. doi:10.1007/s00024-010-0151-8.

(Received January 23, 2015, revised May 4, 2015, accepted June 1, 2015, Published online July 7, 2015)

Pure Appl. Geophys. 173 (2016), 305–319

DOI 10.1007/s00024-015-1114-x

Pure and Applied Geophysics

Statistical Evaluation of Efficiency and Possibility of Earthquake Predictions with Gravity Field Variation and its Analytic Signal in Western China

SHI CHEN,[1] CHANGSHENG JIANG,[1] and JIANCANG ZHUANG[2]

Abstract—This paper aimed at assessing gravity variations as precursors for earthquake prediction in the Tibet (Xizang)-Qinghai-Xinjiang-Sichuan Region, western China. We here take a statistical approach to evaluate efficiency and possibility of earthquake prediction. We used the most recent spatiotemporal gravity field variation datasets of 2002–2008 for the region that were provided by the Crustal Movement Observation Network of China (CMONC). The datasets were space sparse and time discrete. In 2007–2010, 13 earthquakes (>M_s 6.0) occurred in the region. The observed gravity variations have a statistical correlation with the occurrence of these earthquakes through the Molchan error diagram tests that lead to alarms over a good fraction of space–time. The results show that the prediction efficiency of amplitude of analytic signal of gravity variations is better than seismicity rate model and THD and absolute value of gravity variation, implying that gravity variations before earthquake may include precursory information of future large earthquakes.

Key words: Gravity field variation, earthquake prediction, Molchan error diagram, repeated gravity measurement, Chinese mainland.

1. Introduction

During the past three decades, regional gravity variations related to earthquakes have been widely observed. Attempts for searching gravity precursors for empirical earthquake prediction based on regional gravity variations have had encouraging results (CHEN *et al.* 1979; GU *et al.* 1997; KUO and SUN 1993; LIU *et al.* 2002; ZHU *et al.* 2010, 2012, 2013). At present, the gravity variation data comes from the geophysical field observation and has been widely used to analyze the earthquake risk and the earthquake predictions in China. The findings of this annual consultations are published in a report called "The Annual Report of Earthquake Tendency" (China Earthquake Administration 2007, 2008, 2009, 2010). However, the analysis based on the gravity variation is a more experimental option in practice. Though the physical mechanism of earthquake precursor is not fully understood. How to use a sort of the widely accepted method to evaluate the efficiency and possibility of earthquake predictions with gravity field variation is still an important question.

Earthquakes generally reflect active tectonic activities that are associated with the release of large amount of destructive energy depending on the magnitude of an earthquake. Based on the global map of seismicity, we can easily trace the major plate boundaries through the distribution of earthquake epicenters. At present, earthquake forecast is still a major unsolved science problem. However, with the advent of the global seismic monitoring network and advanced computing technology, the mechanics of earthquakes has been fruitfully studied. We know that about 85 % of the global seismic moment release occurs in the subduction zones. More than 95 % is produced by plate boundary earthquakes (SCHOLZ 2002). The potential occurrence for earthquakes along the plate boundaries has been mapped with reasonable success (ENGLAND and JACKSON 2011). The nature of earthquakes that occurred in the continental interiors is less known and still remains a challenge for modern earthquake seismology (LIU *et al.* 2011).

The occurrence probability of intraplate earthquakes in China is high due to the seismic forces caused tectonically by the continent–continent collision of the Indian Plate against the Eurasian Plate and further caused by the subduction of the West Pacific

[1] Institute of Geophysics, China Earthquake Administration, Beijing 100081, China. E-mail: chenshi80@gmail.com
[2] The Institute of Statistical Mathematics, Tokyo 106-8569, Japan.

Plate associated with the lateral compression by Philippine Plate. The tectonic movement is accompanied by crustal deformation and the activity of complex fault systems. This study aimed to assess whether gravity variations can be used as precursors for earthquake prediction in the Tibet (Xizang)-Qinghai-Xinjiang-Sichuan Region, western China. The tectonic setting and fault systems of the study region are complex, as shown in Fig. 1. The Altyn Tagh Fault is in the northwest, which is the tectonic boundary between Tibean Pleateau Block (TPB) and West China Block (WCB), the Karakoram is in the southwest, the Haiyuan fault and the Kunlun fault are in the north, and, in the southeastern part, the Xianshuihe fault zone intersects the southwest end of the Longmen Shan fault to become the Xiaojiang fault. Likewise, the Jiali fault extends southward as the Red River fault. There are three subblocks, namely Lhasa, Qiangtan, and Songpan Ganzi and three sutures, Indus-Zangbo, Bangong Nujiang, and Jinsha.

In the past decades, the Tibetan Plateau, particularly the Longmen Shan fault and Xianshuihe fault were seismically most active in China. We believed that the geophysical field signals can help us to predict earthquakes and would shed new light on the physical mechanism of the seismogenesis, seismo-processes, and earthquake occurrence.

Moreover, with the continual improvement of modern field gravimeter and introduction of absolute gravity observation, high-precision gravity variation dataset has been obtained using a ground-based survey. Gravity variation data can provide information of subsurface deformation as precursors prior to the occurrence of an earthquake. The concept of gravity precursor as a kind of non-catalog-based probabilistic forecasting approach of earthquakes is promising and has gained its entry. Motivated by the Molchan error diagram method (Molchan 1990, 1991, 1997), which has been recommended by the Collaboratory for the Study of Earthquake Predictability (CSEP)

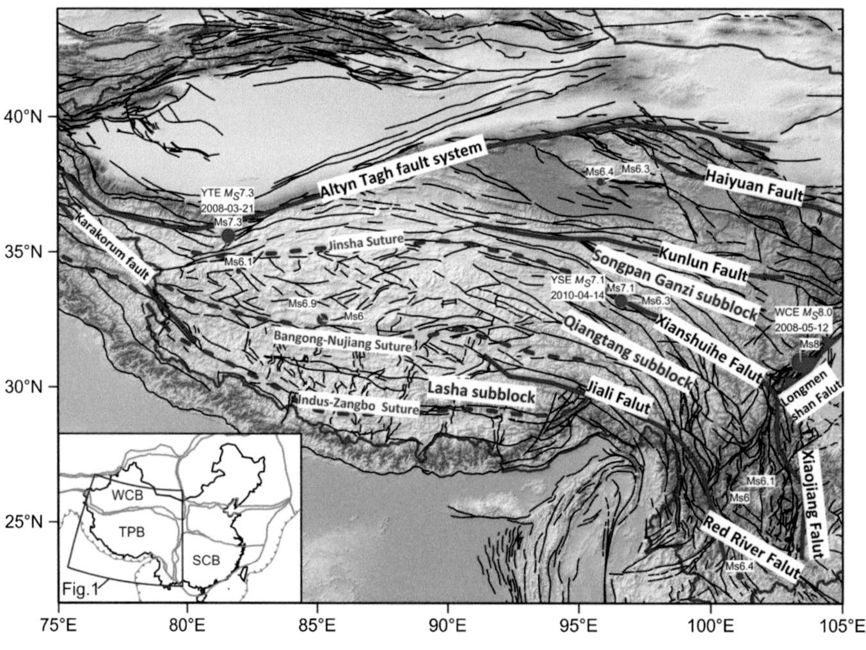

Figure 1

Earthquake locations, active faults and topographic relief in western China. The labels in the figure mean: *WCB* West China block, *TPB* Tibetan Plateau block, *SCB* South China block, *YTE* Yutian Earthquake, *YSE* Yushu Earthquake, *WCE* Wenchuan Earthquake. The boundary of active blocks is shown as *orange line*, active faults as *black lines*, and the national boundary of China as *blue line*. The major faults are also shown as *solid red lines*, and the sutures as *dashed red lines*. The *red circles* mark the epicenters of M 6.0 + earthquakes occurring during 2007 to 2010, with their sizes proportional to their magnitudes. The names of the subblocks of the Tibet (Xizhang)-Qinghai-Xinjiang-Sichuan region are labeled by *blue text*

(Schorlemmer and Gerstenberger 2007; Schorlemmer *et al.* 2010), we aimed to test and evaluate the performance of the gravity precursor in western China.

In this paper, we revisit the gravity variations of the ground-based mobile measurement and take a statistical approach to evaluate the efficiency of earthquake prediction in the Tibet-Qinghai-Xinjiang-Sichuan Region that tectonically includes the Tibetan Plateau Block (TPB) and West China Block (WCB). The gravity field data are observed at the space-fixed stations on 2002, 2005, 2008, respectively. The field work was supported by the project of Crustal Movement Observation Network of China (CMONC). The gravity variations can be easily computed with a 3-year interval in between. The 13 earthquakes of $M > 6.0$ preselected that occurred from 2007 through 2010 in the study region as listed in Table 1. These earthquakes can be divided two sets(No. 1-6 and No. 7-13). The gravity variations before earthquakes as the precursors will be used to evaluate the efficiency of earthquake prediction.

2. Gravity Data and Earthquake Catalogue

There are 13 earthquakes of magnitudes $\geq$6.0, depth $\leq$50 km, occurring in the region of latitude 22°N–44°N, and longitude 75°E–105°E, from 1 January 2005 to 31 December 2010, in the CENC (China Earthquake Network Center) catalog, listed in Table 1. Figure 1 shows the locations of these earthquakes. In the past decade, the overwhelming majority of earthquakes (magnitudes $\geq$6.0) occurred in the Region.

In 1998, Chinese Earthquake Administration (CEA) and other institutions launched Project Crustal Movement Observation Network of China (CMONC) to monitor the dynamics of the crust including mobile gravity surveys of temporal-spatial gravity variations on a nationwide scale that provided the datasets used in this study. These mobile gravity measurements were made along the major roads in the Region and repeated every 2 or 3 years. In addition there were absolute gravity observations as the controls for the mobile gravity observation at the pivotally preinstalled stable stations as shown in Fig. 2. The absolute gravity instrument used was the FG-5 gravimeter. The measured precision of absolute gravity is better than 5×10^{-8} m/s^2 (Zhu *et al.* 2012). The gravimeters for mobile gravity survey used are the LCR-G gravimeters with 3–4 gravimeters read in tandem at each location. The repeated measured time frame has been scheduled on the same month interval to minimize the influence of seasonal and hydrological variations. Before each mobile field survey, all the gravimeters were calibrated for ensuring the reliability each instrument. The

Table 1

Earthquakes (>Ms 6.0) occurring in western China from 2007 to 2010

No.	Date	Time	Location		M_s	GV/μGal	AAS of GV	THD of GV	Location name
			Long.	Lat.					
1	2007-05-05	16:51:42	82.00	34.30	6.1	−7.64	9.81	9.09	Gaizhe, Xizang[a]
2	2007-06-03	05:34:56	101.10	23.00	6.4	9.47	16.75	8.42	Puer, Yunnan
3	2008-01-09	16:26:47	85.20	32.50	6.9	45.19	50.25	33.02	Gaizhe, Xizang
4	2008-01-16	19:54:44	85.20	32.45	6.0	45.82	51.36	32.69	Gaizhe, Xizang
5	2008-03-21	06:33:03	81.60	35.60	7.3	−24.18	27.06	21.51	Yutian, Xinjiang[a]
6	2008-05-12	14:28:04	103.40	30.95	8.0	−0.11	43.75	24.81	Wenchuan, Sichuan
7	2008-08-30	16:30:50	101.90	26.20	6.1	−47.66	45.85	15.88	Panzhihua, Sichuan
8	2008-10-06	16:30:46	90.30	29.80	6.6	6.86	52.47	16.11	Dangxiong, Xizang
9	2008-11-10	09:22:05	95.90	37.60	6.3	11.79	38.42	9.60	Haixi, Qinghai
10	2009-07-09	19:19:13	101.10	25.60	6.0	−38.23	39.68	25.23	Yaoan, Yunnan
11	2009-08-28	09:52:06	95.80	37.60	6.4	14.48	36.00	9.28	Haixi, Qinghai
12	2010-04-14	07:49:38	96.60	33.20	7.1	99.53	59.85	20.36	Yushu, Qianghai
13	2010-04-14	09:25:18	96.60	33.20	6.3	99.53	59.85	20.36	Yushu, Qianghai

GV gravity variation

[a] The epicenters of No. 1 and 5 earthquakes were not well covered by gravity survey lines. The gravity variation and related derivative values were computed by interpolated method

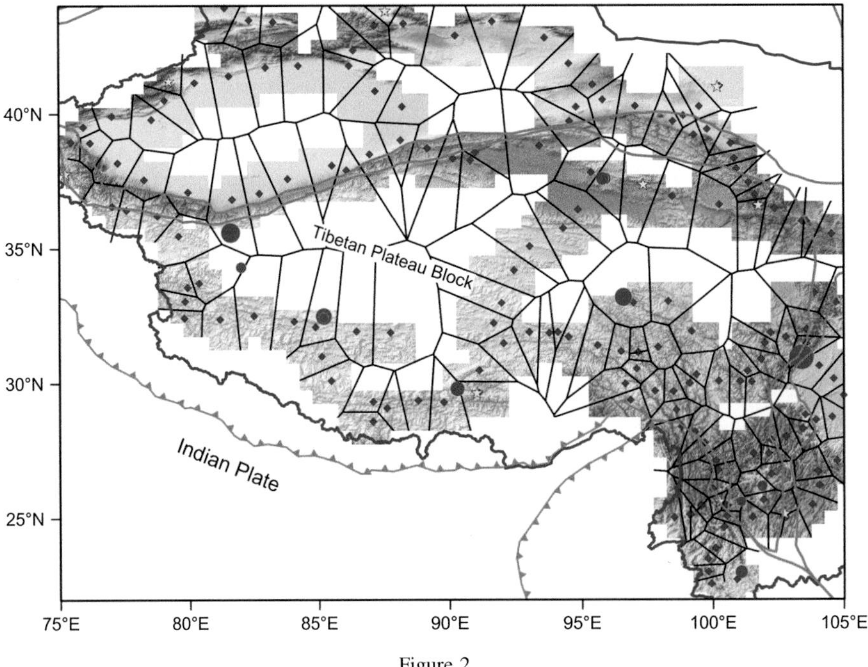

Figure 2
Spatial descretization using the Voronoi tessellation. The *blue squares* represent the gravity observation locations using the relative measurements and the absolute gravity stations are marked by a *yellow star*. The *red circles* show the locations of earthquakes. The blanked region without the topographic relief has no effective gravity variation data. The total number of *blue squares* is 176. The data from the 9 absolute gravity stations are used for adjustment of the relative measurements

precision of average gravity measured values was estimated to be better than 15×10^{-8} m/s^2 (Zhu *et al.* 2012). In this paper, we used the data product from the project of CMONC. However, we blanked out the regions which far away from the gravity survey lines. The well-conditioned regions covered by gravity measurement were shown in Fig. 2.

Figure 2 shows the 167 relative gravity observation locations and 9 absolute gravity stations in the TPB and WCB. Limited by topographic relief, the observation locations are not homogeneously distributed in space. We use the Voronoi tessellation (Aurenhammer 1991) to mesh the monitoring coverage of the gravity survey and define the monitoring ability in a Voronoi cell as proportional to the reciprocal of its area. That is, the larger area of Voronoi cell, the lower monitoring ability. From the definition of the Voronoi tessellation, if an earthquake occurs in a Voronoi cell, the nearest gravity observation location is also located in this cell. We blank the areas which located at the outside of the gravity survey network or where the value of gravity changes can only be acquired by extrapolation. For the interior zone, we blank the areas which far from the survey lines. We choose the 1 degree distance which is far away from the nearest measured station as the blanking rule. In this case, we only consider the areas where the effective gravity change can be observed or acquired by interpolations, as shown with irregular boundaries in Fig. 2, in the evaluation of the precursory significance of the gravity anomalies.

Figure 2 indicates that the monitoring ability of the southern part of North–South seismic belt is better than the interior of the Tibetan plateau as the mobile gravity stations were nearly located at the peripheral of the TPB except one profile across the center of the TPB. There exists mobile gravity points with the epicentral distance less than 50 km for three earthquakes ($M_s > 7.0$) occurring from 2008 to 2010 including the Yutian earthquake, the Wenchuan earthquake and the Yushu earthquake. Therefore, we used three times more gravity data with repeated observation in this region to study the pattern of gravity variation in the context of seismogenesis.

Data reduction and contouring of space sparse and time discrete data in the present case in the Tibet-

Qinghai-Xinjiang-Sichuan Region are very crucial. The threshold precision of each gravity measured values is about 15 to 20 $\times$ 10^{-8} m/s^2. The regional gravity variations as observed in the Tibet-Qinghai-Xinjiang-Sichuan Region ranged from $>+100$ to $<-60 \times 10^{-8}$ m/s^2 that were far above these precision threshold values and believed to be real and significant.

Figure 3 shows two images of gravity field variations from 2002 to 2005 and 2005–2008 in the region, based on gravity variation differencing the repeated observations at each measured point. Since the locations of gravity stations for repeated survey are not uniformly distributed in the study region due to the limitation of natural conditions, we employ a gridding method. In this study, we choose the improved minimum curvature algorithm by Smith *et al.* (1990) to solve the instability problem in the sparse region. This gridding technique is a suitable solution for gridding the potential field data, which solves the modified biharmonic differential equation with tension and generates the target function surface through interpolating the available data (Smith and Wessel 1990). We select grids of 0.5 degree by 0.5 degree on the basis of the average distance of gravity stations.

Then the spatial gravity variation signal with uniform space can be determined. We used the gravity data which are observed repeatedly with an interval of 3 years to determine the gravity variation. The gravity observation years of CMONC are 2002, 2005, and 2008. The earthquakes that occurred during these 3 years after the gravity measurement were also plotted in Fig. 3.

Zhu *et al.* (2012) showed yearly regional gravity variation of tens of microgals (1μGal = 1 $\times$ 10^{-8} m/s^2) and pointed out the gravity variation gradient zones as a precursor for future earthquakes. The zones of high gradient of gravity variation can also be regarded as of high seismic risks. The volume of a high-gradient zone of gravity variation is positively correlated to the magnitude of the future earthquake. The physical mechanism of such gravity field changes can be explained as caused by the underlying mass transportation and the deformation of interfaces and interior of the crust (Chen *et al.* 1979). For example, the Modified Combined Dilatancy Model (MCDM) (Kuo and Sun 1993) is a five-stage hypothesis including the whole possible physical processes during the preseismic, coseismic, and postseismic stages. However, statistical evaluations should be carried out to test whether the gravity change is a kind of precursor and whether it indicates where a large earthquake is likely to occur in next few years.

It can be seen from Fig. 3 that there are no occurrences of earthquakes in the maximum-value or minimum-value zones of gravity variation. However, the gravity variation belts of transition for the gravity variation more likely correspond to a large earthquake. Kuo *et al.* (1999) proposed a hypothesis named the epicentroid model (concept likened to hypocentroid) to explain this phenomena of gravity precursor.

It is difficult to rigorously formulate the quantitative relationship between the magnitudes of the earthquake and the changes value of regional gravity field because the pattern of crustal deformation and the tectonic movement varies from region to region. Zhu *et al.* (2012) summarized the form and magnitude of gravity variations associated with several M_s 6.8+ earthquakes on the continent of China. It has now been confirmed that there exists a positive correlation between the earthquake magnitude and the change value. They also indicated that there is more than 100 $\times$ 10^{-8} m^2/s^2 gravity variations from 1998 to 2005 before the Wenchuan earthquake that occurred in northwest of the Sichuan Basin (Zhu *et al.* 2012). The related high-gradient belt of gravity change is located along Luzhou-Wenchuan-Maerkang, Sichuan. The formation of gravity variation is due to the enhanced movement of the Bayan Kala subblock obstructed by the rigid Sichuan basin.

Based on the above opinions, we mainly focus on the issues of how to quantify gravity variation gradient at each location and of how to test or evaluate the significance of predications made by based on the gravity anomalies.

3. Method

In this section, we introduce the Molchan error diagram (Molchan 2010) (MED) method, which can be used to validate whether there is the precursory

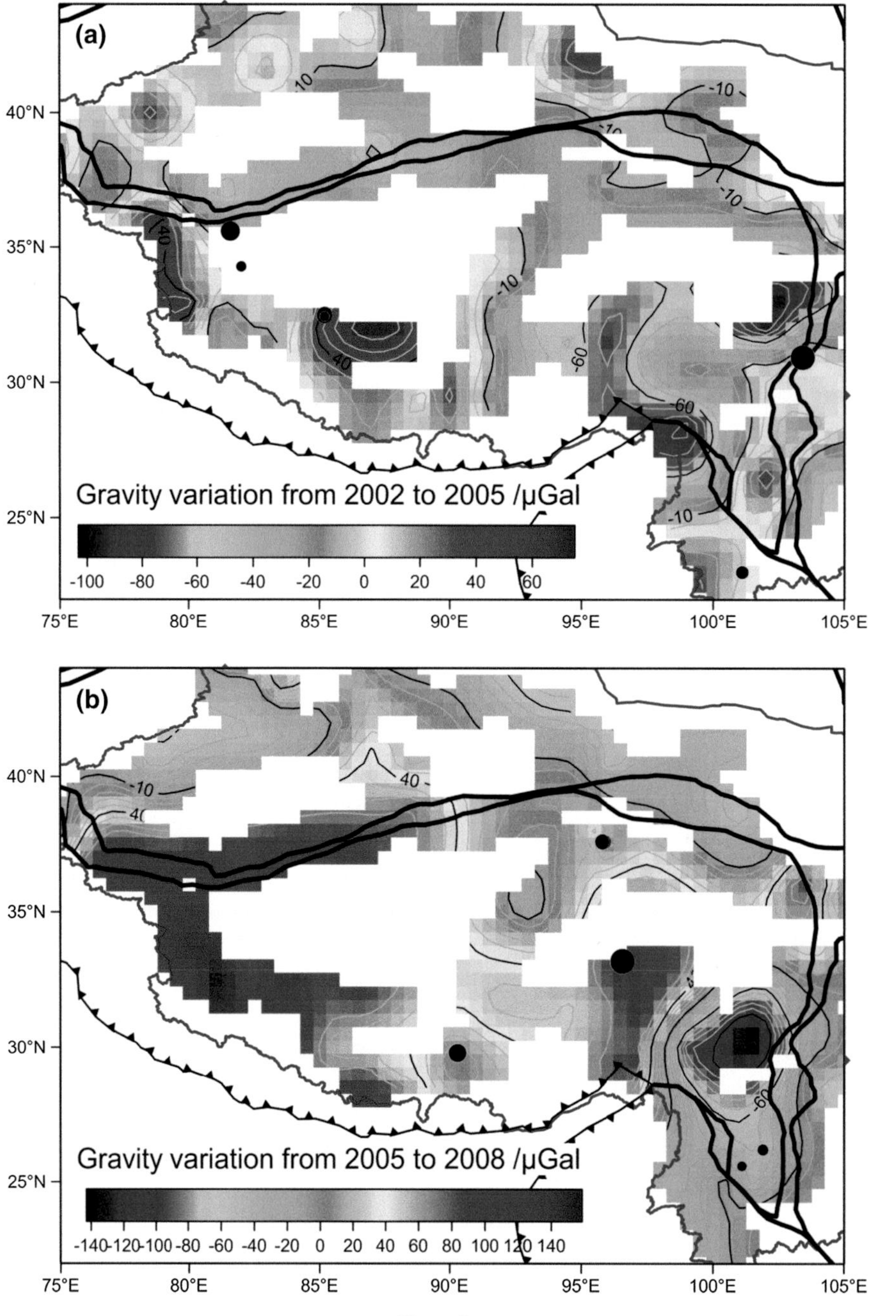

Figure 3
The gravity variations from 2002 to 2008. The *black circles* show the locations of earthquakes. The *small black squares* show the locations of gravity stations. The *colored arrows* show the gravity variation in each period, with *blue color* meaning decrease and *yellow color* meaning increase. The scale of each *arrow* is proportionate to the magnitude of gravity variation at each station along with each survey lines. The gravity variation outside of the effective region has been *blanked*. The *black solid lines* show the boundary of active blocks

information in anomalous signals, by showing how significantly a prediction algorithm is better than the Poisson model (the complete randomness) and the non-homogeneous Poisson model.

3.1. Specification of Alarm Level Function

We use the absolute value of the gravity variation and its gradient as candidates of earthquake precursors. First, we measure the relative gravity repeatedly at each location at different times; after survey adjustment data process, we get the high-accuracy gravity value at each station, relative to the absolute gravity base station. Then, the gravity variation between times t_1 and t_2, is computed using

$$g_v = g_{t_1} - g_{t_2}, \quad (1)$$

where g_{t_1} and $g_{t'_2}$ are the gravity observed values at the same location at t_1 and t_2, respectively. At last, the uniform grid with each same size cell will acquire a gravity variation signal through the gridding algorithm.

Another concept is the gradient of gravity variation. There is no unique definition approach to determine the magnitude of gravity variation gradient at each grid cell.

In our study, we first choose the total horizontal derivative (THD) of gravity data as the metric of the gradient of the gravity change for each cell. This method computed the rate of change of potential field in x and y directions and reconstructed a resultant grid (Cordell and Grauch 1985) in the following form:

$$\mathrm{THD}(x, y) = \sqrt{\left(\frac{\partial g}{\partial x}\right)^2 + \left(\frac{\partial g}{\partial y}\right)^2} \quad (2)$$

Second, we use the amplitude of analytic signal (AAS) as the metric of the gradient of the gravity change for each cell. The AAS is usually used to but not limited to identify the boundary or the margin of causative source bodies in the 3-D space. For example, the analytic signal techniques were applied to interpreted the 2D magnetic by Nabighian (1972, 1974). The amplitude of the analytic signal of the vertical gradient of gravity was defined by Nabighian (1984) in the following form:

$$\mathrm{AAS}(x, y) = \sqrt{\left(\frac{\partial g}{\partial x}\right)^2 + \left(\frac{\partial g}{\partial y}\right)^2 + \left(\frac{\partial g}{\partial z}\right)^2}, \quad (3)$$

where g is the gravitational acceleration. The Laplace's equation is satisfied by the gravitational potential. Generally, the analytic signal and some derived methods can be used to determine the positions of the causative bodies. (Nabighian and Hansen 2001). The amplitude of analytic signal can usually be used to identify the boundary or circumference margin of causative bodies of the 3-D source. The enhancement of gravity anomaly has effectively corresponded to the location of gradient belt by the analytic signal process.

Here, we here also use the absolute value of gravity variation, THD, and AAS as indexing functions (alarm levels) for predicting earthquakes. The higher the alarm level, the higher the earthquakes risk. The comparison results of gravity variation and its THD and AAS will be discussed in 4th section.

3.2. The Molchan Error Diagram

Given the index function, the earthquake alarms can be issued as follows: we first specify a threshold, then issue "Yes" prediction on a cell if the value of the index function is higher than the threshold and, otherwise, issue "No earthquake" prediction on that cell.

The performance of this strategy can be evaluated using the Molchan Error Diagram (MED), which provides a comprehensive summary on the performance of such alarms. Figure 4 illustrates the principle of the Molchan error diagram. We defined the τ as fraction of space–time occupied by alarm, and υ as fraction of earthquakes missed, respectively. If the threshold is changed from the highest to the

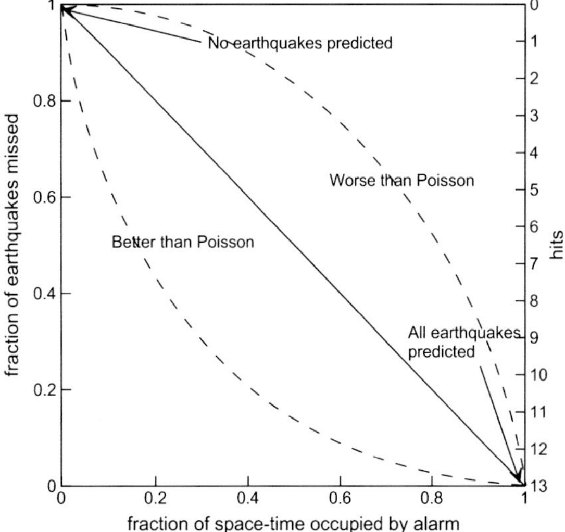

Figure 4
The demonstration of Molchan error diagram method

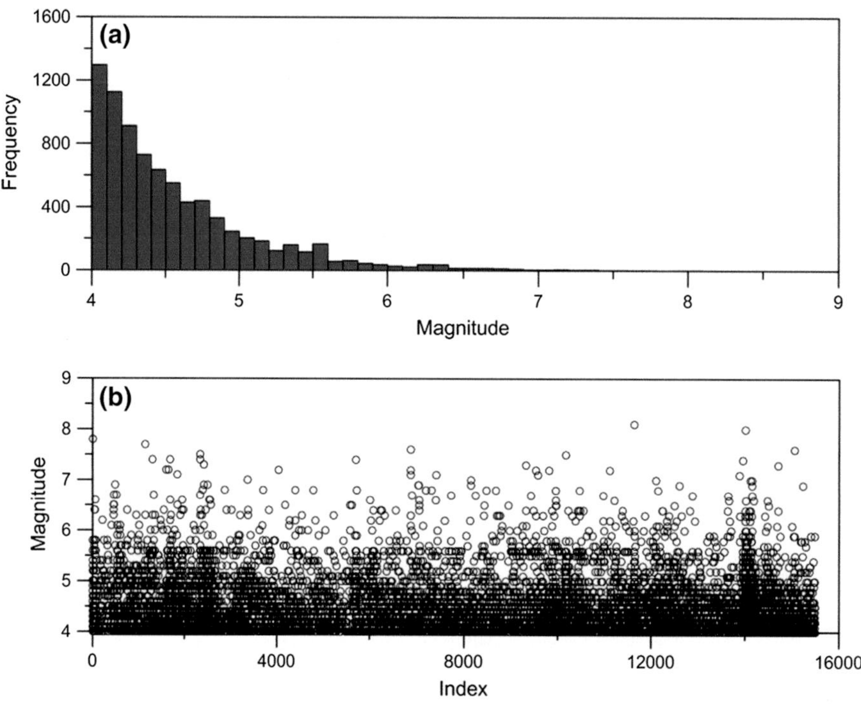

Figure 5
The earthquake catalog which used to obtain seismicity rate started from 1 Jan 1970. **a** Shows the histogram of the magnitudes for events greater than 4.0; **b** shows the magnitudes again sequence numbers

lowest value of the indexing function, τ changes from 0 to 1 and v changes from 1 to 0. Figure 4 gives the curve of τ and υ when the threshold of the indexing function is changed for issuing alarm. In the case of complete random guess, as it is easy to see, $\upsilon = 1-\tau$, the diagram is a straight line joining the points (0, 1) and (1, 0). A better alarm strategy than a completely random guess should have a $\upsilon-\tau$ curve below this straight line while a worse alarm strategy has a curve above it.

3.3. Testing Against Non-Homogeneous Seismicity Rate Model

It can be seen that, in the above evaluation scheme of the Molchan error diagram, the baseline model is complete randomness (the homogeneous Poisson model), which assumes that the spatial locations of earthquake are uniformly distributed in the study region, i.e., each cell has the same probability to have an earthquake occurring inside. However, the seismicity rate in the study research may differ from place to place. To correct this aspect, we weight each cell, say i, by the seismicity rate λ_i in it.

The seismicity rate in each cell can be estimated in the following way: We select the event of magnitude $>M_s$ 4.0 from Jan 1, 1970 to beginning of forecasting time interval. Figure 5 shows the epicenter locations of all selected events used in this estimation. We then fit the model to the proximity to past earthquake (PPE) model (Rhoades and Evison 2004), which has a seismicity rate of

$$\lambda(t,x,y) = \frac{1}{t-t_0} \sum_{i:t_i<t} \left(\frac{a}{d^2 + (x-x_i)^2 + (y-y_i)^2} + s \right) \tag{4}$$

where the point (x_i, y_i) is the epicenter of the i th earthquake; a is a normalization constant; d is a smoothing distance; and s is a small constant to allow for earthquake far from past earthquakes. The parameters (a, d, S) can be computed by fitting the model to the past earthquake catalog. In this work, the learning period is from Jan/01/1970 to Jan/01/1985. Earthquake events (M 4.0+) are used for the calculated conditional intensity, but the likelihood are calculated only for magnitude M 6.0+ events. The model parameters will be estimated by means of maximizing the likelihood function. The model is fitted to the M 6.0+ earthquake in the period from Jan/01/1985 to the beginning of the forecasting period. Figure 6 shows results of the fitted seismicity rate for the two different periods, Jul/1/2005–Jun/30/2008, Jul/01/2008–Jun/30/2011. The major difference between Fig. 6a and b is due to the contribution of the Yutian M_s 7.3 and Wenchuan M_s 8.0 earthquakes, which are listed in Table 1. The seismicity rate is obviously enhanced at the surrounding of the epicenter in Fig. 6b. Finally, we employ the total seismicity rate in each cell before the predicted period as the weight. The fraction of space–time occupied by alarm (τ) of the Poisson and non-homogeneous Poisson models was defined in the form of, respectively,

$$\tau = \frac{\sum_{i=1,n} I(S(i) > slevel) A(i)}{\sum_{i=1,n} A(i)} \quad (5)$$

$$\tau_n = \frac{\sum_{i=1,n} I(S(i) > slevel) \lambda(i) A(i)}{\sum_{i=1-n} \lambda(i) A(i)} \quad (6)$$

The fraction of earthquakes missed (υ) was defined in the form of

$$\upsilon = \frac{\sum_{i=1,n} I(S(i) > slevel) E(i)}{\sum_{i=1,n} E(i)}, \quad (7)$$

where $S(i)$ is the absolute value of gravity variation, THD or AAS value of ith cell, $A(i)$ is the area of ith cell, $\lambda(i)$ is the seismicity rate of i'th cell. and '$s$level' is the threshold value of indexing function.

In this study, we use the MED method to test whether the gravity variation and AAS include precursory information of future large earthquake using the complete randomness and the non-homogeneous Poisson model as the baseline hypothesis. The comparison results will be discussed in next section.

4. Results and Analysis

The gravity variation used by this work is shown in Fig. 3. On the basis of the gravity variation in two phases, the THD and AAS signasl can be computed. Figures 7 and 8 give the THD and AAS signals, respectively. In two figures, the subfigure (a) and (b) show the earthquake precursors on two different temporal intervals, Jul 2005-Jun 2008 and Jul 2008-Jun 2011, respectively. It can be seen that the gradient belt has been enhanced obviously and related to the geometry of major fault systems. The higher-value regions have a gravity variation more intensive than other regions. The earthquakes which occurred within 3 years after the gravity variation are also marked in Figs. 7 and 8, respectively.

First, using the gravity variation and then using the THD and AAS as the indexing function, we apply the Molchan error diagram, which is discussed in Sect. 3, to compute the fraction of alarmed cells and miss rate by adjusting the threshold of the indexing function from the highest to the lowest. The results are shown Fig. 9, in which the red line is the evaluated results by PPE model, which can be regarded as a kind of reference evaluation model. The evaluated result include all 13 events has been calculated in figure.

From Fig. 9, we can see that the prediction efficiency of AAS data (the solid blue line) is better than the THD of gravity variation (the solid cyan line) and absolute value of gravity variation (the solid green line) based on the homogeneous Poisson model. The prediction efficiency of THD data is similar to the absolute value of gravity variation. But only the AAS result is better than PPE model, especially in the section of miss rate value from 0.6 to 0.2. The fraction of space–time occupied by alarm determined by the non-homogeneous model (dashed line) is slightly different from the Poisson model. This result may infer that the gravity variation or its AAS can give some useful information related to the future earthquake risk which is independent of regional seismicity. However, the predication efficiency of AAS data with miss rate <0.2 is lower than PPE model.

Therefore, if we reconsider the causation of regional gravity change, this physical phenomenon can be explained the crustal deformation and underlying mass transportation, etc. The gravity change signal may include many physical factors associated with the complicated preparation process of a large or moderate earthquake, and the gravity variation can

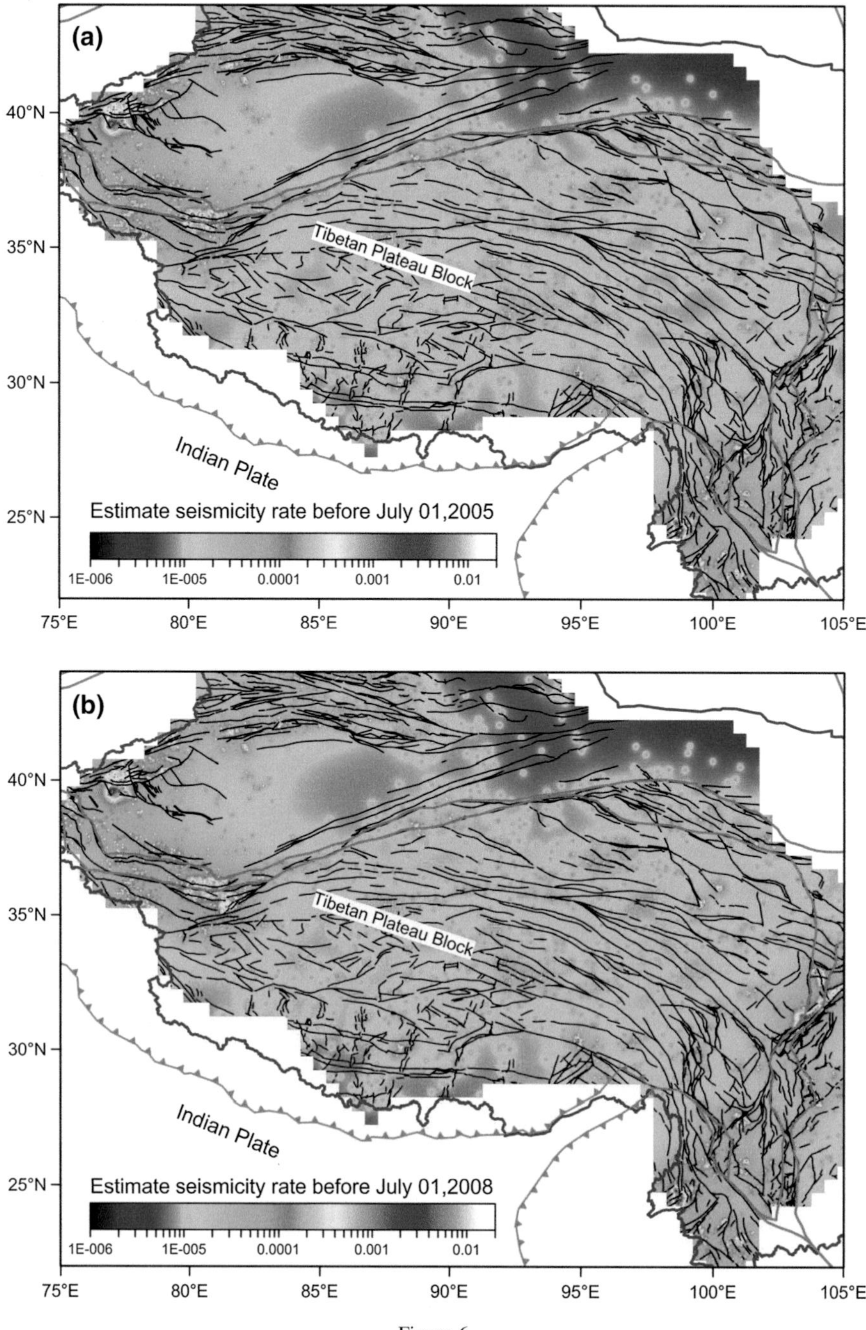

Figure 6

Estimated seismicity rate using earthquake events based on the CENC catalog. The learning period selected from 1970-1-1 to 1985-1-1 (Julian days 5479) based on the PPE model to determine the conditional intensity. **a** The estimated result using data before 2005-July-01 00:00:00; **b** the estimated result using data before 2008-July-01 00:00:00

provide some preseismic information and is independent on the regional seismicity rate.

Moreover, the signal to noise ratio (SNR) of gravity variation can preliminarily satisfy the location predication of large and moderate earthquakes. The gravity variation and the time scale of preparing process of earthquake are comparable. But the data as shown in Fig. 9 show that the predication efficiency

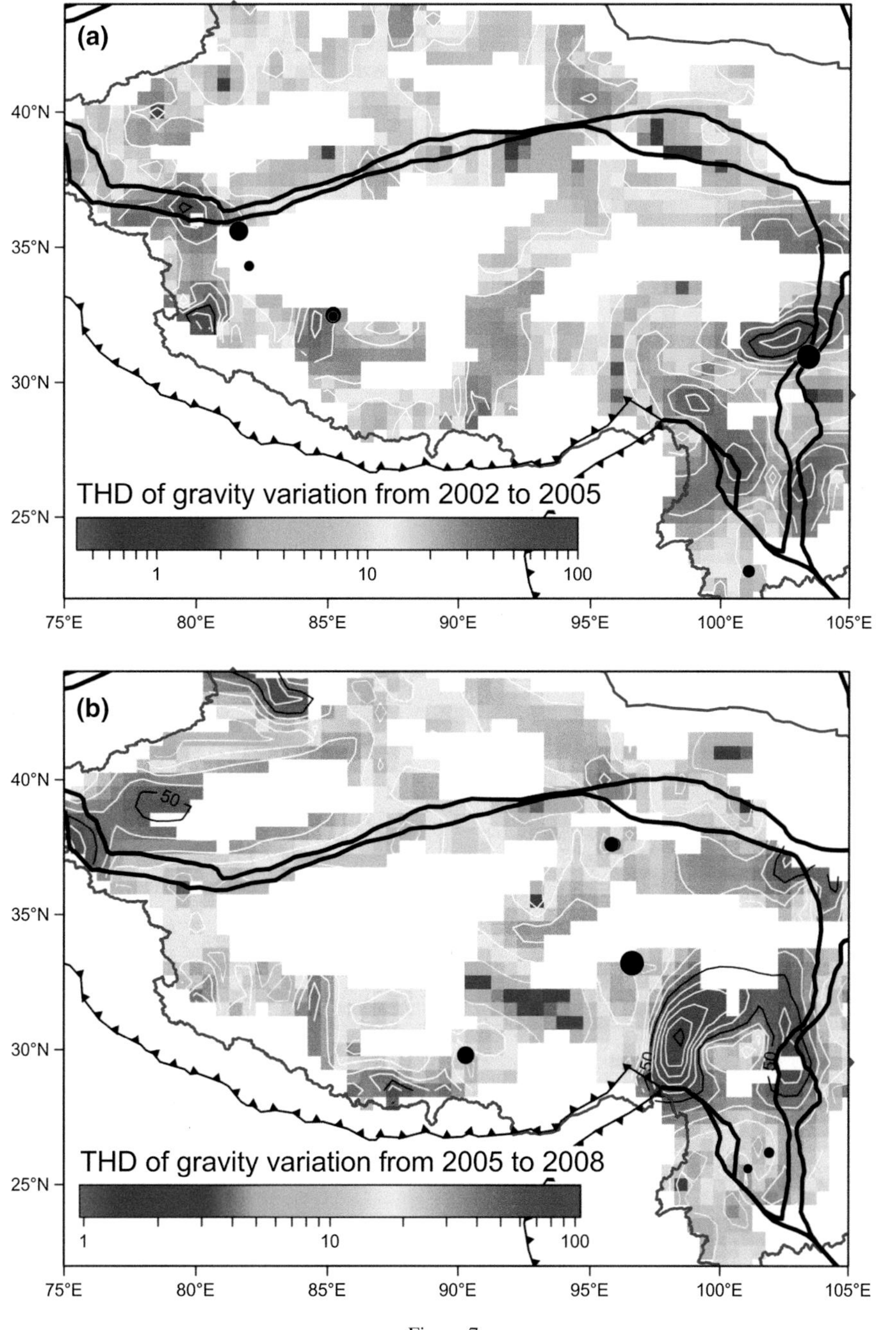

Figure 7
The total horizontal derivative (THD) of gravity variations. The *black circles* show the locations of earthquakes. The *color mapping* used logarithmic scaling function. The *black solid lines* show the boundary of active blocks

may be worse than random guess if only the magnitude of gravity variation is used. That is to say that the maximum region of gravity variation cannot indicate the risk of earthquake occurrences. The explanation on this problem is consistent with and reconfirms the 'epicentroid' hypothesis that the epicenter of the seismic event is not located at the high-value position of gravity field variation (Kuo and Sun 1993). It is difficult to define how much change of the gravity field is enough to trigger an earthquake.

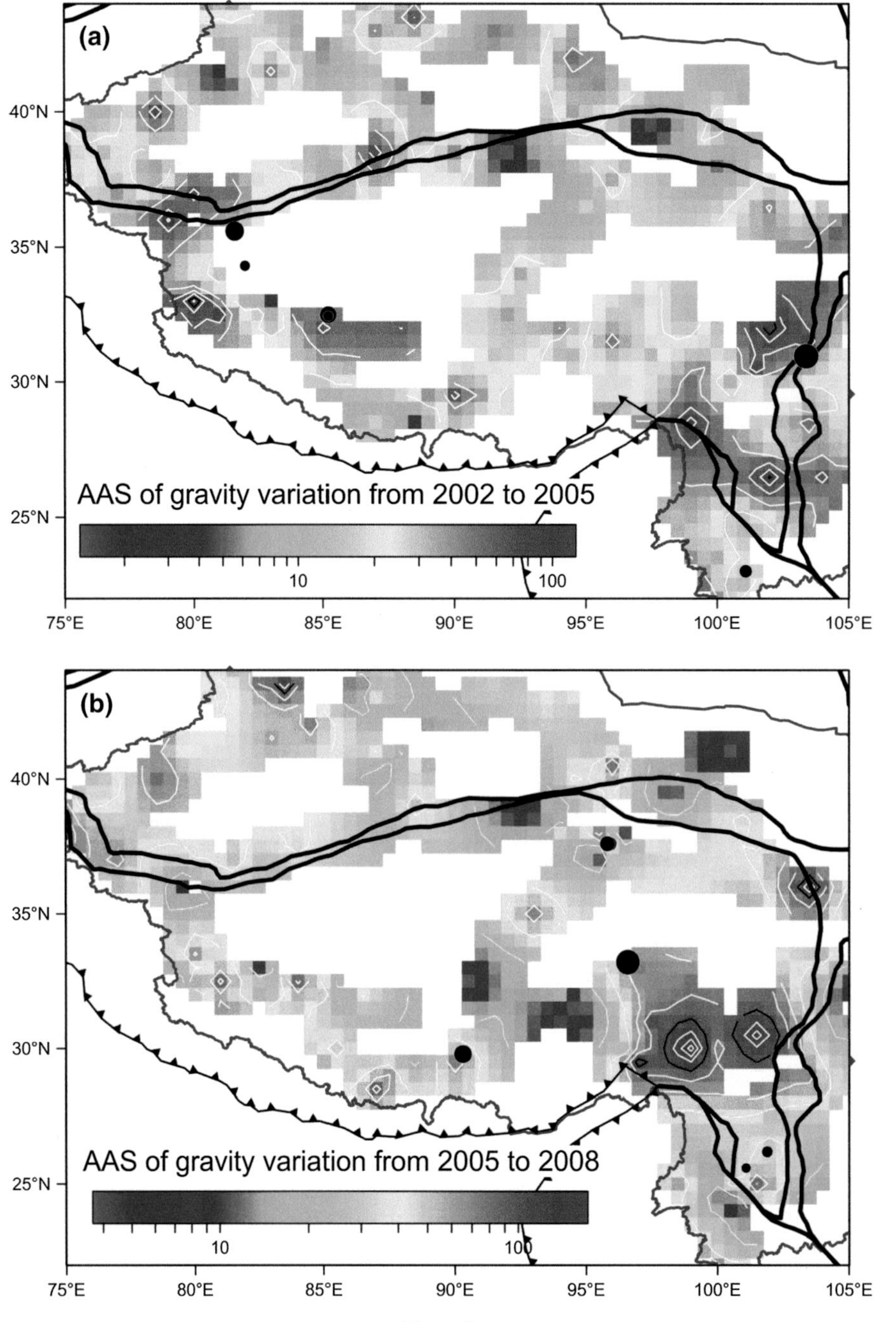

Figure 8
The amplitude of analytic signal (AAS) of gravity variations. The *black circles* show the locations of earthquakes. The *color mapping* used logarithmic scaling function. The *black solid lines* show the boundary of active blocks

However, the AAS data have better prediction ability for the location of earthquake in future than gravity variation data and may be superior to PPE method in case of medium miss rate.

5. Conclusions and Discussions

There are numerous studies published during the past decades on finding precursors in non-seismicity

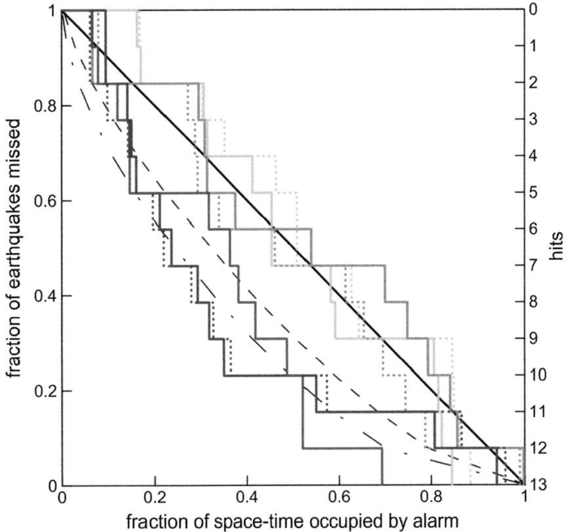

Figure 9
Molchan error diagrams for the absolute value (*green curves*), the AAS (*blue curves*), and the THD signal (*cyan curves*) of gravity variations and the PPE model (*red curve*). The *solid curves* and the *dashed curves* are results when using the homogeneous Poisson model and the non-homogeneous Poisson model (PPE) as the reference models, respectively. The *black dash* and *dash-dot lines* are the 95 and 99 % confidence levels, respectively

observations for the purpose of earthquake prediction, such as geodetic method, GPS, and electric field (Granat 2006; Granat and Donnellan 2002; Ogata 2007; Shen *et al.* 2007; Wang *et al.* 2013; Zhuang *et al.* 2005). Among them, the synthesized gravity variation signals, caused by various geodynamic processes, have also enough SNR to be identified as possible earthquake precursors by the repeated ground-based gravity survey nowadays. On the basis of our work, the conclusions are as follows:

1. The statistical method indicated the AAS as precursor has significant efficiency. This kind of earthquake prediction belongs to the non-catalog-based probabilistic forecasting and is effective compared to the Poisson and non-homogenous Poisson model, proved in this study by the analysis using the Molchan error diagram. Essentially, our results indicate that there is a feasible physical mechanism between earthquake occurrence and the spatiotemporal microgravity variation. The regional gravity change before the occurrence of earthquake is significant. Especially, for some large and moderate earthquakes which occur in the interior of active tectonic plate, the gravity precursors can be identified several years before the occurrence of the earthquake. We have investigated 13 earthquakes occurring in or around the Tibetan plateau block and found that the gravity changes may be correlated to these events. Such geophysical field signals may include information which comes from the interior of crust and the tectonic movement associated with the aseismic events.
2. The regions of high earthquake risk potentials are related to the gradient belt of gravity variation. But, due to the complicated geological settings and tectonic environments in this region, it is difficult to find a criterion to separate strictly the risk region from the safe region. Also, different seismic activity phases in the same region may correspond to the different kinds of earthquake mechanisms and different types of anomalies. Nevertheless, the gradient belt region of gravity variation should been given more attention.
3. Results from analysis on the gravity-based alarm function that are proposed in this study show that the predication based on the gravity data is better than random guess and that the AAS signal is more significant than the absolute value of the gravity change and PPE model. The analytic signal of gravity field has a good property to enhance the boundary imaging of 3-D causative bodies. So the AAS can be used a good tool to estimate the status of gradient belt and evaluate the physical differences underlying the crust.

Needless to say, our evaluation results contain uncertainty and errors, partially derived from the gridding approach applied to inhomogeneous gravity measurements. However, this study mainly focuses on how to evaluate the efficiency and possibility of earthquake predictions with gravity field variation and has obtained encouraging results. If more precise and denser data are observed or new gravity based prediction methods are proposed, this approach for evaluating prediction performance can be still applied in future.

Moreover, the vertical land motion at each station can produce the gravity variation. Generally, the gravity variation derived from the surface elevation change can be computed by $(3.086–0.419\rho)\Delta h$. The Δh is the change of elevation in cm. The ρ is the mass density under the surface. If the 2.6 g/cm^3 is taken for density, the gravity variations due to the 1 cm elevation change are no more than 2×10^{-8} m/s^2. In this case, the precision of gravity measured values is almost 15×10^{-8} m/s^2. This means the uncertainty of gravity measurement is roughly equal to the 7.5 cm elevation change in each period of twice repeated gravity measurement. On the basis of previous literatures (Wang *et al.* 2008; Cui *et al.* 2009), the vertical velocity of ground derived from the GPS and leveling in the Longmen Shan and the eastern margin of Tibet regions does not exceed 1 cm/a. So we consider that the precision of gravity data used in this study is not enough to model the vertical land motion process.

Providing ideal earthquake forecasts is an extremely difficult task. A plausible way to improve earthquake forecast is to incorporate multi-disciplinary geophysical observations and the experiences of retrospective studies on past large earthquake into rigorous probability models. The repeated gravity survey has been carried out for tens of years in Chinese mainland. In recent years, there is a nationwide gravity observation network including the absolute and relative gravity in China. More continuously observed earth tide meters also have been set on the good-condition stations. The rate of gravity change with geological block scale has been achieved. It can provide detailed information which comes from the ground to the deep crust. Then, the spatiotemporal gravity field variation dataset from the Chinese mainland will help us to understand the physics of seismogenesis and seismo-processes and improve our ability to interpret them. It is well known that the precision gravity measurement is a time-consuming task in practice. We wish that these very valuable data can be evaluated using quantificational methods and deeply used into the earthquake consultation. Statistical evaluation approach will be useful to give such standards on the basis of some retrospective study.

Acknowledgments

The authors would like to thank Professors John T. Kuo and M. N. Nabighian for their constructive comments on this manuscript. This work is mainly supported by the National Science-technology Support Plan Projects of the 12th Five-Year-Term of China named “Long-term earthquake risk assessment and the key techniques research for main tendency prediction” with grant No. 2012BAK19B01-05, and also partially supported by the National Natural Science Foundation of China; the grant No. is 41104046. We thank the CSEP testing center in China for their assistance.

References

Aurenhammer F., 1991. *Voronoi Diagrams – A Survey of a Fundamental Geometric Data Structure.* ACM Computing Surveys *23*, 345–405.

Cordell, L and V J S Grauch, 1985. *Mapping basement magnetization zones from aeromagnetic data in the San Juan Basin, New Mexico*, In Hinze, W. J., Ed., *The utility of regional gravity and magnetic anomaly maps.* Soc Explor Geophys, 181–197.

Center for Analysis and Prediction, China Earthquake Administration, 2007. *Study on the Seismic Tendency in China* (for the year 2008) (in Chinese) Seismological Press, Beijing.

Center for Analysis and Prediction, China Earthquake Administration, 2008. *Study on the Seismic Tendency in China* (for the year 2009) (in Chinese). Seismological Press, Beijing.

Center for Analysis and Prediction, China Earthquake Administration, 2009. *Study on the Seismic Tendency in China* (for the year 2010) (in Chinese). Seismological Press, Beijing.

Center for Analysis and Prediction, China Earthquake Administration, 2010. *Study on the Seismic Tendency in China* (for the year 2011) (in Chinese). Seismological Press, Beijing.

Chen, Y.T., Gu, H.D., Lu, Z.X., 1979. *Variations of gravity before and after the Haicheng earthquake, 1975, and the Tangshan earthquake, 1976.* Phys. Earth Planet. Inter. *18*, 330–338.

Cui, D.X., Wang, Q.L., Hu, Y.X., *et al.*, 2009. *Lithosphere deformation and deformation mechanism in northeastern margin of Qinghai-Tibet platea* (in Chinese). Chinese Journal Geophysics, *52*(6): 1490–1499.

England, P., Jackson, J., 2011. *Uncharted seismic risk.* Nature Geosicience *4*, 348–349.

Granat, R.A., 2006. *Detecting regional events via statistical analysis of geodetic networks.* Pure and Appl. Geophys. *163*, 2497–2512.

Granat, R.A., Donnellan, A., 2002. *A hidden Markov model based tool for geophysical data exploration.* Pure and Appl. Geophys. 159, 2271–2283.

Gu, G., Liu, K., Zheng, J., Lu, H., 1997. *Seismogenesis and occurrence of earthquakes as observed by temporally continuous gravity variations in China.* Chinese Science Bulletin (in Chinese) *42*, 1919–1930.

Kuo, J.T., Sun, Y.-F., 1993. *Modeling gravity variations caused by dilatancies.* Toctonophysics *227* 127–143.

Kuo, J.T., Zheng, J.H., Song, S.H., Liu, K.R., 1999. *Determination of earthquake epicentroids by inversion of gravity variation data in the BTTZ region, China.* Tectonophysics *312*, 267–281.

Liu, K.R., Zheng, J.H., Kuo, J.T., Song, S.H., Liu, D.F., Lu, H.Y., Guo, F.Y., 2002. *Mobile gravity survey and modified CDM in the*

BTTZ Region, in: Shu, S. (Ed.), *Advances in Pure and Applied Geophysics* (in Chinese). Meteorological Publishing House, pp. 167–178.

Liu, M., Stein, S., Wang, H., 2011. *2000 years of migrating earthquakes in North China: How earthquakes in midcontinents differ from those at plate boundaries*. Lithosphere, 128–132

Molchan, G.M., 1990. *Strategies in strong earthquake prediction*. Phys. Earth planet. Inter. *61*(1.2), 84–98.

Molchan, G.M., 1991. *Structure of optimal strategies in earthquake prediction*. Tectonophysics *193*(4), 267–276.

Molchan, G.M., 1997. *Earthquake prediction as a decision-making problem*. Pure appl. Geophys. *149*(1), 233–247.

Molchan, G.M., 2010. *Space–Time Earthquake Prediction: The Error Diagrams*. Pure and Appl. Geophys. *167*, 907–917.

Nabighian, M.N., 1972. *The analytic signal of two-dimensional magnetic bodies with polygonal cross-section—Its properties and use of automated anomaly interpretation*. Geophysics *37*, 507–517.

Nabighian, M.N., 1974. *Additional comments on the analytic signal of two-dimensional magnetic bodies with polygonal cross-section*. Geophysics *39*, 85–92.

Nabighian, M.N., 1984. *Toward a three-dimensional automatic interpretation of potential field data via generalized Hilbert transforms - Fundamental relations*. Geophysics, 780–786.

Nabighian, M.N., Hansen, R.O., 2001. *Unification of Euler and Werner deconvolution in three dimensions via the generalized Hilbert transform*. Geophysics *66*, 1805–1810.

Ogata, Y., 2007. *Seismicity and geodetic anomalies in a wide area preceding the Niigata-Ken-Chuetsu earthquake of 23 October 2004, central Japan*. J. Geophys. Res. contributed equally to this, B10301.

Rhoades, D.A., Evison, F.F., 2004. *Long-range earthquake forecasting with every earthquake a precursor according to scale*. Pure and Appl. Geophys. *161*, 47–72.

Scholz, C.H., 2002. *The mechanics of earthquakes and faulting* (2nd edition). Chambridge university press, New York.

Schorlemmer, D., Gerstenberger, M.C., 2007. *RELM Testing Center*. Seismol. Res. Lett. 78.

Schorlemmer, D., Zechar, J.D., Werner, M.J., Field, E.H., Jackson, D.D., Jordan, T.M., 2010. *First results of the regional earthquake likelihood models experiment*. Pure and Appl. Geophys. *167*, 859–876.

Shen, Z.K., Jackson, D.D., Kagan, Y.Y., 2007. *Implications of geodetic strain rate for future earthquakes, with a five-year forecasts of M5 earthquakes in Southern California*. Seismol. Res. Lett. *78*, 116–120.

Smith W.H.F., Wessel, P., 1990. *Gridding with continuous curvature splines in tension*. Geophysics *55*, 293–305.

Wang, Q.L., Cui, D.X., Wang, W.P., *et al*., 2008. *Research for the vertical crustal movement in western Sichuan area nowdays* (in Chinese). Science in China:Earth Sciences *38*(5),589–610.

Wang, T., Zhuang, J., Kato, T., Bebbington, M., 2013. *Assessing the potential improvement in short-term earthquake forecasts from incorporation of GPS data*. Geophysical Research Letters *40*, 1–5.

Zhu, Y., Liang, W., Zhan, F.B., Liu, F., 2012. *Study on dynamic change of gravity field in China continent* (in Chinese). Chinese Journal Geophysics *55*, 804–813.

Zhu, Y., Wen, X., Sun, H., 2013. *Gravity changes before the Lushan, Sichuan, Ms = 7.0 Earthquake of 2013* (in Chinese). Chinese Journal Geophysics *56*, 1887–1894.

Zhu, Y.Q., Zhan, F.B., Zhou, J.C., 2010. *Gravity measurements and their variations before the 2008 Wenchuan earthquake*. Bulletin of the Seismological Society of America *10*, 2815–2824.

Zhuang, J., Vere-Jones, D., Guan, H., Ogata, Y., Ma, L., 2005. *Preliminary analysis of observations on the ultra-low frequency electric field in a region around Beijing*. Pure and Appl. Geophys. *162*, 1367–1396.

(Received October 28, 2014, revised March 18, 2015, accepted May 27, 2015, Published online June 13, 2015)

Printed in the United States
By Bookmasters